ENGINEER'S GUIDE TO HIGH-TEMPERATURE SUPERCONDUCTIVITY

ENGINEER'S GUIDE TO HIGH-TEMPERATURE SUPERCONDUCTIVITY

James D. Doss

A WILEY-INTERSCIENCE PUBLICATION

John Wiley & Sons

New York / Chichester / Brisbane / Toronto / Singapore

Published by John Wiley & Sons, Inc.

Library of Congress Cataloging-in-Publication Data:

Doss, James D.
Engineer's guide to high-temperature superconductivity / by James D. Doss.
p. cm.
"A Wiley-Interscience publication."
Bibliography: p.
Includes index.
1. High temperature superconductivity. 2. High temperature superconductors. I. Title.
QC611.98.H54D67 1989
621.3–dc20 89-34211
ISBN 0-471-51307-5 CIP

Printed in the United States of America

10 9 8 7 6 5 4 3 2 1

This Volume is Dedicated to

Ural and Martha Cobb,
Lynn and Margaret Wiseman,
and
L. V. Schubnikov

CONTENTS

PREFACE

This reference was written for that sizeable group of professionals (particularly engineers) who, as a result of recent developments, are just beginning to work with superconducting materials. While it is hoped that those who have considerable experience with superconductivity will also find this introductory reference useful, it is the engineer-in-transition to whom the book is primarily directed.

The field of superconductivity, which has employed the services of a moderate number of engineers and technicians for several decades, has recently begun to grow in an exponential fashion. The reason for the recent explosion of activity is, of course, the remarkable discoveries since 1986 in the regime commonly referred to as "high-temperature" superconductivity. In a very short period of time, maximum critical transition temperatures increased from 23 K to 35 K and then to 90 K. Indeed, there are now materials that superconduct at 110–125 K and intriguing possibilities of even higher transition temperatures.

It is a challenging period for those working in the field. There is an atmosphere of urgency in the laboratory; a sense that one is living in an extraordinary time in the history of science and technology. Experimental results have attracted so much attention outside the laboratories that reports of progress are often announced in the popular press.

Preprints of journal articles are circulating almost as soon as experiments are completed. Those who work in the field have considerable difficulty keeping up to date with the latest experimental results and theoretical notions; there simply is not enough time to read even a fraction of the new publications. It has been estimated that approximately 4000 papers were published on the subject of high-temperature superconductivity during 1987–1988! The interest in this subject has also led to a variety of excellent news-periodicals dedicated to "high-T_c" superconductivity.

Quite aside from the justifiable excitement among theorists over the possibilities of radical new theories to describe these unexpected phenomena, the potential applications of materials that are superconducting above liquid nitrogen temperature promises economies and simplicity of refrigeration that are not possible with compounds requiring liquid helium.

Even so, experienced practitioners quite properly remind us that it has taken decades to develop the "conventional" superconducting materials to a point where high-field magnets are practical for NMR, particle accelerators, and laboratory measurements. Having thus gotten our attention, they also like to point out that, except for lower transition temperatures, conventional superconductors like niobium and its alloys are superior to the new materials in almost every respect.

If you find this hard reality daunting, there are others who are ready to encourage you with the news that the new high-temperature materials are being improved almost every week. If this is not enough to lift your spirits, there are those incurable optimists who believe we will soon develop materials that superconduct at room temperature and beyond. The potential applications for materials that would operate in the superconducting mode without the need for refrigeration is so enormous that one is awed by the probable consequences.

No one really knows where these new developments will lead, but government agencies and the private sector are increasing their investment in virtually all phases of superconductivity research, from the development and testing of new materials to the investigation of applications suitable to these peculiar ceramics.

As a result of the expanding amount of work to be done, engineers who have not previously worked with superconductors will be called on to apply their skills of design and analysis in this new arena. Some of these individuals will perform measurements of very low-level signals, in a spectrum ranging from dc to several GHz. Others will use the new materials in applications that will be both unfamiliar and challenging. A sound knowledge of the basics of superconductivity, with an emphasis on the special characteristics of the high-temperature compounds, will be required if a first-rate job is to be done.

These individuals will find themselves scouring the scientific journals for recent information on the high-temperature materials and searching the library for texts on the basic phenomena of superconductivity. Some will be introduced, for the first time, to the Meissner effect, critical current density, critical magnetic fields, the London penetration depth, persistent currents, and flux trapping. Already having at least some familiarity with microscopic quantum effects, they will be delighted to learn of the large-scale quantum phenomena that govern the behavior of superconductors.

This reference is intended for *those individuals;* its primary purpose is to provide a lucid introduction to the fascinating properties exhibited by superconductors, followed by a discussion of the most interesting and significant applications.

An introductory chapter provides a history of superconductivity. This is considered helpful, if not essential, since it is difficult to appreciate this field without knowing something about the manner in which it has developed since Onnes' remarkable discovery in 1911.

The chapter on superconducting phenomena will acquaint the reader with the primary behavior of superconductors with an introduction to flux expulsion, penetration depths, coherence lengths, energy gaps, flux quantization, the Josephson effects, and more. The differences in Type I and Type II superconductors are discussed at some length.

The chapter on high-temperature superconductivity introduces the reader to the exciting developments since the Bednorz-Müller breakthrough during 1986. Particular attention is paid to the $YBa_2Cu_3O_x$ ceramic and its rare earth counterparts, since so much data is now available, but the recent developments regarding bismuth- and thallium-based superconductors are also discussed. This chapter lists the most relevant characteristics of these new materials for engineering applications. Where controversy exists regarding particular results, this is also pointed out.

This is followed by a discussion of representative techniques used in processing high-temperature superconductors, both in bulk and thin-film configurations.

A rather extensive chapter deals with applications of superconductors. Rather than describe every application in great detail, the need to limit the book to a manageable size has suggested that a more general treatment is in order. In this spirit, the chapter discusses several general subjects (high current conductors, quenching, junctions, SQUIDs, etc.) that are pertinent to more than one application. Even so, several representative applications are discussed in some detail, with emphasis on computer applications of Josephson junctions, MHD, MAGLEV, *IR* detectors, particle accelerator structures, and magnetic shielding.

Another chapter deals with the questions of cryogenic temperature measurement and superconductor characterization by a variety of techniques, ranging from noncontact eddy current and resonant cavity measurements to thermoluminescence.

The chapter on safety is primarily concerned with hazards associated with the storage and use of cryogens. In addition, the toxicity of thallium is discussed.

The first appendix, a rather large one, is dedicated to a review of relevant electromagnetic phenomena, with particular attention to those subjects of interest to one working with superconductors.

The engineer who uses one of the excellent references on superconductivity is usually faced with Gaussian units. This is generally because these texts and articles have been prepared by physicists for physicists. For the engineer who is startled to see the appearance of c (velocity of light) in unexpected places or 4π as a multiplier of magnetization, an appendix on units has been provided to help minimize this vexation.

There are other appendices dealing with a variety of relevant subjects. These include a glossary of symbols and definitions, and a section on temperature scales. There are also appendices on the subject of perovskites, novel superconductors, temperature conversion, and heat capacity, as well as a listing of commercial sources of superconductor-related products, services, and literature.

To supplement the references placed at the end of each chapter, there is usually a list of recommended reading. Finally, a rather extensive bibliography is provided for the reader who wishes to delve more deeply into a particular subject.

It is hoped that this book will be helpful in introducing the electrical engineer and experimental physicist (and others with similar interests) to the very dynamic field of high-temperature superconductivity. Readers are encouraged to forward their comments and suggestions to me via the publisher.

With knowledge of the risk that I will forget someone, I nevertheless wish to acknowledge assistance, constructive criticism, and encouragement from a variety of helpful souls:

At the Los Alamos National Laboratory:

Bob Newell
Jerry Beery
D. Wayne Cooke
Catherine Mombourquette
M. Maley
Sam Freund
Paul Gaetjens
Ray Gore
Gene Stark
Anton (Tony) Mayer
John Allred
Jim Sturrock

and,

Harold Burnette (Ojo De Dios Book Store)
John Dicello (Clarkson University)
Richard Williams (University of New Mexico)
Helmut Piel (University of Wuppertal)
Lillian Hoddeson (University of Illinois)

finally,

my wife Martha, who spent many hours proofreading.

James D. Doss

Los Alamos, New Mexico
September 1989

ENGINEER'S GUIDE TO HIGH-TEMPERATURE SUPERCONDUCTIVITY

CHAPTER 1

A BRIEF HISTORY OF SUPERCONDUCTIVITY

> An objective history of science is perhaps only possible in very broad outlines and in its relation to the rest of history. When we try to look at a recent event with a microscope, the resolving power may often be insufficient.
>
> (S. A. Goudsmit, 1961)

EARLY INVESTIGATIONS OF METALLIC RESISTIVITY

As the end of the nineteenth century approached, enormous strides had been made in understanding the behavior of electromagnetic fields. Even so, there was no widely accepted theory for the behavior of free electrons in metals, particularly at low temperatures. It was known early in the nineteenth century that the resistivity of metallic electrical conductors decreased with temperature. In the absence of data at the very low temperatures, speculation flourished. James Dewar, for example, suggested that the resistivity of a pure metal would be reduced to zero as temperature approached absolute zero.

Other distinguished theorists, including Lord Kelvin, invoked different models and predicted that the metallic resistivity could be expected to decrease to a minimum level at low temperatures and then *increase* as temperature was lowered still further. The difficulty standing in the way of settling the issue was simple, if daunting: it was impossible to cool samples to sufficiently low temperatures to make the measurements. The cryogenic technology simply did not exist.

Progress must have seemed almost painfully gradual to those who yearned to make the measurements of resistivity at these unprecedented low temperatures. It was probably preordained that the first investigator to solve the cryogenic problems would also settle the question about resistance to electric current flow near absolute zero. The electrical measurements, after all, were straightforward compared to the cryogenic progress that was required.

Developments proceeded at what, especially now, seems a snail's pace. In 1883 Olzewski and Wroblewski succeeded in liquefying oxygen (90.2 K) and nitrogen (77.3 K), repeating work performed earlier in France. James Dewar, 15 years later, managed to liquefy hydrogen, which has a boiling point of approximately 20.5 K. It was important progress, but not nearly enough to solve the problem we are concerned with.

HEIKE KAMERLINGH ONNES

Young Kamerlingh Onnes was admitted to the University of Groningen in 1870, where he made his decision to study physics and mathematics. He later spent time in Heidelberg, studying under such luminaries as Kirchoff and Bunsen. His dissertation, prosaic considering his future contributions, dealt with "new proofs for the axial rotation of the earth." Onnes held positions in Delft and Amsterdam, and in 1882, at the age of 29, was appointed professor of experimental physics at the University of Leiden. Onnes was a hands-on type of fellow who believed that one learns by doing. His motto, which could hardly have been more appropriate, was "to comprehend through measurement."

It was not until July 10, 1908 that Onnes successfully operated the first liquefier of helium (see Figure 1.1), which exhibited a boiling point of approximately 4.2 K at atmospheric pressure. At reduced pressure, temperature dropped to 1.7 K. This accomplishment alone would have assured his place in the history of science. The apparatus produced a scant 280 ml of liquid helium on an hourly basis, and it was natural that the professor dedicated some of this precious fluid to investigate the resistivity of metals at these unprecedented low temperature levels. Onnes planned these measurements with considerable attention to detail; nothing was left to chance. He enlisted the aid of Dr. Cornelius Dorsman, an expert in electrical measurements, to help him plan the experiments. The professor was also assisted by his technician, G. J. Flim, and a Mr. Gilles Holst who made the actual measurements of resistance. Holst, a dedicated graduate student, applied all his skills and concentration to operation of the Wheatstone bridge, while Onnes directed most of his attention to the delicate cryogenic apparatus (de Bruyn Ouboter, 1987). While Onnes obviously hoped to settle the argument about which of the current theories best described metallic conduction at low temperatures, he expected to find that one of the common theories of the day would prevail. Personally, he leaned toward the notion that resistivity would pass through a minimum level and then increase dramatically as the temperature was decreased to lower levels. It was not to be, even though his first experiments seemed to support his ideas.

It was natural to measure elements like gold or platinum, which exhibited a relatively low resistance at room temperature. No one suspected that it would be the metals with relatively poor electrical conductivity that would provide the

FIGURE 1.1 Heike Kamerlingh Onnes (left) and J. D. Van der Waals in Onnes' laboratory in Leiden, 1911. The helium liquefier is in the background. (Provided courtesy of R. de Bruyn Ouboter, Kamerlingh Onnes Laboratorium, Rijksuniversiteit te Leiden)

really spectacular behavior! Furthermore, no one, as far as we know, expected zero dc resistance to occur well above 0 K.

In Onnes' initial experiments with platinum wire, the resistance diminished gradually and then leveled off at 4.3 K. This seemed to fit one of the most widely held theories, but since such a leveling-off was also consistent with impurities, Onnes was cautious. The platinum, he pointed out in his notes, might contain a small amount of impurities. Gold, the experimenters soon learned, displayed similar leveling characteristics. Not very exciting. Not yet!

What elements should he measure to provide a meaningful test of the competing theories? Mercury, after all, was convenient from the point of view of making low-resistance contacts. Since it was a liquid at room temperature, it was also easier to purify (by repeated distillations) than silver or gold. Mercury, Onnes decided, was to be the next test material. The results were completely unexpected. When the resistance suddenly disappeared below 4.2 K, Holst was certain that he had a troublesome short circuit and enlisted the help of Flim. The pair tried in vain to locate and repair it (Hoddeson, et al., 1989). In one sense, a short circuit *had* developed, but not of the conventional type Holst was looking for.

In his notes, Onnes comments (de Bruyn Ouboter, 1987):

> I was formerly of the opinion that the resistance of pure metals reaches a minimum as the temperature is diminished, and then, as the temperature sinks still lower, again begins to increase and becomes infinitely great at the absolute zero; but now it seems to me more probable that, even before the absolute zero is actually reached, the resistance if not zero, has become so extremely small that it practically vanishes, and that this remains the case for further lowering of the temperature.

These tentative comments gave way to assurance as the work progressed:

> . . . at very low temperatures such as could be obtained by helium evaporating under reduced pressure the resistance would, within the limits of experimental accuracy, become zero.

In Onnes' *Communication No. 124c* of November 1911, entitled "On the sudden change in the rate at which the resistance of mercury disappears," the cautious investigator ushered in the age of superconductivity:

> Cooling the resistance to 4.2 K, somewhat below the boiling point of helium, it had become 500 times less than that of the solid wire at the melting point of mercury. At this point within some hundredths of a degree came a sudden fall, not foreseen by the vibrator theory of resistance that I had framed, bringing the resistance at once to less than a millionth of its original value at the melting point . . .

Onnes presented a report of his work with mercury at the first Solvay congress, which met in Brussels during October 1911. If one is to judge from the printed proceedings of this meeting, which included questions from the floor following each presentation, the significance of Onnes' work was apparently not recognized (Dahl, 1984).

A representative plot of Onnes' results with mercury is shown in Figure 1.2. He followed this work by showing that indium, tin, and lead became superconducting at 3.4, 3.7, and 7.2 K respectively. During the next several decades, a wide variety of metallic elements and alloys were found to be

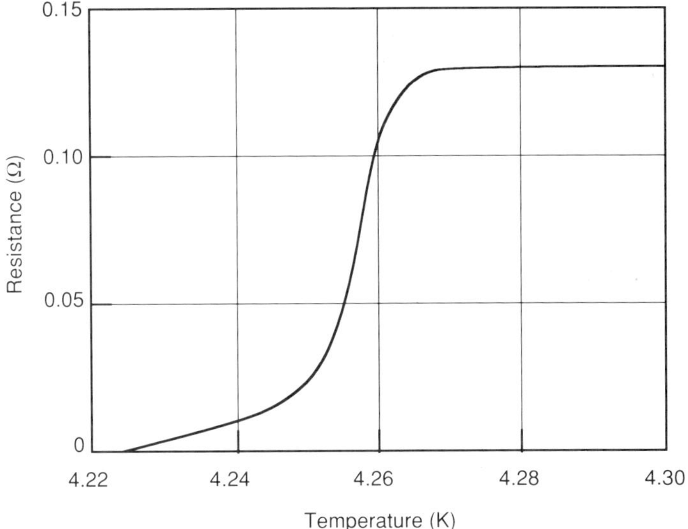

FIGURE 1.2 Kamerlingh Onnes' experiment revealed a rapid decrease in the resistance of mercury at 4.25 K, with the level becoming vanishingly small below 4.2 K. (After Fritz London, *Superfluids, Macroscopic Theory of Superconductivity*, Vol. 1, John Wiley & Sons, New York, 1950, pp. 3–4.)

superconducting, with Nb_3Sn exhibiting a critical temperature (T_c) of 17.9 K (see Figure 1.3). For his work in "researches on the properties of matter at low temperatures," H. Kamerlingh Onnes received the Nobel prize in physics in 1913.

For the reader who is interested in the fate of those faithful graduate students who toil tirelessly in pursuit of their advisor's goals, it is a pleasure to report that G. Holst received his doctorate in Zurich in 1914. Holst eventually went on to direct research at the huge Philips organization (Hoddeson, 1989b). One wonders whether any event in his future career provided as much satisfaction for Holst as the observations he made as a student while nulling the Wheatstone bridge for Professor Onnes.

Early Expectations, 1911–1913

Well-informed science writers of his day, not unlike media enthusiasts of our own time, predicted that Onnes' discoveries would promptly lead to the development of unprecedented magnetic fields in solenoids constructed of small-diameter superconducting wires. Onnes himself was interested in "producing intense magnetic fields with the aid of coils without iron cores" (de Bruyn Ouboter, 1987).

The same technology would, in the view of the time, result in heavy electric machinery that would operate essentially without loss. There was, at first, no reason to doubt this grand view of the future.

It was only two years after his original discovery that the interest in large-scale commercial applications was beginning to wane. Onnes by then had learned that the superconducting behavior was quenched in the presence of small current densities and quite ordinary levels of magnetic fields. Onnes, faithful to his motto on measurements, had seen the "appearance of resistance in supraconductors which are brought into a magnetic field, at a threshold value of the field" (Onnes, 1913).

The disappointment that followed, though based on hard facts, was as premature as the earlier euphoria. What no one realized at the time was that Onnes had only investigated the so-called soft elemental superconductors; these were later classified by investigators as "Type I." Decades would pass before the high-field capabilities of "Type II" materials, like niobium and its alloys, would be fully appreciated. These later developments would justify much of the enthusiasm of 1912.

During the first world war, Onnes had to abandon all research that required liquid helium (Gorter, 1964). It was during this dry period for Onnes that F. B. Silsbee, a physics assistant at the U.S. National Bureau of Standards, suggested that the quenching of superconductivity at higher current densities was due to the generation of a magnetic field at the surface of the superconductor:

> the threshold value of the current is equal to that value at which the magnetic field which caused the current is equal to the magnetic threshold value.
>
> (Silsbee, 1917)

While this insight may not seem so incredible today, the Silsbee hypothesis represented a milestone in the way scientists thought about superconductivity.

Like that aged prophet who would be denied the opportunity to place his sandals upon the soil of the Promised Land, the old professor from Leiden was to proceed no further. Kamerlingh Onnes died in 1926, well before his dream of high-current superconductors would be realized. Nevertheless, Onnes had established the foundation for that which was to come.

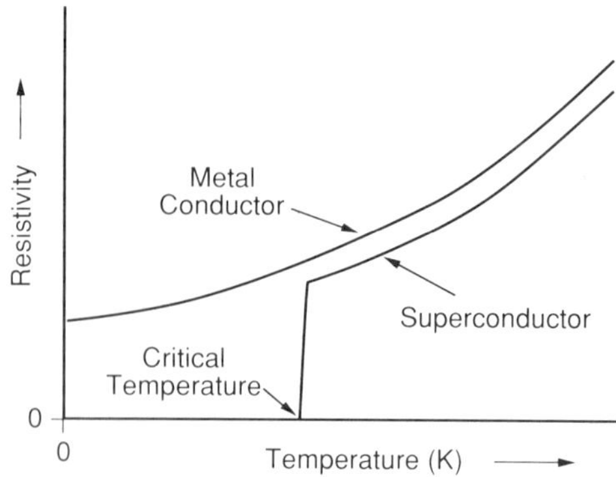

FIGURE 1.3 A generalized comparison of resistivity for normal metals and superconductors as a function of temperature.

H-FIELD EXPULSION; TWO-FLUID MODEL

In 1933, Walter Meissner and Robert Ochsenfeld discovered that magnetic flux was not only excluded, but actually *expelled* from superconductors. There was surely no development that was more important in the progress of the understanding and application of superconductors. This was the first recognition that superconductivity was, most fundamentally, a *magnetic* phenomenon. Even though the electric and magnetic effects are intimately connected, one might exaggerate and suggest that the exhibition of zero dc resistance be considered a fortuitous secondary effect! In any case, prior to the Meissner-Ochsenfeld discovery, it had been wrongly assumed that the fundamental property of superconductors was zero resistance; magnetic properties were considered to be the secondary effect (Dahl, 1986).

This demonstration of perfect diamagnetism in superconductors is commonly known as the Meissner effect. (If an equitable allotment of credit was as important as economy of expression, diamagnetism in superconductors would be widely known as the "Meissner-Ochsenfeld" effect.) This breakthrough was followed by the London phenomenological theories of superconductivity in 1935, and the landmark proposal by F. London (London, 1950) that superconductivity could be understood as a quantum-mechanical phenomenon. More about the London brothers will follow shortly.

Another important development in the thirties was the 1934 Gorter-Casmir phenomenological representation of a superconductor as a mixture of superconducting and normal electrons, with the proportion of superconducting electrons ranging from none at the onset of superconductivity, to 100% at 0 K. While not nearly so accurate or elegant as later theories, this "two-fluid" model is still extremely useful as a tool for modeling superconductor behavior.

ÉMIGRÉ SCIENTISTS

Mendelssohn

Kurt Mendelssohn could hardly contain his excitement when the first helium derived from gas wells arrived from America. Helium gas was scarce and precious, usually laboriously extracted from Monazite sand. He went to the customs house in Berlin to receive the crated cylinder, and had some difficulty convincing the customs official that the cylinder did not contain liquor. As Mendelssohn was leaving the university library in Breslau one day in the early thirties, he remembers "ducking bullets." This herald of the approaching dominance of Hitler's anti-Semitic hooligans was enough to encourage a variety of Germany's best scientific minds to leave for more congenial surroundings. Several of the most gifted departed for England (Mendelssohn, 1964).

Germany was not the only country where dark forces were on the move in those years. In the Soviet Union, a superb program of research on superconductivity at the Kharkov laboratory was ended when L. V. Schubnikov was arrested during the Stalinist purges. He was never heard from again.

THE LONDON BROTHERS

Phenomenological Theories. In 1933, Mendelssohn left Breslau for Oxford, where, with financial aid from the Rockefeller Foundation, he installed the first liquefier of helium in England. The device was a miniature liquefier developed in Germany. After arriving at Oxford, Mendelssohn was eventually joined by two other refugees from Nazi Germany, the brothers Fritz and Heinz London, whose names would become forever linked to their studies of the electromagnetic properties of the superconductive state. Heinz London had also worked at Breslau when Mendelssohn was there, and had begun work on a thesis that dealt with the losses due to alternating currents in superconductors. This particular research direction was, at least in part, due to industrial interests (Siemens) in the high-frequency application of superconductors.

It is recognized that not all scientists exhibit the same expertise in the laboratory; some are considered an outright hazard and encounter pleas from their colleagues to leave forthwith. Heinz London was limited not only by poor cryogenic facilities in Breslau, but Mendelssohn politely suggests that Heinz was also hampered by "a certain lack of manual dexterity" (Mendelssohn, 1964).

These limitations may have caused Heinz to become more than a little frustrated. Probably to everyone's relief, he decided to depart from the world of valves, tubing, cryostats, and such. Instead of breaking glassware, he used his mind and pencil to attack the problems in a theoretical fashion. This departure from the laboratory led to a significant contribution to the sphere of theoretical physics. In a similar situation today, Heinz would probably be appointed to an administrative post to while away his productive years in the preparation of grant proposals.

Heinz London soon realized that the alternating current must penetrate a short distance into the superconducting metal, and derived the correct formula for this depth. He was unaware that these results for penetration depth had been published earlier by several investigators, but he did go much further than his predecessors in combining his findings with the two-fluid model. He correctly predicted that, at sufficiently high frequencies, the superconducting electrons would not shield their normal counterparts, and Joule heating would be the result. The appropriate short-wave electronics was unavailable to Heinz London, and a successful experiment to prove these predictions correct was delayed until 1940.

During this same period, most investigators were attempting to use Maxwell's equations alone to explain the Meissner effect, that is, the expulsion of magnetic fields by the superconductor. This was certainly a reasonable approach; Maxwell's collection of expressions had always proved sufficient when dealing with electromagnetic phenomena, at least in normal materials. Superconductors, however, were anything but "normal"; attempts to use the classical approach were continually frustrated. Fritz London, using his brother's results, came to the astonishing conclusion that superconductivity required a new relationship between the superconducting current flow and the magnetic field. The result was his derivation of the now familiar form:

$$-\Lambda \text{ curl } \mathbf{J} = \mathbf{H} \tag{1.1}$$

While this expression will be dealt with in some detail in the chapter on superconducting phenomena, it is sufficient here to point out that the factor Λ is related to the penetration depth of the magnetic field into the superconductor. In recognition of their contribution, the penetration depth is often referred to simply as the "London depth." The London brothers' work was presented to the British Royal Society in 1934 and published in March 1935. The Londons were also the first to suggest the existence of an energy gap between the superconducting ground state and the lowest-level excited states. Heinz London, while measuring the surface resistance of tin at 1.46 GHz, also discovered the "anomalous skin effect." This effect is a higher than expected resistance at microwave frequencies when the electron mean free path becomes longer than the skin depth (Hoddeson, et al., 1989).

The Quantization of Flux. In a book published in 1950, Fritz London would suggest rather casually, in a footnote, that flux in a superconductor was quantized to a value (using Gaussian units) of hc/e, where h is Planck's constant, c is the velocity of light, and e is the electron charge (London, 1950). In SI units, the Londons' "fluxoid" would be h/e.

His suggestion was not so much ignored as uncommunicated; it was not widely known to the community of theorists and experimentalists who were concerned with such matters. This was somewhat academic, since instrumentation for measuring such a small value of flux would be unavailable for a decade. When the measurements were made, the flux quantum turned out to be exactly half the London prediction, since the superconducting carrier is a *pair* of electrons. If London had known this, he would have predicted the correct value for the flux quantum, that is, in SI units,

$$\Phi_0 = \frac{h}{2e} \tag{1.2}$$

PIPPARD; NONLOCAL EFFECTS

World War II, like the previous world war, had been a period of slow growth for the science of superconductivity. Nevertheless, the importance of radar in the late war had led to the rapid development of microwave technology; this enabled several groups to make precise measurements of microwave absorption and reflection in superconductors, with emphasis on the relation to the energy gap.

This was also the period, in the early fifties, of A. Brian Pippard's important contributions to an understanding of the nonlocality of electromagnetic field effects in superconductors, including his introduction of the concept of a coherence length, ξ.

In a normal conductor when frequency is increased to a point where wavelength is comparable to the mean free path of an electron, one sees deviations from the usual skin effect, that is, the "anomalous" skin effect. In this situation, the electrons which contribute to the current at a particular point have momentum acquired from the field at other locations and earlier times. Therefore, the current at a point is determined not only by the field at that same location (as is the case for lower frequencies), so one must attempt to account for nonlocal effects (Van Duzer & Turner, 1981).

Superconductivity presents a somewhat analogous situation. Since the superconducting current is comprised of pairs of electrons, one must be concerned about the extent of nonlocal effects due to the pair "size," that is, the range over which a pair can remain coherent. The effect on the pair current density at a specific point depends on the vector potential **A**, which is defined by its relation to the induction **B**:

$$\mathbf{B} = \nabla \times \mathbf{A} \qquad 1.3$$

Pippard's integral expression for current density **J**, where r represents the distance from the point of interest in the superconductor is (using his own formalism for a volume integral):

$$\mathbf{J} = -\frac{3}{4\pi\xi_0\Lambda}\int \frac{\mathbf{r}(\mathbf{r}\cdot\mathbf{A})e^{-r/\xi}}{r^4}\,d^3r \qquad 1.4$$

where Λ is the same variable (related to penetration depth) used in the London theory, ξ_0 is the coherence length in the clean limit, and ξ is the more general coherence length for impure superconductors. This expression is very similar to an earlier one known as "Chamber's formula" for normal metals. Pippard recalls "slipping in a free path-dependent coherence length, ξ" and "guessing" that

$$\frac{1}{\xi} = \frac{1}{\xi_0} + \frac{\alpha}{\iota} \qquad 1.5$$

where $\alpha \sim 1$ and ι represents the mean free path. We should all guess with such style. He points out that this guess "fitted the results quite well" (Pippard, 1987). Essentially, Pippard's work revealed that the current at a specific point in the superconductor was a function of the vector potential **A** in a region *near* the point. This nonlocal effect falls off exponentially as one moves away from the point, with a characteristic decay length ξ.

This new understanding of nonlocality in the behavior of the electromagnetic field in the superconductor was a crucial step in the advances that were to be seen shortly. It was just the beginning of great progress during the decade of the fifties. The greatest advance in the understanding of superconductivity, without equal since Gilles Holst had measured Onnes' frozen mercury, would take place on the prairies of Illinois.

BCS THEORY: THE GREAT ENLIGHTENMENT

During the period from approximately 1930 to 1950, a considerable effort was expended by theorists in their attempts to understand the mechanism of superconductivity, and a variety of talented individuals attempted to explain the strange behavior with the quantum mechanical models. They applied their efforts to explanations that used single-electron phenomena, an approach that had been extremely successful in dealing with other important problems. This was done in spite of the fact that most of these scientists realized that superconductivity involved the interactions between enormous numbers of electrons, that is, it was a many-body problem. Before 1950, the methods needed for dealing with many-body problems, or "collective phenomena" simply did not exist (Hoddeson, et al., 1989). When you don't have the right instruments at your disposal, you often make do with what is at hand. On some occasions, it is not enough.

It was during the autumn of 1955, seven years after the publication of London's remarkable flux quantum prediction in *Superfluids*, that Bardeen, Cooper, and Schrieffer began final work on a microscopic theory of superconductivity based on a "condensation" of electron pairs into "Cooper Pairs" (Bardeen, et al., 1957). While this remarkable theory seemed to appear on the scene almost overnight, John Bardeen had actually been laying the groundwork for a period of years.

By 1955, Bardeen was in need of someone steeped in the currently fashionable theoretical techniques of Feynman diagrams and renormalization methods. In his quest for the right talent, Bardeen contacted Frank Yang at Princeton, who promptly recommended Leon Cooper. Cooper had landed a post-doctoral appointment at the Institute for Advanced Studies at Princeton, following his Ph.D. work at Columbia. He recalls that he first met John Bardeen in the spring of 1955. The time was easily remembered as the end of an era: Albert Einstein had just died. This was the first time that Cooper had heard of superconductivity, since Columbia did not offer a course in solid state physics. There had been a brief mention of the subject in Zemansky's *Heat and Thermodynamics*, but he does not remember this section being assigned reading in his course on thermodynamics. Cooper recalls that while he did not absorb much about the problems of superconductivity in his first meeting with Bardeen,

> . . . since at the time prospects for progress in field theory seemed rather discouraging, I decided to take the plunge.
>
> (Cooper, 1987)

Leon Cooper had been lured away from the "true religion" of high-energy physics, but his work would enervate the work at the University of Illinois. He arrived in Champaign-Urbana during September 1955 and was assigned a wooden desk in Bardeen's office that had served David Pines and would later be used by Gerald Rickayzen. He remembers that Robert Schrieffer, one of Bardeen's graduate students, was in an attic office. A sign on Schrieffer's door proclaimed:

Institute for Retarded Studies

Bardeen had assigned superconductivity as the subject for Schrieffer's Ph.D. thesis. While Schrieffer was destined to make crucial contributions to the BCS theory, there were periods of doubt. He recalls one instance in 1956 that was particularly daunting, when Richard Feynman made approximately the following remark after failing in his own efforts to solve the riddle of the superconducting mechanism:

> What we must do is not to compute anything . . . but simply to guess what makes the ground state special. The only reason we can't do this problem is that *we don't have enough imagination*.

This observation by one so brilliant as Feynman was enough to demoralize even the most gifted graduate student. Schrieffer began to wonder if he would ever make enough progress for a thesis on his assigned subject.

> Well that's a very discouraging thing if *Feynman* doesn't have enough imagination . . . So I began working on itinerant ferromagnetism without Bardeen knowing about it. I figured I'd better hedge my bets here; I think: 'we're in some deep, deep water.' Fortunately, Bardeen convinced me in December of 1956 to not give up on the problem; he said to 'spend one more month.' He went off to Japan to a meeting.
> (Schrieffer, 1988)

John Bardeen may have been distracted by other matters; the professor had a pressing appointment in Stockholm, where he would receive the 1956 Nobel Prize in physics for his role in the invention of the transistor (Hoddeson, et al., 1989).

Before his month was up, Schrieffer was also traveling, but not to the rarified atmosphere of Stockholm. Not yet. During January of 1957, the graduate student journeyed to the fair city of Hoboken, New Jersey to attend a meeting that dealt with physics of the many-body problem. He listened to a variety of learned presentations from big names in the field; statistical mechanics was a common feature of the discussions. Inspired, Schrieffer decided to try a new direction. He wrote down a variational form of the wave function for the superconducting ground state that took electron pairing into account (Hoddeson, et al., 1989). This new approach led to an essential breakthrough. In this model, the electron-pairing mechanism was exponentially stronger than had been seen in earlier theoretical attempts. Schrieffer had succeeded in "embodying the pairs in a wave function that satisfied the Pauli principle" (Cooper, 1987).

Leon Cooper met his colleague at the airport, and shared Schrieffer's excitement over this encouraging development. Within minutes of inspecting his graduate student's work, Bardeen expressed his conviction that this was the solution to the problem. The trio then worked furiously for several days to construct the complete framework of their theory and compare calculated quantities to experimental results. It was not a time to tarry; they suspected that the formidable Feynman, who was still applying his imagination to the problem, might be "hot on the trail" (Hoddeson, et al., 1989).

The bottom line was that superconductivity results from an attraction between electron pairs; this attraction is a result of electron coupling to phonons (lattice vibrations). A letter describing their accomplishment was mailed to *Physical Review* on February 15, 1957. The "official" announcement of the breakthrough was made via a pair of post-deadline papers at the March 1957 meeting of the American Physical Society in Philadelphia, just one week before publication of the *Physical Review* letter. Bardeen was concerned that his presence might prevent Cooper and Schrieffer from getting their due credit, so he did not attend. Since Schrieffer did not hear about his expected attendance in time to be in Philadelphia, Cooper had to present both papers (Hoddeson, et al., 1989).

This elegant theory, while difficult to understand, was extremely successful, and has been referred to as "the great enlightenment" (Pippard, 1987). An example of the enormous success of this theoretical framework was the detailed BCS explanation of the isotope-mass effect on critical temperature and critical magnetic field.

> . . . most calculations were as simple as for the normal state—even simpler. The whole thing was analytic and you could predict things and you got quite detailed agreement with most experiments . . .
>
> (Schrieffer, 1988)

Experimentalists, who were delighted with its predictive power, readily accepted the BCS theory. Some theorists, on the other hand, were not so eager to accept this new theory without some resistance. This is not inappropriate; it is a hallmark of genuine science to test any new idea with appropriate criticism. Pippard, however, suggested that there may have been some emotion involved as well, particularly by those who had already considered and rejected the electron-phonon interaction:

> . . . anyone who had travelled that road and decided that it led nowhere was understandably irritated by the suggestion that he had missed discovering the theory of superconductivity . . .
>
> (Hoddeson, et al., 1989)

Eventually, every reasonable criticism was answered. Bardeen, Cooper, and Schrieffer were awarded the 1972 Nobel Prize in physics for their landmark accomplishment. As often happens after a great success, some investigators subsequently felt that virtually everything of fundamental importance was now known about superconductivity; it was only necessary to work out a few details. Some of the "details" would prove to be astonishing.

AT LAST: PRACTICAL HIGH-FIELD SUPERCONDUCTORS

It was actually before the development of BCS theory that great progress was made in what might be termed the "engineering" of useful superconductors.

Great progress was made in the Soviet Union during this period. N. V. Zavaritski, who was in the habit of preparing thin films of pure metals, altered his process one day and generated a rather amorphous structure. The Ginzburg–Landau relationship between critical magnetic field and film thickness did not appear to hold for this peculiar product. This was noticed by a young theoretician, Alexei Abrisokov, who proposed in 1952 that a new type of superconductor, characterized by a negative surface energy, should be recognized as a distinctly different class from the pure metallic superconductors, which exhibited a positive surface energy. This group of alloys and "dirty" superconductors was subsequently known by several names, but most commonly today as "Type II." "Type I" is used to specify the pure metallic superconductors that had been the subject of so much study by Onnes and his contemporaries. It was the breakthrough that Onnes had not lived to see; the Type II materials would prove to be capable of transporting enormous currents and operation in high magnetic fields. It is also noteworthy that, up to the time Abrisokov formulated his theory, all previous theories had considered only pure metals.

Much of the scientific progress in the Soviet Union during the early 1950s, including the important theoretical advances in superconductivity, were late in reaching the United States. The hysteria of the McCarthy era had peculiar effects on virtually every American institution; in some universities it was considered risky to wear a red tie! Until this absurdity subsided, American scientists were relatively isolated from their Soviet counterparts. Nevertheless, progress continued.

Early in the 1950s, George Yntema investigated the behavior of niobium wire on an iron core; he was able to generate fields up to about 7 kgauss (0.7 tesla (T)) (Yntema, 1987). John Hulm was able to confirm these results and proceeded to test air-core windings of niobium. This work was continued by Stan Autler of Lincoln Labs, who achieved fields of 4 kgauss with niobium solenoids.

The technology of superconductors for practical applications was given a significant boost by Eugene Kunzler and his colleagues, who performed extensive tests on an alloy of tin and niobium (Nb_3Sn) developed by Bernd T. Matthias and Ted Geballe (Matthias, et al., 1954; Kunzler, et al., 1961). This was another of the materials with a crystalline structure known as "A15" by the chemists, a class of materials investigated earlier by George Hardy and John Hulm at the University of Chicago. Kunzler recalls the "day of discovery," December 14, 1960:

> The first sample . . . was still superconducting at 88 kgauss even with a substantial current passing through the sample . . . we first doubted our measurements—was the magnet on? Was the current going through the sample?
>
> (Kunzler, 1987)

The magnet *was* on, the current *was* flowing. The age of high-field superconductivity was at hand. A second alloy, niobium-titanium, exhibited similar characteristics and was ductile enough for making wires. Current densities of 100,000 A/cm^2 at fields of 88,000 gauss had been achieved. Kamerlingh Onnes' dream had come true, but it had taken half a century. Even in the late 1980s, it is these A15 materials, used in high-field magnets, that are responsible for the bulk of applications in superconductivity.

IVAR GIAEVER; SINGLE-ELECTRON TUNNELING

The tunneling of electrons through a potential barrier, a quantum-mechanical phenomenon, was predicted by theorists as early as 1928 and was immediately used to explain such diverse phenomena as field emission from cold metals and α-decay (Esaki, 1974). By the mid-1950s, Leo Esaki had developed the p-n junction tunnel diode, and engineers began to think of ways to use the negative-resistance characteristic of this device in high-frequency oscillators. The individual who would play a key role in the investigation of tunneling in superconducting junctions was in Norway, struggling with his studies. He found games more interesting than serious study.

In 1952, Ivar Giaever (Figure 1.4) received his degree and began what promised to be a typical career in mechanical engineering. The young man and his wife made a decision to emigrate from Norway to Canada, where the housing shortage was not so serious. He found a job with Canadian General Electric and signed up for their three-year course in engineering and applied mathematics. Giaever, now more serious, realized that he had not applied himself in Trondheim; now he "really studied hard for a few years" (Giaever, 1974). The family then moved to GE's headquarters in Schenectady, New York, where he decided to take the opportunity to "learn some physics." Giaever soon found himself working with John Fisher.

Fisher, aware of Esaki's tunneling experiments, was interested in experiments with thin films that might lead to useful devices. Fisher and Giaever put in long hours attempting to produce unambiguous tunneling through thin insulating films; the experiments were plagued with pinholes through the insulator. These sometimes led to shorts and false indications of tunneling that were

FIGURE 1.4 Ivar Giaever, a young mechanical engineer from Norway, had the insight to relate tunneling in superconducting films to the energy gap. Giaever (shown above), Brian Josephson (Cambridge University), and Leo Esaki (IBM) were awarded the 1973 Nobel Prize in physics for their various contributions to the understanding of electron tunneling. (Photograph provided by Rensselaer Polytechnic Institute)

responsible for a missed opportunity later. Of these early experiments, Giaever recalls that they were all considered to be "miracles," because none could be repeated!

The first important advance made by Fisher and Giaever resulted from the realization that they could form a thin, pinhole-free insulator by growing metal oxide layers. By 1959, they were routinely producing tunneling effects with aluminum films separated by oxide insulators less than 50 Å thick. These measurements, of course, were performed with the junction at room temperature.

The critical moment came while Giaever, always taking opportunities to "learn more physics," was listening to a lecture on solid state phenomena when the professor, discussing superconductivity, mentioned the energy gap. That, Giaever reports, "really caught my attention." He realized that, when the voltage across the junction was low enough, the "forbidden energy gap" should prevent tunneling of electrons whose energy was in this prohibited region. Such a phenomenon should be easy to see, since the plot of tunneling current versus voltage was linear when the metal was in the normal state.

This "gap" (usually represented by the symbol Δ) in the spectrum of electron energy had been the subject of debate for several years. Evidence was seen for the gap in the exponential behavior of the specific heat of tin for temperatures below $\sim T_c/2$. The gap was not merely of theoretical interest; it would prove to be of fundamental importance in determining the thermal properties as well as the high-frequency electromagnetic response of superconductors (Lynton, 1962).

While most early theorists rejected the idea on the basis of gauge invariance, BCS theory revealed a ground state that "breaks" local gauge theory (Anderson, 1969).

With a lot of help from his friends at GE, Giaever managed to put the cryogenics system together. He used a junction of lead and aluminum separated by the familiar aluminum oxide. Initial experiments failed because the oxide layers were too thick; when he reduced the thickness to about 30 Å, "the current-voltage characteristic changed markedly when the lead changed from the normal state" (Giaever, 1974). Figure 1.5 illustrates how the linear current-voltage characteristic of normal-state tunneling is depressed until a critical voltage is seen across the insulator. Giaever had demonstrated a simple means for measuring the energy gap, which was proportional to the "critical voltage" where electrons began to tunnel. When he was able to cool the system to a temperature where the aluminum became superconducting ($\sim$1.2 K), Giaever saw evidence (which he had predicted) of negative resistance in the I-V curve.

When the young mechanical engineer published his findings in August 1960 (Giaever, 1960), the importance of the discovery to basic science was already apparent. Some scientists speculated the negative resistance characteristic seen in Giaever's tunneling junctions would lead to important applications in amplifiers and switches. In fact, the developing technology of semiconductors would prove to be far superior to superconductor electronics for the great proportion

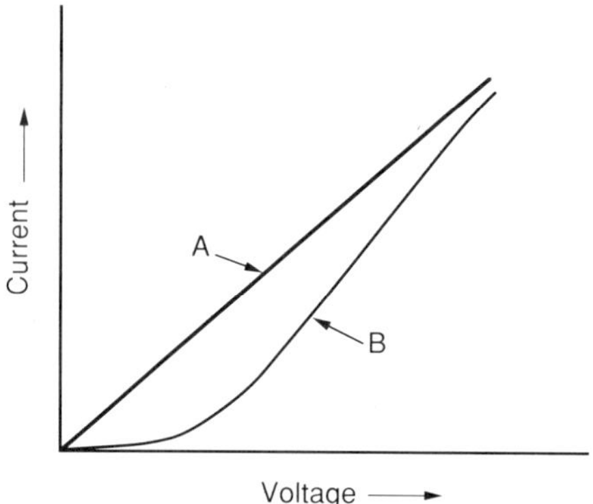

FIGURE 1.5 Ivar Giaever's measurements of electron tunneling through a thin sandwich made of aluminum/aluminum oxide/lead exhibited normal tunneling (A) when the lead was held in the normal state by an applied magnetic field of 2700 oersteds. When the magnetic field was removed (B) at a temperature of 1.6 K, the lead became superconducting and electron tunneling was greatly attenuated until the injection energy exceeded the energy gap Δ. This is a far simpler technique than *IR* transmission methods. (After I. Giaever, "Energy gap in superconductors measured by electron tunneling," *Phys. Rev. Lett.*, Vol. 5, No. 4, Aug. 15, 1960, pp. 147–148.)

of practical applications. Nevertheless, the discovery was remarkable in that it made the determination of the superconducting energy gap a rather simple matter compared to the earlier infrared measurements (Glover & Tinkham, 1957). Nothing so important with such straightforward experimental technique had been seen in experimental superconductivity since Kamerlingh Onnes and his graduate student applied the Wheatstone bridge to the sample of mercury.

In the words of one of his contemporaries at General Electric: "the discovery was elegant and had the aesthetic simplicity that makes a scientist wonder why it had not been made before" (Schmitt, 1961). There were several reasons, but perhaps a major factor was that Giaever, at this early stage in his career, was unaware of all the subtle "reasons" why the experiment was considered unlikely to succeed.

Aside from his major discoveries, some other very odd things were happening during the experiments with the superconducting junctions; Giaever occasionally noticed these rather peculiar phenomena but did not consider them to be of any special significance. This time, the young man from Norway missed something of great importance. We will come back to that story shortly.

BRIAN JOSEPHSON; PAIR TUNNELING

This work by Giaever was followed, in 1962, by a startling theory that originated on the other side of the Atlantic. Brian Josephson, a brilliant graduate student at Cambridge University, was working at the Royal Society Mond Laboratory, supervised by Brian Pippard. Josephson recalls his own early difficulties with the subject:

> . . . my first acquaintance with the theory of superconductivity was with a formidable book "A New Method in the Theory of Superconductivity" by Bogoliubov, Tolmachjev and Shrikov . . . which it was suggested I try to understand. I found this book almost totally incomprehensible . . .
>
> (Josephson, 1986)

Josephson had learned of Giaever's work on electron tunneling. Giaever had considered the possibility of superconductor-pair tunneling but, like others, decided that such behavior was unlikely to be measurable since the probability of two electrons tunneling together would be vanishingly small.

Josephson continued to think about the problem and, like Giaever, gained his initial inspirations in the classroom. During a series of lectures by P. W. Anderson, where the subject of broken symmetry attracted his attention, he began to wonder if the conventional wisdom was to be trusted. In contrast to Giaever's largely experimental approach, Josephson solved his problems by reasoning and calculation. He considered the *coherence* of the electron pair as he attempted his difficult calculations. Eventually, his expression for current flow through a superconductor-insulator-superconductor junction predicted a current that would flow when the voltage across the junction was zero (Josephson, 1962). Furthermore, if a small dc voltage (V) was applied across the junction, the current would oscillate at a frequency equal to $2eV/h$! No one, of course, had expected such peculiar behavior in the simple junction structure.

Josephson reports that there was an "embarrassing feature" in his theory. The current levels predicted were simply too large. The unknown graduate student, under these circumstances, was cautious. His theory, he thought, might turn out to be "of mathematical interest only" (Josephson, 1974).

Josephson made the following entry in his notebook:

> Meanwhile we turned our minds to the question of transport properties of superconducting boundaries. A self-consistent field equation for quasi-particles, treating them as two-component electron-plus-hole wave functions, was developed and the consequences explored. However, curious difficulties arise when treating SIS or SIN junctions. An approximate treatment was thought up and seems to imply oscillatory effects. It seems to be rather difficult to persuade experimentalists that these are more than a mathematical fiction, but we hope to publish a paper on these effects, and then someone may try the experiments . . .
>
> (Josephson, 1986)

The title of his publication in the July 1, 1962, issue of *Physics Letters* was "Possible New Effects in Superconductive Tunneling." The skepticism which met his ideas seemed to justify the cautious title. Common sense suggested that such a large effect would have already been seen in experiments such as those performed by Giaever. There was also the misunderstanding about the probability of coincident tunneling of two electrons mentioned earlier. However, when the coherence of the paired electrons is considered, the actual probability of pair-tunneling is much higher (Tinkham, 1985).

A third factor may have been the fact that Josephson's thoughts were not easy to follow. Giaever remembers that someone told him about Josephson's short publication. He recalls his reaction after reading the report: "What did I think? Well, I did not understand the paper." Giaever also remembers that, after meeting Josephson soon afterwards, he "came away impressed."

The fact is, Giaever and his colleagues *had* seen the currents predicted by Josephson, and other investigators had even published I-V curves that demonstrated their existence. Recall the shorts through the insulating junctions that had plagued Giaever and Fisher. The Josephson current, along with any other unexplainable phenomena, was typically ascribed to shorts through the insulator. Since the shorts were not present in every junction, it seemed a reasonable conclusion. In Giaever's own words,

> . . . all the samples we made showing the Josephson effect were discarded as having shorts . . . Later I have been asked many times if I feel bad for missing the effect? The answer is clearly no, because to make an experimental discovery it is not enough to observe something, one must also realize the significance of the observation . . .
>
> (Giaever, 1974)

The reason that the supercurrents were not always present in the junctions, as P. W. Anderson realized, was their exquisite sensitivity to magnetic fields. Most were extinguished by the earth's field. Happily, Josephson's predictions were confirmed in experiments by John Rowell and P. W. Anderson at Bell Laboratories only a year later (Fitzgerald, 1988). The predictions by Josephson, which seemed so unlikely, are now the very heart of a variety of practical superconducting devices used for ultrasensitive magnetic field measurements and voltage standards, as well as for computer logic gates and memories.

Ivar Giaever and Brian Josephson each received the Nobel prize in physics for their work. Giaever, with characteristic humility, noted the fact that an Oslo newspaper headlined the story of their native son roughly as:

Master in billiards and bridge,
almost flunked physics
—gets Nobel Prize

Anxious to correct any misconceptions that this story might leave about his academic career, Giaever quickly pointed out that he had "almost flunked mathematics as well."

Josephson, a theorist to his marrow, delivered a Nobel address that, while generally fascinating for its history, included arcane material that was almost as difficult to follow as his initial paper.

Before leaving these pioneers, so different in their approach and yet so complementary in their respective contributions of practical experiment and deep theoretical insight, I cannot resist the temptation to compare Giaever and Josephson with two other individuals who exhibit similar qualities in their approach. The plain-speaking fisherman, while writing a letter to folk with backgrounds similar to his own, refers to a brilliant Talmudic scholar who is able to perceive truth at a deeper level:

> . . . Paul, according to the wisdom given to him, has written to you, as also in all his epistles, speaking in them of these things, in which are some things hard to understand . . .
>
> (2 Peter, 3:15–3:16, New King James Version)

COMPUTER APPLICATIONS

Once the work of Josephson and Giaever was widely appreciated, it became apparent that switches, perhaps even logic gates, could be developed using the superconducting junctions. One could switch between the superconducting "Josephson" supercurrent flow at zero junction voltage to the "Giaever" operating point at 2–3 mV where unpaired electrons (quasiparticles) would provide the current flow.

This type of gate would not only be fast, but due to the intrinsically small currents and voltages involved, it would consume very little power compared to semiconducting switches. Time was not wasted; by 1966, Juri Matisoo had demonstrated a device that he called a "tunneling cryotron," in reference to the relatively slow cryotron switch (Matisoo, 1980). This logic gate was the basis for a superconducting "Josephson" computer!

IBM corporation, in particular, has devoted an enormous effort over a 14-year period to the development of Josephson technology for computer logic and memory applications. Although Big Blue made considerable progress in solving problems with lead-alloy junctions for operation at liquid helium temperature, the effort was finally halted in 1983 after the apparent advantages in performance over more conventional (nonsuperconducting) technologies appeared to be narrowing. It was evidently determined that it would be more prudent to invest in gallium arsenide and high-speed silicon semiconductor technology (Fitzgerald, 1988).

It has also been reported that the project was abandoned because of technical difficulty in fabricating the junctions reliably enough for mass production (Garwin & Campbell, 1987). The cache memory needed for a Josephson computer was a very difficult design problem (Hayakawa, 1986). It remains to be seen whether recent developments with copper-oxide superconductors will eventually reverse this trend . . . but that is getting ahead of our story.

PEROVSKITE-RELATED STRUCTURES

A 1975 report by Sleight, Gillson, and Bierstedt of superconductivity at 13 K in $BaPb_{1-x}Bi_xO_3$ was primarily of interest to theorists, since better practical superconductors at higher temperatures were already in use. The report is now considered quite significant in the history of superconductivity, since the material had "perovskite related structures" (Sleight, et al., 1975). It was a hint of remarkable developments to come. (See Appendix F for a brief discussion of perovskite characteristics.)

HIGH-TEMPERATURE SUPERCONDUCTIVITY; BEDNORZ AND MÜLLER

The modern age of the so-called high-T_c superconductors began in April 1986 with a report by IBM-Zurich's K. Alex Müller and J. Georg Bednorz of superconductivity in lanthanum-barium copper oxide at 30 K.

Georg Bednorz had a long-standing interest in the characteristics of perovskite crystals. He had worked with Gerd Binnig in a study of the superconducting properties of $SrTiO_3$. Together, they increased the T_c of this perovskite from 0.3 K to 1.2 K by doping the crystal with niobium. Bednorz found this experimental effort exciting, but Binnig eventually lost interest and began work with Heinrich Rohrer on a scanning tunneling microscope (STM). While Bednorz was disappointed to lose his worthy collaborator, the new direction was a fortunate one for Binnig; he and Rohrer won the 1986 Nobel prize in physics for development of the STM.

Bednorz did not lose his interest in oxide superconductivity. When he was asked by Alex Müller to collaborate in a search for new superconducting oxides, Bednorz recalls that Müller was "knocking on a door already open" (Bednorz & Müller, 1988). Both were aware of Arthur Sleight's 1975 discovery of 13 K T_c in $BaPb_{1-x}Bi_xO_3$; the challenge was to find oxides with even higher critical temperatures. After several false starts and disappointments, the turning point came in late 1985 when Bednorz was alerted to a publication by the French team of C. Michel, L. Er-Rakho, and B. Raveau who reported a metallic-like conductivity in La-Ba-Cu-O in the range from −100°C to +300°C. The French team had not been investigating superconductivity, but their results raised a flag for Bednorz. The material was a perovskite, and the metallic behavior was enough to raise the curiosity of the duo in Zurich.

An eager Bednorz immediately went to the ground-floor laboratory and prepared to mix the appropriate ingredients. By the next day, the synthesis was complete but the resistance measurement had to be postponed. The Director of Research was visiting IBM's Zurich laboratory, and such events are much like an admiral's inspection of a craft at sea. The brass is polished to a mirror finish, the decks swabbed down, and all hands are presented on deck for examination. Such matters as charting the course into the future must wait until another day!

Happily, Bednorz and Müller had their day. The former recalls that, as initial resistivity tests were being performed on the sample, the atmosphere in the laboratory was a mixture of anticipation and anxiety.

> My inner tension, always increasing as the temperature approached the 30 K range, started to be released when a sudden resistivity drop of 50% occurred at 11 K. Was this the first indication of superconductivity?
>
> (Bednorz & Müller, 1988)

Indeed it was. In subsequent refinements during the next two weeks, T_c was increased to 30 K. Bednorz and Müller were painfully aware that earlier reports of high-temperature superconductivity by capable investigators had turned out to be false alarms. The team worked furiously to eliminate the possibility that the transition might be something other than superconductivity. Could the sharp drop in resistance be a metal-to-metal transition? This seemed unlikely, since the transition shifted to lower temperatures as sample current was increased. This would not be expected with a metal-to-metal transition, but was a hallmark of superconductor quenching behavior.

They had everything they needed to prove the case except the most important evidence: a shift to diamagnetic behavior at temperatures below the resistivity transition. Bednorz and Müller did not have access to a sensitive magnetometer at this early stage. Even so, they were convinced of their findings and decided to submit a paper for publication before the DC SQUID magnetometer arrived. Bednorz and Müller approached a friend on the editorial board of *Zeitschrift für Physik*, a journal with a relatively limited circulation. Eric Courtens, in spite of the absence of magnetization data, agreed to submit the paper for publication. Georg Bednorz recalls that this service "did involve some gentle persuasion on our part." Like Brian Josephson's paper, the title was cautious:

> "Possible High T_c Superconductivity in the Ba-La-Cu-O System"

Since there had been several earlier reports of superconducting transition temperatures well over 25 K that were subsequently not verified, the Müller-Bednorz publication did not initially generate a great deal of excitement (Bednorz & Müller, 1986). Bednorz recalls that as he and Müller began giving talks about their discovery, they were "met with a skeptical audience" (Bednorz & Müller, 1988).

The skepticism was to be short-lived. Following verification by Shoji Tanaka's group at Tokyo University, there was a surge of interest from the scientific community (Müller & Bednorz, 1987). Shortly after this, confirmation of the IBM-Zurich findings were reported by Bertram Batlogg and his colleagues at AT&T Bell Laboratories as well as by the group headed by Paul Chu in Houston.

Once accepted, the Müller-Bednorz discovery led to interest in a variety of new compounds that, like La-Ba-Cu-O, are structurally similar to a class of ceramics known as perovskites.

The new developments also suggested that superconductivity could occur in a manner that was not predictable (or perceived to be predictable) from the classic BCS theory that had elegantly explained the phenomenon at lower temperatures. In addition, the possible commercial application of materials with T_c at $\sim$ 30–35 K was a distinct improvement over the best to date for niobium-tin alloys at about 23 K. This was just the beginning; the news was to keep getting better. But first, back to the pioneers.

In October 1987, after a brief period unprecedented in the history of the Nobel prize in physics, Müller and Bednorz (see Figure 1.6) were awarded the Nobel prize for their discovery. Each reported that they would use their $140,000 checks for mortgage payments and continue their work to determine precisely why these particular copper oxides are capable of superconductivity. Many other investigators, no doubt, will also be spending their careers trying to gain a similar understanding, but a substantial number of large mortgages now outstanding will probably be paid in full before the phenomenon is understood in detail.

Late in 1986, Paul Chu and his colleagues at the University of Houston demonstrated that, under tremendous pressure, La-Ba-Cu-O showed unmistakable signs of superconductivity well above 40 K. The T_c, in fact, increased at about 1 K per kilobar, reaching 57 K at 12 kbar. This work suggested to the groups at Houston and Alabama that the barium should be replaced with smaller strontium, to simulate the effect of high pressure. The substitution of strontium worked; the onset of superconductive behavior was as high as 42.5 K. At about the same time (December 1986) a team at AT&T Bell Laboratories at Murray Hill, New Jersey had also demonstrated superconductivity in La-Sr-Cu-O at 36 K (Fitzgerald, 1988).

It thus became apparent that La-Ba-Cu-O might be only one member of an entire family of perovskites that would superconduct well above 30 K.

HITTING THE JACKPOT IN ALABAMA

In Huntsville, M. K. Wu and his two graduate students, Jim Ashburn and Chuan-Jue Torng, made several attempts to improve T_c in La-Sr-Cu-O. Various substitutions for the barium (in La-Ba-Cu-O) were attempted. Magnesium was a total failure. Calcium substitutions resulted in a superconductor, but T_c was lower than that for La-Ba-Cu-O.

On January 17, Jim Ashburn performed some calculations on the reverse side of a sheet of paper containing a homework problem (Pool, 1988). He was considering substitutions for *both* La and Sr in La-Sr-Cu-O. Considering effects on size of the unit crystal, Ashburn generated a chart that suggested yttrium was an appropriate element to substitute for lanthanum; likewise barium for strontium. The fact that the Y-Ba-Cu-O ceramic superconductor can now be produced almost routinely in high school science laboratories leads many to believe that once the notion of substituting yttrium for lanthanum was in hand,

FIGURE 1.6 J. Georg Bednorz (left) and K. Alex Müller of IBM's Zürich Research Laboratory were awarded the 1987 Nobel Prize in physics for their milestone discovery (in 1986) of a La-Ba-Cu-O ceramic that was superconducting at 30 K. This remarkable achievement suggested an unexpected class of superconducting mechanisms and paved the way for development of additional copper oxides with even higher critical temperatures. (Photo provided by IBM-Zürich Research Laboratory)

making the 90 K superconductor was almost automatic. It is not so. Several savvy individuals had thought of simply replacing lanthanum with yttrium, but the breakthrough was not so easy.

The process of invention involves much more than simply having an idea; one must also prove its efficacy. Other than those trivial cases where suggested performance is obvious (like putting wheels on a suitcase so that it can be pulled instead of carried), the inventor is expected to be able to demonstrate that the invention will perform the intended function. This is, in fact, a fundamental principle of law; a concept known as "reduction to practice" by the patent examiner.

In any case, there it was: Y-Ba-Cu-O, or more precisely in the Alabama plan, $Y_{1.2}Ba_{0.8}CuO_{4-\delta}$. The stoichiometry was not yet optimized, but the right ingredients were in the pellets and the required phase would come out in the processing schedule that followed.

There was a hitch; Professor Wu had no yttrium and it was hardly a time to wait weeks for delivery. Fortunately, the Alabama group was able to obtain a modest supply from friends at the Marshall Space Center in Huntsville. The reader should know that this was not an experiment supported by a lavishly equipped laboratory. On the contrary, one of the principals reports that

> . . . the first Y-Ba-Cu-O samples were made with borrowed chemicals, weighed in a borrowed balance, pressed into pellets with a borrowed die, and fired in a borrowed furnace.
>
> (Ashburn, 1989)

The fundamental scientific idea driving this experiment, the essential element, was not borrowed.

By January 28, Ashburn and Chuan-Jue Torng had prepared a mixture of Y-Ba-Cu-O. That evening, the new compound went into the oven for overnight processing. The next day, the black and green charcoal-like materials were removed from the oven and made ready for a test of resistance versus temperature. If one considers the experiment and the possibilities, it could have been Leiden, the year could have been 1911.

The resistivity measurements were displayed on a computer screen, point by point, in real time as the sample was cooled. Ashburn recalls that the experimentalists would watch the computer screen as the sample was cooled, "looking for any hint of a downward turn" in resistivity. Occasional noise, if it had the appearance of a sharp decrease in resistance, would startle the investigators:

> . . . the occasional bad data point would make for a good laugh as we would start counting the points (about one every few seconds) marking a possible transition . . . But on that January afternoon, as we watched every point as it was plotted, and counted, as point piled upon point, *we stopped laughing*.
>
> (Ashburn, 1989)

On January 29, the first resistivity measurement was completed at precisely 2:08 p.m. The results were far beyond the expectations of the investigators; Wu recalls that this first sample tested began to exhibit superconductivity at 90 K. The atmosphere was electric; an elated Ashburn invited a friend to watch the measurements and pointed out that the visitor was "looking at history." The comment was hardly an exaggeration. (See Figure 1.7.)

Even so, the investigators realized, colleagues might not accept such a remarkable measurement uncritically. The resistance measurement was rock-solid, a standard procedure. Would the thermometry be questioned? During the previous week, the investigators had experienced some trouble with the thermocouple reference junction; this had caused an offset in previous temperature measurements. The experimenters had no liquid nitrogen in the laboratory, a substance that was about as useful as ice water for cooling conventional superconductors. They literally ran down the hall to find some in another laboratory. Within minutes, the ceramic sample was moved from liquid helium to liquid nitrogen, where the temperature could not drop below 77 K. By 2:22 p.m. the verdict was in: the material was definitely a superconductor in liquid nitrogen. Ashburn remembers that, when this experiment was finished "the doubts vanished" (Ashburn, 1989). There could be no doubt that a historic breakthrough had been made (Wu, et al., 1987). Following the experiment, Wu was so shaken that his hands were trembling. Later that day, Wu called his colleague in Houston, Paul Chu. Wu's message was direct: "We've hit the jackpot" (Pool, 1988).

On January 30, the group in Houston, using a sample hand-carried to Houston by Professor Wu, reproduced the results from Alabama. The Alabama groups earlier conclusion was unmistakable; the sample exhibited a 24% Meissner effect when ac susceptibility was measured. Subsequently, they realized that their intended preparation of $Y_{1.2}Ba_{0.8}CuO_{4-\delta}$ was actually composed of mixed phases, that is, several distinct chemical compounds.

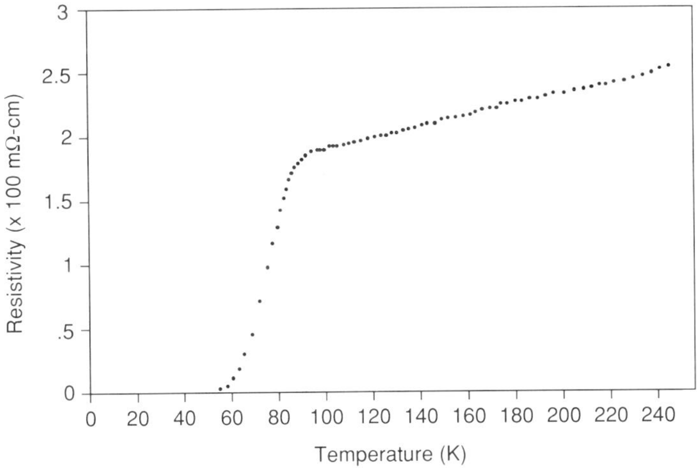

FIGURE 1.7 Resistance vs. temperature for the world's first sample of Y-Ba-Cu-O superconductor from the Wu-Ashburn-Torng team at the University of Alabama in Huntsville. (Figure provided by University of Alabama in Huntsville)

Paul Chu provided a sample to Robert M. Hazen, an experimental mineralogist at the Geophysical Laboratory of the Carnegie Institution of Washington. Hazen found that the historic sample was actually composed of two major phases (Hazen, 1988b). The largest proportion consisted of a transparent, emerald-green crystalline structure, which had 2-1-1 ratios of yttrium, barium, and copper. This material (Y_2BaCuO_5) is now commonly referred to as "that green stuff" (Hazen, 1988a).

The remainder was an opaque, black structure. Since all known superconductors are opaque, the black material was suspected of harboring the superconducting properties. It did. The proportion of yttrium to barium to copper in this orthorhombic crystalline structure was 1-2-3; a label that has stuck to this remarkable material. Hazen reports that he and his colleagues were both delighted and puzzled with the result; they had "never encountered a perovskite with such a low ratio of oxygen atoms to cations." When this oxygen-deficient perovskite was isolated, T_c proved to be approximately 93 K.

It is hard to exaggerate the excitement following the *New York Times* announcement on February 16, 1987, that the groups led by Paul C. W. Chu (Houston) and Mau-Kuen Wu (U. Alabama–Huntsville) had discovered a superconducting ceramic with a T_c above 90 K. Most accounts in the media have overlooked the fact that the material was produced by Wu and his two graduate students in Alabama (Pool, 1988). (See Figure 1.8.)

Several groups immediately produced their own compounds and verified the report that $YBa_2Cu_3O_x$ (where x has a range from approximately 6.4 to 7.0), did, indeed, superconduct at unprecedented high temperatures. Furthermore, analogous compounds with rare earths substituted for yttrium exhibited the same remarkable behavior.

While the previous announcement by Müller and Bednorz of a 30 K T_c had created interest in the advancement of superconductor applications, the news of superconductivity above 90 K galvanized the attention of the technological and commercial sectors. The "old" superconductors required expensive liquid helium for cooling, but these new 1-2-3 materials could be cooled (for some applications) with relatively inexpensive liquid nitrogen at 77 K. The potential advantages in simplicity of refrigeration and lower costs were enormous.

At the New York meeting of the American Physical Society on March 18, 1987, which was immediately dubbed the "Woodstock of Physics," scientists, pressed into the lobby by the crowd, gathered around television monitors until past midnight to hear the speakers describe the incredible new ceramics.

Quite apart from the supercharged atmosphere that had been generated among the condensed-matter specialists who were primarily interested in the science, it seemed probable that a wide variety of instruments and products which had been too costly to operate with the old technology could now be considered with the new materials. If, that is, these odd ceramics could be manufactured with reproducible characteristics and in appropriate shapes (wires, films, etc.). As measurements on the earliest samples were made, it became clear that this would be no easy task. The 1-2-3 materials, like all

FIGURE 1.8 On January 29, 1987, the team at the University of Alabama/Huntsville discovered the 90 K, Y-Ba-Cu-O superconductor. The investigators, from left to right, are Jim Ashburn, C. J. Torng, and M. K. Wu. (Photo provided by University of Alabama in Huntsville)

ceramics, are brittle. Worse still, the current limitations like those Onnes had seen 75 years earlier with Type I superconductors had reappeared in the new materials. In spite of the fact that the critical fields were too high to measure directly ($>$ 100 T), the critical current levels were disappointingly low ($\sim$1000 A/cm^2), in the polycrystalline samples.

Furthermore, the magnetic and current-transport properties of single-crystal samples varied with direction of current flow, with anisotropies as high as 1000:1 eventually reported. If this was not enough, the earliest thin films made with the 1-2-3 compound lost their superconducting characteristics in a few days and samples from different laboratories exhibited wildly different characteristics.

As time went on and much effort was devoted to experiments with variations in processing, the situation gradually began to look better. The stability of thin films was improved. Several groups fashioned wires that had promising characteristics. Even so, processing schedules and techniques, which are the key to optimum performance, appear to be as much an art as a science. One distinguished scientist recalled hearing the following observation, a commentary on the different results from similar Y-Ba-Cu-O processing "recipes" (Teller, 1987):

> This is certainly not physics. It may be chemistry.
> It looks more like *cooking*!

BISMUTH AND THALLIUM COMPOUNDS

During early 1988, when it seemed that the revolutionary developments had reached a plateau, most physicists, chemists, and engineers had settled down to the tedious task of producing better 1-2-3 materials. The theorists, somewhat less subdued, were publishing literally hundreds of papers asserting their varied speculations about the mechanism of superconductivity in these oxygen-deficient perovskites.

Then, when matters seemed almost back to normal, Hiroshi Maeda's group at Tsukuba Laboratories in Japan (see Figure 1.9) reported the bismuth-strontium-calcium-copper oxides, which began superconducting at $\sim$110 K, reaching zero dc resistance at about 80 K (Maeda, 1988). The discovery at Tsukuba Laboratories on December 25, 1987, was the result of persistence and hard work.

These early bismuth compounds contained at least two superconducting phases, with the high-T_c component having a projected zero resistance at approximately 105 K. More ductile and far more stable than the $YBa_2Cu_3O_x$ (1-2-3) compounds, the Bi-Sr-Ca-Cu-O compounds exhibit several superconducting phases (see Figure 1.10) that are not trivial to separate (Maeda, et al., 1988). Even this remarkable report about the bismuth-based 100 K+ materials was not to be the last word in new superconductors.

FIGURE 1.9 Hiroshi Maeda (shown) of Tsukuba Laboratories headed the team, including Yoshiaki Tanaka, Masao Fukutomi, and Toshihisa Asano, that was responsible for the discovery of the bismuth-strontium-calcium-oxide ceramics, with superconductivity above 100 K. (Photograph provided by Tsukuba Laboratories)

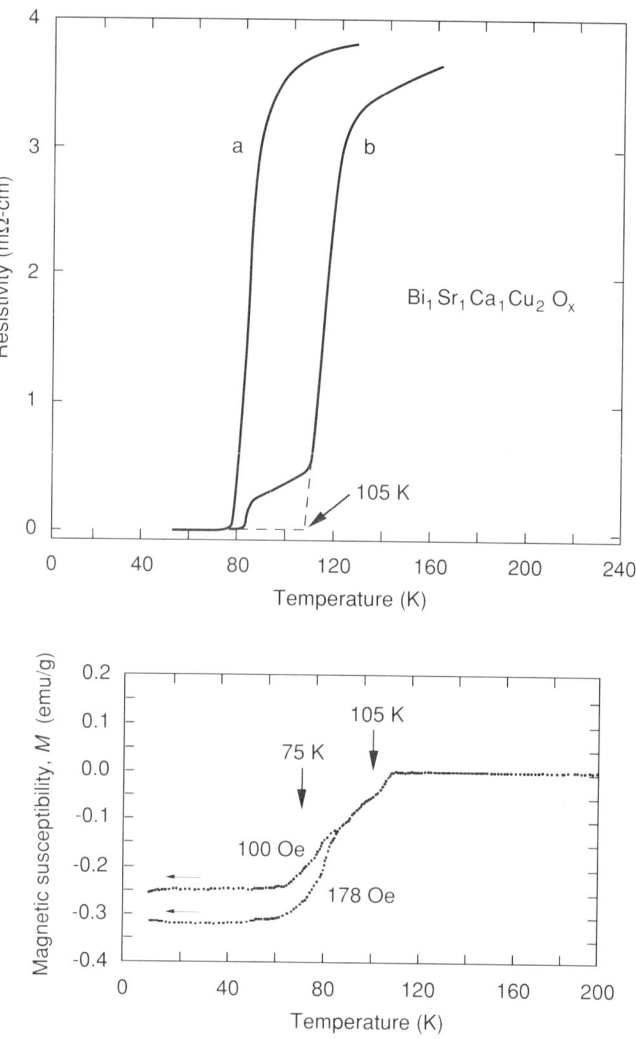

FIGURE 1.10 Resistivity and magnetic susceptibility for the first superconducting Bi-Sr-Ca-Cu-O compound discovered by Maeda & colleagues at the National Research Institute for Metals, Tsukuba Laboratories. (Original figures provided by Tsukuba Laboratories)

The next big surprise came at the World Congress on Superconductivity in Houston during late February 1988 (Khurana, 1988). On Monday, February 22, Zhengzhi Sheng and Allen Hermann (see Figure 1.11) from the University of Arkansas presented a poster that electrified the other investigators. Sheng and Hermann reported success with a thallium compound (Tl-Ca-Ba-Cu-O) that showed an onset of superconducting behavior at ~120 K, and zero dc resistance at 106 K. Sheng and Hermann had made an announcement several days earlier in Arkansas, but their discovery went virtually unnoticed. Following

FIGURE 1.11 Z. Z. Sheng (left) and A. M. Hermann of the University of Arkansas announced their discovery of the 120 K thallium-based superconductor (Tl-Ba-Ca-Cu-O) on February 15, 1988. (Provided by the University of Arkansas)

their poster report, the claim was immediately confirmed by S. Parkin at IBM-Almaden. As with the bismuth compounds, there were a variety of superconducting phases present. A big plus was that there was no troublesome lower-T_c phase that had to be separated, like the 85 K phase that plagued the bismuth compounds (Sheng, et al., 1988).

While the fact that thallium is extremely toxic makes processing of these compounds a task to be treated with appropriate caution, the thallium system shows much promise for applications. Not only are the thallium compounds more stable than 1-2-3 materials; D. S. Ginley and his colleagues at Sandia National Laboratory in Albuquerque have reported that transport current densities are high enough for many applications requirements.

Applications that had looked marginal with 1-2-3 compounds were suddenly more interesting with the new bismuth and thallium-based superconductors. Even so, there were developments at lower temperatures that might eventually prove to be just as significant.

HIGH-TEMPERATURE SUPERCONDUCTIVITY WITHOUT COPPER

While the new high-temperature perovskites based upon yttrium, bismuth, or thallium exhibit intriguing differences in structure and behavior, they all have copper-oxygen planes and exhibit highly anisotropic behavior at the single-crystal level. This anisotropy, a result of the fact that superconductivity occurs in the copper-oxygen planes, is the source of some concern since virtually all existing applications are easier to accomplish with isotropic materials. At the "bulk" or polycrystalline level, this anisotropy is not apparent unless the crystals are oriented.

What it seems to boil down to is that you can have isotropy and low current density in bulk 1-2-3 samples, or high current density in textured films that are anisotropic. This does not sound like a fundamental problem since, in most applications, current does not need to flow in all directions. On the other hand, the production of a practical magnet wire, where the crystals must all be oriented in a particular manner, is hardly a trivial task. Ask someone who makes high-field magnet wire with the "conventional" superconducting materials; it is already a difficult problem, even with isotropic, ductile materials that support high current density.

What we would like to have is high-temperature superconduction in a material that was isotropic in its electric and magnetic properties. A current density of at least 10^6 A/cm^2 at 77 K would also be very welcome. In addition, this current density should not be greatly affected by magnetic fields up to several T.

It is too early to be sure, but there are already signs that something quite like this, while not yet here, may yet be possible. Early in 1988, a group including Robert Cava and Bertram Batlogg at AT&T Bell Laboratories reported superconductivity at about 30 K in a material ($Ba_{0.6}K_{0.4}BiO_3$) that, as indicated, has no copper (Cava, et al., 1988a). While 30 K T_c will no longer make headlines, one can imagine the excitement that would have been generated by this report if it had occurred about two years earlier. But there is more. This compound, unlike the ceramic copper oxides, exhibits superconductivity in *three dimensions*, rather than being restricted to a plane in the unit crystal. It is entirely possible that this work will lead to an entirely new class of isotropic high-temperature superconductors.

Pb-Sr-(RE)-Cu-O: SUPERCONDUCTING SMORGASBORD?

The same group at Bell Laboratories has discovered a new family of compounds that superconduct near 70 K (Cava, et al., 1988b). This material contains a combination of lead, strontium, copper, and oxygen, with another element that may be one of a variety of rare earth metals. The general formula, where (RE) represents the rare earth substitution, is:

$$Pb_2Sr_2(RE)\,Cu_3O_{8+\delta}$$

Since RE may represent a wide variety of rare earth substitutions (which may themselves be combined with strontium or calcium divalent ions on the lanthanide site), the variety of superconductors that may be realized is so large that the group has been referred to as a "superconducting smorgasbord" (Science News, 1988). There is hope that with such a large number of possible combinations, superconducting properties may be markedly improved over those initially measured.

RESEARCH: GOALS AND SUPPORT

Governments of industrialized nations have been quick to take notice of the potential of the new high-temperature superconductors. German physicists are working on better methods for applying films and coatings to metallic surfaces. Japan has launched a major effort into the development of thin films; the electronics applications have obvious commercial potential. The Japanese Science and Technology Agency has plans to build several high-field magnets for measurement of the new superconductors. The United Kingdom's Science and Engineering Research Council is funding several existing programs and is evaluating applications for a new "Center for Excellence" to be dedicated to superconductivity research.

Virtually every U.S. government agency with an interest in technology has begun some level of research on the high-T_c superconductors. Several national laboratories have been given special responsibilities, ranging from materials studies to applications and technology transfer to the private sector. The U.S. Department of Energy has established a Center for Superconductivity Applications at Argonne National Laboratory, a Center for Basic Scientific Information at the Ames Laboratory, and the Computer Data Base on Superconductivity at its own Office of Scientific and Technical Information. The Los Alamos National Laboratory has been directed to develop ways for the D.O.E. labs to form constructive partnerships with the private sector. The National Bureau of Standards, funded by the Department of Commerce, plans a "Superconductivity Center." The National Aeronautics and Space Administration and the National Science Foundation are pursuing similar efforts to fund research and support selected applications.

Following a presidential request, the U. S. Patent Office has given superconductor patent applications a special status to reduce the normal two-year processing time by about six months. While the commercial sector is somewhat more cautious, almost two dozen established U.S. companies are involved in superconductor research, and there are another nine startups (Fitzgerald, 1988). The U.S. Congress' Office of Technology Assessment (OTA) is concerned not only that government funds are not being apportioned in an appropriate manner (OTA wants more funding for the National Science Foundation), but also expresses a concern that the private sector, with the notable exception of a few large corporations, is not placing a significant investment in superconductor research (Lindley, 1988).

The view that Japan will dominate the development and applications of high-temperature superconductivity is widespread. This concern has lead to suggestions, from those who normally prefer to keep government at arms length from the private sector, that the U.S. government should assist private firms in an effort to establish a competitive position in commercial aspects of the new technology (Joyce, 1988).

THE LONG VIEW

At the time of this writing, there is some controversy about the future of the new high-T_c superconductors. Those who might be depicted as moderately pessimistic express the view that, while the recent discoveries are quite important for the advancement of condensed-matter physics, significant applications of the new ceramics are at least a decade away. The reasonably optimistic

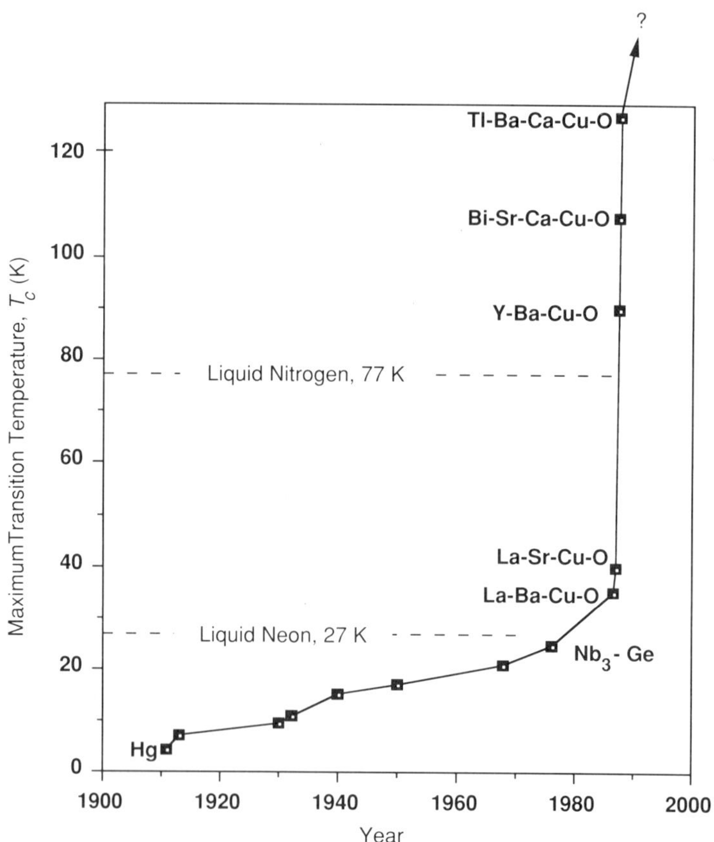

FIGURE 1.12 In spite of the fact that similar plots have appeared *ad nauseum* since the summer of 1987, the author cannot resist the temptation to present this illustration. Regardless of its familiarity, the plot serves to demonstrate the dramatic process triggered by Bednorz and Müller. We may hope that this version will soon be obsolete.

expect significant applications in small-scale electronics by the early 1990s. Almost no one, since the euphoria that characterized 1987, cares to speculate about development of magnetically levitated trains based on high-temperature superconductors.

Whatever the future of these materials may be in applications, there is a great interest in understanding the precise mechanism of operation of the new class (or classes?) of superconductors. This interest in theory has practical as well as academic implications. An accurate, detailed theoretical framework would be very helpful, perhaps even crucial, in support of a systematic development of materials with higher transition temperatures and, perhaps most important, bulk-material critical current densities high enough to meet the most demanding requirements.

It seems safe to predict continued growth in the field of high-temperature superconductivity, with useful applications at 77 K now at hand and materials operating well above 130 K not too far away. On the other hand, there are those reliable voices from the past:

> I'm asking you now and I want to know, how many of these *wrong* predictions are we supposed to have?
>
> (Bernd T. Matthias, 1973)

REFERENCES

Anderson, P. W., (1969). "Superconductivity in the Past and the Future," in *Superconductivity*, R. D. Parks, ed., Vol. 2, Marcel Dekker, New York, pp. 1343–1366.

Ashburn, J. R., (1989). Personnal communication, January 8.

Bardeen, J., L. N. Cooper, and J. R. Schrieffer, (1957). "Theory of Superconductivity," *Phys. Rev.*, Vol. 108, No. 5, December 1, pp. 1175–1204.

Bednorz, J. G., and K. A. Müller, (1986). "Possible High T_c Superconductivity in the Ba-La-Cu-O System," *Z. Phys. B—Condensed Matter 64*, pp. 189–193.

Bednorz, J. G., and K. A. Müller, (1988). "Perovskite-type oxides—The new approach to high-T_c superconductivity," *Rev. Mod. Phys.*, Vol. 60, No. 3, July, pp. 585–600.

Cava, R. J., B. Batlogg, J. J. Krajewski, R. Farrow, L. W. Rupp, Jr., A. E. White, K. Short, and T. Kometani, (1988a). "Superconductivity near 30 K without copper: $Ba_{0.6}K_{0.4}BiO_3$ perovskite," *Nature*, Vol. 332, No. 6167, April 28, pp. 814–816.

Cava, R. J. , B. Batlogg, J. J. Krajewski, L. W. Rupp, L. F. Schneemeyer, T. Siegrist, R. B. vanDover, P. Marsh, W. F. Peck, Jr., P. K. Gallagher, S. H. Glarum, J. H. Marshall, R. C. Farrow, J. V. Waszczak, R. Hull, and P. Trevor, (1986b). "Superconductivity near 70 K in a new family of layered copper oxides," *Nature*, Vol. 336, No. 6196, November, pp. 211–214.

Cooper, L. N., (1987). "Origins of the Theory of Superconductivity," *IEEE Trans. Magn.*, Vol. MAG-23, No. 2, March, pp. 376–389.

Dahl, P. F., (1984). "Kammerlingh Onnes and the discovery of superconductivity: The Leyden years, 1911–1914," *Historical Studies in the Physical Sciences*, Vol. 15, Part 1, pp. 1–37.

Dahl, P. F., (1986). "Superconductivity after World War I and circumstances surrounding the discovery of a state $B = 0$," *Historical Studies in the Physical and Biological Sciences*, Vol. 16, Part 1, pp. 1–58.

de Bruyn Ouboter, R., (1987). "Superconductivity: Discoveries During the Early Years of Low Temperature Research at Leiden, 1908–1914," *IEEE Trans. Magn.*, Vol. MAG-23, No. 2, March, pp. 355–370.

Esaki, L., (1974). "Long journey into tunneling," *Rev. Mod. Phys.*, Vol. 46, No. 2, April, pp. 237–244.

Fitzgerald, K., (1988). "Superconductivity: Fact vs. Fancy," *IEEE Spectrum*, Vol. 25, No. 5, May, pp. 33–41.

Garwin, L., and P. Campbell, (1987). "New superconductors in perspective," *Nature*, Vol. 330, December 17, pp. 611–614.

Giaever, I., (1960). "Energy gap in superconductors measured by electron tunneling," *Phys. Rev. Lett.*, Vol. 5, No. 4, August 15, pp. 147–148.

Giaever, I., (1974). "Electron tunneling and superconductivity," *Rev. Mod. Phys.*, Vol. 46, No. 2, April, pp. 245–250.

Glover, R. E., III, and M. Tinkham, (1957). "Conductivity of superconducting films for photon energies between 0.3 and 40 kT_c," *Phys. Rev.*, Vol. 108, No. 2, October 15, pp. 243–256.

Gorter, C. J., (1964). "Superconductivity until 1940 in Leiden and as seen from there," *Rev. Mod. Phys.*, Vol. 36, January, pp. 3–7.

Goudsmit, S.A., (1961). "Pauli and nuclear spin," *Physics Today*, June, pp. 18–21.

Hayakawa, H., (1986). "Josephson computer technology," *Physics Today*, March, pp. 46–52.

Hazen, R. M., (1988a). *The Breakthrough—The Race for the Superconductor*, Summit Books, New York, p. 145.

Hazen, R. M., (1988b). "Perovskites," *Scientific American*, Vol. 258, No. 6, June, pp. 74–81.

Hoddeson, L., (1989). University of Illinois, personal communication, January 23.

Hoddeson, L., G. Baym, S. Heims, and H. Schubert, (1989). "Collective Phenomena," Chapter 8 in *Out of the Crystal Maze: A History of Solid State Physics, 1900–1960*, L. Hoddeson, E. Braun, J. Teichmann, and S. Weart, eds., Oxford University Press, New York.

Josephson, B. D., (1962). "Possible new effects in superconductive tunneling," *Phys. Lett.*, Vol. 1, No. 7, July 1, pp. 251–253.

Josephson, B. D., (1974). "The discovery of tunneling supercurrents," *Rev. Mod. Phys.*, Vol. 46, No. 2, April, pp. 251–254.

Josephson, B. D., (1986). Personal communication based on talk delivered to the History of Physics Group of the Institute of Physics at University College Cardiff, November 12.

Joyce, C., (1988). "Superconductors: America stalls," *New Scientist*, Vol. 118, No. 1617, June 16, pp. 36–38.

Khurana, A., (1988). "The T_c to Beat is 125 K," *Physics Today*, Vol. 41, No. 4, April, pp. 21–25.

Kunzler, J. E., (1987). "Recollection of Events Associated with the Discovery of High Field-High Current Superconductivity," *IEEE Trans. Magn.*, Vol. MAG-23, No. 2, March, pp. 396–402.

Kunzler, J. E., E. Buehler, F. S. L. Hsu, and J. H. Wernick, (1961). "Superconductivity in Nb_3Sn at high current density in a magnetic field of 88 kgauss," *Phys. Rev. Lett.*, Vol. 6, No. 3, February 1, pp. 89–91.

Lindley, D., (1988). "Superconductor pessimism in the US," *Nature*, Vol. 333, No. 6176, June 30, pp. 789.

London, F., (1950). *Superfluids, Macroscopic Theory of Superconductivity*, Vol. 1, John Wiley & Sons, New York, pp. 3–4.

Lynton, E.A., (1962). *Superconductivity*, Methuen, London, pp. 107–127.

Maeda, H., (1988). Personal communication, November 8.

Maeda, H., Y. Tanaka, M. Fukutomi, and T. Asano, (1988). "A New high-T_c Oxide Superconductor Without a Rare Earth Element," *Jap. J. Appl. Phys.*, Vol. 27, No. 2, February, pp. L209–L210.

Matisoo, J., (1980). "Overview of Josephson Technology Logic and Memory," *IBM J. Res. Dev.*, Vol. 24, No. 2, March, pp. 113–129.

Matthias, B. T., (1973). "High T_c's—The Low and High of It," in *The Science and Technology of Superconductivity*, W. D. Gregory, W. N. Mathews, Jr., and E. A. Edelsack, eds., Vol. 1, Plenum Press, New York, p. 288.

Matthias, B. T., T. H. Geballe, S. Geller, and E. Corenzwit, (1954). "Superconductivity of Nb_3Sn," *Phys. Rev.*, Vol. 95, No. 6, September 15, p. 1435.

Mendelssohn, K., (1964). "Prewar work on superconductivity as seen from Oxford," *Rev. Mod. Phys.*, Vol. 36, January pp. 7–12.

Müller, K. A., and J. G. Bednorz, (1987). "The Discovery of a Class of High-Temperature Superconductors," *Science*, Vol. 237, September 4, pp. 1133–1139.

Onnes, K., (1913). "Report on the Researches Made in the Leiden Cryogenic Laboratory between the Second and Third International Congress of Refrigeration," Suppl. No. 34b. (See *Superconductivity—Selected Reprints*, American Institute of Physics, New York, 1964, pp. 8–23, and de Bruyn Ouboter (1987).)

Pippard, A. B., (1987). "Early Superconductivity Research (Except Leiden)," *IEEE Trans. Magn.*, Vol. MAG-23, No. 2, March, pp. 371–375.

Pool, R., (1988). "Superconductor credits bypass Alabama," *Science*, Vol. 241, No. 4866, August 5, pp. 655–657.

Schmitt, R. W., (1961). "The discovery of electron tunneling into superconductors," *Physics Today*, Vol. 14, No. 12, December, pp. 38–41.

Schrieffer, J. R., (1988). "Theoretical Ideas Relating to Superconductivity," Videotape, Los Alamos National Laboratory Colloquium, December 13.

Science News, (1988). "A superconducting smorgasbord," Vol. 134, No. 22, November 26, p. 348.

Second Epistle of Peter, 3:15–3:16, *The Holy Bible*, New King James Version, Thomas Nelson, Nashville, 1982.

Sheng, Z. Z., W. Kiehl, J. Bennett, A. El Ali, D. Marsh, G. D. Mooney, F. Arammash, J. Smith, D. Viar, and A. M. Hermann, (1988). "New 120 K Tl-Ca-Ba-Cu-O superconductor," *Appl. Phys. Lett.*, Vol. 52, No. 20, May 16, pp. 1738–1740.

Silsbee, F. B., (1917). "Note on electrical conduction in metals at low temperatures," Vol. 14, No. 307, Scientific Papers, U.S. Bureau of Standards, pp. 301–306.

Sleight, A. W., J. L. Gillson, and P. E. Bierstedt, (1975). "High-Temperature Superconductivity in the $BaPb_{1-x}Bi_xO_3$ System," *Solid State Commun.*, Vol. 117, pp. 27–28.

Teller, E., (1987). Comment at Los Alamos National Laboratory Colloquium, Summer.

Tinkham, M., (1985). *Introduction to Superconductivity*, Robert E. Krieger, Malabar, FL.

Van Duzer, T., and C. W. Turner, (1981). *Principles of Superconductive Devices and Circuits*, Elsevier, New York.

Wu, M. K., J. R. Ashburn, C. J. Torng, P. H. Hor, R. L. Meng, L. Gao, Z. J. Huang, Y. Q. Wang, and C. W. Chu, (1987). "Superconductivity at 93 K in a New Mixed-Phase Y-Ba-Cu-O Compound System at Ambient Pressure," *Phys. Rev. Lett.*, Vol. 58, No. 9, March 2, pp. 908–910.

Yntema, G. B., (1987). "Niobium Superconducting Magnets," *IEEE Trans. Magn.*, Vol. MAG-23, No. 2, March, pp. 390–395.

SUGGESTIONS FOR FURTHER READING

Larbalestier, D., G. Fisk, B. Montgomery, and D. Hawksworth, "High-Field Superconductivity," *Physics Today*, March 1986, pp. 24–33.

Parks, R. D., "Quantum effects in superconductors," *Scientific American*, Vol. 213, No. 4, October 1965, pp. 57–67.

Rowell, J. M., "Superconducting Tunneling Spectroscopy and the Observation of the Josephson Effect," *IEEE Trans. Magn.*, Vol. MAG-23, No. 2, March 1987, pp. 380–389.

Tanaka, S., "Research on high-T_c superconductivity in Japan," *Physics Today*, December 1987, pp. 53–57.

CHAPTER 2

THE PHENOMENA OF SUPERCONDUCTIVITY; FUNDAMENTAL CONSIDERATIONS

INTRODUCTION

When one considers the exceptional range of phenomena exhibited by superconductors, it is not surprising that several decades have been required to gain even a rudimentary understanding of their behavior. The concept of zero dc resistance, while hardly commonplace, was a notion considered by theorists long before Kamerlingh Onnes measured his mercury in 1911. While Onnes soon learned (to his intense disappointment) that modest magnetic fields could destroy the superconducting phenomena, it was more than two decades later before anyone realized that superconductors, behaving in an opposite fashion to ferromagnetic materials, actually *expelled* magnetic flux (see Figure 2.1). It must have seemed that the list of peculiar phenomena was near completion, but there was a good deal more to come. There was large-scale quantum behavior, including flux quantization, exotic tunneling phenomena, flux tubes that made practical high-field superconductors possible, and more.

In this chapter, these varied and peculiar phenomena will be presented. To prevent the discussion from becoming overly tedious, lengthy discussions and derivations are avoided. Even so, it is not possible to appreciate the phenomena of superconductivity without some discussion of matters that are, to say the least, not intuitive. That, of course, is generally the case when quantum effects govern a phenomena.

A variety of metals, intermetallic compounds, alloys, semiconductors and, as discovered more recently, ceramic-like copper oxides are superconductors when the thermal and magnetic environments are appropriate. This discussion will be devoted to a description of the characteristics of "conventional" metallic

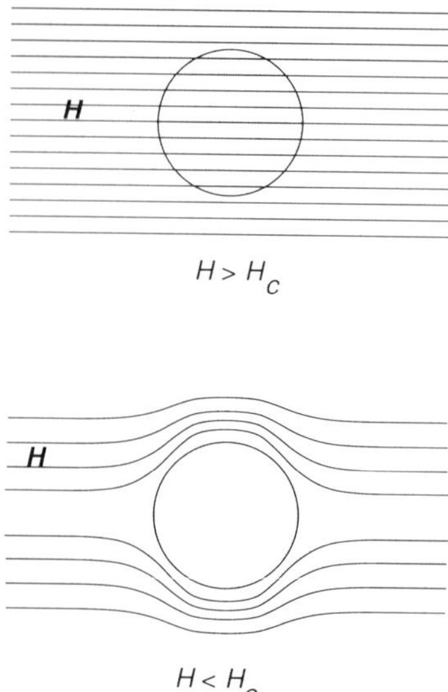

FIGURE 2.1 The earliest investigators of superconductivity believed that the primary phenomena was zero dc resistivity. A more fundamental phenomena, seen in all superconductors, is the exclusion of magnetic flux. This diamagnetism in superconductors is known as the "Meissner effect."

and metal-alloy superconductors, much of which will apply to the copper oxide ceramics. The recently discovered "high-temperature" superconductors will be the subject of a separate chapter, primarily because the copper oxides appear to comprise a different class of superconducting materials.

ZERO RESISTIVITY, MEISSNER EFFECT

All superconductors exhibit zero (dc) electrical resistivity and pronounced diamagnetic properties below a critical transition temperature, T_C. The abrupt change from normal conduction to superconduction occurs at a thermodynamic phase transition determined not only by the temperature but also by the magnetic field strength (H) at the surface of the specimen. There is a critical magnetic field H_C, which, when exceeded, will destroy superconductivity. If a sufficient current density (J_C) is transported through the superconductor, this also causes quenching of the material. It is not surprising that these three parameters, that is, temperature, magnetic field and current density are closely coupled in determining the state of the superconductor.

Zero dc resistivity is, of course, impossible to measure in an absolute sense. Perhaps the most sensitive technique involves the induction of a "persistent" direct current in a superconducting loop. Persistent currents have been measured after circulating for more than a year with no indication of decay (see

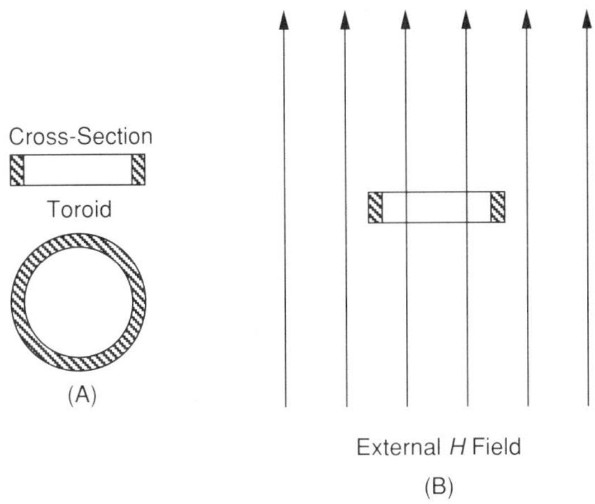

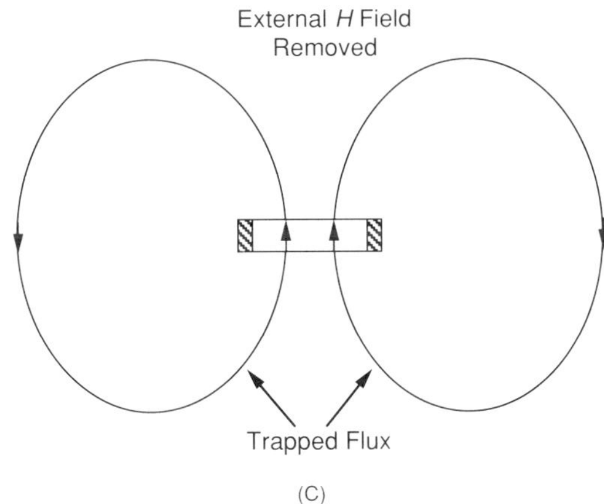

FIGURE 2.2 When a superconducting toroid (A) is cooled in an externally applied magnetic field (B), flux will be expelled from the material. When the magnetic field is removed (C), trapped flux is supported by a persistent current flow in the toroid. (After K. R. Atkins, "Superfluids," in *Handbook of Physics*, E. U. Condon and H. Odishaw, eds., McGraw-Hill, New York, 1958.)

Figure 2.2). Nuclear magnetic resonance (NMR) determination of the static H field associated with the persistent currents has led to the conclusion that the decay time is at least 100,000 years!

There are two fundamental facts to remember about superconductors operating well below critical temperature T_c and critical magnetic field H_c:

H fields are expelled from the superconductor (the Meissner-Ochsenfeld effect)

The internal static electric field (**E**) is zero

Another important basic fact is that superconductivity is not at all the same phenomena as negligible resistivity in a normal conductor; indeed, the latter behavior would preclude the Meissner effect of flux expulsion. Consider the case of a long cylinder, arranged with its axis parallel to a magnetic field. Maxwell's induction equation requires that the field in the cylinder behave according to:

$$\nabla \times \mathbf{E} = -\frac{\partial \mathbf{B}}{\partial t} \tag{2.1}$$

Infinite conductivity would require that $\mathbf{E} = 0$, which would lead to:

$$\frac{\partial \mathbf{B}}{\partial t} = 0 \tag{2.2}$$

This is equivalent to stating that the magnetic induction **B** inside the "perfect conductor" could not change; that is, any magnetic field present would be trapped when the material became perfectly conductive. Experiments show that superconductors *do not* behave in this manner. In contrast, the superconductor not only excludes external magnetic fields, it *expels* fields that had been established inside the conductor at higher (than critical) temperatures (see Figure 2.3).

An example of this peculiar behavior involves the quenching of the superconducting state by application of an external field. As the material becomes normal due to the application of a magnetic field exceeding the critical field

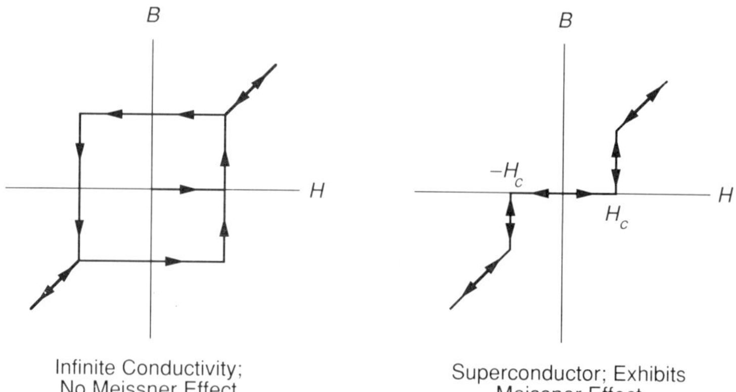

FIGURE 2.3 The *B-H* plot on the left illustrates the behavior of a "perfect conductor," that is, a medium that exhibits infinite conductivity but does not expel flux. In this fictitious material, flux would be frozen in place. In superconductors, as illustrated on the right, flux is expelled as long as the field is below a critical level H_C. (After K. R. Atkins, "Superfluids," in *Handbook of Physics*, E. U. Condon and H. Odishaw, eds., McGraw-Hill, New York, 1958.)

for the material, an internal induction **B** is established. However, when the external **H** field is lowered to the critical level, the superconductor *expels* the internal magnetic flux lines! The actual proportion of the field expelled and the precise mode of expulsion depends on a number of factors, including the type and shape of the superconducting specimen, but more will be said about that later.

The essential point is that the hallmark of superconductivity is expulsion of the internal **B** field in an applied magnetic (**H**) field, unless the applied field exceeds a critical level. This is the reason that reports of superconductivity in new compounds or alloys must meet the Meissner effect criteria prior to acceptance by the scientific community. Since it is impossible to measure absolute zero resistivity, the Meissner effect is considered a definitive test to differentiate between large discontinuities in the temperature-resistivity relation and true superconductive behavior. This consideration is particularly relevant when so many investigators are seeing tantalizing evidence of "room-temperature" superconductivity in variants of the copper oxide materials.

TWO BROAD CATEGORIES OF SUPERCONDUCTORS

Superconductors are generally divided into two categories. While these categories have been given a variety of names, they are most commonly referred to as Type I and Type II.

The first are always rather pure metallic elements, easily quenched in magnetic fields less than ~1000 gauss. Examples of Type I superconductors include mercury, lead, and tin. Such materials were investigated by Onnes and proved to be impractical for operation except in relatively low-level magnetic fields. The Type I materials must exclude virtually all of an applied magnetic field to remain superconducting. This is a problem, since the superconducting state has less free energy than the normal state (Tinkham, 1986). Since transport currents generate **H** fields, a low tolerance for **H** fields naturally leads to a low tolerance for current transport. See Figure 2.4 for H versus T for a Type I superconductor.

Systems have a tendency to seek their lowest energy state. The superconductor is no exception: just below T_c, the lowest energy state is the superconducting state. As temperature is lowered farther below T_c, the difference in energy between the superconducting and the normal state increases. If a magnetic field is applied, energy is required *from the superconductor* for the expulsion of flux. Once the superconductor's free-energy "budget" is used up, the material can no longer remain in the superconducting state. This quenching occurs at a thermodynamic critical field, H_c, which is a function of the composition, temperature and shape of the sample. The energy cost of removing flux, sometimes referred to as the condensation energy of the superconducting state, is $H_c^2/8\pi$ per unit volume. This value is simply the difference in the energy of the normal and the superconducting state in zero magnetic field.

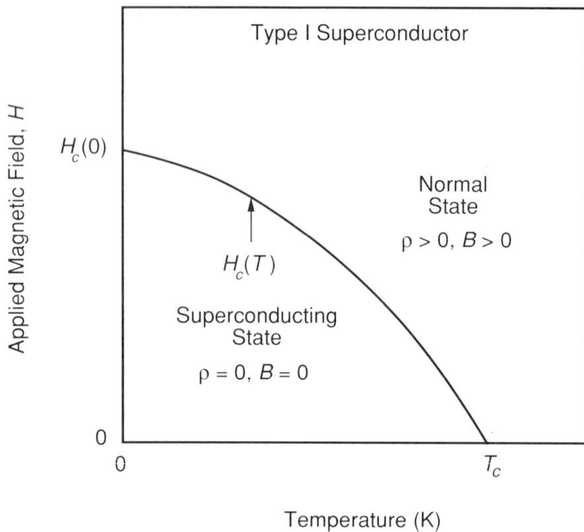

FIGURE 2.4 Both temperature and magnetic field must be below a critical level to prevent a superconductor from returning to the normal state. In a Type I superconductor, which is always a rather pure metal, normal and superconducting regions may coexist, a condition (dependent on geometry) called the "intermediate state."

If there was some way (and, fortunately, there is!) for the magnetic field to remain in the body of the superconductor, so that energy would not be required for its expulsion, it would be possible for the material to remain in a superconducting state at much higher levels of applied magnetic field and transport current. Fortunately, the Type II materials exhibit just such a behavior.

The Type II materials are usually alloys or compounds (niobium and vanadium are exceptions to this rule), and are characterized by being able to retain their superconductive characteristics in rather intense magnetic fields. This class of superconductors accomplishes this feat by a remarkable means: rather than using the energy required to completely expel magnetic fields, the fields are confined to an internal array of normal-state flux tubes called "vortices" since they are surrounded by a circulating supercurrent. Certain Type II materials, since they are capable of carrying relatively large current densities, have a variety of commercial applications. The differences between the Type I and Type II materials will become more apparent as the characteristics of superconductors are discussed. See Figure 2.5 for H versus T for a Type II superconductor.

TWO-FLUID MODELS

The "supercurrent" flow in superconductors is comprised of electron pairs, as opposed to the flow of individual electrons in the normal metallic state.

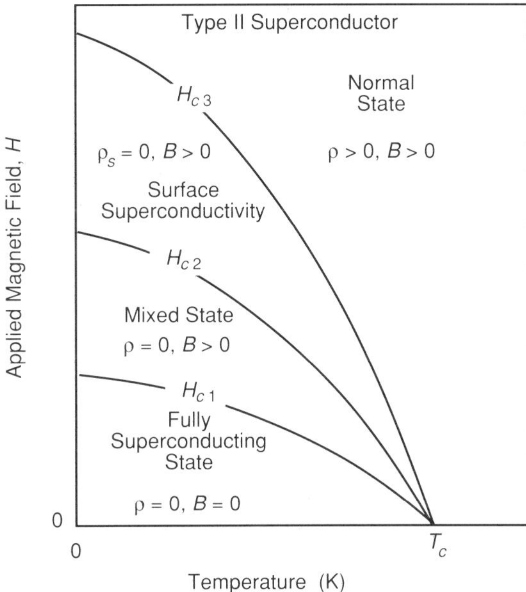

FIGURE 2.5 In a Type II superconductor, there exists a "mixed state" for magnetic field levels between H_{c1} and H_{c2}. The mixed state is characterized by the penetration of quantized "fluxoids" through the material.

More will be said of this electron pairing later, but first let us consider a few fundamental facts about current flow in the superconductor. One might guess that, since all of the current is carried by unpaired normal electrons when the conductor is in its normal state, that all of the current would be carried by paired electrons in the superconducting state. It is a reasonable guess, but not a correct one. In fact, the current transport in the superconducting state is a mixture of paired electrons that comprise the "supercurrent" and unpaired electrons that behave very much in the normal fashion. The superconducting and normal current densities are generally identified by the symbols J_s and J_n, respectively. There are very useful engineering models of superconducting behavior, where each component is considered to be a "fluid," that is, the two-fluid models.

Such two-fluid models reflect the fact that, at nonzero temperatures, not all conduction electrons participate in the superconducting behavior. The two-fluid models, while somewhat artificial, are nevertheless of great practical utility in modeling the behavior of superconductors.

As a superconductor is cooled below T_c, ordinary electrons "condense" into superconducting pairs. These pairs consist of electrons with opposite spins and (where transport current flow and applied magnetic fields are zero) opposite and equal momenta so that the net momentum and the net spin of the pair are zero (Clarke, 1988).

As temperature approaches 0 K, the proportion of the free electrons drafted into the pairing mode increases. Therefore, the number of unpaired electrons will diminish as temperature decreases. The ratio of density of normal (n_n) to superconducting (n_s) electrons decreases dramatically as temperature falls below T_c, with the normal electron density disappearing at $T = 0$ and superconducting electrons vanishing at $T > T_c$ (see Figure 2.6).

A variety of two-fluid models have been proposed. As Lynton points out, there are two assumptions common to all (Lynton, 1969):

a. The entropy of the superconducting system is entirely due to the individual (i.e., noncondensed) charge carriers, which behave essentially as they would for $T > T_c$, and
b. The part of the system responsible for superconductivity is an ordered or condensed state, represented by an order parameter taken as unity at $T = 0$ K and zero at $T = T_c$.

If the two-fluid model is to be useful in engineering analysis, one needs to know precisely how the density of superconducting (and normal) charge carriers vary with temperature.

The formulation of C. J. Gorter and H. B. G. Casmir in 1934 develops a fractional variation of density which is $(1 - t^4)$, where $t = T/T_c$, that is, the "reduced temperature" (Pippard, 1987). If the density of superconducting

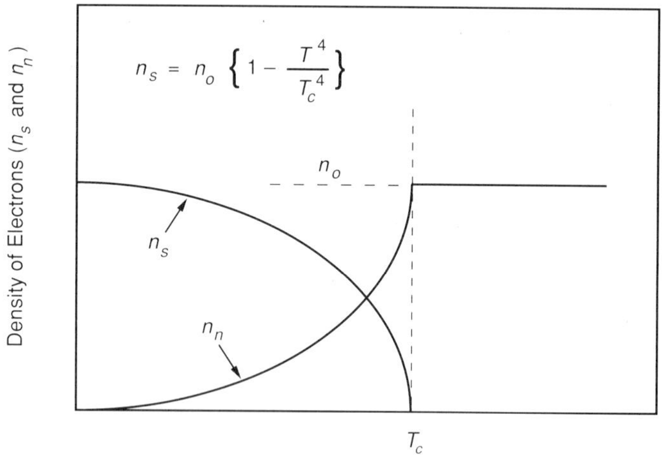

FIGURE 2.6 In the two-fluid model, the density of paired electrons (n_s) increases as temperature drops below T_c, and this is at the expense of the density of unpaired electrons (n_n). It is the unpaired electrons that contribute to losses in the presence of high-frequency alternating electromagnetic fields; the paired electrons provide the supercurrent flow.

electrons is n_o at 0 K, the density n_s at other temperatures below T_c is:

$$n_s = n_o\{1 - t^4\} \tag{2.3}$$

If the fraction of superconducting electrons is $(1 - t^4)$, the fraction of normal electrons is $1 - (1 - t^4)$, or t^4 for values of $T \leq T_c$.

The total current flowing in the superconductor, according to the two-fluid model, is simply the sum of the normal and superconducting components:

$$\mathbf{J} = \mathbf{J}_s + \mathbf{J}_n \tag{2.4}$$

The normal electron current obeys Ohm's law, that is,

$$\mathbf{J}_n = \sigma\mathbf{E} \tag{2.5}$$

where σ is conventional temperature-dependent electrical conductivity (London, 1950).

If one considers an electron with mass m and charge e in an applied electric field $\mathbf{E}$, the force relationship (in the absence of a collisional relaxation time for velocity $\mathbf{v}$) is:

$$\mathbf{F} = m\frac{d\mathbf{v}}{dt} = e\mathbf{E} \tag{2.6}$$

The current density $\mathbf{J}$ is simply $ne\mathbf{v}$, where n is the density of electrons so that

$$\frac{d\mathbf{v}}{dt} = \frac{1}{ne}\left(\frac{d\mathbf{J}}{dt}\right) = \frac{e}{m}\mathbf{E} \tag{2.7}$$

and

$$\frac{m}{ne}\frac{d\mathbf{J}}{dt} = e\mathbf{E} \tag{2.8}$$

so that

$$\frac{d\mathbf{J}}{dt} = \left(\frac{ne^2}{m}\right)\mathbf{E} = \left(\frac{1}{\Lambda}\right)\mathbf{E} \tag{2.9}$$

This is the first of the London equations (proposed by F. and H. London), which describes the supercurrent flow. Incidentally, this equation describes a pure reactance; a fact that will be used later to develop a simple inductance-resistance equivalent circuit for representation of the two-fluid model. The second London equation was assumed by Fritz London to have the form

$$\text{curl } \mathbf{J}_s = -\frac{\mathbf{H}}{\Lambda} \tag{2.10}$$

where Λ, from the development above, is defined as:

$$\Lambda = \frac{m}{n_s e^2} = 4\pi\lambda \tag{2.11}$$

where m is the effective mass of the superconducting charge carrier (electron), n_s is the electron density, e is the electron charge, and λ is a characteristic penetration depth that is crucial in the operation of superconductors. It makes no difference in the definition of Λ whether one assumes a single electron as the charge carrier (as London and all his contemporaries did) or a paired electron carrier is used. If, for the paired charges, $m^* = 2m$, $e^* = 2e$, and $n^* = n_s/2$, then

$$\frac{m^*}{n^*(e^*)^2} = \frac{2m}{(n_s/2)(2e)^2} = \frac{m}{n_s(e)^2} \tag{2.12}$$

The London equations may be rewritten:

$$\mathbf{H} = -4\pi\lambda^2 \text{ curl } (\mathbf{J}_s) \tag{2.13}$$

and

$$\mathbf{E} = 4\pi\lambda^2 \frac{\partial \mathbf{J}_s}{\partial t} \tag{2.14}$$

It is often pointed out that the latter equation effectively replaces Ohm's law for superconductors.

Note that when the current density J_s is *not changing with time*, the field **E** is required to be zero. Since static supercurrents can flow with a zero electric field, the expression properly describes the superconductive state. Those who are familiar with the theory of lumped electrical circuits will note the similarity to the expression for voltage V_L across a lumped inductance L as a function of current i:

$$V_L = L\frac{di}{dt} \tag{2.15}$$

The equation for **E** as a function of the partial time derivative of $\mathbf{J}_s$ is, similarly, a representation of a pure reactance. There is a 90° difference in phase between the supercurrent $\mathbf{J}_s$ and field **E**, and, subsequently, a 90° phase difference between $\mathbf{J}_s$ and $\mathbf{J}_n$, the normal current in the two-fluid model. $\mathbf{J}_n$ is, of course, in phase with **E**.

It is equally important to note that a finite field **E** is required to produce time-varying supercurrents, hence the alternating-current impedance of a superconductor, while it may be extremely low, is never quite zero. A time-varying magnetic field ($\mathbf{B}_s$) in the region that penetrates λ into the surface will generate an orthogonal surface electric field $\mathbf{E}_s$. This surface electric field is not shunted to zero, since the superconducting electron pairs have a finite inertia. The surface $\mathbf{E}_s$ field accelerates the "normal" electrons and produces losses. The power loss is proportional to E_s^2, which is, in turn, proportional to the square of frequency:

$$P_s \propto \sigma_n E_s^2 \propto \omega^2 B_s^2 \qquad 2.16$$

where σ_n is the conductivity component due to the normal electrons in the two-fluid model. This fact has important implications for alternating-current applications, since the power loss in superconductors increases more rapidly with increasing frequency than is the case for normal conductors.

It is worth noting that the London equations, which predict a rather total Meissner effect, are quite useful for superconductors in small magnetic fields. Conversely, the London equations are not applicable to either the mixed state seen in Type II materials or the intermediate state of Type I materials, where magnetic fields may penetrate deeply into the superconductor. (The mixed and intermediate states will be discussed later in this chapter.)

CHARACTERISTIC LENGTHS (λ and ξ)

When the London equations are combined with Maxwell's equations, one obtains the following result for static conditions:

$$\nabla^2 \mathbf{H} = \frac{\mathbf{H}}{\lambda^2} \qquad 2.17$$

This relationship implies that the magnetic field decays exponentially with depth from a parallel field $\mathbf{H}_s$ at the surface of a superconductor. The field **H** at depth z is:

$$\mathbf{H} = \mathbf{H}_s \epsilon^{-z/\lambda} \qquad 2.18$$

Even for high-quality superconductors, the expulsion of flux is not quite complete. Fields remain in a thin surface layer whose depth is $\sim \lambda$ (see Figure 2.7). For a magnetic field less than H_c for a Type I material, the surface field H_s is exponentially attenuated as it penetrates into the surface of the material, with the characteristic depth λ. The value of λ varies with temperature and may be related to its zero temperature value λ_o by recalling that λ is proportional to $n_s^{-1/2}$, that is,

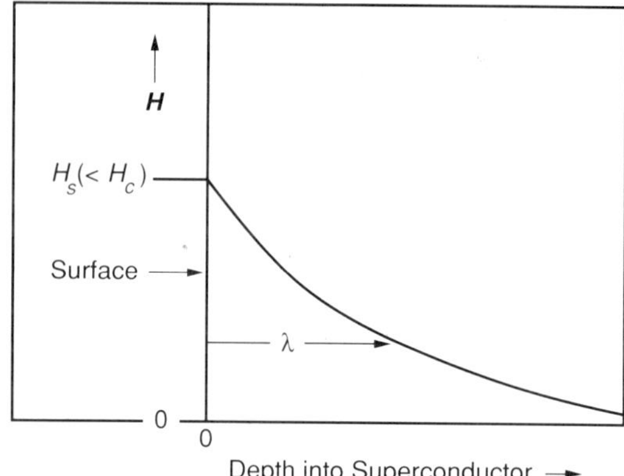

FIGURE 2.7 While virtually all magnetic flux is expelled from a superconductor operated well below its transition temperature and critical magnetic field, there is a slight penetration (λ) in a very thin layer near the surface. This "London" penetration depth ranges from approximately 100 to 1500 Å for metallic superconductors, with somewhat larger values for the high-temperature copper oxides.

$$\lambda = \sqrt{\frac{m}{4\pi n_s e^2}} \qquad 2.19$$

Recall also, the Gorter-Casmir relation (introduced earlier) of superconducting electron density (n_s) as a function of temperature and the maximum density n_o at 0 K,

$$n_s = n_o\{1 - t^4\} \qquad 2.20$$

where $t = T/T_c$. This leads quite naturally to the following expression for London penetration depth as a function of temperature, $\lambda(T)$ (see Figure 2.8), which is useful for temperatures near T_c:

$$\lambda(T) = \lambda_o\left[1 - \left(\frac{T}{T_c}\right)^4\right]^{-1/2} \qquad 2.21$$

where λ_o is the penetration depth (or, alternatively, the London depth, λ_L) at 0 K, and is an intrinsic property of the material. Note that the value of λ approaches infinity as $T \rightarrow T_c$. For temperatures *very* near (and just below) T_c, the expression above is sometimes replaced by

$$\lambda(T) = \frac{\lambda_o}{2\sqrt{1 - (T/T_c)}} \qquad 2.22$$

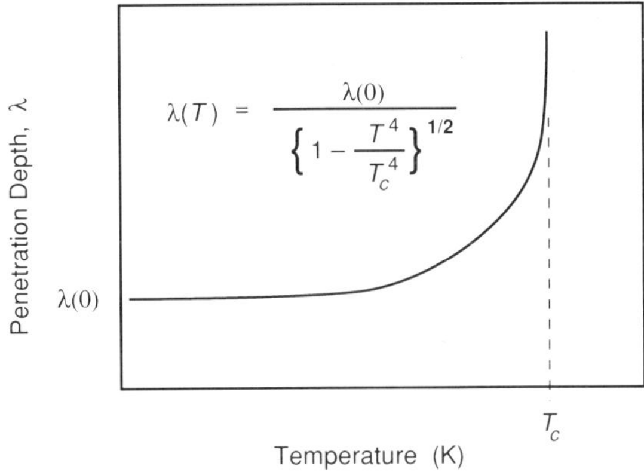

FIGURE 2.8 Penetration depth (λ) is at its minimum value at 0 K, but increases rapidly as the critical transition temperature (T_c) is approached.

The penetration depth for pure superconductors is relatively small, typically in the 500–2000 Å range. Approximate values of λ_o for several elements are given in Table 2.1. For nonstatic fields, the field penetration is also determined by λ. At temperatures above T_c, where conduction becomes normal, penetration is a function of conventional electromagnetic skin depth, which is proportional to $\omega^{-1/2}$. (At sufficiently high frequencies and low temperatures, the "anomalous" skin depth varies as $\omega^{-1/3}$. This occurs where the conduction electron's mean free path ι becomes comparable to skin depth.)

In regard to penetration depth, Type II superconductors behave somewhat differently than Type I, particularly when the mean free path ι becomes comparable to the Pippard coherence length, ξ_o (Solymar, 1972) ($\xi_o = 0.18\ hv_F/2\pi kT_c$, where v_F is the Fermi velocity). The coherence length may be thought of as a measure of the maximum distance between members of a Cooper pair. Furthermore, the amount of impurity doping for effective flux

TABLE 2.1 London penetration depth for selected metallic superconductors. (After E. A. Lynton, *Superconductivity*, Methuen, London, 1969.)

Element	λ_0 (Å)
Al	500
Cd	1300
Hg	380–450
In	640
Nb	470
Pb	390
Sn	510
Tl	920

pinning has a direct effect on the mean free path. In particular, the relation between $\lambda(T)$, the temperature-dependent penetration depth in a "dirty superconductor," and $\lambda(0)$, the penetration depth at 0 K in the absence of impurity scattering, is

$$\lambda(T) = 0.615\lambda_0\sqrt{\frac{\xi_o}{\iota(1-t)}} \qquad 2.23$$

where $t = T/T_c$, the reduced temperature.

In 1953, Brian Pippard's measurements of microwave surface impedance on superconductors led him to introduce a "coherence length" that was a measure of the superconducting electron's nonlocality. This distance, the "Pippard coherence length," is represented by the symbol ξ_o and may extend to the order of 10^{-4} cm for very pure metals (Edelsack, 1973), but is only on the order of 10^{-6} cm for transition metals like niobium and tantalum. For compounds such as Nb_3Sn, ξ_o is on the order of 5×10^{-7} cm (Al'tov, 1977). Irregular lattice structures or imperfections will decrease the magnitude of ξ_o. The Pippard coherence length is closely related to the Ginzberg-Landau parameter $\xi(T)$.

Within the coherence length, the electrons are said to be "correlated," and one may think of ξ_o as being the "size" of the Cooper pair. For very pure metals, the coherence length is usually specified as ξ_o, with the symbol ξ defining the coherence length for impure metals, alloys, and so forth. The symbol ξ_ι is usually reserved for calculation of coherence length as a function of the mean free path ι. The symbol ξ_o is used by Lynton to denote the range of coherence (between paired electrons) in a material where the mean free path of electrons is "unlimited." ξ_o has the value

$$\xi_o = \frac{hv_F}{2\Delta\pi^2} \qquad 2.24$$

where h is Planck's constant, v_F is the electron velocity at the Fermi surface for the normal state of the material, and Δ is the superconductor energy gap (to be discussed later in this chapter). An alternate expression, used in the discussion of Type II penetration depth, is

$$\xi_o = a\frac{hv_F}{2\pi kT_c} \qquad 2.25$$

where hv_F is the energy at the Fermi surface and the constant a is 0.18 in Bardeen-Cooper-Schrieffer (BCS) theory (Bardeen, et al., 1957). Pippard empirically determined a value of 0.15 for this same parameter (Tinkham, 1985). See Table 2.2 for the values of ξ_o for three Type I superconductors.

For impure metals where the mean free path of electrons becomes comparable to ξ_o, the following empirical relationship applies:

TABLE 2.2 Values of ξ_o for representative Type I superconductors. (After Lynton, *Superconductivity*, Methuen, London, 1969.)

Element	ξ_o (Å)
Aluminum	18,000
Indium	4400
Tin	2300

$$\xi_\iota^{-1} = \xi_o^{-1} + (\alpha\iota)^{-1} \quad 2.26$$

where $\alpha \approx 0.8$ and ι is the mean free path of electrons. The same relationship may be expressed as follows:

$$\xi_\iota = \xi_o\left(\frac{1}{1 + (\xi_o/\alpha\iota)}\right) \quad 2.27$$

It can be seen that as the mean free electron path ι diminishes (at higher temperatures), ξ_ι will also decrease.

As mentioned above, the Ginzberg-Landau (GL) theory of superconductivity introduces a related dimension, known as the temperature-dependent or Ginzberg-Landau coherence length, that is, $\xi(T)$ or $\xi(GL)$. This parameter diverges as $T \rightarrow T_c$, while $\xi(GL) \rightarrow\sim \xi_o$ as $T \rightarrow 0$:

$$\xi(T) \equiv \xi_{GL} = 0.74\xi_o\sqrt{\frac{1}{1 - (T/T_c)}} \quad 2.28$$

or, when mean free path ι is comparable to ξ_o, as (Solymar, 1972)

$$\xi(T) = 0.85\sqrt{\frac{\xi_o\iota}{1 - (T/T_c)}} \quad 2.29$$

The GL theory kappa, κ, is the ratio of penetration depth to coherence length, that is,

$$\kappa = \frac{\lambda}{\xi_{GL}} \quad 2.30$$

The value of κ is relevant to characterizing a superconducting material as Type I or Type II. More specifically, Alexei Abrikosov showed that a value of κ greater than $(2)^{-1/2}$ can be expected to be associated with a Type II superconductor, while if $\kappa < (2)^{-1/2}$ the superconductor is Type I. The relationship between $\xi(T)$ and λ, which distinguishes Type I and Type II superconductors, will be discussed later in this chapter.

CRITICAL TEMPERATURE, T_c

It is important to realize that T_c is, by convention, defined for a zero value of applied magnetic field.

In a common convention used for superconductors where the transition region is quite narrow, T_c is simply the transition temperature. Where the transition width is larger, this convention is not used. Instead, one attempts to define onset temperatures, the temperature for zero dc resistivity, and finally, T_c itself.

In a common convention, used widely to discuss the new ceramic superconductors, T_c is the temperature where the resistivity has decreased to half its value at the "onset" temperature (T_{co}) (see Figure 2.9). Onset temperature for superconducting phenomena is the level where the resistivity versus temperature curve just begins to depart from the normal to superconducting behavior. Since the slope of resistance versus temperature plots (near T_c) may vary widely between dc and ac measurements, there will be corresponding differences in the measured value of T_c. In a typical case, dc resistance of a sample may drop abruptly to zero at 90 K, but significant ac resistance will be measured down to 60–80 K.

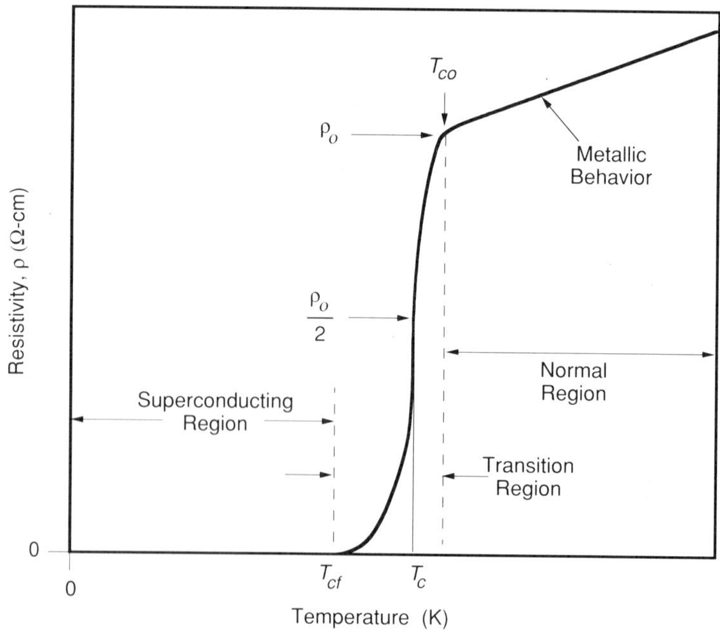

FIGURE 2.9 Characteristic shape of the resistivity vs. temperature response for a superconductor in a small H field.

Measurements of parameters like resistivity (ρ) or magnetic susceptibility (χ) are typically used to determine T_c. The accuracy of these techniques is acceptable if the transition region is narrow; a long "resistive tail" below T_c can complicate definition of the critical temperature, or perhaps suggest that the sample has multiple superconducting phases, each exhibiting a different critical temperature.

If the dc resistivity at T_{co} is ρ_o, the dc resistivity at T_c is $\rho_o/2$. A third temperature, T_{cf}, is the highest temperature where the material is considered to be fully superconducting, that is, dc resistivity is zero. The "transition width" is often defined as the temperature range where resistivity varies between 10% and 90% of ρ_o. A high-purity sample will have a narrow transition width. In practice, the T_{co} and T_{cf} temperatures often may be determined only approximately.

While the transition widths may be as little as a few thousandths of a degree in very pure, homogeneous superconductors like lead or tin, the presence of thermodynamic fluctuations always leads to an intrinsic (albeit very narrow) transition width. Such fluctuations are responsible for the presence of finite (but extremely small) resistance below T_c, and also for the presence of slight superconductive effects up to about $2T_c$ (Tinkham, 1985). In complex superconductors (like the copper oxides), the transition width ($T_{co} - T_{cf}$) is typically on the order of 2–10 degrees K.

The need to standardize the determination of T_c for samples of the new ceramic superconductors (where the transition is often rather broad) has led to an interesting proposal that the normalized temperature derivative of dc magnetic susceptibility χ be used to define the value of T_c (Datta, et al., 1988). By this approach, the temperature where the maximum value of the derivative $d\chi/dT$ occurred would be defined as the value of T_c. This approach has considerable merit and might be adopted if more investigators routinely measured dc susceptibility. In the meantime, measurements of dc resistance are likely to continue to be the basis of T_c determinations.

The essential point here is that the definitions of T_c vary considerably and are often influenced by a desire to quote as high a value as possible. Indeed, investigators sometimes quote the value of T_{co} as their reported value for T_c. A supposed justification is that the value of T_c onset may be associated with the true T_c of an ideally stoichiometric portion of the superconducting sample.

FLUX QUANTIZATION

Let us, for a moment, consider the most simple model for the expulsion of magnetic fields by a superconductor, that is, the Meissner effect. In this explanation, the external **H** field causes a flow of electron current on the surface of the superconductor. This "shielding" current flow has precisely the appropriate magnitude and direction to generate a magnetic field that cancels effects of the external **H** field, leading to a zero **B** field inside the superconductor. As Parks has pointed out, this simple model is hardly sufficient to explain other magnetic phenomena, such as the behavior of magnetic fields trapped in a superconducting ring (Parks, 1965).

Fritz London was apparently the first to realize that a specific superconducting charge carrier could not move independently from other equivalent supercurrent carriers. This led directly to the concept of a macroscopic quantum phenomena that involved all of the superconducting charge carriers and a more detailed understanding of persistent currents in a ring: if an external **H** field is removed from a superconducting ring, the ring will trap the internal field by the flow of a persistent current.

In 1950, London published the bold assertion that this trapped flux would be quantized—it could only exist in discrete units! This remarkable prediction attracted little attention when it was made, perhaps because London mentioned the notion rather casually in a footnote in one of his books (London, 1950):

> We note that in order for U to be a single-valued function, as required by quantum mechanics, it is necessary that the moduli of c fulfill a kind of quantum condition . . . This means that there exists a universal unit for the fluxoid:
>
> $$\Phi_1 = \frac{hc}{e} \sim 4 \cdot 10^{-7} \text{ gauss-cm}^2$$

Even after the prediction became common knowledge, confirmation was delayed because the flux quanta predicted by London was so small that measurement was not possible with the instrumentation available. When the initial measurements were finally made in 1961, the results, while not precisely what London had expected, amply confirmed his footnote prediction. Since, unknown to London, the charge carrier consists of a *pair* of electrons, the quantization was $h/2e$, exactly half London's prediction. (We drop the c, for consistency with SI units.)

The question of flux quantization typically is related to a specific sample shape; it is illuminating to consider a hollow superconducting cylinder. The superconducting state is associated with a macroscopic wave function Ψ that has a specific phase and amplitude. The phase is restricted to changes that are an integral multiple of 2π for any closed path in the superconductor sample, including a path around the hole through the cylinder. The hole through the cylinder cannot contain an arbitrary amount of flux, but must contain integral multiples of the flux quantum or "fluxon" above, i.e.:

$$\Phi_o = \frac{h}{2e} = 2.07 \times 10^{-15} \text{ Wb (SI units)}$$

$$\Phi_o = \frac{hc}{2e} = 2.07 \times 10^{-7} \text{ gauss-cm}^2 \text{ (cgs units)}$$

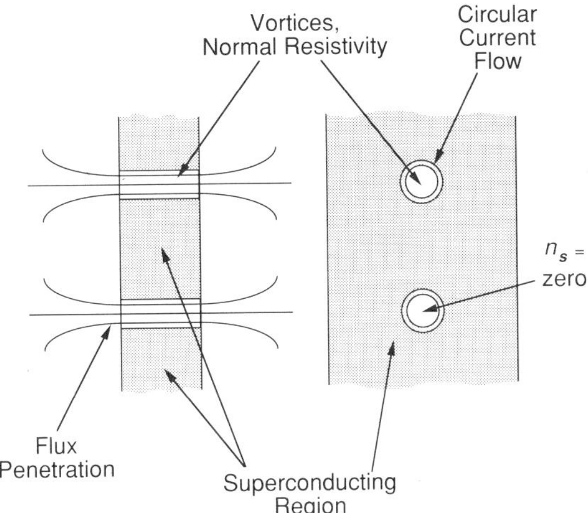

FIGURE 2.10 A Type II superconductor in the "mixed" or "Schubnikov" phase creates normal-state flux tubes or "vortices" surrounded by material that remains in the superconducting state. Flux penetrating these vortices is quantized at levels of $h/2e$. Since the Type II material need not expend the energy to expel this flux, superconductivity is maintained in much higher ambient H fields than can be withstood by Type I superconductors.

where h is Planck's constant, c is the velocity of light, and e is the electron charge.

Experiments designed to measure Φ_o, while difficult, are useful in determining whether a particular superconducting mechanism uses electron pairing. Such measurements have proven valuable in basic research on the new high-T_c superconductors as will be pointed out in the chapter on that subject.

The flux quantization applies not only to macroscopic holes in superconductors, rings, and so forth, but also the the flux vortices in Type II superconductors (see Figure 2.10). In the latter instance, while it is possible to have multiples of the fluxoid ($n\Phi_o$), the single-fluxoid condition (Φ_o) is energetically far more likely.

See Figures 2.11 and 2.12 for a description of Parks' vortex experiments in thin films.

GINZBURG-LANDAU WAVE FUNCTION

According to the theory proposed by Vitaly Ginzburg and Lev Landau in 1950, all of the electron pairs are condensed into a macroscopic quantum state with a wave function $\Psi(\phi)$, which has the same phase (ϕ) throughout the superconductor! Such quantum behavior on a large scale has a certain aesthetic appeal; this cooperative behavior of a large congregation of paired electrons is often poetically compared to a perfectly orchestrated "dance" of particles with opposite spin.

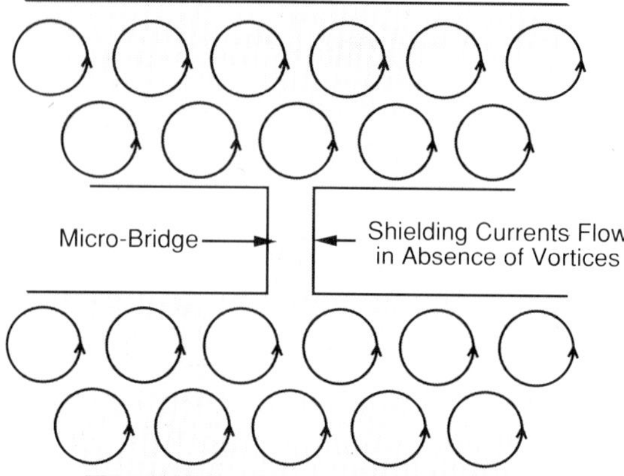

FIGURE 2.11 In an experiment that was both direct and elegant, R. D. Parks and his colleagues tested the vortex theory by use of a very narrow tin bridge connecting a pair of superconducting films. If each vortex actually confines a single flux quanta, then the number of vortices (and their size) must be a function of applied magnetic field. If the applied field is sufficiently low (as illustrated) so that the "size" of a vortex is larger than the width of the bridge, vortices will be unable to form on the bridge. In this situation the bridge expels the flux, resulting in a flow of shielding or Meissner current that can be detected. See Figure 2.12 for the instance where the magnetic field is increased. (After R. D. Parks, "Quantum effects in superconductors," *Scientific American*, Vol. 213, No. 4, Oct. 1965, pp. 57–67.)

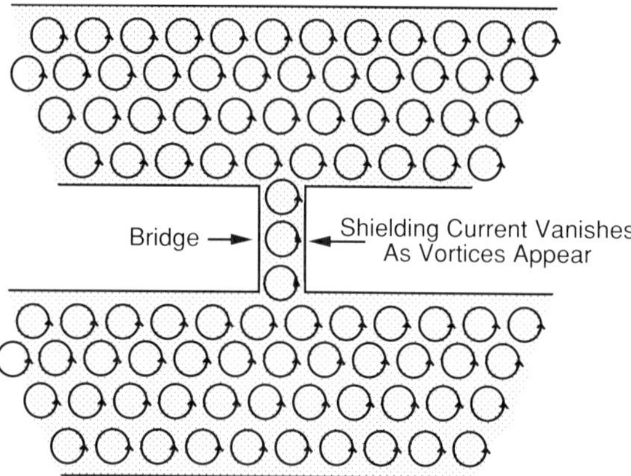

FIGURE 2.12 In this case the field is increased, resulting in an increase in the density of vortices and a reduction in their size. Here, the vortices are small enough to appear on the bridge and the Meissner currents decay rather abruptly. (After R. D. Parks, "Quantum effects in superconductors," *Scientific American*, Vol. 213, No. 4, Oct. 1965, pp. 57–67.)

Aside from its elegance, this macroscopic quantum phenomena is quite a useful feature. Properties that lead to important applications are demonstrated when the long-range phase coherence is interrupted either by insulating junctions or weakly-superconducting regions called, appropriately enough, "weak-links." These subjects will be discussed in more detail near the end of this chapter when Josephson tunneling junctions are introduced. Their importance to electronic instrumentation will be dealt with in the chapter on applications.

PAIRED ELECTRONS, ISOTOPIC MASS

Prior to the development of the BCS theory, there was evidence to point to the fact that although superconductivity is essentially an electronic phenomenon, the isotopic mass M of the specific material had a surprising effect on the magnitude of the critical temperature. Measurements revealed that T_c and H_c for a given elemental superconductor are proportional to $M^{-\alpha}$, which suggests a mass-on-a-spring model. While α is usually close to the BCS prediction of 1/2, this is not true in every instance; see Table 2.3 and the comments from Matthias quoted in the chapter on high-temperature superconductivity. Nevertheless, this dependence on isotopic mass suggested a coupling between phonons (quantized lattice vibrations) and electrons as the mechanism for superconductivity.

Classic superconductor behavior is explained in BCS theory as an interaction between phonons and (Cooper) pairs of electrons that have spins of equal

TABLE 2.3 The value of the exponent α in the expression $T_c = K/M^{\alpha}$ where K is a constant and M is the isotopic mass. The isotopic mass effect in a superconductor is considered a certain indication that lattice vibrations (phonons) are involved in the superconducting behavior. Original BCS theory predicts a value of 0.5 for α, but other factors (such as coulomb effects associated with the electrons) tend to modify α.

Superconductor	Isotope-Effect Exponent (α)
Zn	0.45
Cd	0.32
Sn	0.47
Hg	0.50
Pb	0.49
Tl	0.61
Ru	0.00 ± 0.05
Os	0.15
Mo	0.33
Nb_3Sn	0.08 ± 0.02
Mo_3Ir	0.33
Zr	0.00 ± 0.05

magnitude but opposite polarity. There is the usual coulomb repulsion between Cooper pairs, but an attraction between the electrons is introduced by the phonon-electron interaction at sufficiently low temperatures.

If the net result between these two forces is attractive, there is a strong correlation between electron pairs with opposite spin and opposite (but equal magnitude) momenta for energies within the range of $k_B\Theta_D$ (i.e., $3.53k_BT_c$) about the Fermi level, where Θ_D is the Debye temperature and k_B is Boltzmann's constant. (The value of Θ_D is typically determined from measurements of specific heat.)

The net attractive force is rather weak, but is effective over a rather large distance for a quantum phenomena.

ENERGY GAP, Δ

The first hints of a "forbidden" energy range or energy gap in the superconducting state came from measurements of an exponential specific heat; the presence and magnitude of the gap was confirmed by measurements of electromagnetic absorption. Perhaps the most straightforward method of measuring the energy gap is by determination of the "critical gap voltage" (V_g) for quasiparticle tunneling through a Giaever junction. More will be said about this in the section on quasiparticle (electron) and Cooper-pair tunneling.

It is now well established that the energy spectrum of Cooper-pair electrons has a gap of $\pm\Delta$ around the Fermi level E_F, that is, a width specified as 2Δ, typically on the order of 10^{-3} eV (electron-volts) (see Figure 2.13). The gap is reduced to zero as temperature approaches T_c, but is maximum at 0 K (see Figure 2.14). This energy gap at zero absolute temperature is:

$$2\Delta \approx 4k\Theta_D\epsilon^{(-1/N(0)V)} \qquad 2.31$$

where $N(0)$ is the density of electronic states at the Fermi level and V is the net electron-electron coupling constant.

It is interesting that there are superconductors that do not exhibit an energy gap. Since these are generally of less interest to the engineer than are conventional superconductors, a brief discussion of novel superconducting phenomena has been placed in Appendix G.

RELATIONSHIP BETWEEN TRANSITION TEMPERATURE (T_c), DEBYE TEMPERATURE (Θ_D), AND ENERGY GAP (Δ)

The superconducting transition temperature T_c is, according to BCS theory, related to the energy gap by

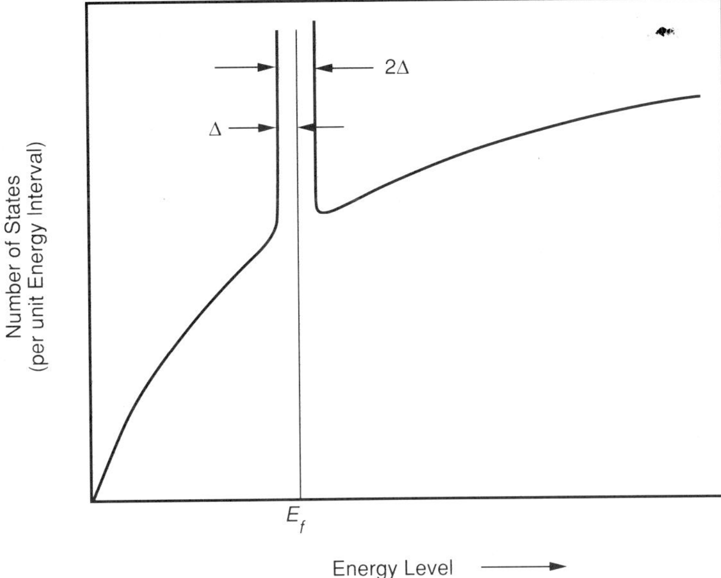

FIGURE 2.13 The density of electron energy states vs. energy level for a superconductor differs from that of a normal metal in an important aspect: the presence of an energy gap positioned symmetrically about the Fermi level.

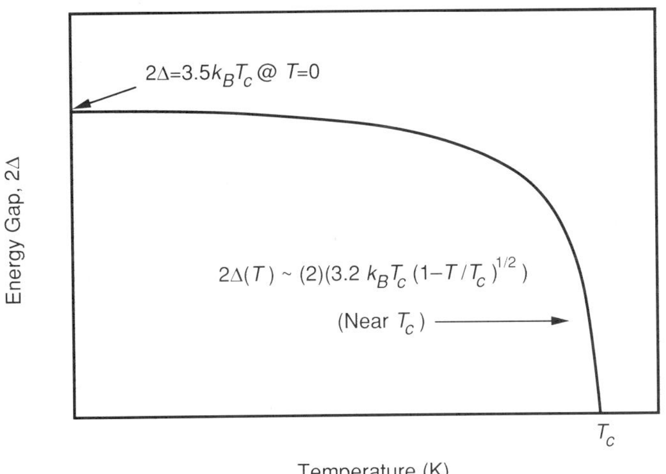

FIGURE 2.14 Superconductors that conform to the BCS model have an energy gap that increases from zero at $T = T_c$ to a maximum level at 0 K.

$$T_c \approx \frac{2\Delta(0)}{3.53k_B} \quad 2.32$$

where $\Delta(0)$ is the energy gap at zero temperature (see Table 2.4). The energy gap as a function of temperature $\Delta(T)$ is $\sim \Delta(0)$ for temperatures from 0 K up to about $T_c/2$.

For temperatures close to T_c, the temperature-dependent energy gap $\Delta(T)$ varies approximately as (Van Duzer, 1981)

$$\Delta(T) \approx 3.2k_B T_c \sqrt{1 - \frac{T}{T_c}} \quad 2.33$$

The $2\Delta = 3.53k_B T_c$ relationship, when combined with the expression relating the 0 K energy gap to the Debye temperature Θ_D, yields

$$T_c \approx 1.14\Theta_D \epsilon^{(-1/N(0)V)} \quad 2.34$$

For example, the Debye temperature for Sn is 195 K, and the $N(0)V$ product is 0.25. The estimated value for T_c from the above expression is 4.07 K, compared to a measured value of 3.75 K. Values of Θ_D, T_c, and the $N(0)V$ product are given for representative metallic elements in Table 2.5.

TABLE 2.4 Values of the energy gap and $2\Delta/k_B T_c$ for several elemental superconductors. According to BCS theory, $2\Delta/k_B T_c$ should equal 3.53. This excellent agreement between theory and a very complex phenomenon for a variety of elemental superconductors is one of the hallmarks of BCS theory success.

Element	Energy Gap @ $T = 0$ K (10^{-4} eV)	$2\Delta/k_B T_c$ (BCS value = 3.53)
V	16	3.4
Al	3.4	3.3
Zn	2.4	3.2
Ga	3.3	3.5
Nb	30.5	3.8
Mo	2.7	3.4
Cd	1.5	3.2
In	10.5	3.6
Sn	11.5	3.5
La	19	3.7
Ta	14	3.6
Hg	16.5	4.6
Ti	7.35	3.57
Pb	27.3	4.38

TABLE 2.5 Values for Debye temperature Θ_D, critical temperature T_c, and the $N(0)V$ product for several metals. (After De Gennes, P. G., *Superconductivity of Metals and Alloys,* trans. by P. A. Pincus, W. A. Benjamin, New York, 1966.)

Metal	T_c (K)	$N(0)V$	Θ_D (K)
Cd	0.56	0.18	164
Zn	0.9	0.18	235
Al	1.2	0.18	375
Tl	2.4	0.27	100
In	3.4	0.29	109
Sn	3.75	0.25	195
Hg	4.16	0.35	70
Pb	7.22	0.39	96

Using the Debye temperature Θ_D relationship to the Debye vibrational frequency ω_D, this expression may be rewritten as:

$$T_c \approx \frac{1.14 h \omega_D}{2\pi k_B} \epsilon^{(-1/N(0)V)} \quad 2.35$$

The Debye vibrational frequency ω_D is proportional to $M^{-1/2}$, an expression of the isotopic mass effect on T_c.

CRITICAL THERMODYNAMIC MAGNETIC FIELD, H_c

Recall from the introductory statements that the superconductor's free energy is utilized in the expulsion of flux; and once the free energy can no longer be expended without destroying the superconducting state, the material "quenches," that is, converts to the normal state of finite dc resistance. Quenching occurs at a thermodynamic critical field (H_c) that is a function of the material, temperature, and shape of the specimen.

As the applied magnetic field increases, temperatures lower than T_c are required to maintain the superconductive state. The variation of H_c with temperature (for $T < T_c$) is empirically shown to approximate a parabola. If $H_c(0)$ is the critical field at absolute zero, a useful approximation of H_c at a temperature T is:

$$H_c = H_c(0)\left[1 - \left(\frac{T}{T_c}\right)^2\right] \quad 2.36$$

If the applied field is greater than $H_c(0)$, it is impossible to produce a superconducting state even at absolute zero temperature. Consideration of magnetic field strength in the vicinity of a superconductor is of considerable practical importance, since many applications of superconductors require relatively high current densities (which produce high **H** fields) and other applications require operation in high ambient magnetic fields.

$H_c(0)$ may be estimated by extrapolation from measurements of critical fields rather close to 0 K. In instances where $H_{c2}(0)$ is so high that obtaining a sufficiently intense magnetic field for direct measurement is impractical, the following estimate is made using the incremental slope of H_c (vs. temperature) at the critical temperature T_c:

$$H_{c2}(0) \approx 0.69 T_c \left| \frac{dH_{c2}}{dT} \right|_{T=T_c} \qquad 2.37$$

TABLE 2.6 T_c and $H_c(0)$ for selected Type I superconductors. (After Lynton, *Superconductivity*, Methuen, London, 1969.)

Element	T_c(K)	$H_c(0)$ (Oe)
Aluminum	1.19	99
Cadmium	0.56	30
Gallium	1.09	51
Indium	3.404	293
Iridium	0.14	~20
Lanthanum-α	~5.0	—
Lanthanum-β	6.0	1600
Lead	7.18	803
Mercury-α	4.153	411
Mercury-β	3.95	340
Niobium	9.46	1944
Osmium	0.66	65
Protactinium	1.4	—
Rhenium	1.698	198
Ruthenium	0.49	66
Tantalum	4.482	830
Technetium	7.75	1410
Thallium	2.39	171
Thorium	1.37	162
Tin	3.722	309
Titanium	0.39	100
Tungsten	0.012	1070
Uranium-α	0.68	~2000
Uranium-β	1.80	—
Vanadium	5.414	1370
Zinc	0.875	53
Zirconium	0.546	47

This is a fairly common practice with the 1-2-3 copper oxide superconductors, where $H_{c2}(0)$ is evidently well in excess of 100 T. The upper critical field for Type II superconductors may also be expressed in terms of the Ginzberg-Landau coherence length ξ_{GL} and the flux quantum Φ_o:

$$H_{c2} = \frac{\Phi_o}{2\pi\mu_o\xi_{GL}^2} \qquad 2.38$$

where μ_o is the permeability of free space (Van Duzer, 1981).

Values of T_c and $H_c(0)$ for several superconducting elements are listed in Table 2.6.

CRITICAL CURRENT DENSITY, J_c

Along with T_c and H_c, the critical direct-current density, J_c, forms the third leg of the fundamental triad of superconductor characteristics that are the most significant for engineering purposes. The operating region for a superconductor may be considered as lying below a temperature-current-density-magnetic field surface; see Figure 2.15.

The current density that, when exceeded, causes the material to "quench," that is, lose its superconducting properties, is J_c. In certain geometries, it is approximately correct to consider the critical current (I_c) as that current which

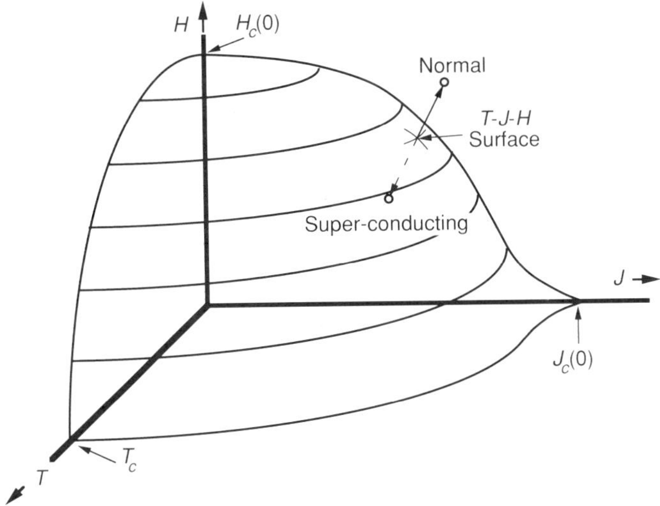

FIGURE 2.15 For a material to remain superconducting, a particular temperature/H-field/current density environment must exist. The material must operate below a surface similar to the one shown in this illustration.

develops the critical magnetic field H_c on the Type I conductor surface. This approach is useful for understanding the phenomena of critical current density, even if it is limited in its application for precise calculation of J_c.

If the superconductor is a pure metal (not an alloy or compound) and sample geometry is not complex, I_c can be estimated as the current level that will create surface magnetic fields just sufficient to exceed H_c. This estimation technique is known as he "Silsbee approximation," in reference to suggestions made by F. B. Silsbee. In actual practice, the Silsbee critical current estimate is generally rather higher than the actual critical current that can be realized.

For example, consider a long cylindrical superconductor of radius R carrying a total current I amperes. The surface field is then

$$H = \frac{I}{2\pi R} \tag{2.39}$$

Thus, the current required to develop a field H_c at the conductor surface is simply

$$I_c = 2\pi R H_c \tag{2.40}$$

(Other references, where cgs or Gaussian units are used, will present this relationship as $I_c = RH_c/2$.)

It is important to note that, even if the estimate for I_c was accurate, the critical current density for the wire is *not* simply $I_c/(\pi R^2)$ or $2H_c/R$, since, in Type I superconductors, the current is concentrated in the surface layer, with a characteristic penetration depth λ (see Figure 2.16).

In the case of Type II superconductors, operating in the mixed state, current flow is essentially a bulk phenomenon (Lynton, 1969). In this latter instance, supercurrent flow is confined to superconducting regions that are mixed with normal-resistance regions. The picture that is often invoked (Bean, 1962; Kim, et al., 1963) to describe dc current distribution in a Type II material is

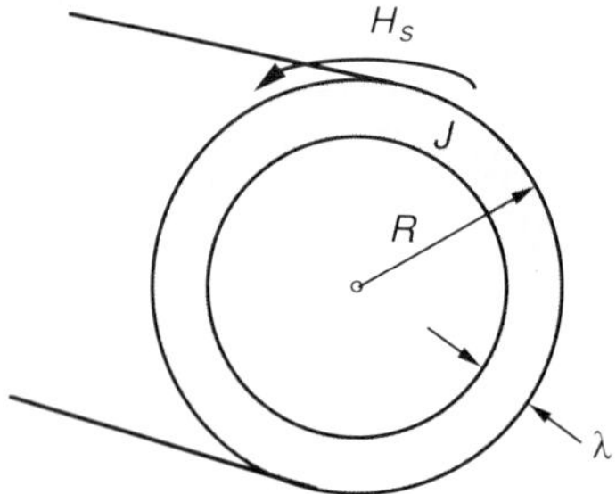

FIGURE 2.16 In a pure, elemental (Type I) superconductor, the magnetic field at the surface determines the maximum critical current density that can be maintained.

that of a network or mesh of superconducting filaments surrounding normal regions. This same concept is also expressed by comparing the superconducting region in the hard superconductors to a sponge. Since this model was proposed by K. Mendelssohn in 1935, it is usually referred to as the "Mendelssohn sponge." See Figure 2.17 for an idealized comparison of current transport in normal conductors and superconductors.

With these significant differences in current transport in Type I and Type II superconductors, one important result is that (surface) critical current flow (I_c) in a Type I wire is proportional to the circumference of the wire while the value of I_c in a Type II wire is proportional to the cross-sectional area (Blatt, 1964).

A word of caution: This approach for determination of the critical current is based on these assumptions:

1. The material is a perfect superconductor, that is, it is completely diamagnetic below T_{cf}.
2. The relevant dimensions of the sample (a wire or radius R in the instance just described) are large compared to the London penetration depth.

In practice, if very pure samples of tin are used to construct 1 mm diameter wires, the measured value of I_c will be on the order of half of the calculated value based on the assumption of a current that produces H_c at the conductor surface. In addition, recall that H_c is a function of $H(0)$, T, and T_c, and that H_c is zero at $T = T_c$. To determine the approximate value of H_c at a particular temperature (less than T_c), use the expression introduced earlier, that is,

$$H_c = H_c(0)\left[1 - \left(\frac{T}{T_c}\right)^2\right] \qquad 2.41$$

The more general point is this: to calculate the critical current in a Type I superconductor by this approximate method, one must first know the critical field H_c. The relationship between J_c and the critical current must take the actual geometry of current flow into account. For type I superconductors this is a surface sheath of depth $\sim\lambda$. In the case of Type II materials, the picture is more complex, but we will get to that subject directly.

Type II Critical Current Density; Bean's Critical State Model

Type II superconductors behave like Type I materials for magnetic fields below a critical level H_{c1}. When fields are increased above H_{c1}, however, Type II materials allow bulk penetration of magnetic flux. This penetration continues until the magnetic field is increased above the upper critical level, H_{c2}. This region where $H_{c1} < H < H_{c2}$ is the mixed state where microscopic normal

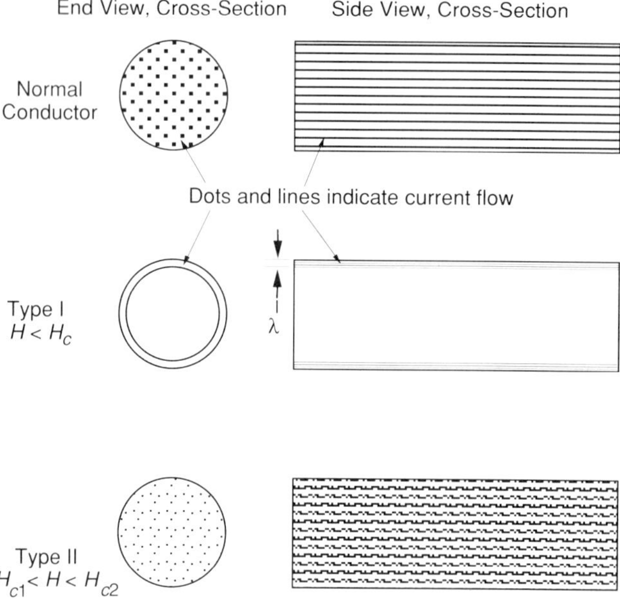

FIGURE 2.17 The current flow pattern compared in normal, Type I, and Type II cylindrical conductors, with the latter in the mixed state.

regions are microscopically mingled with superconducting regions. Since the superconductor's energy is not utilized in expelling flux, the critical current density is generally much higher in Type II materials than in Type I.

It is, of course, very important to be able to determine the level of these relatively high values of critical current density. When it is convenient to attach leads to a superconductor sample, the magnitude of critical current may be measured in a direct fashion. One simply increases the level of current flow through the sample until the superconducting state is destroyed. The loss of superconductivity is demonstrated by the appearance of voltage across the contacts where the current is injected. This rather straightforward technique is often referred to as the "transport" method of determining the level of I_c.

In some instances, for example, the characterization of very small crystalline samples, the transport method may not be practical due to the difficulty of attaching electrical leads. Since the value of critical current depends on the criteria for measurement of a finite voltage across the sample, variations in this voltage (due to poor contacts) may be attributable to the choice of threshold voltage selected by the experimenter. To minimize this effect, voltage is often measured across an additional set of contacts so that the voltage created by poor contacts where current is injected is less of a factor.

A standard noncontact method for determining critical current density in Type II superconductors involves the measurement of magnetization hysteresis as a function of applied H field. The width of the resultant hysteresis loop is proportional to the magnetization current J_M and, thus, to J_c. This result is based upon the work by C. P. Bean in development of his "critical state model" to explain the behavior of Type II superconductors (Bean, 1962). Essentially, Bean defines a parametric field H^*, where the critical current density J_c is flowing through the entire volume of the superconductor sample. The relationship between J_c and H^* for a magnetization M, in the Gaussian units employed by Bean, is:

For a slab of thickness D:

$$J_c = \frac{5H^*}{\pi D} = \frac{5[-8\pi M]}{\pi D} = -\frac{40M}{D} \qquad 2.42$$

For a cylinder of radius R:

$$J_c = \frac{10H^*}{4\pi R} = \frac{10[-12\pi M]}{4\pi R} = -\frac{30M}{R} \qquad 2.43$$

The negative sign is of no consequence when one merely wishes to know the magnitude of J_c, and is usually dropped. The reader is referred to the article by Bean in *Reviews of Modern Physics* (Bean, 1964) for a thorough treatment of this approach. More will be said about Bean's critical state model later in this chapter. See Figure 2.18 for distributions of J and H in the Bean model.

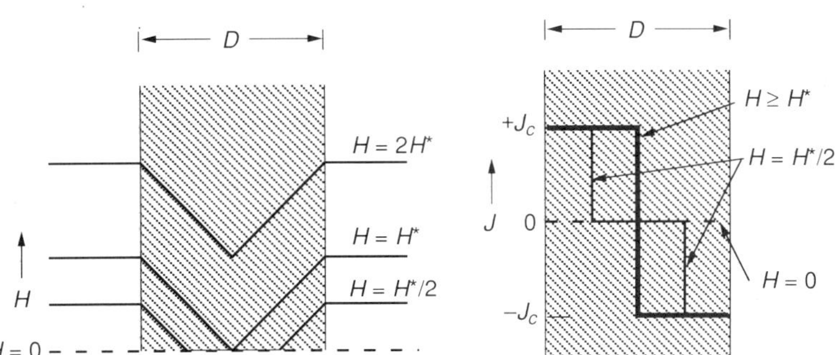

FIGURE 2.18 Distribution of magnetic field (H) and the current density (J) in a Type II superconducting slab. The quantity H^* (in the Gaussian units used by Bean) is $\pi J_c D/5$ for a slab thickness D and $4\pi J_c R/10$ for a cylinder of radius R. (After C. P. Bean, "Magnetization of high-field superconductors," *Rev. Mod. Phys.*, Vol. 36, January 1964, pp. 31–39.)

The basic relationship (in SI units) for the relationship between dc magnetization M and magnetization current density J_M for a specimen of thickness D is:

$$J_M = \frac{2\Delta M}{\mu_o D} \tag{2.44}$$

where ΔM is the width of hysteresis in the major magnetization curve (Ni, et al., 1988) This expression may be converted to Gaussian units for consistency with expressions commonly used in the physics literature. (See the conversion table for units in Appendix B.) Gaussian units for magnetization are in common use in the current literature on superconductivity.

Use of magnetization and the size parameter (D) may vary depending on the system under consideration. For example, in the case of granular superconductors where the typical particle radius (in cm) is R, and the hysteresis is ΔM gauss (or emu/cm^3), critical current J_c in A/cm^2 is related to hysteresis (see Figure 2.19) and average grain size by (Willis, et al., 1988)

$$J_c = \frac{15\Delta M}{R} \tag{2.45}$$

In another example (Dinger, et al., 1987), an equivalent form is used by investigators to estimate the J_c of single crystals:

$$J_c = \frac{30M}{R} \tag{2.46}$$

where M (in this instance) represents the magnetization (i.e., $\Delta M/2$) in emu/cm^3 (gauss), and R represents a disk radius in cm, resulting in units of A/cm^2 for J_c. In this latter case, since the crystals were not disks, the value of R was taken as the geometric mean of half of the sample dimensions perpendicular to the field. (This high-T_c Y-Ba-Cu-O sample was known to be anisotropic, so the field direction had to be taken into consideration.)

In low magnetic fields, values for J_c on the order of 10^6–10^7 A/cm^2 have been obtained for certain Type II metallic alloys. Levels on the order of at least 10^5 A/cm^2 are generally considered necessary to exploit a wide variety of superconductor applications and, if this includes high field magnet design, this J_c magnitude should be obtainable in a field of several tesla.

Flux Pinning

The lowest energy state (and therefore the most probable state) for the fluxon lattice is a triangular array of flux tubes, each vortex of supercurrent encompassing a single fluxon Φ_o. It is important that this array of fluxons remain

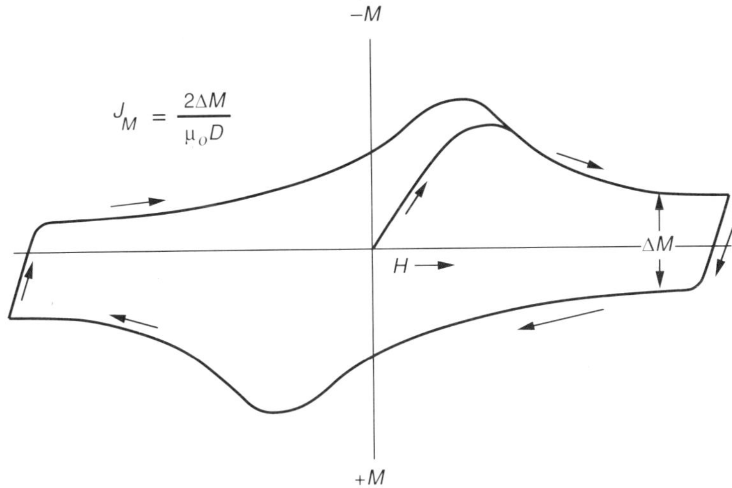

FIGURE 2.19 Typical (diamagnetic) magnetization curve for Type II superconductor. The hysteresis (ΔM) at a given field H is a measure of the critical current density, J_c. The D in the expression above is a measure of sample thickness, or, in the case of polycrystalline ceramics, represents grain size.

stable, since motion results in dissipation by normal currents induced in the fluxon cores. Fluxon stability and critical current density can be increased by the presence of impurities or defects, which tends to stabilize or "pin" the fluxons.

Such useful defects may include holes, precipitate particles, grain boundaries or a variety of other heterogeneities. Defects introduced by radiation damage (e.g., neutron irradiation) may also serve to pin the vortices found in Type II materials. Effective fluxon pinning is achieved when the separation between pinning locations is on the order of one to five coherence lengths (i.e., ξ to 5ξ) (Larbalestier, et al., 1986). In typical metallic alloys where the feature responsible for pinning is the grain boundary, grain size must therefore be relatively small. Effective grain sizes for pinning may be on the order of 0.1 μm or less.

It is notable that while T_c and H_c may generally be considered intrinsic properties of the specific material, J_c is largely a function of the microstructure and can vary by several orders of magnitude as a function of material processing techniques.

It is helpful, in understanding the difference of current transport in Type I and Type II superconductors, to consider the current flow in a wire. The discussion that follows applies for static conditions, that is, constant, direct current flow.

As pointed out earlier, in a type I material, current is restricted to a thin surface sheath of depth λ, as long as the magnetic field (externally applied or due to the transport current) is less than the critical thermodynamic field H_c. While the penetration depth λ is very small near 0 K, λ increases dramatically as temperature approaches T_c. Above H_c, of course, the Type I superconductor is quenched, allowing complete penetration of magnetic fields.

Below H_{c1}, the Type II superconductor behaves like a Type I material: current flow and magnetic field penetration is restricted to the thin surface sheath of depth λ. As the field is increased above H_{c1}, however, flux begins to penetrate the surface in thin tubes with diameters less than λ. As mentioned earlier, the configuration of the superconducting region in this mixed state has been compared to a mesh or sponge.

In the case of circumferential H fields generated by the surface current transported by a superconducting Type II wire, the flux tubes are rings that enter at the surface of the wire. Even though the flux-tube circles are created at the surface, they tend toward their lowest energy state, which requires a decrease in radius. If this decrease in radius is not checked by effective pinning, the circular flux tubes shrink toward the center of the wire where they disappear (De Gennes, 1966). This movement of the shrinking circular flux tube through the wire represents a loss, equivalent to a nonzero resistance at fields above H_{c1}.

If Type II superconductors are to be practical for large current transport densities, it is necessary to prevent this movement of flux tubes across the conductor cross-section. As indicated above, this is accomplished by pinning the flux tubes to defects. The pinning force (α) at a defect site has a limiting or threshold value, which we will call α_t.

In actual superconductors, the induced field B is never completely homogeneous, and a variation in density (B/Φ_o) of flux tubes creates the equivalent of a pressure (p) that tends to cause expansion of flux tubes from higher to lower-density regions in the superconductor. Fundamentally, the pressure is caused by the repulsion between flux lines. Consider a gradient in pressure along a direction z; it is necessary (for effective pinning) that the pressure at any location in the sample remain less than the pinning threshold force α_t, that is,

$$\alpha_t \geq \frac{\partial p}{\partial z} \qquad 2.47$$

More On The Critical State Model

The general basis for the critical-state model is that the superconductor is capable of supporting a maximum *local* current density of J_c (Fietz, et al., 1964). Bean's assumption is that there exists a relationship for J_c as a function of induction **B**, so that the critical current J_c is determined only by the value of **B** at the point under consideration. Any transport current, however small, will cause J_c to flow in some small region of the superconductor. Thus, the value of J_c is not uniform throughout the superconductor, that is, it depends on the induction **B** at each location. As transport current flow is increased, the critical current will flow in additional regions. When transport current has increased to a level where this critical current $J_c(B)$ flows at *every location* in the sample, the specimen is in the "critical" state. In this critical state, any additional transport current will necessarily lead to an increase over J_c in some region, leading to quenching of the superconductor.

The superconductor can, of course, be perturbed by other means than by an increase in transport current. If the flux-lattice pressure gradient becomes larger than the pinning force *in some region* of the superconductor, flux lines will move toward a region of lower flux density until the threshold condition above is again met throughout the sample.

A useful analogy for the dynamics of the Bean critical state used by De Gennes is that of a hill of sand which, at some location on its surface, exceeds a critical slope. At this location, sand will avalanche down the hill until the original slope is again critical. The analogy is particularly appropriate, since flux tubes in regions where the critical state is exceeded tend to move toward lower-density regions in groups (avalanches) of at least 50 flux lines (De Gennes, 1966).

An important result of Bean's approach was the prediction that magnetization of the filamentary structure depended on the macroscopic dimensions of the sample under consideration; this size-dependent magnetization was supported by experiments with solid cylinders of Nb_3Sn (Bean, 1962). Bean's calculations also accounted for hysteresis in the magnetization curve, which had not been explained by previous theories of Type II behavior. As pointed out earlier, this magnetization can be used to make indirect measurements of the value of J_c.

Kim and his colleagues used the critical state equation (Kim, et al., 1963)

$$J_c = \frac{\alpha}{B + B_o} \qquad 2.48$$

in their description of the behavior of hollow, superconducting cylinders. B_o, in this instance, corresponds approximately to the thermodynamic critical field. Generally, α and B_o are obtained from experiments.

Flux Creep

When flux lines move in a Type II superconductor, energy is invariably dissipated. The dissipation is generally distinguished as related to "flux creep" when the pinning forces dominate, and as "flux flow" when the Lorentz forces are dominant. If optimum critical current transport is to be realized, the movement of flux must be minimized. The flux decay measurements of Kim, Hempstead, and Strnad previously mentioned were interpreted by Anderson's flux-creep theory (Anderson, et al., 1964; Anderson, 1962). Essentially, flux pinned by vortices, or microscopic persistent current loops jumps barriers by a thermal activation process. This leads to a logarithmic decay of flux with time (t) (Mohamed, et al., 1988):

$$\partial B \propto k_B T \ln(t) \tag{2.49}$$

While pinning centers tend to prevent the movement of flux tubes, as long as the pressure gradient due to the gradient in flux density is less than the pinning force, there is some tendency for flux to jump across pinning barriers. This phenomena is generally termed "flux creep." This phenomena can be measured by placing a magnetic field sensor on the inside of a hollow superconducting cylinder. As the external field (normal to the cylinder axis) is increased, no field is seen inside the cylinder until the critical field is reached. If flux creep did not exist, the field inside the cylinder would then vary in proportion to the external field as it was increased above the critical level. In fact, when the external field is held constant, the internal field tends to gradually increase as flux tubes jump across pinning barriers.

Flux creep can be quantified by measurement of the emf that is created in a current-carrying conductor by the drift of flux vortices (De Gennes, 1966). Such experiments have shown that the flux creep velocity (v) has a relationship that suggests an activation energy (E_a):

$$v = v_o \epsilon^{-E_a/k_B T} \tag{2.50}$$

where v_o is typically on the order 1000 cm/sec.

Surface Barriers to Flux Tubes

There are interactions between the surface of a superconducting sample and the flux tube, usually referred to as surface barriers (see Figure 2.20). These barriers operate in both directions, that is, both to prevent entry of a flux tube and to prevent its exit from the sample. In some samples, there is a hysteresis in the magnetization curve that can be explained, in part, by a difference in the magnitude of the surface barrier, depending on whether the flux tube is entering or leaving the sample.

The surface barrier has been modeled by consideration of a single flux tube and its interaction with the surface of the superconductor (Bean & Livingston, 1964). Two competing forces act on the flux tube normal to the specimen surface:

a. The image force. Surface boundary conditions are met by adding an image flux tube, of opposite direction (sign), outside (and also normal to) the surface. This results in an attractive force between the internal

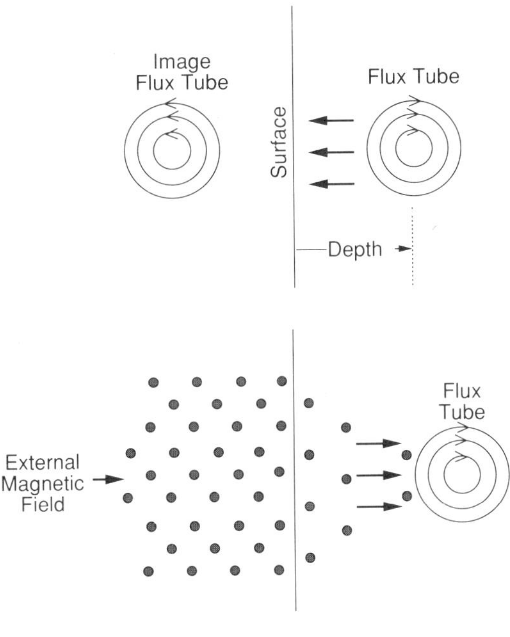

FIGURE 2.20 The surface barrier to the movement of flux into or out of the superconductor is illustrated schematically above. At the top of the figure, the image flux line, since it has a sign opposite to the internal (actual) flux tube, tends to attract the flux tube toward the surface. An opposing force is generated by an external magnetic field that has the same sign as the flux tube, tending to force the flux tube away from the surface. Except at very low applied fields, the repulsive force tends to dominate the surface barrier effect. (After C. P. Bean and J. D. Livingston, "Surface barrier in type II superconductors," *Phys. Rev. Lett.*, Vol. 12, No. 14, January 6, 1964, pp. 14–16.)

flux tube and the surface. For depth (z) much greater than the coherence length ξ, this attractive force vanishes as $\epsilon^{(-2z/\lambda)}$.

b. External field penetration into the specimen, if of the same sign as the flux tube, causes a repulsive force that tends to push the flux tube away from the surface.

Since the repulsive force due to the penetration of external field will dominate the (attractive) image term for depths much greater than λ, the attractive effect is important only in very low fields. At higher fields, domination by the repulsive surface effect leads to hysteretic phenomena.

This simple picture, however, is complicated by surface irregularities. Flux may enter at surface irregularities and either spread into the sample or block flux escape (at the irregularity) as field level decreases. Concave irregularities on the surface, by producing lowered surface fields, may reduce the barrier to flux escape.

Type I and Type II Superconductors Compared

We have already been referring to superconductors as being subdivided into two main categories, referred to either as Type I or Type II, and these are also identified as "first kind" and "second kind," respectively. To add to the confusion, Type I may be referred to as "Pippard" or "soft" and Type II as "London" or "hard," particularly in the older literature.

Aside from the behavior of the magnetization curve, one of the more distinguishing aspects of Type I and Type II materials is the relative value of coherence length ξ and penetration depth λ. As indicated in the following list of characteristics, when ξ is greater than λ, the material is Type I. This is a consequence of the fact (dealt with in the Ginzberg-Landau equations) that the surface energy of an interface between superconducting and normal regions is positive. For ξ less than λ, the surface energy for the same surface is negative; in this instance, a Type II superconductor exists (Yntema, 1971; Rickayzen, 1965).

It was pointed out earlier that the Ginzberg-Landau (GL) theory yields a dimensionless parameter κ that may be defined in terms of the GL coherence length ξ_{GL} and the London penetration depth λ_L (Lynton, 1969; Tinkham, 1985):

$$\kappa = \frac{\lambda_L}{\xi_{GL}} \qquad 2.51$$

This parameter defines the difference in Type I and Type II superconductors in a particularly simple fashion.

For $\kappa < 2^{-1/2}$, the superconductor is Type I, exhibiting a normal Meissner effect and flux exclusion for fields below a critical level H_c.

For $\kappa > 2^{-1/2}$, a magnetic field greater than a critical level (H_{c1}) will penetrate the material in quantized flux tubes. Each flux tube (or vortex) is confined by a current vortex and carries the single flux quantum $h/2e$.

In addition, when comparing a Type I and a Type II superconductor that have the same thermodynamic critical field H_c, the upper critical field for the Type II material is given by

$$H_{c2} = \sqrt{2}\,\kappa H_c \qquad 2.52$$

Characteristics of each type are summarized in Table 2.7. When not specified otherwise, λ will be understood to be the London penetration depth λ_L, and ξ will be the GL characteristic length ξ_{GL}.

TABLE 2.7 Summary of the characteristics of Type I and Type II.

Type I (First Kind, Pippard, Soft)	Type II (Second Kind, London, Hard)
Thermodynamic Field: H_c	Critical Fields: H_{c1}, H_{c2}, H_{c3}
Exhibits Intermediate State	Quantized Vortex Lines, Mixed State
Generally Pure Elements	Generally Compounds (Except Nb & V)
Lower Critical Currents	High Critical Currents*
Lower Critical Fields	Higher Critical Fields
$\xi \gg \lambda$	$\xi \ll \lambda$
Surface Current Flow for $H < H_c$	Bulk Current Flow for $(H_{c2} > H > H_{c1})$

* The high-T_c copper oxide superconductors are Type II materials, but unlike most Type II superconductors, these ceramics have relatively low values of J_c in bulk samples. This is apparently due to grain-boudary effects, the presence of impurities, and other effects that depend largely upon processing. As single crystals or oriented films, the copper oxides do have fairly high values of J_c. See the following chapter for details.

High-purity elemental superconductors, if sufficiently free from mechanical strain on the crystal lattice, are generally Type I. (Vanadium and Niobium are important exceptions to this rule.) If the mean free path ι of conduction electrons is sufficiently shortened by defects in the crystal lattice or by chemical impurities, a Type II superconductor is the result. In fact, the difference between H_{c2} and H_{c1} increases as the electron mean free path decreases.

Type I Characteristics Summarized. Type I (first kind) materials of certain geometries are characterized by an abrupt change from normal magnetic permeability above T_c to total diamagnetism below T_c. Type I superconductors exhibit a single (thermodynamic) critical magnetic field level, H_c. Below H_c, (for temperatures below T_{cf}) Type I materials exhibit zero resistivity (ρ) and zero internal field **B** in the presence of an applied external field **H**, except for an interesting condition called the "intermediate state." This occurs when the expulsion of flux creates distortions at specific regions on the surface of the material which results in the field in those regions exceeding H_c. This occurs, for certain shapes, even when the external applied field is less than H_c. Since the material in those regions must remain nonsuperconducting, the result is a coexistence of macroscopic superconducting and normal regions (see Figure 2.21). As mentioned earlier, there is an energy "cost" paid by the Type I superconductor in expelling the field; it is

$$\frac{\text{Energy}}{\text{Unit} - \text{Volume}} = \frac{H_c^2}{8\pi} \qquad 2.53$$

It is not surprising that the intermediate state is highly dependent on sample geometry (see Figure 2.22). In the case of a sphere, the intermediate state exists for ambient fields in the range

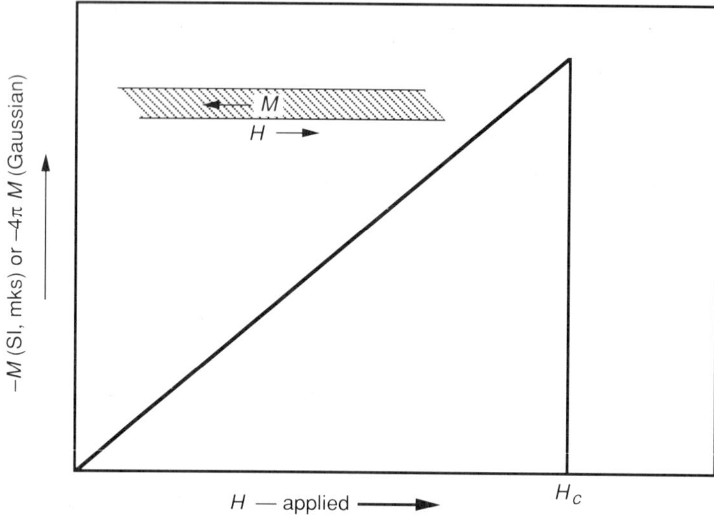

FIGURE 2.21 In a long, thin, Type I superconductor placed in an axial magnetic field, behavior is almost perfectly diamagnetic until the critical thermodynamic field H_c is reached. Above H_c, the field collapses into the material (i.e., $M \rightarrow 0$) as it reverts to the normal state.

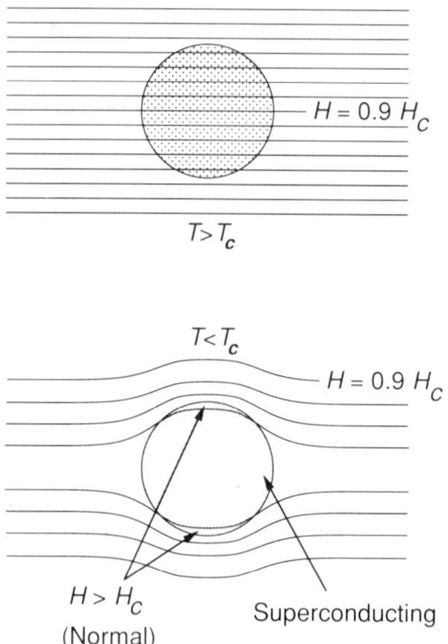

FIGURE 2.22 In contrast to the special case illustrated in Figure 2.21, a more typical Type I superconductor configuration has a tendency to distort an ambient field that is slightly less than H_c, leading to a field slightly greater than H_c in some parts of the specimen. This is the "intermediate" state, where part of the specimen is normal and the balance is superconducting.

$$H_c > H_{\text{sphere}} > \frac{2}{3}H_c \qquad 2.54$$

For a thin slab of Type I material, the intermediate state exists for all magnetic fields (normal to the slab surface) less than H_c, that is,

$$H_c > H_{\text{slab}} > 0 \qquad 2.55$$

For a cylindrical conductor (where length is much greater than radius) in an **H** field parallel to its axis, the intermediate state is absent or minimal since expulsion of flux creates a minimal distortion of the external field.

Type II Characteristics Summarized. Type II materials show a more gradual change in magnetic properties, regardless of sample geometry, and have three critical magnetic field levels: H_{c1}, H_{c2}, and H_{c3}. In the range $0 \rightarrow H_{c1}$ (Perfectly Diamagnetic State), H_{c1} is the level below which magnetic permeability is zero (for certain sample geometries), and is equivalent to H_c for a Type I material. In the range $H_{c1} \rightarrow H_{c2}$ (Mixed State), H_{c2}, the upper critical field, is that field strength where electrical resistivity will abruptly change from zero to its "normal" level. Type II materials are in a mixed state between H_{c1} and H_{c2}, where, although resistivity is zero, the internal magnetic field **B** is nonzero and confined to an array of internalized flux tubes (see Figure 2.23).

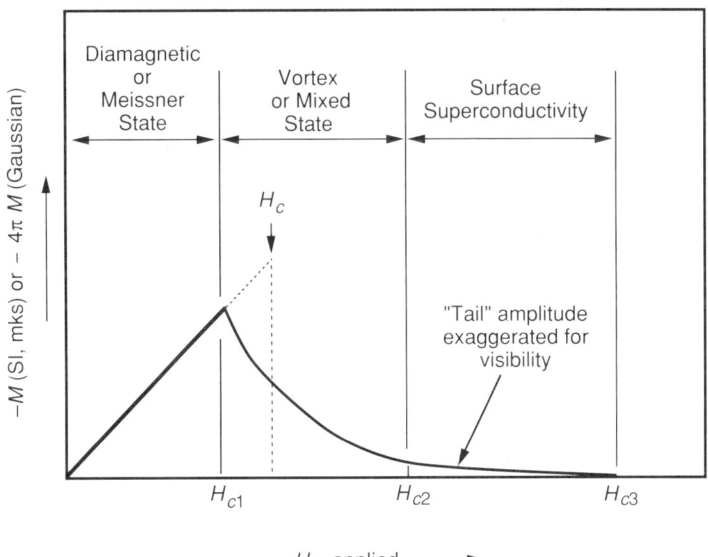

FIGURE 2.23 In a Type II superconductor, the material is in a diamagnetic or Meissner state below H_{c1}, and in the mixed or vortex state between H_{c1} and H_{c2}. The "tail" between H_{c2} and H_{c3} represents surface superconductivity.

In the mixed state, a portion of the magnetic field normal to the surface penetrates Type II materials in individual flux quanta Φ_o. In this model, first proposed by Alexei Abrikosov, the term "vortex state" is often used to describe the phenomenon.

Of considerable practical importance is the phenomenon of flux pinning that is seen in Type II materials in the mixed state. If a Type II material is cooled in a magnetic field, field lines normal to the surface can become trapped or "pinned" in the vortex tubes. The stabilization of the pinning sites is important for increasing the value of critical current J_c. As discussed elsewhere in this chapter, defects play a major role in the stabilization of pinning sites and enhancement of J_c.

The geometry for vortex distribution which has the lowest free energy is a triangular one, but the presence of impurities tends to generate a somewhat random distribution of pinning sites (Tinkham, 1985).

The value of H_{c2} is much higher than H_c, that is, the thermodynamic critical field. Using the Ginzberg-Landau kappa (κ), the penetration depth (λ), and the coherence length as a function of temperature ($\xi(T)$); the relationship between H_{c2} and H_c is

$$H_{c2} = \sqrt{2}\,\kappa H_c = \sqrt{2}\left(\frac{\lambda}{\xi_{GL}}\right)H_c \qquad 2.56$$

Since λ is much larger than ξ_{GL}, H_{c2} is much larger than H_c.

In the range $H_{c2} \rightarrow H_{c3}$, at field strengths between H_{c2} and H_{c3}, there remains a superconducting current "sheath" on the surface (Van Duzer, 1981). A useful approximation for H_{c3} is

$$H_{c3} \approx 1.69 H_{c2} \qquad 2.57$$

AC LOSSES IN SUPERCONDUCTORS

At frequencies below about 1 MHz, losses in superconductors are due largely to the hysteretic motion of vortices through the material.

At higher frequencies, the two-fluid model may be used to account for losses due to the presence of "normal" electrons that have not condensed into pairs.

Recall that the normal-state surface resistance (Ohms (Ω) per square) of a normal conductor at high frequency is a function of skin depth and conductivity σ, that is,

$$R_{\text{surface}} = \frac{1}{\sigma\delta} = \sqrt{\frac{\omega\mu}{2\sigma}} \qquad 2.58$$

Superconductors *do not* exhibit zero resistivity to alternating currents. While the superconductor has a much lower ac resistance below T_c than above, like normal conductors, superconductor ac resistance increases with frequency.

Recall that a superconductor can be represented (by the two-fluid model) as having both a superconducting current density J_s and a "normal" current density J_n, defined by:

$$\mathbf{E} = 4\pi\lambda^2 \frac{\partial \mathbf{J}_s}{\partial t} \qquad 2.59$$

which, since

$$\Lambda = \frac{m}{ne^2} = 4\pi\lambda^2 \qquad 2.60$$

is equivalent to

$$\frac{\partial \mathbf{J}_s}{\partial t} = \frac{1}{\Lambda}\mathbf{E} = \frac{n_s e^2}{m}\mathbf{E} \qquad 2.61$$

where a finite value of $\mathbf{E}$ is clearly required to produce an alternating current flow. The normal component $\mathbf{J}_n$ is defined by

$$\frac{\partial \mathbf{J}_n}{\partial t} + \mathbf{J}_n\left(\frac{1}{\tau}\right) = \frac{n_n e^2}{m}\mathbf{E} \qquad 2.62$$

where τ represents the momentum relaxation time (Hartwig & Passow, 1975).

Consider a superconductor that has dimensions much larger than the penetration depth λ, operating at a temperature $T \ll T_c$. Consider also the same conductor operating in its "normal" mode (quenched by a magnetic field) with a "normal" homogeneous conductivity σ_n that is sufficiently large so that the skin depth δ is much smaller than the sample dimensions. If the alternating-current resistance in the normal mode is R_n, the ac resistance R_s at superconducting temperatures is (Herring, 1960; London, 1950):

$$R_s = \frac{R_n \lambda}{\delta}\left\{\frac{\sigma_n}{\sigma}\right\}^{1/2} \sqrt{\frac{\sqrt{1 + (\lambda/\delta)^4} - 1}{1 + (\lambda/\delta)^4}} \qquad 2.63$$

The skin depth δ varies with frequency ω, conductivity σ, and magnetic permeability μ in the following manner (if the mean free path ι of the electrons is smaller than δ):

$$\delta = \sqrt{\frac{2}{\omega\mu\sigma}} \qquad 2.63$$

See Appendix A on electromagnetism for a discussion of anomalous skin depth and surface resistivity for $\iota \gg \delta$. Also see the references (Reuter, 1948; Pippard, 1987) for a thorough discussion of the interaction between λ and δ.

At superconducting temperatures, the surface alternating electric field $\mathbf{E}_s$ at radial frequency ω rad/s penetrates the conductor attenuated exponentially with a characteristic depth $1/K$, that is,

$$\mathbf{E} = \mathbf{E}_s \epsilon^{-Kz} \tag{2.65}$$

The parallel **H** field and surface current behave in the same manner:

$$\mathbf{H} = \mathbf{H}_s \epsilon^{-Kz} \tag{2.66}$$

and

$$\mathbf{J} = \mathbf{J}_s \epsilon^{-Kz} \tag{2.67}$$

The parameter K is a function of penetration depth (λ) and skin depth (δ):

$$K = \frac{1}{\lambda\sqrt{2}} \{ \sqrt{m+1} + j\sqrt{m-1} \} \tag{2.68}$$

where

$$m = \sqrt{1 - \frac{4\lambda^4}{\delta^4}} \tag{2.69}$$

which reduces the expression for K to:

$$K = \frac{1}{\lambda}\sqrt{1 + j2\left(\frac{\lambda}{\delta}\right)^2} \tag{2.70}$$

As is usual in engineering notation,

$$j = \sqrt{-1} \tag{2.71}$$

This expression for K is also often given in the following form, where δ_L has been substituted for δ in this instance:

$$K = \frac{1}{\lambda}\sqrt{1 + j\left(\frac{\lambda}{\delta_L}\right)^2} \tag{2.72}$$

The difference in London's expression is that

$$\delta_L = \frac{\delta}{\sqrt{2}} \tag{2.73}$$

The parameter δ_L is not the skin depth (δ) familiar to electrical engineers, but a related "small length" defined by London (London, 1950).

Engineers are familiar with the effects of skin depth in limiting the flow of current to the surface of a conductor at higher frequencies (see Figure 2.24). How do the λ and δ parameters interact? It is clear that the parameter λ performs a similar function to δ over a wide range of frequencies. The effects of λ certainly predominate at most frequencies of interest, that is, λ determines the penetration of electromagnetic fields into the material. It is clear that, for a specific superconductor, there is relatively high frequency where the skin depth δ and London penetration depth λ will have the same value. As frequency increases above that level, skin depth will continue to decrease while λ does not, so that δ becomes the predominant factor in determining electromagnetic penetration and surface resistivity. The relevant relationship is (London, 1950)

$$\frac{|J_n|}{|J_s|} = \left(\frac{\lambda}{\delta_L}\right)^2 = \sigma\Lambda\omega \tag{2.74}$$

where σ is the conductivity associated with the "normal" electrons, and $\Lambda = m/ne^2$ has a value typically on the order of 10^{-31} s^2 (London, 1950).

To summarize, at frequencies of general interest, ($\omega < 1/(\sigma\Lambda)$), the penetration of surface currents is controlled by λ, which is much smaller than δ_L. For high frequencies ($2\pi f > 1/(\sigma\Lambda) \approx 10^{11}$ rad/s, i.e. ~16 GHz) skin effects gradually begin to predominate over λ. For *much* higher (optical) frequencies, the "superconductor" behaves essentially as a normal conductor.

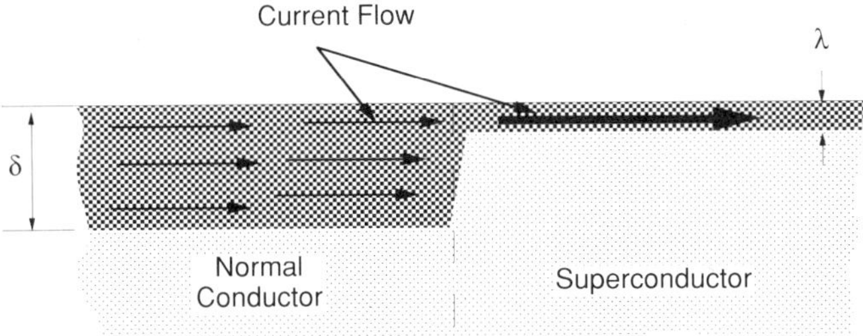

FIGURE 2.24 In normal conductors, microwave currents flow on the surface with a skin depth (δ) that is a function of frequency and conductivity. In a superconductor, shown here connected to the normal metal, microwave penetration is limited to the London penetration depth (λ). When frequency (ν) of operation becomes sufficiently high that the proton energy ($h\nu$) exceeds the superconductivity energy gap (2Δ), electron pairs are disrupted and the superconductor reverts to normal metal behavior.

When the photon frequency ν becomes high enough so that the photon energy $h\nu$ is comparable to the superconductor energy gap Δ ($2\Delta \sim 3.53k_BT_c$), the material behaves like a normal conductor (see Figure 2.25). This conversion to normal behavior at very high frequencies is associated with the observation that, at optical frequencies, ($\omega \approx 10^{14}$ rad/s), the coefficients of reflection and absorption do not vary for $T > T_c$ and $T < T_c$. This is related to the fact that a superconductor does not have different visual appearances, that is, reflection or absorption of light, in the superconducting and the normal state.

Experiments in resonant cavities at temperatures well below $T_c/2$ yield a useful relationship for surface resistance R_s (which is in agreement with BCS theory),

$$R_s = \frac{A\omega^2}{T}\epsilon^{-\alpha T_c/T} + R_{\text{res}} \tag{2.75}$$

where A is a constant for the specific conductor, $\alpha = 1.76$, and R_{res} is a "residual resistance" with a typical range of 10^{-7} to 10^{-9} Ω for niobium, and is probably related to a variety of surface defects as well as trapped flux. When the residual resistance is due to normal materials on the surface, it will generally not show a marked temperature dependence. On the other hand, a strong temperature dependence in R_{res} may indicate a degraded superconductor with a reduced energy gap (Braginski, et al., 1988). In practice, the exponent on ω may be slightly less than 2.

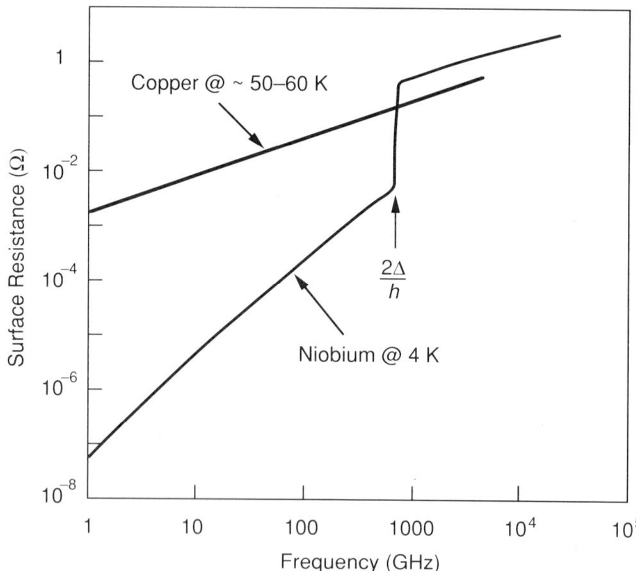

FIGURE 2.25 A comparison of the radio-frequency surface resistance for cryogenic copper and superconducting niobium. Note the abrupt change in niobium surface resistance when the photon energy exceeds the Nb energy gap.

Helmut Piel has suggested the following simplified approach to account for the ω^2 dependence of superconductor surface resistance (Piel, 1988).

First, consider the alternating magnetic field H_s at the surface of a superconductor, which is parallel to the surface. Recall that, in light of the Faraday induction law, the magnitude of the induced surface field E_s is proportional to the product of ω and B_s:

$$E \propto \omega B_s \tag{2.76}$$

The electric field generates losses in the fraction of normal electrons that still exist above 0 K. The "normal" conductance associated with these unpaired electrons is represented here by σ_e, which is a function of the density of normal electrons n_e, the mean free path ι, and their mean velocity v_o:

$$\sigma_e = \frac{n_e e^2 \iota}{2m v_o} \tag{2.77}$$

where m and e are the mass and charge, respectively, of the electron.

One can express the surface power density (P_s) over an area A for an incremental volume dV in terms of E_s:

$$P_s = \frac{\sigma_e E_s^2}{2}(dV) \propto \frac{\sigma_e E_s^2}{2}(A\lambda) \tag{2.78}$$

The "2" appears in the denominator to account for the fact that E and H represent peak values of a sinusoidal field. The rms value, required for power calculations, is peak divided by $(2)^{1/2}$. Since E_s is proportional to ωB_s (or ωH_s), it is also true that

$$\frac{P_s}{A} \propto \frac{\sigma_e (\omega H_s)^2}{2}(\lambda) \tag{2.79}$$

but, P_s may be expressed in terms of surface resistance R_s and surface field H_s by

$$\frac{P_s}{A} = \frac{R_s H_s^2}{2} \tag{2.80}$$

It follows that

$$R_s \propto \sigma_e \omega^2 \tag{2.81}$$

For normal-state conductors, R_s has the following relationship to frequency depending on the electron mean free path ι (which increases as temperature decreases).

For ι small compared to skin depth:

$$R_s \propto \omega^{1/2} \tag{2.82}$$

For ι comparable to skin depth (an admittedly unusual case):

$$R_s \propto \omega^{1/3} \tag{2.83}$$

For ι large compared to skin depth:

$$R_s \propto \omega^{2/3} \tag{2.84}$$

It is noteworthy that R_s is proportional to $\omega^{1/2}$ for normal conduction, and to $\omega^{1/3}$ or $\omega^{2/3}$ for cryogenic cavities at sufficiently high frequencies (where ι becomes comparable to or larger than δ), while superconductors have a surface resistance that varies as ω^2, except for the residual component for $T \ll T_c$.

One consequence of this relationship is that superconducting niobium cavities (for particle accelerator applications, for example) have decreasing advantages over cryogenic copper cavities as frequency increases, or even normal (Cu @ 300 K) at a sufficiently high frequency. Generally, in the design of microwave cavities, normal metals have a Q that increases with frequency as $\omega^{+1/2}$ while superconducting cavities exhibit a Q proportional to $\omega^{-1/2}$ (Van Duzer, 1981). Clearly, the operating frequency is a major factor in determining whether a particular cavity design should be of superconducting or nonsuperconducting materials.

CONDUCTIVITY; BY THE TWO-FLUID MODEL

Our earlier invocation of the Gorter-Casmir two-fluid model dealt with the case where momentum relaxation time τ could be ignored; in this instance τ is necessary to illustrate the development of complex conductivity for superconductors. First, one considers the force equations for the normal electrons, with average velocity $\mathbf{v}_n$, and the superconducting electrons, with velocity $\mathbf{v}_s$:

$$m\frac{d\mathbf{v}_n}{dt} + m\frac{\mathbf{v}_n}{\tau} = -e\mathbf{E} \tag{2.85}$$

$$m\frac{d\mathbf{v}_s}{dt} = -e\mathbf{E} \tag{2.86}$$

The current densities are simply the products of charge density, charge, and velocity:

$$\mathbf{J}_n = n_n(-e)\mathbf{v}_n \tag{2.87}$$

$$\mathbf{J}_s = n_s(-e)\mathbf{v}_s \tag{2.88}$$

As usual, in the two-fluid model, the total current ($\mathbf{J}_t$) is simply the sum of the currents due to the "normal" and "superconducting" electrons, and these components may be expressed in terms of real and imaginary conductances, σ_1 and σ_2:

$$\mathbf{J}_t = \mathbf{J}_n + \mathbf{J}_s = (\sigma_1 - j\,\sigma_2)\mathbf{E} \qquad 2.89$$

Assuming a sinusoidal form for the electric field (i.e., $E_o \sin \omega t$), one can derive the admittances σ_1 and σ_2 for the form of 2.89:

$$\sigma_1 = \frac{n_n \tau e^2}{m(1 + \omega^2\tau^2)} = \sigma_{1n} \qquad 2.90$$

and

$$\sigma_2 = \frac{n_n e^2(\omega\tau)^2}{m\omega(1 + \omega^2\tau^2)} + \frac{n_s e^2}{m\omega} = \sigma_{2n} + \sigma_{2s} \qquad 2.91$$

Note that while σ_1 is associated only with the normal electron density n_n, σ_2 is a function of both normal and superconducting carrier density. This is clarified in the expression

$$\mathbf{J}_t = \mathbf{J}_n + \mathbf{J}_s = (\sigma_{1n} - j\,[\sigma_{2n} + \sigma_{2s}])\mathbf{E} \qquad 2.92$$

The expressions for real and imaginary admittance are simplified considerably if we consider only the frequency range below about 100 GHz, where $(\omega\tau)^2$ is much less than 1. Recalling that $\lambda_L^2 = m/(\mu_o n_s e^2)$, the expression for total current finally reduces to:

$$\mathbf{J}_t = \mathbf{J}_n + \mathbf{J}_s \approx (\sigma_n - j\,\sigma_s)\mathbf{E} \approx \left(\frac{n_n e^2 \tau}{m} - j\left[\frac{1}{\mu_o \omega \lambda_L^2}\right]\right)\mathbf{E} \qquad 2.93$$

This expression suggests a simple equivalent circuit for the two-fluid model that is shown in the figure. This rather straightforward model not only has practical use in predicting the behavior of superconductors at temperatures well above 0 K; the equivalent circuit is also useful in helping one appreciate the power of the two-fluid model (see Figure 2.26).

Consider the expression for surface impedance Z_s on a conducting plane with thickness large compared to skin depth δ:

$$Z_s = R_s + jX_s = \frac{1 + j}{\sigma\delta} = (1 + j)\sqrt{\frac{\omega\mu}{2\sigma}} \qquad 2.94$$

By using the value of the two-fluid complex conductivity in an expression for

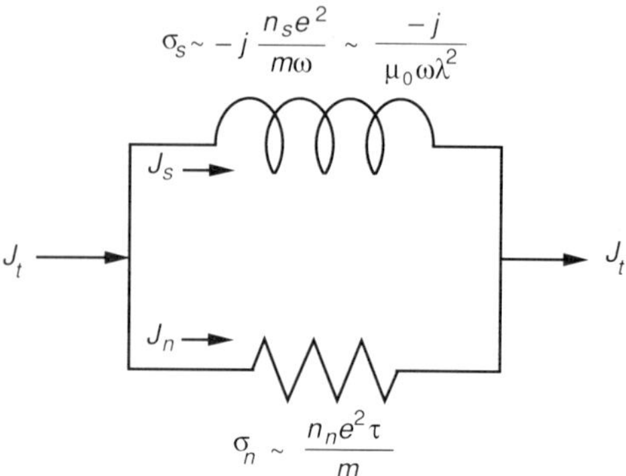

FIGURE 2.26 In this low-frequency equivalent admittance circuit based on the two-fluid model, a representation of normal and superconducting current flow is illustrated. As frequency approaches zero, virtually all current flows through the inductive branch, which represents the supercurrent flow. As frequency increases, voltage increases across the inductive branch, forcing more current through the resistive "normal" branch. At higher frequencies, the model is improved by adding a second inductive element to the "normal" branch. See the text for a detailed explanation. (See also T. Van Duzer and C. W. Turner, *Principles of Superconductive Devices and Circuits*, Elsevier, New York, 1981, p. 127.)

surface impedance, it is shown, at least to first order, that the surface resistance is proportional to ω^2 and λ^3, that is, (Van Duzer, 1981):

$$Z_s = R_s + j X_s \approx \frac{\omega^2 \mu_o^2 \lambda^3 n_n \sigma_n}{2 n_s} + j\, \omega \mu_o \lambda \qquad 2.95$$

This expression is consistent with experimental results that show that R_s is approximately proportional to ω^2. The dependence of R_s on λ^3 is particularly relevant to the new high-temperature ceramics, where λ is anisotropic in the Y-Ba-Cu-O unit crystals.

ELECTRON (GIAEVER) TUNNELING

Generally, tunneling is a quantum-mechanical effect that deals with very small electron current flow through barriers; flow that would not be permitted by the rules of classical physics. If two conductors are separated by a sufficiently thin insulating barrier, (the combination usually referred to as a "junction") there is a finite probability that a free electron will "jump" the barrier, that is, tunnel through it. One may consider (a) junctions with normal metals on each side of a thin insulator, (b) superconducting and normal conductors separated by the insulating barrier (SIN), or (c) superconductors separated by the barrier. While the latter superconductor-insulator-superconductor (SIS) configuration is of most interest here, it was the SIN barrier that Ivar Giaever first investigated.

For the instance where a pair of superconductors are separated by an insulating film with a thickness on the order of 50 Å or less, single electrons (quasiparticles) will tunnel through the barrier in the presence of a relatively small electric field. This tunneling exhibits a characteristic voltage threshold V_g (where single-electron current increases dramatically) that is related to the superconductor energy gap Δ in a straightforward manner. Electron tunneling which begins at V_g is commonly referred to as *Giaever* tunneling after its discovery by Ivar Giaever in 1960 (Foner & Orlando, 1988). The reason that electron tunneling does not occur until the particle energy is greater than the gap is simply that electrons with energy in the "forbidden" range are rejected by the superconductor. One simply cannot inject electrons unless the electron energy is greater than the forbidden value (Schmitt,1961).

Giaever realized the fortunate consequence that there is a simple relationship between the voltage V_g across an SIN barrier, the electron charge e, and the superconductor energy gap Δ, that is,

$$V_g = \frac{\Delta}{e} \qquad 2.96$$

For an SIS junction, with superconductors a and b exhibiting energy gaps Δ_a and Δ_b respectively, the equivalent relationship would be

$$V_g = \frac{\Delta_a + \Delta_b}{e} \qquad 2.97$$

This relationship has reduced the effort required to determine the energy gap to a relatively straightforward electronic procedure, largely supplanting more tedious techniques where specific heats or absorption characteristics of thin superconducting films at far-infrared wavelengths were measured for the same purpose. (Nevertheless, far-IR absorption techniques for determination of the energy gap still have application to certain high-T_c materials where the construction of reliable tunneling junctions for measurement of V_g is nontrivial.)

PAIR TUNNELING; THE DC JOSEPHSON EFFECT

On the contrary, *no potential difference* is required for tunneling of superconducting electron *pairs*, as predicted by Brian Josephson in 1962. Current flow must remain below a critical level (I_c) at zero gap voltage; if current is increased beyond I_c, a voltage is developed across the gap and normal (quasiparticle) current begins to flow. The quasiparticle current has rather small values at low gap voltages, but will increase when a threshold voltage (V_g) is reached, that is, where the Giaever normal-electron tunneling current increases rapidly. This very nonlinear relationship between gap current and applied dc voltage is useful in a number of applications.

In computer applications of Josephson junctions (see Figure 2.27), switching is generally accomplished by moving the operating point between the

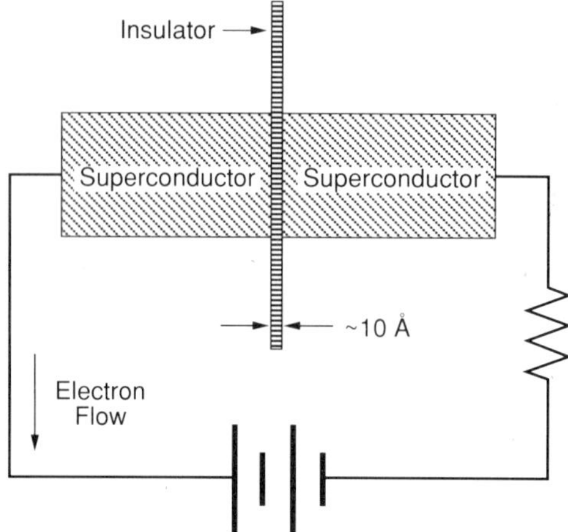

FIGURE 2.27 A Josephson junction is constructed by placing a very thin insulator (typically less than 20 Å) at the junction between two superconductors.

Josephson operating point ($V \sim 0$) and the Giaever operating point ($V \sim V_g$). More will be said about this useful feature of Josephson junctions in the chapter on applications (see Figure 2.28).

The level of current flow through the junction is limited to a peak value that cannot exceed the critical current I_c, and has a phase (ϕ) that is the difference between the phases (ϕ_1 and ϕ_2) of the (Ginzburg-Landau) macroscopic quantum wave function on each side of the junction:

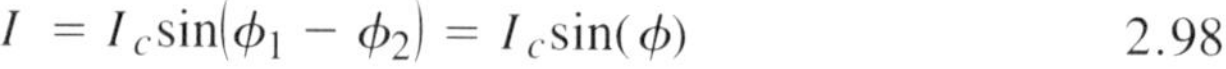

$$I = I_c \sin(\phi_1 - \phi_2) = I_c \sin(\phi) \qquad 2.98$$

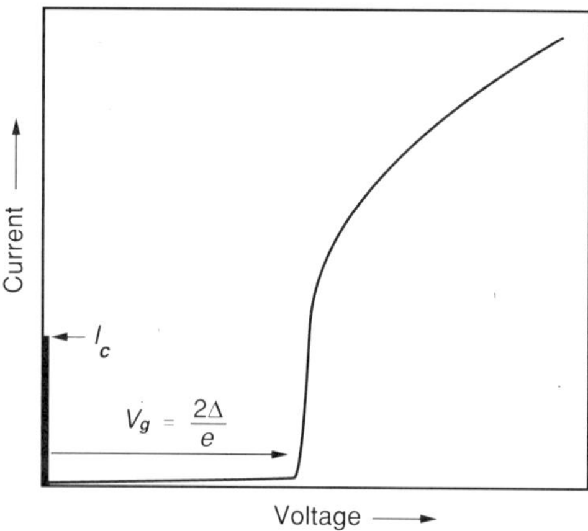

FIGURE 2.28 In addition to their utility in determining the energy gap of superconducting materials, Josephson junctions are used to construct extremely sensitive detectors of magnetic fields and for design of very fast switching and memory elements in digital computer circuitry.

These junctions are exquisitely sensitive to magnetic fields. In the presence of an H field, current is reduced to minimal levels when the junction is intersected by integral units of the flux quantum Φ_o introduced earlier:

$$\Phi_o = \frac{h}{2e} = 2.07 \times 10^{-15} \text{ Wb (SI units)} \qquad 2.99$$

$$\Phi_o = \frac{hc}{2e} = 2.07 \times 10^{-7} \text{ gauss-cm}^2 \text{ (cgs units)} \qquad 2.100$$

As a point of reference, the earth's magnetic field is on the order of 0.5 gauss.

THE AC JOSEPHSON EFFECT

The Cooper-pair current flow is proportional to the sine of the phase differences between the individual wave functions of each conductor. A consequence of this relationship between current flow and phase difference is that the direction of current flow reverses when the phase difference is changed by π radians.

This results in an oscillatory current flow at rather high frequencies. The oscillatory frequency is linearly proportional to the voltage V applied to the gap:

$$f = \frac{2eV}{h} \text{ Hz} \qquad 2.101$$

where e is the electron charge and h is Planck's constant. This amounts to a frequency shift of 4.83×10^{14} Hz/V or 483 MHz/μV. The frequency can be measured by applying a microwave signal to the junction to mix with the ac supercurrent. At integer multiples of the microwave frequency, steps in the current-voltage characteristic are produced. These steps provide the basis for very precise voltage standards, since the applied voltage can be determined to the resolution and accuracy of the measured frequency of the applied microwave signal.

JOSEPHSON EQUATIONS

There are five differential equations derived by Josephson which relate pair phase ϕ (recall that ϕ is the difference in the superconductor macroscopic wave-function phase across the insulating gap) and its derivatives to other barrier parameters. In the two-dimensional case, let the sides of the junction be in the x-y plane and the thickness (τ) of the junction be measured along the z-axis. Let d be the sum of the thickness and the penetration depth on each side of the barrier, that is,

$$d = \tau + 2\lambda \qquad 2.102$$

The Josephson equations are listed (Petley, 1971):

$$\frac{\partial \mathbf{H}_y}{\partial x} - \frac{\partial \mathbf{H}_x}{\partial y} = \mathbf{J}_z + C\frac{\partial \mathbf{V}}{\partial t} \qquad 2.103$$

This is essentially the Maxwell equation (in two dimensions),

$$\mathrm{Curl}(\mathbf{H}) = \mathbf{J}_z + \frac{\partial \mathbf{D}}{\partial t} \qquad 2.104$$

The remaining four Josephson equations are

$$\frac{\partial \phi}{\partial x} = 2\pi d\frac{B_y}{\Phi_o} \qquad 2.105$$

$$\frac{\partial \phi}{\partial y} = 2\pi d\frac{B_x}{\Phi_o} \qquad 2.106$$

$$\frac{\partial \phi}{\partial t} = \frac{4\pi eV}{h} \qquad 2.107$$

$$J_z = J_1 \sin(\phi) \qquad 2.108$$

where B_y, B_x, H_y, and H_x are magnetic field components at the gap, Φ_o is the flux quantum $h/2e$, V is the gap voltage, C is the capacitance per unit area of the gap, and J_1 is a current density proportional to K, a constant for the particular junction ($K = 0$ for zero coupling across the junction):

$$J_1 = \frac{4\pi K}{h}\sqrt{n_a n_b} \qquad 2.109$$

where h is Planck's constant and n_a and n_b are electron densities on either side (a and b) of the junction (Feynman, et al., 1965).

These equations may be used to derive the general "Barrier Phase Equation,"

$$\nabla^2\phi - \frac{1}{v^2}\frac{\partial^2 \phi}{\partial t^2} = \frac{1}{\lambda_J^2}\sin(\phi) \qquad 2.110$$

While the previous differential equation is perfectly general, it is nonlinear and thus requires a numerical approach for solutions. For small values of ϕ, (where $\sin(\phi) \sim \phi$) the linearized version may be used for analytical solutions:

$$\nabla^2\phi - \frac{1}{v^2}\frac{\partial^2\phi}{\partial t^2} = \frac{\phi}{\lambda_J^2} \qquad 2.111$$

where the velocity term is defined as

$$v = \sqrt{\frac{1}{\mu_o C d}} \qquad 2.112$$

and the Josephson penetration depth is

$$\lambda_J = \sqrt{\frac{h}{4\pi \mu_o J_1 e d}} \qquad 2.113$$

A Josephson junction, much like a waveguide, has a lower cutoff frequency ω_o, that is a function of λ_J and drift velocity v (Petley, 1971):

$$\omega_o = 2\pi f_o = \frac{v}{\lambda_J} = \sqrt{\frac{4\pi J_1 e}{hC}} \qquad 2.114$$

Typical values for f_o are on the order of a few GHz.

Among other applications, the Josephson junction is used as an extremely sensitive detector of magnetic fields, including the tiny fields developed by currents in neurons and nerve fibers. Josephson junctions also find important applications in SQUID devices used in switching circuitry for advanced digital computing circuits.

SUMMARY

All superconductors can be expected to exhibit zero resistivity and the Meissner effect, that is, the expulsion of magnetic fields, when operated below their critical temperature T_c. At temperatures far below T_c, the expulsion of low-level magnetic fields is complete except for a very thin surface layer with a characteristic thickness λ, which is on the order of a few hundred Å. A superconductor, regardless of how low the temperature may be, can be forced into its normal state by the application of a sufficiently intense magnetic field at the surface; this field can be generated by transport current flow through the superconductor.

The expulsion of **H** fields is a consequence of the fact that superconductors are not merely materials with zero resistivity; they may be considered (as in the two-fluid model) as having both a normal ($\mathbf{J}_n$) and a superconducting ($\mathbf{J}_s$) current density, the latter being composed of electron pairs which experience a weak net attractive force as a result of phonon-electron interactions.

Two basic types of materials (Types I and II) exhibit different behavior, the former characterized by an intermediate state where macroscopic normal and superconducting regions can coexist, the latter by a mixed state characterized by the penetration of flux quanta in isolated, microscopic vortices. Type II materials, because of their relatively high critical current densities and high critical magnetic fields strengths, are favored for most applications.

With the exception of a few types of so-called gapless materials, superconductors have an energy gap in the spectrum of the conduction electrons that is centered around the Fermi level and has a typical magnitude of about 10^{-3} eV.

Resistivity is zero only for dc current flow; ac resistance, while it may be several orders of magnitude smaller than the normal state impedance, is finite and increases with frequency. At low frequencies, losses are primarily due to flux creep, hysteresis, and related magnetic phenomena. At microwave frequencies, surface resistance R_s typically increases as the square of frequency and decreases with temperature, reaching a small residual value near 0 K. As frequency approaches the optical range, the material gradually loses its superconducting behavior. This occurs when the photon energy $h\nu$ is approximately equal to the full energy gap 2Δ of the superconductor, that is, $3.53k_B T_c$.

Tunneling through a very thin insulator can involve single electrons (quasiparticles) as well as Cooper pairs. Junctions based on such tunneling phenomena have been used to measure the energy gap, construct exquisitely sensitive magnetic field detectors, and realize superconducting digital logic circuits with unprecedented speed.

REFERENCES

Al'tov, V. A., V. B. Zenkevich, M. G. Kremlev, and V. V. Sychev, (1977). *Stabilization of Superconducting Magnetic Systems*, Trans. by G. D. Archard, Plenum Press, New York, p. 14.

Anderson, P. W., (1962). "Theory of flux creep in hard superconductors," *Phys. Rev. Lett.*, Vol. 9, No. 7, October 1, pp. 309–311.

Anderson, P. W., and Y. B. Kim, (1964). "Hard superconductivity: Theory of the motion of Abrikosov flux lines," *Rev. Mod. Phys.*, Vol. 36, January, pp. 39–43.

Bardeen, J., L. N. Cooper, and J. R. Schreiffer, (1957). "Theory of Superconductivity," *Phys. Rev.*, Vol. 108, No. 5, December 1, pp. 1175–1204.

Bean, C. P., (1962). "Magnetization of hard superconductors," *Phys. Rev. Lett.*, Vol. 8, No. 250, March 15, pp. 250–253.

Bean, C. P., (1964). "Magnetization of high-field superconductors," *Rev. Mod. Phys.*, Vol. 36, January, pp. 31–39.

Bean, C. P., and J. D. Livingston, (1964). "Surface barrier in type II superconductors," *Phys. Rev. Lett.*, Vol. 12, No. 14, January 6, pp. 14–16.

Blatt, J. M., (1964). *Theory of Superconductivity*, Academic Press, New York, pp. 348–349.

Braginski, A. I., M. G. Forrester, J. Talvacchio, and G. R. Wagner, (1988). "Prospects for thin-film electronic devices of high-T_c superconductors," Proc. 5th Int. Workshop on Future Electron Devices—High Temperature Superconducting Devices, June 2–4, pp. 171–179.

Clarke, J., (1988). "Small-scale analog applications of high-transition-temperature superconductors," *Nature*, Vol. 333, No. 6168, May 5, pp. 29–35.

Datta, T., C. P. Poole, Jr., H. A. Farach, C. Almasan, J. Estrada, D. U. Gubser, and S. A. Wolf, (1988). "New approach to characterizing the high-temperature superconducting transition," *Phys. Rev. B*, Vol. 37, No. 13, May 1, pp. 7843–7845.

De Gennes, P. G., (1966). *Superconductivity of Metals and Alloys*, Trans. by P. A. Pincus, W. A. Benjamin, New York.

Dinger, T. R., T. K. Worthington, W. J. Gallagher, and R. L. Sandstrom, (1987). "Direct observation of electronic anisotropy in single-crystal $Y_1Ba_2Cu_3O_{7-x}$," *Phys. Rev. Lett.*, Vol. 58, No. 25, June 22, pp. 2687–2690.

Edelsack, E. A., (1973). "Fundamentals of Superconductivity," in *The Science and Technology of Superconductivity*, Gregory, W. D., W. N. Mathews, Jr., and E. A. Edelsack, eds., Vol. I, Plenum Press, New York, pp. 306–310.

Feynman, R. P., R. B. Leighton, and M. Sands, (1965). *The Feynman Lectures on Physics, Quantum Mechanics*, Vol. III, Addison-Wesley, Reading, MA, pp. 21–7, 21–18.

Fietz, W. A., M. R. Beasley, J. Silcox, and W. W. Webb, (1964). "Magnetization of superconducting Nb-25%Zr wire," *Phys. Rev.*, Vol. 136, No. 2A, October 19, pp. A335–A345.

Foner, S., and T. P. Orlando, (1988). "Superconductors: The Long Road Ahead," *Technology Review*, Vol. 91, No. 2, February-March, pp. 36–47.

Hartwig, W. H., and C. Passow, (1975). "RF Superconducting Devices," in *Applied Superconductivity*, Vol. II, V. L. Newhouse, ed., Academic Press, New York, pp. 542–639.

Herring, C., (1960). in *Fundamental Formulas of Physics*, D. H. Menzel, ed., Vol. 2, Dover Publications, New York, p. 597.

Kim, Y. B., C. F. Hempstead, and A. R. Strnad, (1963). "Magnetization and critical supercurrents," *Phys. Rev.*, Vol. 129, No. 2, January 15, pp. 528–535.

Larbalestier, D., G. Fisk, B. Montgomery, and D. Hawksworth, (1986). "High-Field Superconductivity," *Physics Today*, March, pp. 24–33.

London, F., (1950). *Superfluids, Macroscopic Theory of Superconductivity*, Vol. 1, John Wiley & Sons, New York, pp. 27–52.

Lynton, E. A., (1969). *Superconductivity*, Methuen, London, pp. 20, 52, 84.

Mohamed, M. A-K., W. A. Miner, J. Jung, J. P. Franck, and S. B. Woods, (1988). "Decay of trapped flux in the high-T_c superconducting compound $Y_1Ba_2Cu_2O_{6.5+\delta}$," *Phys. Rev. B*, Vol. 37, No. 10, April 1, pp. 5834–5836.

Ni, B., T. Munakata, T. Matsushita, M. Iwakuma, K. Funaki, M. Takeo, and K. Yamafuji, (1988). "AC inductive measurements of intergrain and intragrain currents in high-T_c superconductors," *Jap. J. Appl. Phys.*, Vol. 27, No. 9, September, pp. 177–181.

Parks, R. D., (1965). "Quantum effects in superconductors," *Scientific American*, Vol. 213, No. 4, October, pp. 57–67.

Petley, B. W., (1971). *An Introduction to the Josephson Effects*, Mills & Boon, London, pp. 23, 27.

Piel, H., (1988). Private communication.

Pippard, A. B., (1987). "Early Superconductivity Research (Except Leiden)," *IEEE Trans. Magn.*, Vol. MAG-23, No. 2, March, pp. 371–375.

Reuter, G. E. H., and E. H. Sondheimer, (1948). "The theory of the anomalous skin effect in metals," *Proc. R. Soc. A*, Vol. A195, pp. 336–364.

Rickayzen, G., (1965). *Theory of Superconductivity*, Wiley-Interscience, New York, p. 11.

Schmitt, R. W., (1961). "The discovery of electron tunneling into superconductors," *Physics Today*, Vol. 14, No. 12, December, pp. 38–41.

Solymar, L. (1972). *Superconductive Tunneling and Applications*, Wiley-Interscience, New York, p. 18.

Tinkham, M., (1986). "Superconductivity, 75th Anniversary," *Physics Today*, March, pp. 22–23.

Van Duzer, T., and C. W. Turner, (1981). *Principles of Superconductive Devices and Circuits*, Elsevier, New York, pp. 68, 129, 138, 317, 319.

Willis, J. O., M. E. McHenry, M. P. Maley, and H. Sheinberg, (1988). "Magnetic shielding by superconducting Y-Ba-Cu-O hollow cylinders," presented at 1988 Applied Superconductivity Conference, San Francisco, August 21–25, (MT-3), Abstracts, p. 76.

Yntema, G.B., (1971). "Superconductivity," in *McGraw-Hill Encyclopedia of Science & Technology*, Vol. 13, 3rd. Ed., pp. 308–317.

SUGGESTIONS FOR FURTHER READING

Atkins, K.R., "Superfluids," in *Handbook of Physics*, E. U. Condon and H. Odishaw, eds., McGraw Hill, New York, 1958.

Chaddah, P., G. Ravi Kumar, A. K. Grover, C. Radhakrishnamurty, and G. V. Subba Rao, "Critical state model and the magnetic behavior of high-T_c superconductors," presentation at International Conference on Critical Currents in High-T_c Superconductors, Snowmass Village, Colorado, August 16–19, 1988.

Gregory, W. D., W. N. Mathews, Jr., and E. A. Edelsack, *The Science and Technology of Superconductivity*, Vol. I, Plenum Press, New York, 1973, pp. 306–310.

London, F., and H. London, "The Electromagnetic Equations of the Supraconductor," *Proc. R. Soc. A*, Vol. A149, 1935, pp. 71–88.

Tinkham, M., *Superconductivity—Documents on Modern Physics* (series), Gordon and Breach, New York, 1965.

Whitcomb, R. T., "Superconductivity," in *Encyclopedia of Science & Technology*, Vol. 13, McGraw-Hill, 1982, pp. 348–357.

CHAPTER 3

HIGH-TEMPERATURE SUPERCONDUCTORS; SPECIAL CONSIDERATIONS

. . . there are many different mechanisms leading to superconductivity.
Bernd T. Matthias, 1973

. . . The river of God is full of water;
David, King of Israel, Psalm 65, ~950 B.C.

INTRODUCTION

This chapter begins with a statement by Bernd Matthias that, with a modest stretch of the imagination, can be interpreted as a prediction of superconductors that are quite different from anything in the past. This same account ends with a reminder from the same observer that one should exercise caution in making predictions based more on enthusiasm than hard facts.

History, in each case, has proven Matthias' assertions to be ones that we should not ignore. The story of the development of superconductivity reveals a series of cyclic events, beginning with Kamerlingh Onnes' first great expectations for high-current machinery and subsequent disappointments. As recorded in the earlier chapter on history, Onnes' dreams were met after his death, with the new understanding and development of the "A15" high-field, high-current superconductors. There have been several similar cycles: the 1980s began with great enthusiasm for superconducting Josephson-junction computers that, at least for IBM, soon ended in termination of the program. In Japan however, enthusiasm for a similar project continues as improved niobium-nitride SQUIDs are developed which are far superior to the troublesome lead technology abandoned by IBM.

As this account is being written, the cycles, which formerly had a period of several years, can now be measured in months, and peak may follow peak with no time for a valley! The discovery in Huntsville of the 90 K Y-Ba-Cu-O superconductor was accompanied by speculation in the popular press of liquid-nitrogen technology for MAGLEV trains as a near-term prospect. Once it became clear that polycrystalline samples of this material exhibited a rather low critical current, skepticism became the fashion of the day. Within months, the development of high-quality epitaxial Y-Ba-Cu-O films with critical currents in excess of 10^6 A/cm^2, along with the discovery of superconducting bismuth and thallium-based compounds with critical temperatures above 100 K, has given cause for new optimism. Perhaps the biggest difference in the field of superconductivity is that events are moving (and cycling) much faster than ever before.

We must not be overly influenced by either the peaks or the valleys—the field of high-temperature superconductivity is here to stay, whether significant applications occur within five years or 25 years. To date, most of the disappointments are the result of an oversell in the popular media about applications that verge on the realm of science fiction. The media, of course, have heard much of this hyperbole from scientists and sincere fund-raisers who may have been somewhat more optimistic than the facts would justify.

DEFINITIONS

A careful reader might begin by asking what we mean by the term "high-temperature" in respect to superconductors. This question would not have seemed so relevant immediately after the discovery of ceramics with a T_c on the order of 90 K, but we now have materials that exhibit superconductivity well above 100 K and any convention adopted now may be obsolete within another year. Nevertheless, the term must have some meaning, so we will refer to any material with superconducting behavior above ~25 K as belonging to the "high-T_c" category. Most of the time, of course, the T_c's associated with the compounds of the greatest interest to us will far exceed this arbitrary temperature level.

Let us first consider the compound that was the beginning of the current revolution in the science and technology of superconductivity, the inspiration for compounds that attracted the attention of the technologists.

LANTHANUM-BARIUM-COPPER OXIDE

La-Ba-Cu-O was the prototype "high-T_c" superconductor, the forerunner of more remarkable advances that were yet to come. In the initial report, with a caution that is reminiscent of Brian Josephson's first report of his predictions on superconductive tunneling, Bednorz and Müller describe "Possible high-T_c superconductivity in the Ba-La-Cu-O system" (Bednorz & Müller, 1986). The precise formulation of the polycrystalline samples described in this landmark publication was $Ba_xLa_{5-x}Cu_5O_{5(3-y)}$, where x varied from 0.75 to 1.0, with $y > 0$.

Measurements of resistivity versus temperature yielded a typical metallic behavior, that is, a linear drop in resistivity as temperature was decreased, down to about 80–100 K. At this temperature, resistance began to increase in an exponential fashion as temperature was lowered further. Then, when temperature reached the 25–30 K range, resistance dropped dramatically by as much as three orders of magnitude (see Figure 3.4). More significant for proving that the phenomenon was indeed superconductivity, a small diamagnetic effect was also seen.

The highest onset of superconductivity (T_{co}) was "observed in the 30 K range" (Müller & Bednorz, 1987). The previous transition temperature record for superconductivity was 23.3 K; virtually overnight the level was increased by almost 30%! Significantly, the authors reported that the phenomenon might be related to "2D superconducting fluctuations of double perovskite layers of one of the phases."

Let's back up for some perspective. Superconducting oxides, while not common, had been seen earlier. The first superconducting metallic oxide discovered, $SrTiO_3$, had a T_c of only 0.3 K (Müller & Bednorz, 1987). The fact that they were dealing with a perovskite structure was not lost on the experimenters; they noted that Sleight and his colleagues had discovered superconductive behavior in another perovskite ($BaPb_{1-x}Bi_xO_3$) a decade earlier (Sleight, et al., 1975). If this latter development did not attract the wide recognition that it now appears to deserve, it is probably because there were more than a thousand compounds known to be superconductive, and many were based on novel mechanisms.

Back to the IBM-Zurich discovery: Work continued on the La-Ba-Cu-O compounds, with substitutions of strontium for barium and variations on the oxygen stoichiometry. There were also experiments with the La-Ba-Cu-O compound under high pressure to help illuminate the mechanism of superconductivity and to direct the way to compounds with even higher T_c.

La-Ba-Cu-O Pressure Effects

While the effects of high pressure on superconductors may interest the engineer only insofar as the instrumentation is concerned, theorists are particularly interested in shifts on T_c induced by pressure. Early in 1987, Paul Chu and the group at Houston reported on experiments with La-Ba-Cu-O under pressure. The two phases tested were

1. $La_{1-x}Ba_xCuO_{3-y}$
2. $(La_{1-x}Ba_x)_2CuO_{4-y}$

where x was either 0.20 or 0.15 and y was not yet determined. As pressure was increased up to levels in excess of 13 kbar, the onset temperature (T_{co}) of superconductivity increased. At zero pressure, T_{co} was approximately 32 K; at 13 kbar, T_{co} had increased to 40 K! Alas, the sample broke at ~16 kbar. There were other experiments; the general result was that pressure increased T_c in the La-Ba-Cu-O system at a rate of at least 10^{-3} K/bar. In these experiments, T_c could be increased up to 57 K, with "zero resistance" temperature raised to 40 K (Wu, et al., 1988).

La-Sr-Cu-O

When a complex superconducting compound is discovered, investigators almost immediately begin substituting other elements. This process not only yields important information about the nature of the superconductor, but may lead to improved compounds. In fact, a substitution of strontium for barium in La-Ba-Cu-O was notably successful.

Early in 1987, R. J. Cava and his colleagues at AT&T Bell Laboratories reported superconductivity at 36 K in $La_{1.8}Sr_{0.2}CuO_4$. There was an encouraging development: the transition, beginning at 36.2 K, was relatively sharp compared to La-Ba-Cu-O. The total transition from normal to superconducting for $La_{1.8}Sr_{0.2}CuO_4$ required a transition "width" of only 1.4 K. In contrast to earlier oxides that contained relatively small concentrations of the superconducting phase, these new samples containing strontium showed Meissner effects on the order of 60–70% of the level expected for perfect diamagnetism

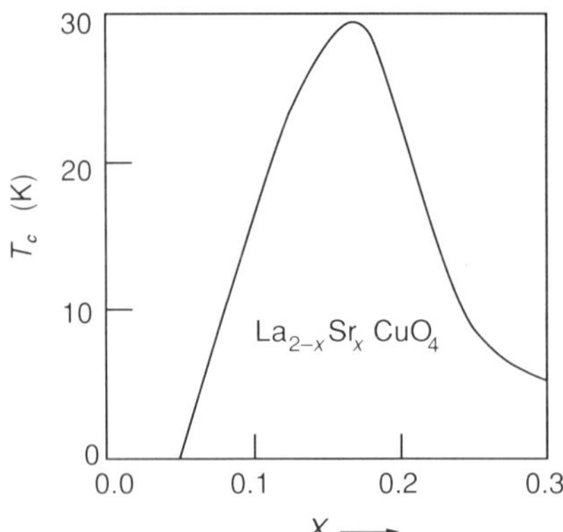

FIGURE 3.1 The critical temperature (T_c) for the $La_{2-x}Sr_xCuO_4$ superconductor peaks for a strontium stoichiometry of approximately 0.17 to 0.18. (After S. J. Rothman, J. L. Routbort, L. J. Nowicki, K. C. Goretta, L. J. Thompson, and J. N. Mundy, "Oxygen diffusion in high-T_c superconductors," Proc. DIMETA-88 [Metals Science Forum], International Conference on Diffusion in Metals and Alloys, Balatonfüred, Hungary, September 5–9, 1988.)

(see Figure 3.1 for the effects of Sr stoichiometry on T_c). The authors, on the basis of estimating density of states from resistivity and critical magnetic field data, felt that conventional BCS phonon-mediated phenomena could account for the superconductive phenomena in this compound.

The ceramics were clearly Type II superconductors. Measurements of the slope of H_{c2} as a function of temperature (near T_c) suggested that the value of H_{c2} at 0 K ($H_{c2}(0)$) was on the order of "at least 100–150 kOe." Such a high value of critical field was totally unexpected, and led to premature speculation in many circles that operation of liquid-nitrogen cooled magnets at unprecedented high fields might soon be possible.

Within less than three months this same group reported on more detailed measurements of upper and lower critical fields (H_{c1} and H_{c2}), specific heat, and Meissner effect for $(La,Sr)_2CuO_4$. While H_{c1} was on the order of 150–200 Oe, The material exhibited "extreme Type II behavior with large upper-critical-field slopes" (Batlogg, et al., 1987a).

The Sommerfield constant (γ) was 5.3 mJ/mole-K^2. This was derived from the specific heat anomaly at T_c ($\Delta C/T_c$ was 7.6 mJ/mole-K^2). (See Appendix H for a discussion of specific heat in superconductors.) Since the critical temperature for this La-Sr-Cu-O compound (and $Ba(Pb,Bi)O_3$) was approximately three times higher than that of other superconductors with similar values for γ, the authors suggested these results empirically distinguished both as belonging to "a separate class of superconductors."

The assertion of novel behavior was not made without the presence of a differing view, however. In the very same issue of *Physical Review B*, another paper presented data on $La_{1.85}Sr_{0.15}CuO_4$, including specific heat. The authors reported their view that "phonon-mediated superconductivity can explain the high transition temperatures in these materials" (Kwok, et al., 1987).

Regardless of the mechanism of superconductivity (which would turn out to be intimately related to the copper oxide *planes* (see Figure 3.3) in this and other ceramic perovskite structures), there was considerable interest in the unprecedented high values for the upper critical field. By using a combination of H_{c2} slope estimates (near T_c) and pulsed fields at 45 T, it was reported that there was a possibility of an $H_{c2}(0)$ as high as 140 T (Orlando, et al., 1987).

By the middle of 1987, there were reports of measurements on thermopower (TEP) and Hall constant (Hundley, et al., 1987). As would be expected for a true superconductor in its ground state, both dropped essentially to zero below T_c. In another report, pressure had little effect on T_c for $La_{1.8}Sr_{0.2}CuO_4$ (Przyslupski, et al., 1987).

YTTRIUM-BARIUM-COPPER OXIDE AND RELATED COMPOUNDS

The great success with the La-Ba-Cu-O and the La-Sr-Cu-O compounds naturally led to the investigation of other perovskites in the hope that even higher T_c's could be found. A major goal was to break the 77 K barrier, since this

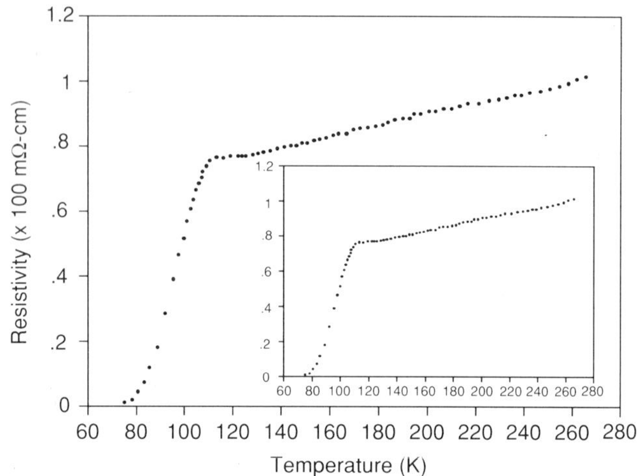

FIGURE 3.2 This sample (Batch 2, #3) of $YBa_2Cu_3O_x$ from Wu's group at Huntsville was measured twice on January 29, 1987. The larger plot was made with liquid helium, while the inset data was taken with liquid nitrogen as the cryogen. (Figures provided by the University of Alabama in Huntsville)

would allow the use of liquid nitrogen for operation of the superconductor. Liquid nitrogen is not only much easier (and less expensive) to produce than liquid helium, its heat transfer characteristics are also far superior. But again, one must be cautious: for effective operation at 77 K, T_c should be well above this temperature, since operation at $\sim T_c/2$ is required for most applications. More will be said later about the use of liquid nitrogen as a cryogen for the new ceramic superconductors.

The group led by M.K. Wu at the University of Alabama decided that a compound of $Y_{1.2}Ba_{0.8}CuO_{4-\delta}$ should have the appropriate characteristics. The Huntsville team produced the first samples that showed resistance transitions above 80 K. Superconductive behavior was immediately confirmed by the group in Houston when they tested samples from the Huntsville laboratory (see Figure 3.2). Magnetic susceptibility measurements revealed a 24% Meissner effect; this was solid proof of true superconductivity above liquid nitrogen temperature.

This original sample was also (like many) a mixed-phase material. As has been pointed out in the earlier chapter on history, the phase responsible for superconductivity was eventually proven to be the now well-known 1-2-3 material: $YBa_2Cu_3O_{7-\delta}$, where δ ranges from 0 to $\sim$ 0.6, with the highest T_c being associated with the lowest value for δ. In another common convention, the 1-2-3 material is given as $YBa_2Cu_3O_x$ where x varies from 6.4 to 7.0, with the highest T_c being at $x = 7.0$. More will be said about the importance of oxygen stoichiometry below, but let us first consider substitutions for yttrium.

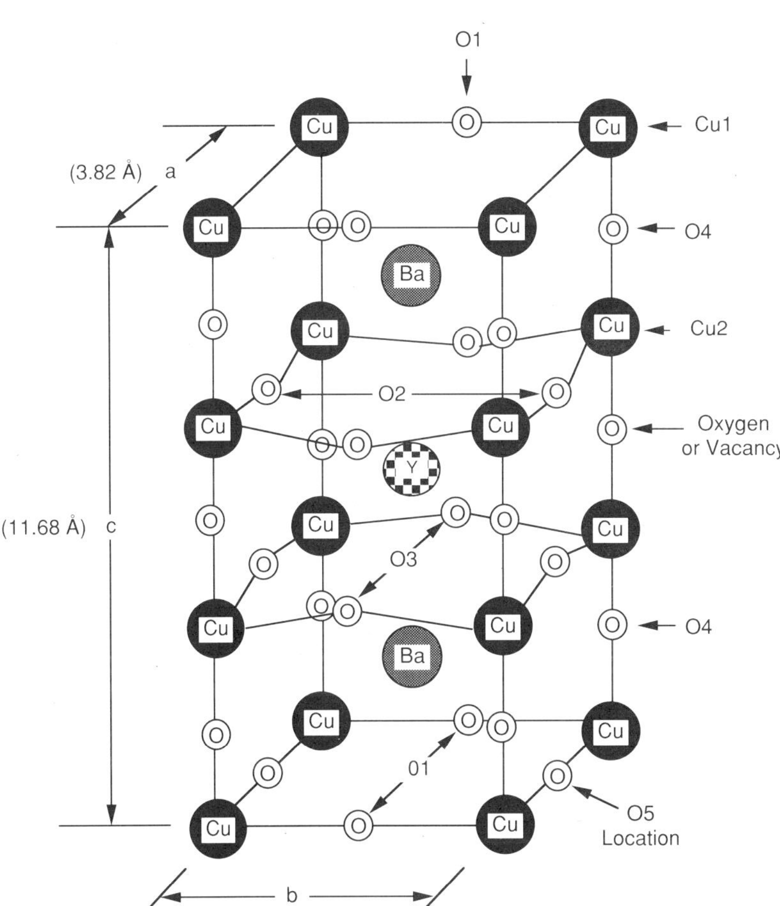

FIGURE 3.3 The molecular structure of the oxygen-deficient perovskite unit crystal $YBa_2Cu_3O_{7-\delta}$.

Rare Earth Substitutions for Yttrium

It was considered remarkable when a determination was made that rare earths could be substituted for yttrium without destructive effects on superconducting characteristics of $YBa_2Cu_3O_{7-\delta}$, since such substitutions in "conventional" BCS superconductors would eliminate or severely reduce superconducting phenomena (Fisk, et al., 1987). In the conventional case, the large magnetic moments of some of the rare earth ions tend to interfere with Cooper-pair conduction. This was further evidence that the new ceramic superconductors were not, strictly speaking, BCS-like in their behavior.

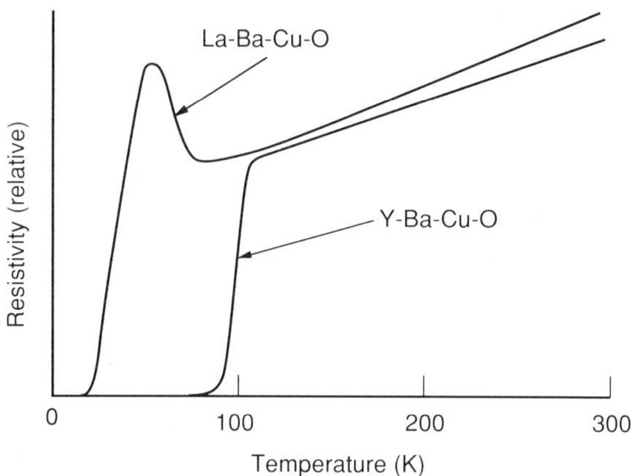

FIGURE 3.4 While response varies considerably with processing schedules, resistivity vs. temperature curves for Y-Ba-Cu-O and La-Ba-Cu-O in the early experiments appeared essentially as shown above. The peak prior to the transitions on the La-Ba-Cu-O suggests extraneous phases or semiconductivity, while the linear slope on the Y-Ba-Cu-O is typical of metallic behavior. (After J. G. Bednorz and K. A. Müller, "Possible High-T_c Superconductivity in the Ba-La-Cu-O System," *Z. Phys. B—Condensed Matter 64*, 1986, pp. 189–193, and M. K. Wu, J. R. Ashburn, C. J. Torng, P. H. Hor, R. L. Meng, L. Gao, Z. J. Huang, Y. Q. Wang, and C. W. Chu, "Superconductivity at 93 K in a New Mixed-Phase Y-Ba-Cu-O Compound System at Ambient Pressure," *Phys. Rev. Lett.*, Vol. 58, No. 9, March 2, 1987, pp. 908–910.)

It suffices to say that most of the statements to be made about $YBa_2Cu_3O_x$ apply to $(RE)Ba_2Cu_3O_x$ where "RE" simply denotes a rare earth substituted for yttrium. Rare earths have atomic numbers 57 to 71; some characteristics when substituted for yttrium in the $YBa_2Cu_3O_{7-\delta}$ compound are shown in Table 3.1.

Not all of the rare earths have been successfully substituted for yttrium. It should be said that yttrium, even though it is typically found with the rare earths (in the minerals gadolinite, samarskite, etc.) and has similar properties, is not itself a rare earth. The rare earths from Ho to Lu are sometimes referred to as "yttrium earths" because of their resemblance to yttrium.

It has become apparent that the substitution of rare earths for yttrium may have consequences beyond theoretical interest; yttrium-based copper oxide superconductors appear to be less stable under repeated heating at the 400–500°C level than several of the rare earth substitutes such as Gd and Eu.

Structure; Oxygen Stoichiometry of $YBa_2Cu_3O_{7-\delta}$

As mentioned previously, the oxygen stoichiometry is particularly important in determining the 1-2-3 superconductor's characteristics (or whether there are any superconducting phenomena to measure!).

TABLE 3.1 Comparison of characteristics of (Y or RE)$Ba_2Cu_3O_{7-\delta}$, primarily after Fisk, Tamegai, and Xiao.* Since data are drawn from several sources, there may be some inconsistencies due to variations in samples or measurement techniques.

Element (Y or RE)	T_c (K)	$-dH_{c2}/dT$ @ T_c (kOe/K)	$H_{c2}(0)$ (est) (kOe)
yttrium (Y)	~90	30	1860
Rare Earths			
cerium earths:			
lanthanum (La)	62.5	21	906
cerium (Ce)	—	—	—
praseodymium (Pr)	—	—	—
neodymium (Nd)	88–93.6	20	1290
promethium (Pm)	—	—	—
samarium (Sm)	93.5–94.2	21	1365
terbium earths:			
europium (Eu)	93.5–93.9	20	1296
gadolinium (Gd)	93.5–93.6	29	1873
terbium (Tb)	—	—	—
dysprosium (Dy)	91.8–92	27	1714
yttrium earths:			
holmium (Ho)	91.7–92.5	26	1659
erbium (Er)	89.5–91.4	20	1261
thulium (Tm)	90.2–91	21	1318
ytterbium (Yb)	88.7–89.5	24	1482
lutetium (Lu)	85		

* See: Fisk, Z., J. D. Thompson, E. Zirngiebl, J. L. Smith, and S-W. Cheong, "Superconductivity of Rare Earth-Barium-Copper Oxides," *Solid State Commun.*, Vol. 62, No. 11, 1987, pp. 743–744. Tamegai, T., A. Watanabe, I. Oguro, and Y. Iye, "Structures and upper critical fields of high T_c superconductors (RE)$Ba_2Cu_3O_x$," *Jap. J. Appl. Phys.*, Vol. 26, No. 8, August 1987, pp. L1304–1306. Xiao, G., F. H. Streitz, A. Gavrin, and C. L. Chein, "Magnetic characteristics of superconducting $RBa_2Cu_3O_{6+y}$ (R = Nd, Sm, Eu, Gd, Dy, Ho, Er, Tm and Yb)," *Solid State Commun.*, Vol. 63, No. 9, 1987, pp. 817–820.

Neutron scattering experiments indicate that $YBa_2Cu_3O_{7-\delta}$ materials have molecular structures that are typically either orthorhombic or tetragonal, depending on oxygen stoichiometry (see Figure 3.3). For values of δ less than ~0.6, $YBa_2Cu_3O_{7-\delta}$ is orthorhombic, but when δ increases beyond ~0.6, a transition to a tetragonal structure occurs. The tetragonal state, where $0.6 < \delta < 1.0$, is not superconducting. In fact, as δ approaches 1.0, ($YBa_2Cu_3O_{7-\delta} \rightarrow YBa_2Cu_3O_{6.0}$) the ceramic becomes a semiconductor at low temperatures. (See Figure 3.5 for the effect of oxygen stoichiometry on Y-Ba-Cu-O T_c.)

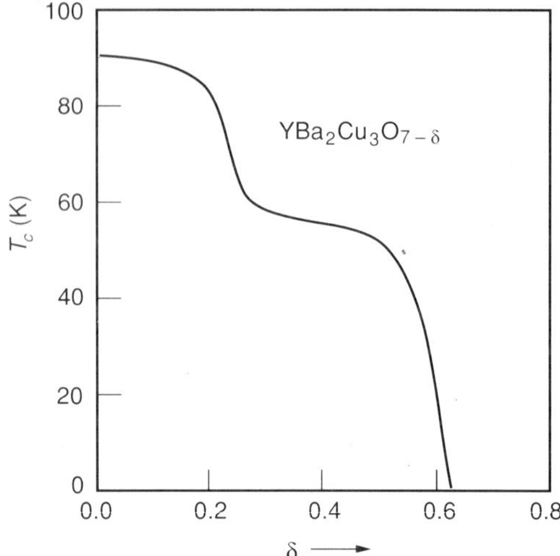

FIGURE 3.5 The effect of oxygen stoichiometry on $YBa_2Cu_3O_{7-\delta}$ critical temperature. (After S. J. Rothman, J. L. Routbort, L. J. Nowicki, K. C. Goretta, L. J. Thompson, and J. N. Mundy, "Oxygen diffusion in high-T_c superconductors," Proc. DIMETA-88 [Metals Science Forum], International Conference on Diffusion in Metals and Alloys, Balatonfüred, Hungary, September 5–9, 1988.)

It has also been determined that the value of T_c can be lowered to about 60 K in homogeneous polycrystalline samples of $YBa_2Cu_3O_{7-\delta}$ where δ has values ranging from 0.3 to 0.4 (Cava, et al., 1987). Achievement of a structure with sufficient homogeneity to see a sharp transition at 60 K relied on a Zr gettering technique during processing in the 360–520°C range.

Energy Gap, Δ(0)

In testing competing theories for mechanisms of high-temperature superconductivity, it becomes necessary to distinguish between strong and weak coupling of the charge carriers. Measurement of the energy gap Δ is crucial in this regard, as is the dependence on temperature of Δ. Unless stated otherwise, when referring to the energy gap, we always have in mind the value of Δ for for $T \ll T_c$, that is, where its value is not significantly different from $\Delta(0)$.

Recall from the previous chapter, that BCS theory predicts the following relationship between critical temperature (T_c) and the energy gap Δ:

$$\frac{2\Delta}{k_B T_c} = 3.53 \qquad 3.1$$

Even when dealing with conventional superconductors, it is not uncommon to see variations on the order of 30% from this predicted value of $2\Delta/k_B T_c$, but the determination of this ratio is nevertheless a valuable tool for testing whether a particular superconductor has BCS-like behavior. A value near 3.5 indicates weak coupling, while values substantially above this level are evidence of strong coupling.

If weak coupling is involved for a 90 K superconductor like Y-Ba-Cu-O, one would expect an energy gap (2Δ) of $(3.53k_B) \times (90)$, or

$$2\Delta \sim (3.53) \times (8.625 \times 10^{-5}\ \text{eV/K}) \times (90\ \text{K}) \sim 27 \times 10^{-3}\ \text{eV}$$

This corresponds to an expected value of $\sim 13.5 \times 10^{-3}$ eV for Δ.

The energy gap is usually measured either by the tunneling method or by measurement of far-infrared reflection. The tunneling approach essentially involves measurement of the voltage required for initiation of quasiparticle flow across a superconductor-insulator-normal (SIN) or a superconductor-insulator-superconductor (SIS) junction. Tunneling measurements, while very simple in principle, tend to give values of $2\Delta/k_B T_c$ that are too low. This problem is related to the extremely short coherence lengths in the high-temperature ceramics. The short coherence lengths lead to anomalous behavior at surfaces and interfaces, such as those in superconductor-normal interfaces used in tunneling experiments. For this reason, tunneling measurements are not considered as trustworthy as other techniques in determining the energy gap in ceramic superconductors (Bednorz & Müller, 1988).

The optical technique involves the difference in reflectance (at far infrared wavelengths) in the normal and the superconducting state of the specimen. In the superconducting state, there is little or no absorption of the electromagnetic waves until the photon energy is high enough to break the Cooper pairs.

A third technique for energy gap measurement is Andreev reflection, which deserves a brief description. Andreev reflection measurements are made with a rather small point contact on the sample (van Bentum, et al., 1988). Electrons are injected from the point into the sample. An electron with an energy less than Δ will not enter the superconductor as a quasiparticle because there are no states below the energy gap value. Instead, the electron tends to condense with a second electron (in the point contact) to form a Cooper pair. The resultant hole in the normal region of the point contact will move in the direction where the original electron originated. This "reflection" results in a lower contact resistance when the injected electron energy is below Δ.

While measurement of the tunneling voltage is straightforward for conventional superconductors, one encounters considerable difficulty in preparing high-quality SIN or SIS barrier junctions of the copper oxide ceramics. Not only do processing problems make it difficult to prepare a high-quality, thin, insulating layer between layers of 1-2-3 material, but there is the problem previously mentioned, related to the very short coherence length ξ.

In essence, it is necessary that good superconducting properties exist to within a few coherence lengths of the barrier surface (Gray, 1988). The combination of two factors, that is, a coherence length that is only on the order of the unit crystal dimensions and the common existence of an insulating layer on $YBa_2Cu_3O_7$, makes formation of Josephson junctions extremely difficult. Tunneling experiments with $YBa_2Cu_3O_7$, for this reason, are generally of the point-contact type, where the point penetrates the insulating layer. Tunneling with such weak links does not provide the clear determination of Δ that is possible with high-quality, thin-film Josephson junctions. In general, the value of Δ provided by tunneling measurements is believed to underestimate the $2\Delta/k_BT_c$ ratio, suggesting a weak coupling that does not exist. As will be seen, however, not all tunneling measurements are consistent with weak coupling.

This difficulty with tunneling determinations of Δ makes infrared (IR) reflection a more appropriate approach for energy-gap measurement on high-T_c ceramics. Representative measurements of the energy gap will be given as follows for comparison of results.

In one instance (Ye, et al., 1987), IR measurements for $YBa_2Cu_3O_{7-\delta}$ have yielded a value for the energy gap of

$$2\Delta = 15.5 \times 10^{-3} \text{ eV} \tag{3.2}$$

at temperatures far below T_c, which yields a value of $2\Delta/k_BT_c$ of ~2.0 for a T_c of 90 K.

Such results have been interpreted, along with the unexpected high values of T_c, as an indication that the copper oxides are not "conventional" BCS superconductors. In fact, results from various investigations yield values of $2\Delta/k_BT_c$ that are either well above or below the value of 3.53 predicted by BCS theory.

A determination of the energy gap made by analysis of surface impedance at 600 MHz yielded a value of 9.5 ± 0.9 for Y-Ba-Cu-O $2\Delta/k_BT_c$, while the same experiments with La-Sr-Cu-O resulted in 3.5 ± 0.3 for $2\Delta/k_BT_c$ (Andrienko, et al., 1988).

On the other hand, a recent report on measurements of 1-2-3 material electron tunneling over a wider temperature range (4.2–110 K) yields quite a different result (Sera, et al., 1988). In this instance, the value of $2\Delta/k_BT_c$ has a value ranging between 3 and 4, bracketing the value (~ 3.53) that would be in agreement with BCS theory. This latter investigation, unlike many earlier results with tunneling, also found that the variation of the energy gap with temperature was consistent with predictions of BCS theory.

Generally, one expects the temperature-dependent energy gap $\Delta(T)$ to vary approximately as (Van Duzer, 1981)

$$\Delta(T) \approx 3.2k_BT_c\sqrt{1 - \frac{T}{T_c}} \tag{3.3}$$

for temperatures just below T_c. Below about $T_c/2$, there is little variation in the energy gap, which is approximately equal to $\Delta(0)$.

One group of investigators has measured the energy gap using all three of the techniques just listed (van Bentum, et al., 1988). Their results are listed in the following table:

Method	*Energy Gap Δ (meV)*
Tunneling	14.0 ± 2
Far-Infrared	14.5 ± 1
Andreev reflection	12.5 ± 2

The authors point out that, for the samples tested, the average value of $2\Delta/k_B T_c$ was 3.6 ± 0.6, rather close to the "weak-coupling" BCS prediction of 3.53.

Whether the energy gap of 1-2-3 compounds is consistent with BCS theory has not been finally determined, at least to the satisfaction of most of those involved in the investigations. There is probably a significant difference in the composition of the samples measured by various groups and there is certainly a variation due to measurement techniques.

Characteristic Lengths (ξ and λ)

Coherence length (ξ) is relatively small in the 1-2-3 ceramic superconductors, typically being reported to be on the order of 10 Å or less compared to values on the order of 1000 Å for conventional Type II superconductors. Since ξ may be considered as the distance over which coherence can be maintained across a normal region, it is a measure of the scale at which lattice interruptions can disrupt the supercurrent. The result is that even grain boundaries or twinning planes may have sufficiently large dimensions to act as weak links in the superconducting state, and this may partially explain the relatively low values of J_c typically seen in bulk ceramic superconductors. A potential advantage of the short coherence length is that the level of J_c may be enhanced by tight pinning vortices, which reduces losses induced by Lorentz forces (Garwin, 1987).

One group, taking the anisotropy of $YBa_2Cu_3O_{7-x}$ into account, has made determinations of the coherence lengths for the c-axis (intralayer) and copper-oxygen a-b plane (interlayer) separately (Oh, et al., 1988). These lengths, $\xi_c(0)$ and $\xi_{ab}(0)$, are approximately 2 Å and 13 Å, respectively. Another investigation is in good agreement, with values of 2 ± 1 Å and 16 ± 2 Å, respectively (Kapitulnik, 1988).

In a third study, coherence length and penetration depth for both the a-b plane and along the c-axis of the $YBa_2Cu_3O_{7-x}$ unit crystal have been estimated to have approximately the values indicated in the following table (Braginski, et al., 1988):

Characteristic Length	*Parallel to c-Axis*	*Parallel to a-b plane*
λ	1800 Å	270 Å
ξ	2–4 Å	16–30 Å

These extremely short coherence lengths, on the order of the size of the Y-Ba-Cu-O unit cell, make this superconductor quite sensitive to local defects

(Deutscher, 1988). See Table 3.2 for a comparison of characteristic lengths for a variety of conventional superconductors and representative copper oxide materials.

Pressure Effects on 1-2-3 T_c

As indicated earlier in this chapter, the original work on La-Ba-Cu-O compounds by Bednorz and Müller determined a temperature level for the onset of superconductivity (T_{co}) on the order of 30 K. Shortly thereafter, Chu et al. reported an onset temperature for the same compound that was above 40 K. The difference was pressure—a great deal of pressure.

A diamond anvil in an optical cryostat is typically used in such experiments; a ruby fluorescence method is used to measure the high pressures. In a typical La-Ba-Cu-O sample, T_{co} increases from ~ 32 K at zero applied pressure up to 40.2 K at a pressure of 13 kbar, yielding an average slope of 0.63 K/kbar. This is a rather large value for dT_{co}/dp. The general effect of the applied pressure is to drive the lattice planes closer together.

When Y-Ba-Cu-O is tested, the results are rather different from those for La-Ba-Cu-O. Pressures up to 19 kbar on Y-Ba-Cu-O show only slight pressure effects (Hor, 1987) and experiments with pressures up to 170 kbar yield levels for dT_{co}/dp of only 0.043 K/kbar (Driessen, 1987). Pressure reduces the value of the lattice parameters (i.e., dimensions of the unit crystal: a, b, and c) and increases the ratio of Cu^{3+} to Cu^{2+}.

TABLE 3.2 A comparision of characteristic lengths for a variety of conventional superconductors and representative copper oxide materials. Values of T_c are approximate; values of characteristic lengths are for temperatures far below T_c. Values given for several copper oxides, i.e., $L\rightarrow/L\uparrow$, reflect single-crystal anisotropy. $L\rightarrow$ is parallel and $L\uparrow$ is perpendicular to the copper oxide plane of the crystal. In the case of the thallium compound, values are average for polycrystalline material. (After J. Talvacchio, "Electrical contact to superconductors " in press, *IEEE Trans. CHMT*, 1989.)

Material	T_c (K)	Penetration Depth λ (nm)	Coherence Length ξ (nm)
Pb	7.2	40	90
Nb	9.2	60	40
NbTi	9.2	60	40
NbN	16	200	4
Nb_3Sn	18	80	3
Nb_3Ge	23	100	3
$La_{1.85}Sr_{0.15}CuO_4$	40	80/430	3.7/0.7
$YBa_2Cu_3O_7$	95	27/180	3.1/0.4
Bi-Sr-Ca-Cu-O	85–115	300/40	3.1/0.4
Tl-Ba-Ca-Cu-O	125	<220>	<2.6>

For a detailed discussion of some of the theoretical implications of high-pressure studies on these ceramics, the reader is directed to other references (Griessen, 1987; Kaneko, 1987).

Isotope Effect: La-Sr-Cu-O and Y-Ba-Cu-O

Recall that the isotopic-mass effect seen in most metallic superconductors was explained in elegant fashion by BCS theory, which predicted that T_c would be proportional to $1/M^{\alpha}$, where $\alpha = 1/2$. In alloys and compounds superconductors, α may take on a range of values, including zero. Measurement of any isotopic mass effect in the ceramics is of considerable interest to theorists, since the presence of a large effect would indicate that lattice vibrations (phonons) are of importance to the mechanism involved in high-temperature superconductivity.

(Note: It may seem a trivial point, but the sign used for reporting of values of α varies depending on whether a particular author is using a $T_c \sim M^{\alpha}$ convention or the $T_c \sim 1/M^{\alpha}$ expression used above. In the first instance, the BCS value for α is $-1/2$, while in the second it will be given as $+1/2$.)

La-Sr-Cu-O. The La-Sr-Cu-O 30 K superconductors exhibit a modest isotope effect, with one investigation finding a value for α of 0.16 ± 0.02 (Batlogg, et al., 1987b). Other investigators, also substituting ^{18}O for ^{16}O, measured a shift in T_c of 0.3 to 1 K, likewise concluding that phonons play a significant role in the superconductivity seen in $La_{1.85}Sr_{0.15}CuO_4$ (Faltens, et al., 1987). In another experiment, substitution of 75% of the ^{16}O by ^{18}O in $La_{1.85}Sr_{0.15}CuO_4$ resulted in a shift in T_c of 1 K, but this was approximately half of the 2.1 K predicted by BCS theory (zur Loye, 1987).

Y-Ba-Cu-0. The major evidence of the difference between classic superconductors and their 1-2-3 counterparts lies in the fact that the $YBa_2Cu_3O_{7-\delta}$ 90 K copper oxides show far less isotope-mass effect than would be expected if phonon-electron interactions were the predominant mechanism.

Some investigators see no measurable isotope-mass effect (Bourne, et al., 1987). In other experiments with $YBa_2Cu_3O_x$ and $EuBa_2Cu_3O_x$, with a substitution of 75% of the usual ^{16}O with ^{18}O, the change in T_c was only 0.25%. This is on the order of 5% of the expected change if phonon-electron interaction was responsible for superconductivity (Batlogg, et al., 1987). A subsequent report shows a stronger effect. Substitution of ^{18}O for ^{16}O in $YBa_2Cu_3O_7$ resulted in a shift of approximately 0.9 K in T_c (zur Loye, 1987). If the material conformed to classic BCS theory one would expect a shift of 5.21 K for the experiment that was performed, where 67% of the ^{16}O was replaced by ^{18}O. Other investigations also see small shifts in T_c (0.3 to 0.5 K) when ^{18}O is substituted for ^{16}O in Y-Ba-Cu-O (Leary, et al., 1987).

The general effect of these results is that, for 1-2-3 compounds, exchange mechanisms involving strong, short-range interelectronic repulsion are favored and mechanisms involving low-energy phonon interactions are considered to be eliminated by some theorists. The possible mechanism of high-energy phonons (i.e., "optical" phonons) are not eliminated by the relatively small isotope effect.

While the evidence that the isotope effect is relatively small for the 1-2-3 copper oxides may prove to be significant, it is worthwhile to recall comments by Bernd Matthias. (Note that these remarks, from a summer course in 1971, predate the discovery of high-T_c superconductivity in copper oxides by 15 years.) Italics are mine:

> When Fröhlich and Bardeen predicted the isotope effect, one over the square root, everybody said "This is the ultimate triumph, now we know it all! We understand everything." I always thought the isotope effect was only a coincidence near 0.5, and could take on any other value. Well, Geballe and I spent one year trying to measure the isotope effect among the transition elements. And, for the first thing we tried, the very first element, ruthenium, the isotope effect was zero; *there was no isotope effect*. Well, for a moment there was some consternation from the theoretical camp when ruthenium didn't show an isotope effect. After a while they recovered and started to predict. Now they predicted that some elements shouldn't have an isotope effect. It should be between 0 and 0.5, until in Los Alamos *the isotope effect of uranium was measured*.
>
> (Matthias, 1973)

While the value of the exponent for the isotope mass effect on T_c had ranged between 0 and +0.5 for elements tested previously, α turned out to be -2.2 for uranium! Matthias' comments serve as a useful reminder at a time when the theories about copper oxide superconductivity are in a very formative state, and new superconductors of unexpected composition are reported almost monthly.

Electron Pairing in 1-2-3 Compounds

However novel the superconducting mechanisms may be in the 1-2-3 copper oxides, there is at least one fundamental similarity to conventional superconductor behavior: pairing of two electron charges. This has been demonstrated by measurement of the superconducting flux-quantum (or fluxoid) which, for the two-electron supercurrent charge carrier, is

$$\Phi_o = \frac{h}{2e} = 2.07 \times 10^{-15} \text{ Wb} \tag{3.4}$$

Results were $(0.97 \pm 0.04)\Phi_o$, rather conclusive evidence for paired electron charges (actually holes in this instance) (Caplin, 1987; Gough, 1987).

Critical Magnetic Fields

Generally, H_{c2} is quite high for $YBa_2Cu_3O_{7-\delta}$ and $(RE)Ba_2Cu_3O_{7-\delta}$. At lower temperatures, in fact, few facilities have the magnetic field intensity required to measure H_{c2} as temperature decreases very far below T_c. For this reason, $H_{c2}(0)$, the upper critical field at zero temperature, is often estimated using the expression introduced earlier:

$$H_{c2}(0) \approx 0.69 T_c \left| \frac{dH_{c2}}{dT} \right|_{T=T_c} \qquad 3.5$$

The absolute value signs reflect the fact that the derivative is typically negative.

For a single-crystal sample of $Eu_1Ba_2Cu_3O_y$, measurements indicate an anisotropy of dH_{c2}/dT of 4 between the c axis and the a-b plane of the crystal (Hikita, 1987). Measured dH_{c2}/dT parameters were -0.7 T/K and -3.0 T/K for the c-axis and parallel to the a-b plane, respectively. Derived $H_{c2}(0)$ values, using the approximation just given, were 45 T and 190 T respectively. Others estimate the value of $H_{c2}(0)$ for 1-2-3 ceramics to be between 250–350 T (Foner & Orlando, 1988).

In what appears to be the highest-field measurement to date, Koichi Nakao and his colleagues have produced pulsed magnetic fields on the order of 100 T and demonstrated that, at a temperature of 6 K, superconductivity in $YBa_2Cu_3O_{7-x}$ was not destroyed by this enormous magnetic field (Nakao, et al., 1988). See Figure 3.6.

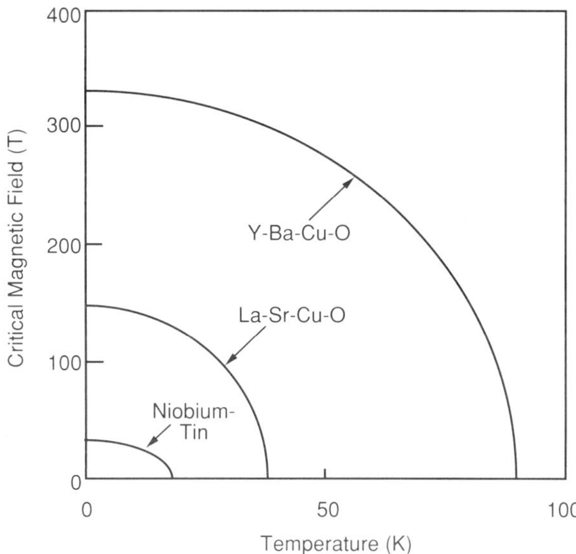

FIGURE 3.6 Approximate H_c vs. T response compared for Nb_3Sn, La-Sr-Cu-O, and Y-Ba-Cu-O superconductors. Even though the high-temperature ceramic superconductors evidently have a very high critical magnetic field and a high intragranular critical current, critical transport current tends to be low unless the anisotropic grains are well-aligned, as in epitaxial films.

It is significant that even with these very high values for $H_{c2}(0)$, some indication of resistance can be measured at 77 K in a 10 T field. This is an order of magnitude lower than the expected $H_{c2}(T)$ at 77 K and is probably another indication of the broadness of the phase transition in the 1-2-3 copper oxides. Again, the anisotropy of the individual superconducting crystals and their separation by nonsuperconducting areas (at least in nonaligned polycrystalline samples) probably contributes to these broad transitions.

Critical Current Density

J_c Dependency on Sample Type. One might say, in regard to the 1-2-3 ceramics, that the good news is that critical transition temperature (T_c) is at sufficiently high levels to be quite interesting. The bad news is that critical current density (J_c) in bulk polycrystalline specimens remains at levels which, to date, eliminate some of the most attractive potential applications of the new high-T_c superconductors. The situation can be somewhat confusing when one hears reports of J_c values varying from 10 to 10^6 A/cm^2.

These variations are a reflection of the type of measurement that was made and/or the specific form of the material that was measured. Polycrystalline films and bulk samples prepared from sintered grains tend to have low critical currents, on the order of 10 to 1000 A/cm^2, and even these values are restricted to small magnetic fields. This relatively low critical current density (compared to niobium, for example) is evidently related to the "weak-link" architecture of sintered preparations. Superconducting domains are connected poorly or even by very thin insulating regions that act as tunneling junctions at low temperatures.

Twinning. When $YBa_2Cu_3O_7$ is being processed at high temperatures, the crystalline structure undergoes a tetragonal-to-orthorhombic phase transition at ~ 700°C. On subsequent cooling, "twin" boundaries are formed which separate the orthorhombic domains (Bednorz & Müller, 1988). These twinning boundaries, given the very short values of the coherence length ξ, form effective intragrain weak-link (Josephson) junctions that severely limit critical current levels.

Twinning junctions are not the only weak links in polycrystalline samples. Weak links occur between grain boundaries and may be exacerbated by the presence of impurity phases.

Specific Sample Types. Single crystals, not surprisingly, tend to exhibit very good critical current characteristics. Actual samples may not always represent the pure, single-phase, stoichiometric samples that are, by definition, single crystals. In addition, those samples presented as single crystals virtually always exhibit twinning.

Regardless of their specific origin, the presence of these Josephson-junction links in granular bulk material leads to an unsettling result: very small magnetic fields tend to quench the bulk ceramic superconductor regardless of the extraordinary high values of *intragranular* $H_{c2}(T)$. Even though the *intragrain* critical current densities are quite high ($\sim 10^6$ A/cm^2), bulk J_c is limited by the Josephson-like weak links, and these are affected by very small values of magnetic field, on the order of 1 gauss or even less. The currents between grains are commonly referred to as *intergrain* as opposed to the relatively high intragrain currents. The intergrain currents are high only when weak links can be eliminated, as in highly textured 1-2-3 films.

Weak Pinning and J_c. There is some concern that there may be *inherent* limitations on maximum attainable current density in the high-temperature superconductors. One suspect parameter is an intrinsic weakness of flux pinning. The argument is made that the high-T_c materials may have weak flux pinning due to a variety of unfortunate parameter combinations, including the fact that T_c is high and carrier density is relatively low (Tinkham, 1988). The notion that the value of J_c is limited by weak flux pinning, leading to flux creep, is supported by evidence of a strong dependence on the temperature of J_c at temperatures below $\sim 0.6T_c$. Tinkham observes that J_c normally varies as the summation of three terms:

$$J_c \approx J_c(0)(1 - \alpha t - \beta t^2) \qquad 3.6$$

where t is the familiar reduced temperature, T/T_c, and $J_c(0)$ is the critical current density at absolute zero temperature. While the third (quadratic) term is a measure of the temperature dependence of thermodynamic quantities, the second (linear) term is related to effects of flux creep. The factor α in the linear term has a value on the order of 0.1 for conventional superconductors, but may have a value on the order of 10 times larger for the high-T_c copper oxide ceramics. At the lowest values of reduced temperature (t), this would lead to a variation of J_c as:

$$J_c \approx J_c(0)(1 - t) \qquad 3.7$$

In some measurements on thin 1-2-3 films below the reduced temperature $t = 0.6$, J_c does show approximately this type of behavior.

Further evidence of the weak-pinning phenomena in limiting J_c is a semilogarithmic decay of persistent currents. In addition to Y-Ba-Cu-O, fairly rapid decays are also seen in thallium superconductor samples, as will be discussed later when those materials are considered.

J_c Measurements. The direct method for measuring critical current density is simply to gradually increase a transport current through a sample (that has a well-defined geometry) until the superconducting characteristics are destroyed. Quenching is generally defined as that state when a finite voltage initially appears across the sample, and the experimenter's decision on the voltage magnitude generally depends on sensitivity of the instrumentation available for the measurement. The transport method is preferable for determining the characteristics of a bulk superconductor sample and is the most appropriate method for determining values for J_c for a variety of practical applications.

The transport method can be difficult and even produce uncertain results, whenever the attachment of test leads to a sample is not straightforward. Lead attachment can be quite difficult on very small samples, such as single crystals. In the case of thin, textured films, attachment of contacts to accommodate the electrical leads may significantly modify the films characteristics. Much more will be said about the problem of contacts in the chapter on measurements.

A second method involves inferences from screening currents associated with measurements of the Meissner effect and these produce relatively large values for J_c when applied to single crystal samples. Using critical field H_c and penetration depth λ, the maximum theoretical critical current density is determined by

$$J_c = \frac{H_c}{3\sqrt{6}\pi\lambda} \tag{3.8}$$

This calculation, using measured values of H_c and λ, produces values on the order of 5×10^6 A/cm^2. Typical values for J_c in $YBa_2Cu_3O_x$ single crystals, representing the upper-limit value that one would expect to see in a bulk or film Y-Ba-Cu-O sample, are on the order of 2×10^4 A/cm^2 at 77 K in zero magnetic field (Jin, et al., 1988).

Indirect measurements of J_c are typically applied to small single-crystal samples of $YBa_2Cu_3O_{7-\delta}$, where it is rather difficult to attach electrical leads for direct measurements. The unit-crystal anisotropy discussed earlier is an important factor whenever single-crystal structures are measured. One effect of this anisotropy is that J_c and the slope of H_{c2} $(-dH_{c2}/dT)$ are dependent on direction of current flow through the crystal. For electric fields aligned with the 1-2-3 crystal's *a-b* plane, J_c is at least an order of magnitude higher than for **E** fields parallel to the *c*-axis. Similar anisotropies are seen when measuring $H_{c2}(T)$ on single crystals; effects on λ and ξ were discussed earlier.

Characterization of a high-purity, single-phase crystal may, at least in a limited sense, indicate the *potential* J_c for a bulk material if problems related to boundaries between polycrystalline sintered pellets could be resolved. Even "single crystals," however, may not be completely free of impurities and may also present weak links at twinning boundaries. Further, the oxygen concentration gradient across a very small crystal may be sufficiently high to produce a superconductor with characteristics that vary with depth into the crystal.

Magnetization-hysteresis measurements are also performed on polycrystalline samples of 1-2-3 superconducting ceramics for the purpose of determining J_c. Intergrain and intragrain supercurrent density may be separated by ac inductive techniques, since there is a large difference in the penetration rate of flux into specimen as a function of field amplitude (Ni, et al., 1988).

As described in the previous chapter, magnetization current J_M ($J_M \sim J_c$) can be related to magnetic hysteresis by the expression for magnetization current density

$$J_M = \frac{2\Delta M}{\mu_o D} \quad 3.9$$

where ΔM is the width of a major hysteresis loop in the M/H curve and D is the thickness of the specimen.

It is encouraging that transport and magnetic inference of J_c on high-quality epitaxial films of $YBa_2Cu_3O_{7-\delta}$ both indicate J_c values in the range of 10^5 to 10^6 A/cm^2, perhaps surprising in light of the lower values seen in single crystals. Again, the specific single crystals samples used in these experiments may not represent the optimum configuration.

J_c levels in excess of 10^4 A/cm^2 at $T = 0.9T_c$ have been measured on polycrystalline films, with J_c proportional to $(1 - T/T_c)^{3/2}$ near T_c and proportional to $(1 - T/T_c)^2$ for temperatures below about $0.8T_c$ (Ogale, et al., 1987). A 1-2-3 (0.7 μm) film where holmium was substituted for yttrium, that is, $HoBa_2Cu_3O_x$, is reported to have a J_c value of 2.54×10^6 A/cm^2 at 77.3 K. In a field of 1 T, the critical current density was over 1.5×10^6 A/cm^2. This value is approaching the maximum theoretical J_c for single-crystal Y-Ba-Cu-O of $\sim 5 \times 10^6$ A/cm^2.

Silver-Enhanced Increase in J_c. There are more recent reports of Y-Ba-Cu-O sintered pellets where J_c has been increased by a factor of five by addition of 4% Ag_2O with the $YBa_2Cu_3O_7$ (Malik, et al., 1988). It should be noted that the current density, while improved by the addition of silver, was nevertheless small, ranging from 40 A/cm^2 for no silver, increasing to 125 A/cm^2 for 4% silver, and falling to 80 A/cm^2 for 6% silver. It is not clear whether this enhancement by the addition of silver is due to improved connections between superconducting grains or some less obvious mechanism. The authors do not rule out the possibility of silver being present in the crystalline lattice.

Silver-encased wires with a diameter of 0.8 mm have been constructed with J_c in excess of 4000 A/cm^2 (Swinbanks, 1987). This value, while not sufficient for most practical applications of high-T_c superconducting wires, nevertheless represents a significant step forward in the technology of applying these ceramics to magnet windings, motors, power transmission, and the like.

Use of Fluorine to Improve 1-2-3 J_c Levels. Fluorine is used in the processing of Y-Ba-Cu-O compounds to increase the level of critical current. It has been reported that while the addition of a limited amount of fluorine does not appear to change the structure of the Y-Ba-Cu-O unit crystal, the fluorine is suspected of strengthening the weak-link connections between the superconducting grains (Hakuraku, et al., 1988). These weak links are suspected of being the primary limitation on the level of critical current in bulk 1-2-3 superconductors.

Melt-Textured Growth; Effect on J_c. Considerable progress in improving the current transport of $YBa_2Cu_3O_x$ has been made by a processing technique called melt-textured growth. Temperatures for melt processing range from 1030°C to 1180°C for production of a liquid + solid phase, and to 1320°C for a single-phase liquid (Jin, et al., 1988). The material is held at these temperature levels for periods ranging from 30 seconds to as long as two hours, then cooled to room temperature in a gradient of approximately 50°C/cm. During this solidification process, needle-like crystals are formed.

In this study, scanning electron microscope examination of the conventional sintered $YBa_2Cu_3O_x$ exhibits a porous, granular structure with randomly oriented grains that have dimensions on the order of 5 μm, with a density that is only ~90% of the theoretical maximum. Of some practical importance to applications, the value of transport critical current density in the sintered material is on the order of 150–600 A/cm^2 in zero **H** field at a temperature of 77 K.

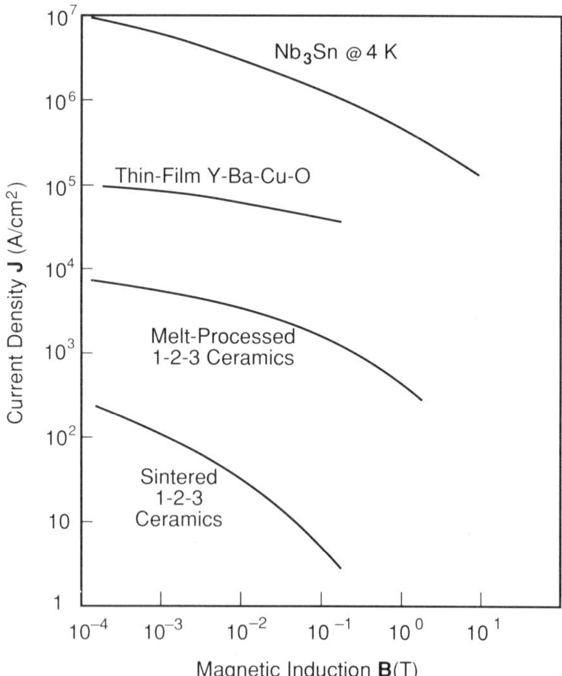

FIGURE 3.7 Representative critical current (J_c) vs. B, comparing Nb_3Sn to Y-Ba-Cu-O (i.e., 1-2-3) in film, melt-processed, and sintered polycrystalline ceramic forms.

The melt-textured material, in contrast, is virtually 100% of theoretical density and exhibits locally textured structures that have the appearance of needles with radii of 2 to 5 μm and lengths up to 600 μm. In addition, the value of transport J_c at 77 K is on the order of 7400 A/cm^2, more than an order of magnitude higher than typical sintered samples. Just as significant, the value of J_c does not decrease as rapidly as that for sintered samples when the magnetic field is increased. At 10,000 gauss, the melt-textured J_c is still on the order of 1000 A/cm^2, while the value for the sintered material has dropped to 1 A/cm^2 or less (see Figure 3.7). The investigators at AT&T Bell Laboratories suggest at least three benefits from the melt-textured growth process:

1. The formation of a more dense structure improves the connections between individual grains,
2. orientation of crystalline structure along the a or b axes, that is, the preferred superconducting axes, and
3. the formation of a cleaner grain boundary due largely to the decomposition (at higher temperature) of the carbonate-type impurities.

However the effect of each of these factors may be weighted, each would contribute to improved current transport by elimination of weak-link junctions between grains seen in typical sintered material. The grain boundary problem is rather a fundamental one in most bulk materials, and will be considered in more detail below.

Grain-Boundary Problems; Contamination. Particularly in polycrystalline bulk samples of $YBa_2Cu_3O_x$, grain boundary effects may dominate behavior over the entire frequency spectrum (see Figure 3.8). At dc and low frequencies, current density is limited by weak-link behavior at the interfaces between grains. These same weak links are, of course, of extreme importance in limiting microwave performance. Also of great significance in high-frequency applications is the presence of any impurities (generally concentrated at the grain boundaries) since these impurities will contribute to the normal-electron losses generated in an alternating field.

Several groups (Verhoeven, et al., 1988) have found carbon contamination in $YBa_2Cu_3O_x$, in one instance despite "the deep calcination process, repeated several times in order to minimize the C in the system (Parmigiani, et al., 1987)." The investigators reported that all samples had a significant amount of carbon. Others have found an excess of yttrium between grains (Mai, et al., 1988). The presence of any contaminant severely restricts the applicability of the material at the higher frequencies.

Adding to these problems is the fact that the surface of $YBa_2Cu_3O_x$, in many instances, may not be superconducting, and may in fact be an insulator (Clarke, 1988). As a result of this phenomenon, it is now a fairly common practice to remove a few Å of the surface prior to applying contacts. While dc and

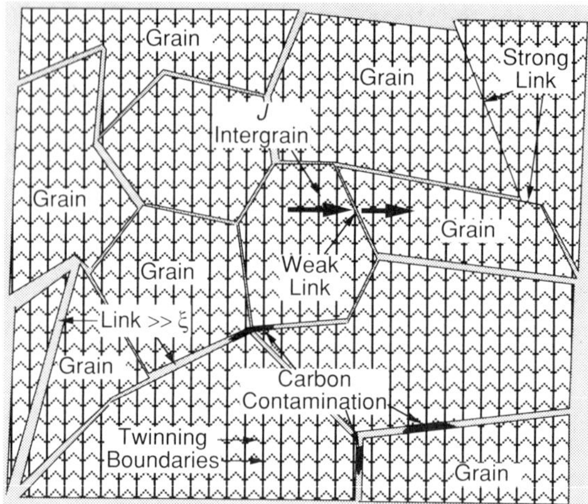

FIGURE 3.8 A rather schematic illustration of grain boundary effects in polycrystalline ceramic superconductors. See the text for details.

many low-frequency ac applications (where the surface layers are penetrated by contacts) may not suffer from an insulating or lossy layer, the problem is of considerable interest to those who have applications at frequencies in the microwave region, since all of the current will flow on the surface of the material. This leads quite naturally into a consideration of how the new materials perform in microwave environments.

Microwave Characteristics

Early reports on surface resistance of bulk samples of the copper oxides at 9.85 Ghz were consistent with the relatively low dc critical current density for such samples; surface resistivity at 70 K was about the same as for copper, with a "residual" resistance at very low temperatures only an order of magnitude lower than for Cu at the same temperatures (Sridhar, et al., 1987; Sridhar & Kennedy, 1988). This less-than-spectacular superconducting performance (particularly when compared to niobium) is probably due to granularity and anisotropic behavior of these ceramics (see Figure 3.9). Single crystals and epitaxial films can be expected to have far superior microwave characteristics, but the problem of contaminants must also be addressed.

Other investigators have found that the absorption ratio (A) of incident fields on $YBa_2Cu_3O_{7-x}$ at 9 GHz has the following empirical relationship for temperatures well below T_c (Tateno & Masaki, 1987):

$$A = 17.6\epsilon^{-0.031(eV)/kT} \qquad 3.10$$

Of considerable interest is a report of field-dependent microwave absorption in $YBa_2Cu_3O_{7-x}$; the nonlinear behavior was markedly different for polycrystalline and single-crystal samples. The authors suggest that microwave absorption as a function of radio frequency (rf) field strength (0–20 gauss) may prove to be a useful diagnostic for these materials (Glarum et al., 1988).

Recall, from discussion in the previous chapter, that the two-fluid relationship for surface impedance of a superconductor is approximated by the expression

$$Z_s = R_s + jX_s \approx \frac{\omega^2 \mu_o^2 \lambda^3 n_n \sigma_n}{2n_s} + j\,\omega\mu_o\lambda \tag{3.11}$$

The interesting point for this particular discussion is that

$$R_s \propto \lambda^3 \tag{3.12}$$

Since the 1-2-3 materials are quite anisotropic, this anisotropy must be taken into account when considering how the films should be constructed to minimize surface resistance R_s. Clearly, the unit-crystal orientation with the smallest λ should have the smallest value of R_s if the new ceramics are governed by this two-fluid model. Since λ_{ab} is on the order 270 Å while λ_c is about 1800 Å ($\lambda_c/\lambda_{ab} \sim 6.7$), one would expect that the *a-b* plane of the 1-2-3 film would have the lowest value of R_s. These numbers suggest that, if the 1-2-3 materials follow this two-fluid prediction, microwave resistivity should be on the order of $(6.7)^3$, that is, 300 times higher for **E** fields parallel to the *c* axis than for **E** fields in the *a-b* plane. Since it is evidently the copper oxygen planes (parallel to the *a-b* plane) that are chiefly involved in transport of the superconductive current flow in the 1-2-3 compounds, this suggestion is not contrary to intuition.

It has also been pointed out that in the Mattis-Bardeen formalism (Braginski, et al., 1988),

$$R_s \propto \frac{\lambda}{\xi} \tag{3.13}$$

or, when $\lambda \gg \xi$, as is the case for 1-2-3 ceramics,

$$R_s \propto \frac{\lambda^2}{\xi} \tag{3.14}$$

For 1-2-3 films this type of behavior would lead to a ratio of surface resistance for the *c*-axis to the *a-b* plane approximated by

$$\frac{R_{s(c)}}{R_{s(a-b)}} \approx \frac{\lambda_c^2/\xi_c}{\lambda_{ab}^2/\xi_{ab}} \approx \frac{(1800)^2/3}{(270)^2/20} = 296.3 \tag{3.15}$$

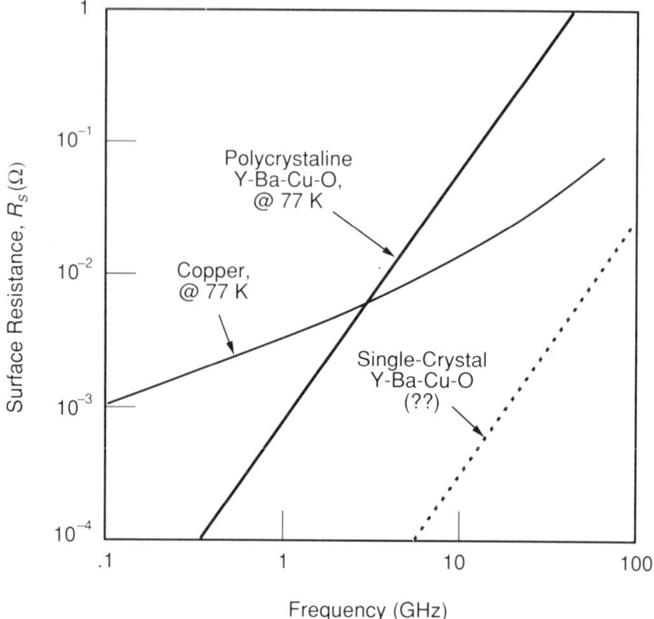

FIGURE 3.9 Surface resistance (R_S): Cu vs. Y-Ba-Cu-O (H. Piel, private communication.)

which is approximately the same (i.e., 300) as for the two-fluid estimate that R_S is simply proportional to λ^3.

Results on microwave measurements of thin films thus far do not clarify the accuracy of these predictions, but it is clearly important for practical applications of these ceramics to understand how single-crystal films should be oriented for minimum microwave surface resistance.

Thermal Expansion, Random and Parallel Cracks

A rather important practical aspect of using superconductors in electronic applications relates to the coefficient of thermal expansion; if the superconductor has a coefficient that is significantly different from its substrate, cooling the system from room temperature (~ 300 K) to operating temperature (~ 77 K?) may prove to be disastrous. A similar mismatch in thermal expansion coefficients can also disrupt films during processing, where temperatures may vary from 20°C to ~ 700°C. The formation of cracks in superconducting films and on the surface of bulk polycrystalline samples is, in fact, a fairly common occurrence.

One study has shown a relation between crack formation and the difference in thermal expansion coefficient coefficients between the superconducting material and particular substrates (Hashimoto, et al., 1988). For example, the average thermal coefficients of expansion (from 30 to 900°C) of $YBa_2Cu_3O_{7-\delta}$ and strontium titanate ($SrTiO_3$) were 1.69×10^{-5}/°C and 1.11×10^{-5}/°C, respectively. It is interesting that the investigators report a significant change in the 1-2-3 compound thermal expansion coefficient at approximately 350°C and 650°C, corresponding respectively to a change in concentration of oxygen and a phase transition.

Other investigators have seen an interruption of the crack propagation when a relatively small percentage of silver was included in the 1-2-3 system. While a silver matrix would not necessarily cause a problem for dc operation, the presence of silver would lead to normal-state losses that would become more significant as frequency was increased.

While randomly oriented cracks limit current transport, cracks that were introduced to be oriented parallel to the direction of current flow would not interfere with current transport, and could serve as pinning centers for flux. Such centers, if effective, would help to increase the low-frequency critical current density. One theorist has estimated that, in thin 1-2-3 films at 77 K, the current density might be as high as 1.7×10^6 A/cm^2 at 5 T for current-parallel cracks on the order of 50 nm apart (Matsushita, 1988).

See the chapter on processing for more information on thermal expansion of ceramic superconductors and substrates.

Radiation Effects; Damage

The question of the effect of ionizing radiation on the superconducting perovskites is naturally of great interest, since a variety of applications require that the material be exposed to a rather large "dose" of radiation. Examples of such applications are particle accelerator bending magnets and magnetic confinement fusion reactors. In addition, the effects of radiation damage may contribute to knowledge of the mechanism of superconductivity in high-temperature ceramics.

Neutron Exposure. Investigators at the Los Alamos National Laboratory have exposed $YBa_2Cu_3O_7$ to fast neutrons at a flux rate of 4×10^{12} neutrons/cm^2 per second at the Omega West reactor (Willis, et al., 1988) (see Figure 3.10). It was determined that the transition temperature of $YBa_2Cu_3O_7$ decreased more rapidly with fast neutron fluence than is the case with A15 superconductors, limiting use of this perovskite to environments where total exposure does not exceed about 2×10^{18} neutrons/cm^2. It was also noted that neutron fluences up to approximately 3×10^{18} neutrons/cm^2 tended to increase the critical current density of $YBa_2Cu_3O_7$ at 7 K (see Figure 3.11). For measurements at 75 K, J_c is enhanced at neutron fluences up to approximately 1×10^{18} neutrons/cm^2, and then begins to decrease for higher exposures. The enhancement of J_c by neutron irradiation is presumably due to the creation of pinning sites produced by localized radiation effects.

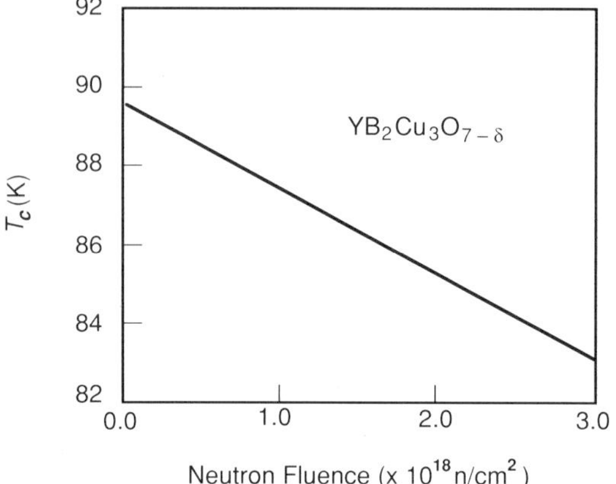

FIGURE 3.10 Approximate effect of fast neutron exposure on Y-Ba-Cu-O critical temperature. (After J. O. Willis, J. R. Cost, R. D. Brown, J. D. Thompson, and D. E. Peterson, "Radiation damage in $YBa_2Cu_3O_7$ by fast neutrons," Proc. 1988 Symposium, Vol. 99, Materials Research Society, pp. 391–394.)

Electron Irradiation. Bulk samples of Y-Ba-Cu-O have also been irradiated, at temperatures below 20 K, with homogeneous doses of 3 MeV electrons (Stritzker, et al., 1987). Exposure to electrons resulted in dramatic decreases in T_c (while the transition width remained approximately constant at $\sim$ 1.5 K) and significant increases in normal-state resistivity. The transition temperature

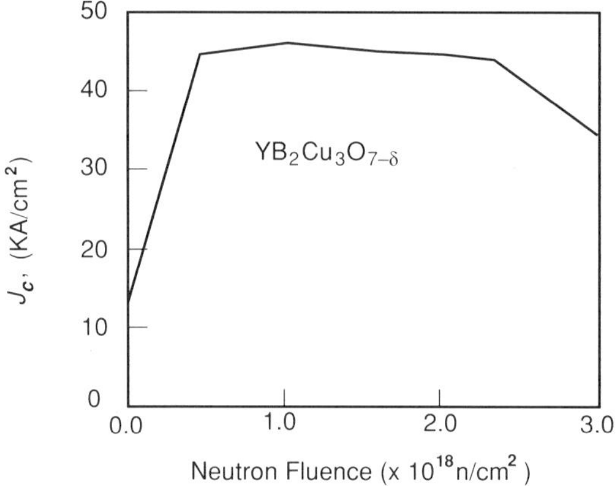

FIGURE 3.11 General behavior of critical current at 7 K as a function of fast neutron fluence in bulk polycrystalline Y-Ba-Cu-O. See reference below for detailed information, including behavior at 75 K. (After J. O. Willis, J. R. Cost, R. D. Brown, J. D. Thompson, and D. E. Peterson, "Radiation damage in $YBa_2Cu_3O_7$ by fast neutrons," Proc. 1988 Symposium, Vol. 99, Materials Research Society, pp. 391–394.)

dropped by 20 K following a total irradiation of 5.7×10^{19} electrons/cm^2. These high-energy electrons cause only single-displacement processes, and effects on either copper or oxygen atoms could be responsible for the sensitivity seen to the irradiation. The investigators felt that, while their results did not lead to firm conclusions about the precise mechanism of damage, the changes in T_c and resistivity were strongly correlated with volume effects.

Proton Exposure. Exposures to 3.1 MeV protons leads to a modest decrease in T_c (to about 0.8 $T_{c(or)}$) for $YBa_2Cu_3O_7$ out to approximately 10^{17} protons/cm^2, where T_c begins to drop dramatically. Here, $T_{c(or)}$ is the original T_c prior to irradiation. The comparable critical 3.1 MeV proton irradiation for Nb_3Sn is an order of magnitude higher (Nikonov, et al., 1988) (see Figure 3.12).

H^+ and He^+ Ion Irradiation. The effects of low-fluence (i.e., less than 10^{14} ions/cm^2) 63 MeV H^+ and 65 MeV He^+ ions on plasma-arc spray $YBa_2Cu_3O_{7-\delta}$ films has been determined to produce a catastrophic loss of superconductivity. The investigators suspect that a "bond percolation threshold" was responsible for the sudden loss of superconducting characteristics, which were more dramatic on the plasma-spray films than for films produced by other methods (Chrisey, et al., 1988).

Oxygen and Arsenic Ion Irradiation. Work has also been performed at the IBM-Watson Research Center to determine the effects of ion implantation on

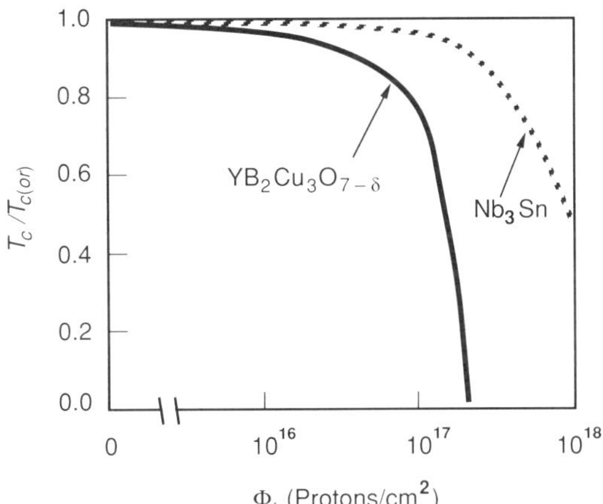

FIGURE 3.12 Effects of 3.1 MeV proton exposure in T_c for Nb_3Sn and Y-Ba-Cu-O. $T_{c(or)}$ represents the value of T_c prior to exposure. (After A. A. Nikonov, G. V. Sotnikov, A. S. Tokarev, I. V. Kurchatov, "Influence of radiation defects on superconductivity in $YBa_2Cu_3O_{7-\delta}$ films," presented at 1988 ASC (*IEEE Trans. Magn.*, Vol. 25, No. 2, 1989, pp. 2579–2582), San Francisco, August 21–25, 1988.

thin films of $YBa_2Cu_3O_7$. Particles used in this study were 500 keV oxygen and 1 MeV arsenic ions; effects on T_c were recorded as a function of ion dose. Zero resistivity temperature was dramatically affected, decreasing to ~ 0 K for approximately 1.1×10^{14} 500-keV oxygen ions/cm^2 (Clark, et al., 1987). Since the onset temperature for superconductivity was not significantly altered, this result amounts to an extreme broadening of the transition as a function of ion doses in this range.

The deposited energy required to change the critical temperature was found to be 0.2 eV per atom. A larger energy (~1–2 eV/atom) was required to significantly modify room temperature conductivity, and ~4 eV/atom was required to render the film amorphous. The precise mechanism for these various thresholds of damage are not yet identified, but the investigators suggest that disruption of the copper-oxygen bonds is probably a major factor at lower doses, while collisional disordering is likely responsible for the conversion of the $YBa_2Cu_3O_7$ films.

Other Y-Ba-Cu-O Superconducting Phases

In addition to $YBa_2Cu_3O_{7-x}$, that is, the well-known 1-2-3 compound with a T_c at approximately 90 K, there are at least two other superconducting Y-Ba-Cu-O compounds. All three phases are listed, along with nominal values for critical temperature.

Compound	*Shorthand Notation*	T_c
$YBa_2Cu_{3.5}O_{7+x}$		~40 K
$YBa_2Cu_4O_{8+x}$	1-2-4	~80 K
$YBa_2Cu_3O_{7-\delta}$	1-2-3	~90 K

The first two phases (with T_c's at 40 and 80 K) are not usually seen in conventionally processed batches of Y-Ba-Cu-O ceramic, because they are not stable under normal processing conditions. Each of these unusual phases is synthesized under rather high oxygen pressure, that is, on the order of 3000 bar O_2 (Karpinski, et al., 1988a, 1988b, 1988c).

BISMITH AND THALLIUM SYSTEMS

High-temperature superconductors based on bismuth and thallium will be discussed below. These two systems, while discovered quite independently, are evidently related, having the general formula (Greaves, 1988)

$$A_2B_2Ca_{n-1}Cu_nO_{4+2n} \qquad 3.16$$

where A is either bismuth or thallium, B is strontium or barium, and n is the number of CuO_2 layers in the unit crystal. Interestingly, T_c increases with n, at least for $n \leq 3$. In the case of the thallium compound, the relationship is given as follows for $n = 1$, 2, and 3:

n	T_c
1	80 K
2	110 K
3	125 K

This behavior has quite naturally led to speculation that significantly higher values of T_c might be achieved by adding more copper-oxide planes to the unit crystals. The picture may not be as simple as such speculation suggests.

Clearly, as n becomes very large, the $A_2B_2Ca_{n-1}Cu_nO_{4+2n}$ approaches $CaCuO_2$, which is an insulator. Nevertheless, one might reasonably expect higher values of T_c for modest increases in n. While values of n up to 3 are commonly produced in the thallium system, larger values have been seen in electron microscopy studies. Even though $CaCuO_2$ is evidently not stable, it can be stabilized with the substitution of small amounts of strontium at the calcium site, producing, for example, $(Ca_{0.86}Sr_{0.14})CuO_2$. This system is under study as the "parent" structure of the bismuth and thallium systems (Siegrist, et al., 1988).

The bismuth and thallium systems will be discussed in more detail in the following section.

Bismuth-Based Superconductors

The first of the new families of superconductors to compete with Y-Ba-Cu-O is a compound of bismuth, strontium, calcium, copper, and oxygen that superconducts at ~ 114 K. The discovery was reported early in 1988, on the compound $Bi_1Sr_1Ca_1Cu_2O_x$ with a T_c of about 105 K and a resistive "tail" extending to 80 K before dc resistivity dropped to zero (Maeda, et al., 1988) (see Figure 3.13).

It has been difficult to isolate the higher-T_c from the lower-T_c phase, but this has been accomplished with magnetron sputtering of thin films where a partially melted underlayer was maintained by controlled substrate heating (Fukutomi, et al., 1988). In another instance, the high-T_c phase was separated by processing mixed nitrates of Bi, Pb, Sr, Ca, and Cu at 830°C under low oxygen pressure (Endo, et al., 1988). In each of these cases, critical temperatures above 100 K were obtained.

For applications, it is significant that the bismuth compounds are moderately ductile, exhibiting a flaky structure not unlike mica. Also encouraging are reports that the bismuth superconductors are resistant to chemical attack; in marked contrast to Y-Ba-Cu-O, bismuth compounds also exhibit stability in water. Moisture, as was mentioned earlier, was generally destructive to early

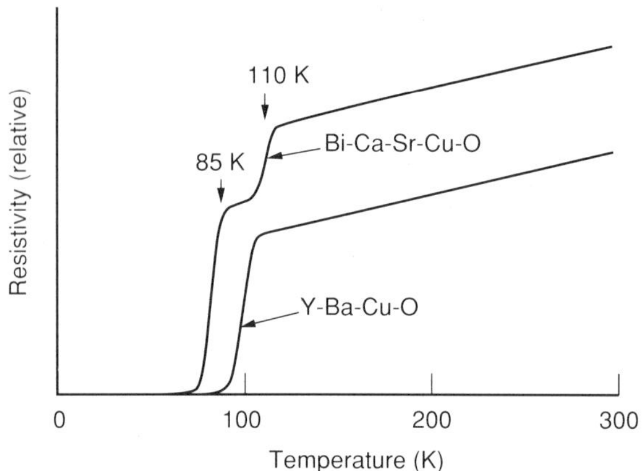

FIGURE 3.13 The early Bi-Sr-Ca-Cu-O system showed an onset of superconductivity near 110 K, and another transition near 85 K. Several investigators have succeeded in separating the 85 K and the 110 K bismuth-system phases. Y-Ba-Cu-O response is shown for comparison.

films of $YBa_2Cu_3O_{x-\delta}$, although moisture degradation is becoming less of a problem as higher quality 1-2-3 films are being developed. The 1-2-3 materials are also quite brittle. Compared to the $YBa_2Cu_3O_{x-\delta}$ ceramics, the bismuth-strontium-calcium-copper oxides are also easier to process.

One investigation (Takayama-Muromachi, et al., 1988) suggests that the superconducting phase is either $Bi_2Sr_3Ca_3Cu_2O_9$ or $Bi_2(Sr,Ca)_{3-x}Cu_2O_{9-y}$ while another (Tarascon, et al., 1988) has isolated a high-T_c phase identified as $Bi_4(Sr,Ca)_6Cu_4O_{16+x}$. Another report indicates that another superconducting phase is $Bi_2Sr_3Ca_3Cu_2O_{8+x}$, where x is much less than 1 (Peterson, 1988).

The superconducting phases of the bismuth compound continue to proliferate. According to a recent report, investigators at the Nagoaka University in Japan have measured five different superconducting phases of Bi-Ca-Sr-Cu-O_x (Nikkan Koygo Shimbun, 1988):

Phase	T_c
$Bi_{1.5}CaSrCu_2O_x$	65 K
$Bi_{1.5}Ca_{1.5}SrCu_2O_x$	63 K
$Bi_{0.5}Pb_{0.5}CaSrCu_2O_x$	71 K
$BiAl_{0.1}CaSrCu_2O_x$	73 K
$BiAl_{0.3}CaSrCu_2O_x$	78 K

It is not uncommon, in the copper oxides, to have several superconducting phases, as the bismuth system demonstrates.

While the bismuth compounds have the copper-oxygen layers that are characteristic of the superconducting ceramics, the copper-oxygen chains found in Y-Ba-Cu-O and La-Sr-Cu-O are not found in Bi-Sr-Ca-Cu-O.

Bismuth Compounds—Critical Current Density. A report from R.B. van Dover and colleagues describes transport measurements of J_c in bismuth single-crystal superconductors ($Bi_{2.2}Sr_2Ca_{0.8}Cu_2O_{8+\delta}$) which exceed 4×10^4 A/cm^2 in zero magnetic field at 77 K (van Dover, et al., 1988). Thus far, the bismuth compounds appear to behave much like yttrium compounds as far as critical current transport is concerned.

Thallium-Based Superconductors

Following on the heels of the successful experiments demonstrating the high T_c in the bismuth superconductors, there were developments involving compounds containing thallium and calcium, which superconduct at 120–125 K (Sheng, et al., 1988; Parkin, et al., 1988) (see Figure 3.14). There are evidently several superconducting phases, including

$$Tl_2Ca_2Ba_2Cu_3O_{10+\delta} \text{ (2223 material)}$$

and

$$Tl_2Ca_1Ba_2Cu_3O_{8+\delta} \text{ (2123 material)}$$

As mentioned previously, the system $Tl_2Ca_{n-1}Ba_2Cu_nO_{4+2n}$ exhibits three different superconducting phases, as a function of n (Feigelson, et al., 1988):

n	T_c
1	80 K
2	105 K
3	125 K

Another report indicates that excellent results were seen with pellets of $TlCaBaCu_2O_x$; the same group at Sandia National Laboratory reports critical current density in *untextured* thallium films that is in excess of 10^5 A/cm^2 (Ginley, et al., 1988). Moreover, these critical currents, unlike those measured in yttrium or bismuth ceramics, were reported to be not significantly affected by the presence of magnetic fields. At this writing, while the J_c results have been repeated in other laboratories, the insensitivity to magnetic fields has not (to the author's knowledge) been verified by other investigators. In fact, it is evidently relatively difficult to produce high-quality, thallium-compound superconducting films. The thallium compounds, while perhaps not quite so stable as the bismuth superconductors, evidently exhibit stability that is superior to 1-2-3 superconductors.

Measurements on toroids of Tl-Ba-Ca-Cu-O superconductors have determined that persistent currents decay in a relatively short time, that is, by 27% in only 3800 s (Grader, et al., 1988). The authors point out that this was a mixed-phase sample, and they estimate an upper limit for dc resistivity of

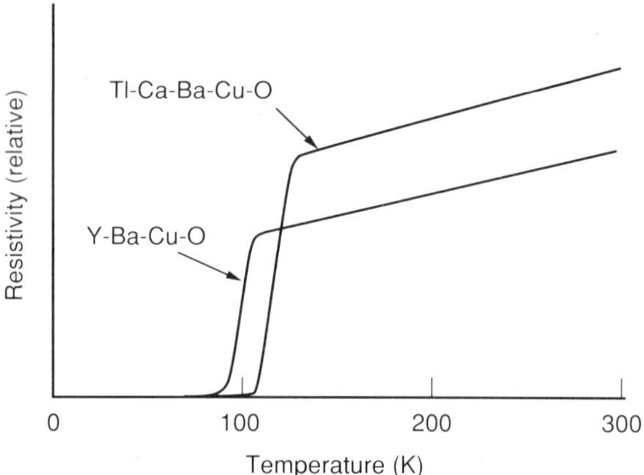

FIGURE 3.14 Tl-Ca-Ba-Cu-O response compared to Y-Ba-Cu-O. Typical thallium-system superconductors show evidence of superconductivity at 120–125 K, and exhibit zero dc resistivity at ~105–107 K.

1.3×10^{-14} Ω-cm at 77.3 K. The investigators believe that the decay was primarily due to flux creep; not good news for alternating-current applications. It is possible, of course, that the sample may have had contaminants that were responsible for this behavior. Samples of greater purity may exhibit much lower losses.

Tl-Strontium-Ca-Cu-O Superconductor. In addition to the Tl-Ba-Ca-Cu-O superconducting system, it has been determined that strontium can be substituted for barium, producing phases with critical temperatures at approximately 50 K and 100 K. The stoichiometry for the 100 K phase is believed to be $Tl_{1.5}Sr_1Ca_1Cu_2O_x$ (Nagashima, et al., 1988).

A Safety Precaution

A word of warning: **thallium is extremely toxic**. Special care should be exercised in handling this element, particularly during preparation and processing procedures. While it is generally believed that bulk and film samples are relatively safe to handle when the thallium is effectively trapped in the matrix, caution should be the rule when compounds containing thallium are heated, ground, sliced, and so forth. Even in bulk samples containing "trapped" thallium, there are potential hazards, such as the formation of thallium-oxide, which is toxic, soluble in water, and can be absorbed through human skin. See the chapters on processing and safety for more information.

Bismuth and Thallium Superconductors: Final Comments

Both Bi-Ca-Sr-Cu-O and Tl-Ca-Ba-Cu-O evidently have a variety of superconducting phases, and exhibit both higher T_c (in selected phases) and better stability than 1-2-3 materials. Since these systems are very complex, even in comparison to 1-2-3 compounds, it is likely that phases with even higher transition onset temperatures than those reported to date (~125 K) will be found.

LIQUID NITROGEN; A SUITABLE CRYOGEN?

It must be pointed out, that for a superconductor to perform well, particularly in high-frequency applications, it is not advisable to operate near T_c. In the case of Y-Ba-Cu-O, with a nominal T_c of 90–95 K, operation at the liquid nitrogen temperature of 77 K may be advisable only for dc applications and only at relatively low transport current density and magnetic field levels. Recall the three–dimensional plot of the relationship of J_c, H_c, and temperature from the previous chapter on superconducting phenomena. Additionally, measurements of ac loss with either superconducting cavities or eddy-current methods (see the chapter on measurements) show the presence of significant losses well below T_c. The presence of excessive ac losses is often referred to as a "tailing" on the loss versus temperature curve. These losses are not generally seen in dc resistivity measurements, since a single, percolative, filamentary path between the measurement electrodes will produce zero resistance to effectively shunt any nonsuperconducting material in the specimen.

Recall that T_c is the temperature where a material just begins to superconduct with low critical current density in zero ambient magnetic fields, that is, where a slight increase in current density or a modest magnetic field may quench superconductivity. Obviously, one does not generally wish to operate a superconductor "on the edge" of quenching. This rule is not absolute: there are applications where such operation is central to the requirement, as with cryotron switches and certain types of infrared detectors.

Nevertheless, it is usually necessary to operate well below T_c to exploit the current-transport and magnetic properties of any superconductor. The approximate "rule of thumb" used by most practitioners of the superconducting art is to operate at one-half the critical temperature (Fitzgerald, 1988). In the case of Y-Ba-Cu-O, or even with the bismuth or thallium systems, this implies an operating temperature on the order of 45–50 degrees K, well below liquid nitrogen temperature (77 K). Seen from another view, operation with liquid nitrogen would normally require a superconductor with a T_c of at least 150–160 K.

HIGH-T_c WITHOUT COPPER

The new high-T_c ceramics appear to have at least one common element: copper-oxygen planes. Associated with superconductivity mediated by the Cu-O_2 planes is a large anisotropy in critical current (J_c) and critical field (H_{c2}).

This anisotropy is, of course, bad news for those who are considering high-field applications for the new compounds, since one must produce materials with highly oriented crystalline structure to achieve high-field, high-current operation. This limitation is quite in contrast with "conventional" superconducting compounds like niobium-tin, which are essentially isotropic in their behavior.

Partially due to these anisotropies, there is considerable interest in a report of a 29.8 K (T_c-onset) superconductor that does not include copper (Cava, et al., 1988). This barium-potassium-bismuth-oxygen ($Ba_{0.6}K_{0.4}BiO_3$) compound supports superconductivity in a *three-dimensional* bismuth-oxygen array, and does not show signs of anisotropy like those associated with the Cu-O planes in the 1-2-3, the bismuth, or the thallium ceramics.

Further work at Argonne National Laboratory, where the relative proportions of barium, bismuth, and oxygen were varied, found the highest T_c at 27 K for a chemical composition identified as $Ba_{0.7}K_{0.3}BiO_3$ (Waldrop, 1988). When the proportion of potassium was reduced to yield $Ba_{0.75}K_{0.25}BiO_3$, the structure of the crystal shifted to a nonsuperconducting phase. While critical currents have not been reported at this writing, tolerance of magnetic fields is apparently not particularly high. Investigators at Bell Labs find that T_c falls by ~4 K in the presence of a 0.2 T field.

It appears that the structure of the superconducting phase is cubic, with dimensions on the order of 4.3 Å. Investigators at Argonne National Laboratory on $Ba_{1-x}K_xBiO_{3-y}$ have performed experiments which show that optimum stoichiometry (i.e., where T_c is maximum) is seen for $x \sim 0.25$. This is approximately the stoichiometry required for a transition to the cubic phase of Ba-K-Bi-O. That is to say, as x is decreased below 0.25, the cubic structure is lost and superconductivity disappears entirely. As x is increased above 0.25, T_c decreases, dropping to about 15 K at $x = 0.4$ (Hinks, et al., 1988).

The fact that T_c's on the order of 30 K can be achieved without copper-oxygen planes may prove to be significant (Rice, 1988). It is perhaps noteworthy that this new compound is similar to the barium-lead-bismuth-oxygen ($T_c = 12$ K) compound reported more than a decade earlier by A. W. Sleight of Dupont.

La_2CuO_4: TRANSIENT SUPERCONDUCTIVITY?

During the summer of 1987, there was an interesting report from the group at IBM-Yorktown Heights concerning superconductivity in La-Cu-O (Grant et al., 1987). The layered perovskite La_2CuO_4 is, after all, the "parent compound" of the $La_{2-x}Ba_xCuO_{4-y}$ 30 K superconductor made famous by Bednorz and Müller.

The discovery of superconductivity was the result of following a clue: a thermopower went to zero at low temperatures even as the resistance *increased*.

The increasing resistance had been seen previously by many investigators; it was consistent with a material that would reach an insulating ground state at very low temperatures. The measurements of a resistance transition near 40 K were "extremely process dependent" and "quite sensitive to stoichiometry."

The authors point out the similarity of these findings to reports of superconductive transitions near 240 K in other compounds. This report of transient evidence of superconductivity at relatively low temperature will take on added significance as we discuss reports of remarkable behavior at much higher temperatures.

ROOM-TEMPERATURE SUPERCONDUCTORS: WILL-O'-THE-WISP?

As this is being written, there is some controversy over reports of evidence for room-temperature superconductivity. Scores of investigators have seen tantalizing resistivity transitions in the range from 200 K to 350 K, and these transitions sometimes have a structure that is quite similar to proven superconducting transitions seen in the range from 80–125 K.

To date, these reports typically describe transient phenomena that have not been reproduced in other laboratories, and often not even by those who made the original observations. Many of these reports are originating from very capable investigators, so there is a strong feeling in the community of condensed-matter science that the level of T_c will continue to be pushed upwards, beyond the ~100–125 K level for the bismuth and thallium systems. This optimism is based not only on the high reputation of many of the scientists who are reporting on the ephemeral high-T_c phenomena, but also on the conviction that the superconducting phenomena in perovskites is a *new* phenomena, representing quite different mechanisms from those found in materials like lead, tin, and niobium.

It may be significant that at least one relatively successful theory of the ceramic superconductor behavior (which assumes that the superconducting behavior is related to magnetic interactions between electrons and pairs of copper atoms, i.e.,"magnon pairing") suggests that T_c for this class of materials is not likely to exceed approximately 225 K (Pool, 1988; Guo, et al., 1988; Chen & Goddard, 1988). On the other hand, equally alluring explanations developed by other theorists do not require such limitations on T_c, and the theoretical basis of the 225 K limitation on T_c is hotly contested by at least one noteworthy theorist (Anderson, 1988).

Some examples of the (near) room-temperature superconductivity investigations follow:

Very large changes in resistivity have been seen at 230–240 K in Y-Ba-Cu-O and Eu-Ba-Cu-O; including suggestive "magnetic anomalies" in europium 1-2-3 samples, but a definitive Meissner effect has not been confirmed (Huang, et al., 1987).

A remarkable effect of temperature cycling to a precise level (239 K) has been reported (Bhargava, et al., 1987). In Y-Ba-Cu-O and Y-Ba-Cu-O:F (fluorine) samples, the level of T_c was increased from 95 K to beyond 130 K simply by cycling the sample temperature between levels below T_c to 238.9 ± 0.2 K, referred to by the investigators as the "superconducting-phase healing temperature" (T_{SPH}). Heating to lower temperatures does not increase T_c, while heating to higher levels tends to reduce T_c back to 95 K. Repeated cycling to the precise T_{SPH} level tend to continue increasing T_c (up to a maximum saturation level for a given sample). The authors believe that a two-dimensional to one-dimensional structural change occurs in this narrow temperature range, and that the nonrepeatability of other experiments showing $T_c > 150$ K may be partially explained by this phenomenon. Other workers have found an anomaly in specific heat at 220 K which may be related to oxygen-ordering at that temperature (Laegreid, 1987).

Other investigators suggest that the presence of coherent twin boundaries in Y-Ba-Cu-O, and other rare earth 1-2-3 structures, are essential to superconductivity at 95 K and have important implications for T_c's up to 240 K (Mueller, et al., 1987).

There have been reports of stable superconductivity with transition temperatures in the range of 228–323 K in a compound of lanthanum-strontium-niobium-oxide. The first published report (Ogushi, et al., 1987) described "sharp resistive transitions" at temperatures above 300 K, and a more recent article (Cambridge, 1988) reports that the compound has been refined ($La_2Sr_1Nb_5O_{10}$ in thin films) and exhibits zero resistance at 320 K. Since zero resistance is impossible to verify, it is more significant that the investigators have reportedly measured a significant Meissner effect at 320 K. While most previous reports of room-temperature superconductivity concerned compounds that were stable for a few days at most, the recent experiments in Japan yielded samples that were stable for at least a month. If these findings are reproduced, and if high values of J_c can be reached, the technological results could be quite remarkable. Curiously, little has been heard of these results since the initial reports.

During the summer of 1988, there were reports of superconductivity at 162 K in a multiphase sample with the "nominal composition" of $TlCa_4Ba_3Cu_6O_x$. The reader is reminded that, during initial experiments, it is extremely difficult to segregate the various phases that are produced during processing of these ceramics. "Zero resistance" at 162 K, seen by R.S. Liu and his colleagues at the Industrial Technology Research Institute in Taiwan, was reported to be transient, while "zero resistance" above 140 K was more repeatable (Liu, et al., 1988). These investigators conclude that the superconducting phase responsible for this behavior was $TlCa_2Ba_3Cu_4O_x$, a tetragonal structure with lattice constants a and b = 3.95 Å and c = 39.5 Å. The investigators were not able to confirm a Meissner effect above 122 K, and suggest that this lack of measurable diamagnetism is probably due to the "filamentary nature of the compound." To date, these results have not been confirmed in other laboratories.

Reports of "superconducting effects" near 300 K continue on almost a weekly basis, but these phenomena are typically either transient or nonreproducible, or both. Rigorous standards must be met, especially when presenting evidence for superconductors with T_c's far in excess of 100 K. The investigators who report these transient results would be among the first to admit (or insist!) that sharp *transitions* in dc resistivity, while intriguing, are not sufficient to prove superconducting behavior. At a minimum, a definite Meissner effect should be demonstrated, and the experiment should be reproducible by other investigators, in other laboratories.

SUMMARY

The copper oxide superconductors exhibit Type-II behavior including extremely high critical (H) field levels. Even though BCS-type theories that invoke electron-phonon interactions appear to be untenable for Y-Ba-Cu-O (since the isotope mass effect is relatively small), there is strong evidence of an electron-charge (hole) pairing mechanism from measurement of the flux quantum. A sound theoretical framework for the copper oxides will be very useful in the effort to understand behavior of the present 1-2-3 materials and design better ones. While the critical current levels are initially not sufficiently high in bulk materials for some important applications, measurements on single-crystal samples and epitaxial films indicate that J_c on the order of 10^5–10^6 A/cm^2 can be achieved if the weak-link problem in bulk material is solved. Another major problem with present samples of 1-2-3 ceramic is the presence of contaminants, particularly at grain boundaries. Microwave applications will require that the weak-link problems at the grain boundaries, as well as the contaminants, be dealt with.

The bismuth and thallium compounds exhibit critical transition temperatures in excess of 100 K, and are more stable than 1-2-3 ceramics. While the critical current problem in bismuth compounds is similar to that seen in 1-2-3 ceramics, there is some evidence that thallium films may support critical currents in the presence of high magnetic fields that are sufficiently large for many applications.

While "room-temperature" T_c's are reported to appear in a transient and nonreproducible fashion *thus far*, one is tempted to predict that critical temperatures well in excess of 200 K will be demonstrated shortly, perhaps in one of the bismuth or thallium compounds.

> We didn't ask for these predictions—we got them whether we wanted them or not. They clutter up the literature, they confuse the mind, and they give all of us a bad image.
>
> (Bernd T. Matthias, 1973)

REFERENCES

Anderson, P. W., (1988). "Public Statement on the "Magnon-Pairing" Theory of High-T_c Superconductors of W. A. Goddard III," available from the author at Princeton University.

Andrienko, A. V., V. I. Ozhogin, L. V. Podd'Yakov, and A. Yu. Yakubovskii, (1988). "Surface impedance measurements of energy gap in high-T_c superconductors," Proc. of Int. Conf. on High Temperature Superconductors, Interlaken, Switzerland, *Physica C*, Vol. 153-155, Superconductivity, June, pp. 665-666.

Batlogg, B., A. P. Ramirez, R. J. Cava, R. B. van Dover, and E. A. Rietman, (1987a). "Electronic properties of $La_{2-x}Sr_xCuO_4$ high-T_c superconductors," *Phys. Rev. B*, Vol. 35, No. 10, April 1, pp. 5340–5342.

Batlogg, B., G. Kourouklis, W. Weber, R. J. Cava, A. Jayaraman, A. E. White, K. T. Short, L. W. Rupp, and E. A. Reitman, (1987b). "Nonzero isotope effect in $La_{1.85}Sr_{0.15}CuO_4$," *Phys. Rev. Lett.*, Vol. 59, No. 8, August 24, pp. 912–914.

Batlogg, B., R. J. Cava, A. Jayaraman, R. B. van Dover, G. A. Kourouklis, S. Sunshine, D. W. Murphy, L. W. Rupp, H. S. Chen, A. White, K. T. Short, A. M. Mujsce, and E. A. Rietman, (1987c). "Isotope effect in the high-T_c superconductors $Ba_2YCu_3O_7$ and $Ba_2EuCu_3O_7$," *Phys. Rev. Lett.*, Vol. 58, No. 22, June 1, pp. 2333–2336.

Bednorz, J. G., and K. A. Müller, (1986). "Possible High-T_c Superconductivity in the Ba-La-Cu-O System," *Z. Phys. B—Condensed Matter 64*, pp. 189–193.

Bednorz, J. G., and K. A. Müller, (1988). "Perovskite-type oxides—The new approach to high-T_c superconductivity," *Rev. Mod. Phys.*, Vol. 60, No. 3, July, pp. 585–600.

Bhargava, R. N., S. P. Herko, and W. N. Osborne, (1987). "Improved High-T_c Superconductors," *Phys. Rev. Lett.*, Vol. 59, No. 13, September 28, pp. 1468–1471.

Bourne, L. C., M. F. Crommie, A. Zettl, H.-C. zur Loye, S. W. Keller, K. L. Leary, A. M. Stacy, K. J. Chang, M. L. Cohen, and D. E. Morris, (1987). "Search for isotope effect in superconducting Y-Ba-Cu-O," *Phys. Rev. Lett.*, Vol. 58, No. 22, June 1, pp. 2337–2339.

Braginski, A. I., M. G. Forrester, J. Talvacchio, and G. R. Wagner, (1988). "Prospects for thin-film electronic devices of high-T_c superconductors," Proc. 5th Int. Workshop on Future Electron Devices—High Temperature Superconducting Devices, June 2–4, pp. 171–179.

Cambridge Report on Superconductivity, (1988). "Is this It? SC Now Reported by Japanese Group at 320 K in Niobium Compound," Vol. I, No. 5, February.

Caplin, D., (1987). "The electrons are paired," *Nature*, Vol. 326, No. 6116, April 6, pp. 827–828.

Cava, R. J., B. Batlogg, C. H. Chen, E. A. Rietman, S. M. Zahurak, and D. Werder, (1987). "Single-phase 60-K bulk superconductor in annealed $Ba_2YCu_3O_{7-\delta}$ $(0.3 < \delta < 0.4)$ with correlated oxygen vacancies in the Cu-O chains," *Phys. Rev. B*, Vol. 36, No. 10, October 1, pp. 5719–5722.

Cava, R. J., B. Batlogg, J. J. Krajewski, R. Farrow, L. W. Rupp, Jr., A. E. White, K. Short, and T. Kometani, (1988). "Superconductivity near 30 K without copper: $Ba_{0.6}K_{0.4}BiO_3$ perovskite," *Nature*, Vol. 332, No. 6167, April 28, pp. 814–816.

Chen, G., and W. A. Goddard III, (1988). "The magnon pairing mechanism of superconductivity in cuprate ceramics," *Science*, Vol. 239, No. 4842, February 19, pp. 899–902.

Chrisey, D. B., G. P. Summers, W. G. Maisch, E. A. Burke, W. T. Elam, H. Herman, J. P. Kirkland, and R. A. Neiser, (1988). "Catastrophic loss of superconductivity in ion-irradiated films of $YBa_2Cu_3O_{y-\delta}$," *Appl. Phys. Lett.*, Vol. 53, No. 11, September 12, pp. 1001–1003.

Clark, G. J., A. D. Marwick, R. H. Koch, and R. B. Laibowitz, (1987). "Effects of radiation damage on ion-implanted thin films of metal-oxide superconductors," *Appl. Phys. Lett.*, Vol. 51, No. 2, July 13, pp. 139–141.

Clarke, J., (1988). "Small-scale analog applications of high-transition-temperature superconductors," *Nature*, Vol. 333, No. 6168, May 5, pp. 29–35.

David, King of Israel, Psalm 65, Vs. 9, *The Holy Bible*, New King James Version, Thomas Nelson, Nashville, 1982.

Deutscher, G., (1988). "Superconducting glass and related properties," Proc. of Int. Conf. on High Temperature Superconductors, Interlaken, Switzerland, *Physica C*, Vol. 153–155, Superconductivity, June, pp. 15–20.

Driessen, A., R. Griessen, N. Koeman, E. Salomons, R. Brouwer, D. G. de Groot, K. Heeck, H. Hemmes, and J. Rector, (1987). "Pressure dependence of the T_c of $YBa_2Cu_3O_7$ up to 170 kbar," *Phys. Rev. B*, Vol. 36, No. 10, October 1, pp. 5602–5605.

Endo, U., S. Koyama, and T. Kawai, (1988). "Preparation of the high-T_c phase of the Bi-Sr-Ca-Cu-O superconductor," *Jap. J. Appl. Phys*, Vol. 27, No. 8, August, pp. L1476–L1479.

Faltens, T. A., W. K. Ham, S. W. Keller, K. J. Leary, J. N. Michaels, A. M. Stacy, H.-C. zur Loye, D. E. Morris, T. W. Barbee III, L. C. Bourne, M. L. Cohen, S. Hoen, and A. Zettl, (1987). "Observation of an oxygen isotope shift in the superconducting transition temperature of $La_{1.85}Sr_{0.15}CuO_4$," *Phys. Rev. Lett.*, Vol. 59, No. 8, August 24, pp. 915–918.

Feigelson, R. S., D. Gazit, D. K. Fork, and T. H. Geballe, (1988). "Superconducting Bi-Ca-Sr-Cu-O fibers grown by the laser-heated pedestal growth method," Science, Vol. 240, No. 4859, June 17, pp. 1642–1645.

Fisk, Z., J. D. Thompson, E. Zirngiebl, J. L. Smith, and S-W. Cheong, (1987). "Superconductivity of Rare Earth-Barium-Copper Oxides," *Solid State Commun.*, Vol. 62, No. 11, pp. 743–744.

Fitzgerald, K., (1988). "Superconductivity: Fact vs. Fancy," *IEEE Spectrum*, Vol. 25, No. 5, May, pp. 33–41.

Foner, S., and T. P. Orlando, (1988). "Superconductors: The Long Road Ahead," *Technology Review*, Vol. 91, No. 2, February-March, pp. 36–47.

Fukutomi, M., J. Machida, Y. Tanaka, T. Asano, T. Yamamoto, and H. Maeda, (1988). "New technique for preparation of BiSrCaCuO thin films with T_c of 100 K and above," *Jap. J. Appl. Phys.*, Vol. 27, No. 8, August, pp. L1484–L1486.

Garwin, L., and P. Campbell, (1987). "New superconductors in perspective," *Nature,* Vol. 330, December 17, pp. 611–614.

Ginley, D. S., E. L. Venturini, J. F. Kwak, R. J. Baughman, M. J. Carr, P. F. Hlava, J. E. Schirber, and B. Morosin, (1988). "A 120 K Bulk Superconductor $(Tl_1Ba_1Ca_1)Cu_2O_x$," to be published in *Physica C*.

Glarum, S. H., J. H. Marshall, and L. F. Schneemeyer, (1988). "Field-dependent microwave absorption in high-T_c superconductors," *Phys. Rev. B.*, Vol. 37, No. 13, May 1, pp. 7491–7495.

Gough, C. E., M. S. Colclough, E. M. Forgan, R. G. Jordan, M. Keene, C. M. Muirhead, A. I. M. Rae, N. Thomas, J. S. Abell, and S. Sutton, (1987). "Flux quantization in a high-T_c superconductor," *Nature*, Vol. 326, No. 6116, April 6, p. 855.

Grader, G. S., E. M. Gyorgy, L. G. Van Uitert, W. H. Gordkiewicz, T. R. Kyle, and M. Eibschutz, (1988). "Persistent currents in Tl-Ba-Ca-Cu-O superconductors," *Appl. Phys. Lett.*, Vol. 53, No. 4, July 25, pp. 319–323.

Grant, P. M., S. S. P. Parkin, V. Y. Lee, E. M. Engler, M. L. Ramirez, J. E. Vazquez, G. Lim, and R. L. Greene, (1987). "Effect for Superconductivity in La_2CuO_4," *Phys. Rev. Lett.*, Vol. 58, No. 23, June 8, pp. 2482–2485.

Gray, K. E., (1988). "Tunneling in high-temperature superconductors," preprint: invited review, *Mod. Phys. Lett. B*, World Scientific Publishing.

Greaves, C., (1988). "Infinite stacks of copper oxide," *Nature*, Vol. 334, No. 6179, July 21, pp. 193–194.

Griessen, R., (1987). "Pressure dependence of high-T_c superconductors," *Phys. Rev. B*, Vol. 36, No. 10, October 1, pp. 5284–5290.

Guo, Y., J-M Langlois, and W. A. Goddard III, (1988). "Electronic structure and valence-bond band structure of cuprate superconducting materials," *Science*, Vol. 239, No. 4842, February 19, pp. 896–899.

Hakuraku, Y., F. Sumiyoshi, and T. Ogushi, (1988). "Role of added fluorine to enhance the electromagnetic properties of superconducting Y-Ba-Cu-O compounds," *Appl. Phys. Lett.*, Vol. 52, No. 18, May 2, pp. 1528–1530.

Hashimoto, T., K. Fueki, A. Kishi, T. Azumi, and H. Koinuma, "Thermal Expansion Coefficients of High-T_c Superconductors," *Jap. J. Appl. Phys.*, Vol. 27, No. 2, February, pp. L214–L216.

Hikita, M., Y. Tajima, A. Katslui, Y. Hidaka, T. Iwata, and S. Tsurumi, (1987). "Electrical properties of high-T_c superconducting single-crystal $Eu_1Ba_2Cu_3O_y$," *Phys. Rev. B*, Vol. 36, No. 13, November 1, pp. 7199–7202.

Hinks, D. G., B. Dabrowski, J. D. Jorgensen, A. W. Mitchell, D. R. Richards, S. Pei, and D. Shi, (1988). "Synthesis, structure and superconductivity in the $Ba_{1-x}K_xBiO_{3-y}$ system," *Nature*, Vol. 333, No. 6176, June 30, pp. 836–838.

Hor, P. H., L. Gao, R. L. Meng, Z. J. Huang, Y. Q. Wang, K. Forster, J. Vassilious, C. W. Chu, M. K. Wu, J. R. Ashburn, and C. J. Torng, (1987). "High-Pressure Study of the New Y-Ba-Cu-O Superconducting Compound System," *Phys. Rev. Lett.*, Vol. 58, No. 9, March 2, pp. 911–912.

Huang, C. Y., L. J. Dries, P. H. Hor, R. L. Meng, C. W. Chu, and R. B. Frankel, (1987). "Observation of possible superconductivity at 230 K," *Nature*, Vol. 328, No. 6129, July 30, pp. 403–404.

Hundley, M. F., A. Zettl, A. Stacy, and M. L. Cohen, (1987). "Transport properties of the superconducting oxide $La_{1.85}Sr_{0.15}CuO_4$," *Phys. Rev. B*, Vol. 35, No. 16, June 1, pp. 8800–8803.

Jin, S., T. H. Tiefel, R. C. Sherwood, R. B. van Dover, M. E. Davis, G. W. Kammlott, and R. A. Fastnacht, (1988). "Melt-textured growth of polycrystalline $YBa_2Cu_3O_{7-\delta}$ with high transport J_c at 77 K," *Phys. Rev. B*, Vol. 37, No. 13, May 1, pp. 7850–7853.

Kaneko, T. H. Yoshida, S. Abe, H. Morita, K. Noto, and H. Fujimori, (1987). "Pressure Dependence of High-T_c Superconducting Oxides," *Jap. J. Appl. Phys.*, Vol. 26, No. 8, August, pp. L1374–L1376.

Kapitulnik, A., (1988). "Properties of films of high-T_c perovskite superconductors," Proc. of Int. Conf. on High Temperature Superconductors, Interlaken, Switzerland, *Physica C*, Vol. 153–155, Superconductivity, June, pp. 520–526.

Karpinski, J., E. Kaldis, E. Jilek, S. Rusiecki, and B. Bucher, (1988a). "Bulk synthesis of the 81 K superconductor $YBa_2Cu_4O_8$ at high oxygen pressure," *Nature*, Vol. 336, December 15, pp. 660–662.

Karpinski, J., E. Kaldis, S. Rusiecki, E. Jilek, P. Fischer, P. Bordet, C. Chaillout, J. Chenavas, J. L. Hodeau, M. Marezio, (1988b). "Two new bulk superconducting phases in the Y-Ba-Cu-O system: $YBa_2Cu_{3.5}O_{7+x}$:(T_c = 40 K) and $YBa_2Cu_4O_{8+x}$:(T_c = 80 K)," Proc. of E-MRS, (& *J. Less-Common Met.*), Strasbourg, November 8–10.

Karpinski, J., E. Kaldis, S. Rusiecki, (1988). "High pressure phase diagrams (1-3000 bar O_2) of the (Y-Ba-Cu-O)-O_2 systems," Proc. of E-MRS, (& *J. Less-Common Met.*), Strasbourg, November 8–10.

Kwok, W. K., G. W. Crabtree, D. G. Hinks, D. W. Capone, J. D. Jorgensen, and K. Zhang, (1987). "Normal and superconducting-state properties of $La_{1.85}Sr_{0.15}CuO_4$," *Phys. Rev. B*, Vol. 35, No. 10, April 1, pp. 5343–5346.

Laegreid, T., K. Fossheim, E. Sandvold, and S. Lulsrud, (1987). "Specific heat anomaly at 220 K connected with superconductivity at 90 K in ceramic $YBa_2Cu_3O_{7-\delta}$," *Nature*, Vol. 330, December 17, pp. 637–638.

Leary, K. J., H.-C. zur Loye, S. W. Keller, T. A. Faltens, W. K. Ham, J. N. Michaels, and A. M. Stacy, (1987). "Observation of an oxygen isotope effect in $YBa_2Cu_3O_7$," *Phys. Rev. Lett.*, Vol. 59, No. 11, September 14, pp. 1236–1239.

Liu, R. S., P. T. Wu, J. M. Liang, and L. J. Chen, (1988). "Zero resistance up to 162 K in a multiphase Tl-Ca-Ba-Cu-O system," preprint provided to author by P. T. Wu, July.

Maeda, H., Y. Tanaka, M. Fukutomi, and T. Asano, (1988). "A New high-T_c Oxide Superconductor Without a Rare Earth Element," *Jap. J. Appl. Phys.*, Vol. 27, No. 2, February, pp. L209–L210.

Mai, Z., L. Chen, X. Chu, D. Dai, Y. Ni, Y. Huang, Z. Xiao, P. Ge, and Z. Zhao, (1988). "The Microregion Compositional Variation in $Y_1Ba_2Cu_3O_{7-x}$ Material," *Phys. Lett. A*, Vol. 127, No. 5, February 29, pp. 297–299.

Malik, M. K., V. D. Nair, A. R. Biswas, R. V. Raghavan, P. Chaddah, P. K. Mishra, G. Ravi Kumar, and B. A. Dasannacharya, (1988). "Texture formation and enhanced critical currents in $YBa_2Cu_3O_7$," *Appl. Phys. Lett.*, Vol. 52, No. 18, May 2, pp. 1525–1527.

Matsushita, T., (1988). "Strong flux pinning by cracks in films of superconducting oxide," *Jap. J. Appl. Phys.*, Vol. 27, No. 9, September, pp. L1712–L1714.

Matthias, Bernd T., (1973). "High T_c's—The Low and High of It," in *The Science and Technology of Superconductivity*, W. D. Gregory, W. N. Mathews, Jr., and E. A. Edelsack, eds., Vol. 1, Plenum Press, New York, pp. 263–288 (emphasis added).

Mueller, F. M., S. P. Chen, M. L. Prueitt, J. F. Smith, J. L. Smith, and D. Wohllenben, (1987). "Coherent Twin Boundaries in 1-2-3 Superconducting Oxides," submitted to *Phys. Rev. Lett.*, November 12.

Müller, K. A., and J. G. Bednorz, (1987). "The Discovery of a Class of High-Temperature Superconductors," Science, Vol. 237, September 4, pp. 1133–1139.

Nagashima, T., K. Watanabe, H. Saito, and Y. Fukai, (1988). "Superconductivity in $Tl_{1.5}SrCaCu_2O_x$," *Jap. J. Appl. Phys.*, Vol. 27, No. 6, June, pp. L1077–L1079.

Nakao, K., N. Miura, K. Tatsuhara, S-I Uchida, H. Takagi, T. Wada, and S. Tanaka, (1988). "Superconductivity in $YBa_2Cu_3O_{7-x}$ in a 100 tesla magnetic field," *Nature*, Vol. 332, No. 6167, April 28, pp. 816–818.

Ni, B., T. Munakata, T. Matsushita, M. Iwakuma, K. Funaki, M. Takeo, and K. Yamafuji, (1988). "AC inductive measurements of intergrain and intragrain currents in high-T_c superconductors," *Jap. J. Appl. Phys.*, Vol. 27, No. 9, September, pp. 177–181.

The *Nikkan Koygo Shimbun*, March 16, 1988, p. 5.

Nikonov, A. A., G. V. Sotnikov, A. S. Tokarev, I. V. Kurchatov, (1989). "Influence of radiation defects on superconductivity in $YBa_2Cu_3O_{7-\delta}$ films," *IEEE Trans. Magn.*, Vol. 25, No. 2, March 1, pp. 2579–2582.

Ogale, S. B., D. Dijkkamp, T. Venkatesan, X. D. Wu, A. Inam, (1987). "Current transport in high-T_c polycrystalline films of Y-Ba-Cu-O," *Phys. Rev. B*, Vol. 36, No. 13, November 1, pp. 7210-7213.

Ogushi, T., G. N. Suresha, Y. Honjo, Y. Ozono, Y. I. Kawano, and T. Numata, (1987). "Possibility of Superconductivity with High T_c in La-Sr-Nb-O System," *J. Low Temp. Phys.*, Vol. 69, Nos. 5/6, pp. 451–457.

Oh, B., K. Char, A. D. Kent, M. Naito, M. R. Beasley, T. H. Geballe, R. H. Hammond, A. Kapitulnik, and J. M. Graybeal, (1988). "Upper critical field, fluctuation conductivity, and dimensionality of $YBa_2Cu_3O_{7-x}$," *Phys. Rev. B*, Vol. 37, No. 13, May 1, pp. 7861–7864.

Orlando, T. P., K. A. Delin, S. Foner, E. J. McNiff, Jr., J. M. Tarascon, L. H. Green, W. R. McKinnon, and G. W. Hull, (1987). "Upper critical fields of high-T_c superconducting $La_{2-x}Sr_xCuO_{4-y}$: Possibility of 140 tesla," *Phys. Rev. B*, Vol. 35, No. 10, April 1, pp. 5347–5349.

Parkin, S. S. P., V. Y. Lee, E. M. Engler, A. I. Nazzal, T. C. Huang, G. Gorman, R. Savoy, and R. Beyers, (1988). "Bulk Superconductivity at 125 K in $Tl_2Ca_2Ba_2Cu_3O_x$," submitted to *Phys. Rev. Lett.*, March 11.

Parmigiani, F., G. Chiarello, N. Ripamonti, H. Goretzki, and U. Roll, (1987). "Observation of carboxylic groups in the lattice of sintered $Ba_2YCu_3O_{7-y}$ high-T_c superconductors," *Phys. Rev. B*, Vol. 36, No. 13, November 1, pp. 7148–7150.

Peterson, I., (1988). "A New Recipe for Superconductivity," *Science News*, Vol. 133, No. 8, February 20, p. 116

Pool, R., (1988). "A testable theory of superconductivity," *Science*, Vol. 242, No. 4875, October 7, p. 31.

Przyslupski, P., J. Igalson, J. Rauluszkiewicz, and T. Skos'kiewicz, (1987). "High-T_c superconductivity in $La_{1.8}Sr_{0.2}CuO_4$ and $Y_{1.2}Ba_{0.8}CuO_4$ systems," *Phys. Rev. B*, Vol. 36, No. 1, July 1, pp. 743–744.

Rice, T. M., (1988). "Clues from copper-free samples," *Nature*, Vol. 332, No. 6167, April 28, p. 780.

Sera, M., S. Shamoto, and M. Sato, (1988). "Electron tunneling studies of high-T_c superconductor $YBa_2Cu_3O_{7-\delta}$," *Solid State Comm.*, Vol. 65, No. 9, pp. 997–999.

Sheng, Z. Z., W. Kiehl, J. Bennett, A. El Ali, D. Marsh, G. D. Mooney, F. Arammash, J. Smith, D. Viar, and A. M. Hermann, (1988). "New 120 K Tl-Ca-Ba-Cu-O superconductor," *Appl. Phys. Lett.*, Vol. 52, No. 20, May 16, pp. 1738–1740.

Siegrist, T., S. M. Zahurak, D. W. Murphy, and R. S. Roth, (1988). "The parent stucture of the layered high-temperature superconductors," *Nature*, Vol. 334, No. 6179, July 21, 1988, pp. 231-232.

Sleight, A. W., J. L. Gillson, and P. E. Bierstedt, (1975). "High-Temperature Superconductivity in the $BaPb_{1-x}Bi_xO_3$ System," *Solid State Commun.*, Vol. 117, pp. 27–28.

Sridhar, S., and W. L. Kennedy, (1988). "Novel technique to measure the microwave response of high T_c superconductors between 4.2 and 200 K," *Rev. Sci. Instrum.*, Vol. 59, No. 4, April, pp. 531–536.

Sridhar, S., C. A. Shiffman, and H. Hamdeh, (1987). "Electrodynamic response of $Y_1Ba_2Cu_3O_y$ and $La_{1.85}Sr_{0.15}CuO_{4-\delta}$ in the superconducting state," *Phys. Rev. B*, Vol. 36, No. 4, August 1, pp. 2301–2304.

Stritzker, B., W. Zander, F. Dworchak, U. Poppe, K. Fischer, (1987). "Electron irradiation of $YBa_2Cu_3O_7$ at low temperatures," Proc. Mat. Res. Soc. Symp, Vol. 99, High Temperature Superconductors, December 30, pp. 491–495.

Swinbanks, D., (1987). "Japanese poised to dominate in superconductors as well?," *Nature* (News section), Vol. 327, No. 6121, June 4. pp. 356–357

Takayama-Muromachi, E., Y. Uchida, A. Ono, F. Izumi, M. Onoda, Y. Matsui, K. Kosuda, S. Takekawa, and K. Kato, (1988). "Identification of the Superconducting Phase in the Bi-Ca-Sr-Cu-O System," (submitted to *Jap. J. Appl. Phys.*).

Tarascon, J. M., Y. Le Page, P. Barboux, B. G. Bagley, L. H. Greene, W. R. McKinnon, G. W. Hull, M. Giroud, and D. M. Hwang, (1988). "Crystal Substructure and Physical Properties of the Superconducting Phase $Bi_4(Sr,Ca)_6Cu_4O_{16+x}$" (submitted to *Phys. Rev.*).

Tateno, J., and N. Masaki, "Measurement of RF Superconductivity in $YBa_2Cu_3O_{7-x}$ with the Standing-Wave Method at 9 GHz," *Jap. J. of Appl. Phys.*, Vol. 26, No. 10, October 1987, pp. L1654-L1656.

Tinkham, M., (1988). "Critical currents in high T_c superconductors," *Helv. Phys. Acta*, Vol. 61, pp. 443–446.

van Bentum, P. J. M., H. F. C. Hoevers, H. Van Kempen, L. E. C. Van De Leemput, M. J. M. F. De Nivelle, L. W. M. Schreurs, R. T. M. Smokers, and P. A. A. Teunissen, "Determination of the energy gap in $YBa_2Cu_3O_{7-\delta}$ by tunneling, far infrared reflection and Adreeev Reflection," Proc. of Int. Conf. on High-Temperature Superconductors, Interlaken, Switzerland, *Physica C*, Vol. 153–155, Superconductivity, June, pp. 1718–1723.

Van Duzer, T. and C. W. Turner, (1981). *Principles of Superconductive Devices and Circuits*, Elsevier, New York.

van Dover, R. B., L. F. Schneemeyer, E. M. Gyorgy, and J. V. Waszczak, (1988). "Critical current densities in single crystal $Bi_{2.2}Sr_2Ca_{0.8}Cu_2O_{8+\delta}$," submitted to *Appl. Phys. Lett.*, April 1.

Verhoeven, J. D., A. J. Bevolo, R. W. McCallum, E. D. Gibson, and M. A. Noack, (1988). "Auger study of grain boundaries in large-grained $YBa_2Cu_3O_x$," *Appl. Phys. Lett.*, Vol. 52, No. 9, February 29, pp. 745–747

Waldrop, M. M., (1988). "New superconductors come through," *Science*, Vol. 240, No. 4859, June 17, pp. 1613–1615.

Willis, J. O., J. R. Cost, R. D. Brown, J. D. Thompson, and D. E. Peterson, (1988). "Radiation damage in $YBa_2Cu_3O_7$ by fast neutrons," Proc. 1988 Symposium, Vol. 99, Materials Research Society, pp. 391–394.

Wu, M. K., J. R. Ashburn, C. J. Torng, P. H. Hor, R. L. Meng, L. Gao, Z. J. Huang, Y. Q. Wang, and C. W. Chu, (1987). "Superconductivity at 93 K in a New Mixed-Phase

Y-Ba-Cu-O Compound System at Ambient Pressure," *Phys. Rev. Lett.*, Vol. 58, No. 9, March 2, pp. 908–910.

Ye, Hong-juan, Wei Lu, Zhi-yi Yu, and Hue-chu Shen, (1987). "Superconducting energy gap of $YBa_2Cu_3O_{9-\delta}$ measured by far-infrared reflection," *Phys. Rev. B*, Vol. 36, No. 16, December 1, pp. 8802–8803.

zur Loye, H.-C., K. J. Leary, S. W. Keller, W. K. Ham, T. A. Faltens, J. N. Michaels, A. M. Stacy, (1987). "Oxygen Isotope Effect in High-Temperature Oxide Superconductors," *Science*, Vol. 238, December 11, pp. 1558–1560.

SUGGESTIONS FOR FURTHER READING, BY CATEGORY:

Review Articles

Chu, C. W., P. H. Hor, R. L. Meng, Z. J. Huang, L. Gao, Y. Y. Xue, Y. Y. Sun, Y. Q. Wang, and J. Bechtold, "New materials and high temperature superconductivity," Proc. of Int. Conf. on High Temperature Superconductors, Interlaken, Switzerland, *Physica C*, Vol. 153–155, Superconductivity, June 1988, pp. 1138–1143.

Geballe, T. H., and J. K. Hulm, "Superconductivity - The state that came in from the cold," *Science*, Vol. 239, No. 4838, January 22, 1988, pp. 367–375.

Khurana, A., "The T_c to Beat is 125 K," *Physics Today*, Vol. 41, No. 4, April 1988, pp. 21–25.

Little, W. A., "Superconductivity at room temperature," *Scientific American*, Vol. 212, No. 2, February 1965, pp. 21-27.

Grain Boundary, Contamination Problems

Braginski, A. I., "Material constraints on electronic applications of oxide superconductors," Proc. of Int. Conf. on High Temperature Superconductors, Interlaken, Switzerland, *Physica C*, Vol. 153–155, Superconductivity, June 1988, pp. 1598–1603.

Flükiger, R., T. Müller, W. Goldacker, T. Wolf, E. Seibt, I. Apfelstedt, H. Küpfer, W. Schauer, "Metallurgy and critical currents in $YBa_2Cu_3O_7$ wires," Proc. of Int. Conf. on High Temperature Superconductors, Interlaken, Switzerland, *Physica C*, Vol. 153–155, Superconductivity, June 1988, pp. 1574–1579.

Kwak, J. F., E. L. Venturini, D. S. Ginley, and W. Fu, "Grain decoupling at low magnetic fields in ceramic $YBa_2Cu_3O_{7-\delta}$," in *Novel Superconductivity*, S. A. Wolf and V. Z. Kresin, eds., Plenum Publishing, 1987, pp. 983–991.

Kwasnitza, K. V. Plotzner, M. Waldmann, and E. Widmer, "Currents, magnetization and ac-losses of $YBa_2Cu_3O_7$ superconductors in rapidly changing magnetic fields," Proc. of Int. Conf. on High Temperature Superconductors, Interlaken, Switzerland, *Physica C*, Vol. 153–155, Superconductivity, June 1988, pp. 1565–1566.

Larbalastier, D. C., S. E. Babcock, X. Cai, M. Daeumling, D. P. Hampshire, T. F. Kelly, L. A. Lavanier, P. J. Lee, and J. Seuntjens, "Weak links and the poor transport critical currents of the 1-2-3 compounds," Proc. of Int. Conf. on High Temperature

Superconductors, Interlaken, Switzerland, *Physica C*, Vol. 153–155, Superconductivity, June 1988, pp. 1580–1585.

Matsuda, M., A. Kikuchi, T. Maeda, M. Ishii, Y. Iwai, M. Takata, and T. Yamashita, "Observation of $GdBa_2Cu_3O_{7-d}$ ceramic microstructure," *Jap. J. Appl. Phys.*, Vol. 27, No. 4, April 1988, pp. L529–L530.

High-Pressure Effects

Neumeier, J. J., M. B. Maple, and M. S. Torikachvili, "Pressure dependence of the superconducting transition temperature of $(Y_{1-x}Pr_x)Ba_2Cu_3O_{7-\delta}$ Compounds; Evidence for 4f electron hybridization," *Physica C*, Vol. 156, 1988, pp. 574–578.

Isotope Effect, 1-2-3

Hidaka, T., T. Matsui, and Y. Nakagawa, "Ba isotope effect in a $Ba_2YCu_3O_{7-\delta}$ superconductor," *Jap. J. Appl. Phys.*, Vol. 27, No. 4, April 1988, pp. L553–L555.

H_{c_2}, 1-2-3

Clougherty, D. P. and K. H. Johnson, "Thermodynamic critical fields in high-T_c superconductivity," Proc. of Int. Conf. on High Temperature Superconductors, Interlaken, Switzerland, *Physica C*, Vol. 153–155, Superconductivity, June 1988, pp. 699–700.

Microwave Characteristics

Peterson, G. E., R. P. Stawicki, and U. C. Paek, "Radio frequency properties of high-T_c superconductors," Proc. 38th Electronic Components Conference, Los Angeles, May 9–11, 1988, pp. 159–167.

Piel, H. M. Hein, N. Klein, U. Klein, A. Michalke, G. Müeller, and L. Ponto, "Superconducting perovskites in microwave fields," Proc. of Int. Conf. on High Temperature Superconductors, Interlaken, Switzerland, *Physica C*, Vol. 153–155, Superconductivity, June 1988, pp. 1604–1609.

Bismuth-Compound Superconductor

Adachi, S., O. Inoue, and S. Kawashima, "Superconducting properties in a Bi-Sr-Ca-Cu-O system," *Jap. J. Appl. Phys.*, Vol. 27, No. 3, March 1988, pp. L344–L346.

Hidaka, Y., M. Oda, M. Suzuki, Y. Maeda, Y. Enomoto, and T. Murakami, "Large anisotropy of the upper critical magnetic field in single crystal Bi-(Sr,Ca)-Cu-O," *Jap. J. Appl. Phys.*, Vol. 27, No. 4, April 1988, pp. L538–L541.

Ishida, T., and H. Mazaki, "Complex susceptibility of the oxide superconductor $BiSrCaCu_2O_x$," *Jap. J. Appl. Phys.*, Vol. 27, No. 4, April 1988, pp. L531-532.

Komatsu, T., R. Sato, K. Imai, K. Matusita, and T. Yamashita, "High-T_c superconducting glass ceramics based on the Bi-Ca-Sr-Cu-O system," *Jap. J. Appl. Phys.*, Vol. 27, No. 4, April 1988, pp. L550–L552.

Kumakura, H., H. Shimizu, K. Takahashi, K. Togano, and H. Maeda, "Upper critical field of new oxide superconductor Bi-Sr-Ca-Cu-O," *Jap. J. Appl. Phys.*, Vol. 27, No. 4, April 1988, pp. L668–L669.

Maeda, A., T. Yabe, H. Ikuta, Y. Nakayama, T. Wada, S. Okuda, T. Itoh, M. Izumi, K. Uchinokura, S-I Uchida, and S. Tanaka, "Physical properties of an 80 K superconductor: Bi-Sr-Ca-Cu-O ceramics," *Jap. J. Appl. Phys.*, Vol. 27, No. 4, April 1988, pp. L661–L664.

Matsumoto, T., H. Aoki, A. Matsushita, M. Uehara, N. Mori, H. Takahashi, C. Murayama, and H. Maeda, "Effect of magnetic field and high pressure on the superconductivity of the new high-T_c oxide Bi-Sr-Ca-Cu-O," *Jap. J. Appl. Phys.*, Vol. 27, No. 4, April 1988, pp. L600–L602.

Ohta, M., K. Takahashi, and M. Kosuge, "Superconductor with a layered structure in the Bi-Sr-Ca-Cu oxides," *Jap. J. Appl. Phys.*, Vol. 27, No. 4, April 1988, pp. L567–L568.

Sumiyama, A., T. Yoshitomi, H. Endo, J. Tsuchiya, N. Kijima, M. Mizuno, and Y. Oguri, "Superconductivity of $Bi_{0.25-y}Sr_{0.25-y}Ca_{2y}Cu_{0.5}O_x$ (y = 0.1, 0.125, 0.15)," *Jap. J. Appl. Phys.*, Vol. 27, No. 4, April 1988, pp. L542–L544.

Syono, Y., K. Hiraga, N. Kobayashi, M. Kikuchi, K. Kusaba, T. Kajitani, D. Shindo, S. Hosoya, A. Tokiwa, S. Terada, and Y. Muto, "An X-ray diffraction and electron microscopic study of a new high-T_c superconductor based on the Bi-Ca-Sr-Cu-O system," *Jap. J. Appl. Phys.*, Vol. 27, No. 4, April 1988, pp. L569–L572.

Takayama-Muromachi, E., Y. Uchida, Y. Matsui, M. Onoda, and K. Kato, "On the 110 K superconductor in the Bi-Ca-Sr-Cu-O system," *Jap. J. Appl. Phys.*, Vol. 27, No. 4, April 1988, pp. L556–L558.

Uehara, M., Y. Asada, H. Maeda, and K. Ogawa, "Magnetic properties of $BiSrCaCu_2O_x$ superconductors," *Jap. J. Appl. Phys.*, Vol. 27, No. 4, April 1988, pp. L665–667.

Thallium & Bismuth-Compound Superconductors

Feigelson, R. S., D. Gazit, D. K. Fork, and T. H. Geballe, "Superconducting Bi-Ca-Sr-Cu-O fibers grown by the laser-heated pedestal growth method," *Science*, Vol. 240, No. 4859, June 17, 1988, pp. 1642–1645.

Thallium-Compound Superconductor

Itoh, T., H. Uchikawa, and H. Sakata, "Synthesis of Tl-Ba-Ca-Cu-O superconductor and its properties," *Jap. J. Appl. Phys.*, Vol. 27, No. 4, April 1988, pp. L559–560.

Kubo, Y., Y. Shimakawa, T. Manako, T. Satoh, and H. Igarashi, "Superconducting phases with T_c above 100 K in the Tl-Ba-Ca-Cu-O system," *Jap. J. Appl. Phys.*, Vol. 27, No. 4, April 1988, pp. L591–593.

Subramanian, M. A., J. C. Calabrese, C. C. Torardi, J. Gopalakrishman, T. R. Askew, R. B. Flippen, K. J. Morrissey, U. Chowdhry, and A. W. Sleight, "Crystal structure of the high-temperature superconductor $Tl_2Ba_2CaCu_2O_8$," *Nature*, Vol. 332, No. 6163, March 31, 1988, pp. 420–422

Zhou, X. Z., A. H. Morrish, Y. L. Luo, M. Raudsepp, and I. Maartense, "Bulk high-temperature superconductivity in the Tl-Ca-Ba-Cu-O system," *J. Phys. D*, Vol. 21, No. 7, July 1988, pp. 1243–1245.

Copperless Superconductor at 30 K

Pool, R., "Copperless Superconductivity," *Science*, Vol. 240, No. 4859, June 17, 1988, p. 1614.

Radiation Effects

Küpfer, H., W. Wiech, I. Apfelstedt, R. Flükiger, R. Meier-Hirmer, T. Wolf, and H. Scheurer, "Influence of fast neutron irradiation on inter– and intragrain properties of ceramic $YBa_2Cu_3O_7$," *IEEE Trans. Magn.*, Vol. 25, No. 2, March 1989, pp. 2303–2306.

General Interest

Allgeier, C., and J. S. Schilling, "Meissner effect in Y-Ba-Cu-O and La-Sr-Cu-O high-temperature superconductors," *Phys. Rev. B*, Vol. 35, No. 16, June 1, 1987, pp. 8791–8793.

Cava, R. J., R. B. van Dover, B. Batlogg, and E. A. Rietman, "Bulk Superconductivity at 36 K in $La_{1.8}Sr_{0.2}CuO_4$," *Phys. Rev. Lett.*, Vol. 58, No. 4, January 26, 1987, pp. 408–410.

Chu, C. W., P. H. Hor, R. L. Meng, L. Gao, Z. J. Huang, and Y. Q. Wang, "Evidence for Superconductivity above 40 K in the La-Ba-Cu-O System," *Phys. Rev. Lett.*, Vol 58, No. 4, January 26, 1987, pp. 405–407.

Cooper, J. R., B. Alavi, L-W Zhou, W. P. Beyermann, and G. Grüner, "Thermoelectric power of some high-T_c oxides," *Phys. Rev. B*, Vol. 35, No. 16, June 1, 1987, pp. 8794–8796.

Marshall, D. B., R. E. DeWames, P. E. D. Morgan, and J. J. Ratto, "Flux penetration in high-T_c superconductors: Implications for magnetic suspension and shielding," Preprint: submitted to *Phys. Rev. Lett.*

Rothman, S. J., J. L. Routbort, L. J. Nowicki, K. C. Goretta, L. J. Thompson, and J. N. Mundy, "Oxygen diffusion in high-T_c superconductors," Proc. DIMETA-88 (Metals Science Forum), International Conference on Diffusion in Metals and Alloys, Balatonfüred, Hungary, September 5–9, 1988.

Talvacchio, J., "Electrical contact to superconductors," to appear in *IEEE Trans. CHMT* (1989).

van Bentum, P. J. M., H. van Kempen, L. E. C. van de Leemput, J. A. A. J. Perenboom, L. W. M. Schreurs, and P. A. Teunissen, "High-field measurements on the high-T_c superconductors $La_{1.85}Sr_{0.15}Cu0_{4-\delta}$ and $YBa_2Cu_3O_{7-\delta}$," *Phys. Rev. B*, Vol. 36, No. 10, October 1, 1987, pp. 5279–5283.

CHAPTER 4

APPLICATIONS OF SUPERCONDUCTORS

> We can even imagine riding on magnetic skis down superconducting slopes . . . many fantastic things would become possible.
>
> (W.A. Little, 1965)

INTRODUCTION

In this chapter, a wide variety of superconductor applications will be discussed. In the interest of providing as much useful information as possible while maintaining the chapter at a reasonable size, certain subjects will be covered in some detail while others will be treated in a more general fashion. In general, low-power applications will receive the more detailed, individual attention while those applications that depend on high-field magnets will, for the most part, be covered in a more general fashion. Even so, a variety of applications of high-field inductors (generators, SMES, MAGLEV, marine propulsion, etc.) will receive special attention. Some other subjects, including superconducting shielding, *IR* detection, and antennas will be dealt with separately.

Rather than deal with the potential high-T_c applications separately, the subject of the possible use of the new ceramics will be integrated into each discussion of "conventional" applications. Since there is, for the present, a deficiency of proven applications for the high-temperature ceramics, this seemed to be the most sensible approach. One may hope that a future edition will require a chapter devoted entirely to the subject of high-T_c applications.

It is a reasonable generalization to state that applications of superconductors can be considered as falling into two primary realms that are identified as much by physical size as by the level of energy involved.

The first category is that of lower-power electronic applications. This broad area of applications typically involves superconductors that are plated as thin films on dielectric substrates to form Josephson junctions (and variants of this important device). These junctions are used for SQUIDs and a variety of other applications involving both analog and digital circuitry. While the current carried in a typical low-power electronic conductor may be only on the order of a few mA, the current density can be quite high. Typical values of current density and magnetic field environment for several applications are shown in Table 4.1.

The applications of Josephson junctions in computers, detectors, mixers, and voltage standards will be discussed in some detail, as will the earlier development of cryotron switches. Even though the cryotron has not found wide applications of the thin-film Josephson junction, it is used for relatively slow switching applications and is an excellent vehicle for a demonstration of the importance of relative branch inductances in general superconductor switching applications. Far infrared detection using superconducting devices will also be discussed; this is a particularly important subject for radio astronomy as well as for military applications.

In the second category of superconductor applications, one uses bulk superconducting material to form conductors with relatively large cross-sections. These conductors generally support a high current density, and must often

TABLE 4.1 Approximate combinations of current density (J) and magnetic field (H) required for typical applications of superconductors.

Application	H (T)	J (A/cm^2)
Interconnects	0.1	5×10^6
AC Transmission Lines	0.2	10^5–10^6
Power Transformers	0.3^a–3^b	10^5
DC Transmission lines	0.2	2×10^4
SQUIDs	0.1	2×10^2
SQMEs (Energy Storage)	2.5^a–5^b	5×10^5
Motors, Generators	2.5^a–5^{b*}	4×10^4–10^{5b*}
Magnetic Separation	2–5	3×10^4
MAGLEV	5–6	4×10^4
Fusion	10–15	10^5
Fault Current Limiters	> 5	$> 10^5$

*Figure for AC Generators

Compiled from information in:

[a] Wolsky, A. M., R. F. Giese, and E. J. Daniels, "The new superconductors: Prospects for applications," *Scientific American*, Vol. 260, No. 2, February 1989, pp. 60–69

and:

[b] Schlabach, T. D., "High-T_c superconductor wire fabrication and prospects for use in large-scale applications," ASME Ann. Mtng., November 30, 1988, in *Superconductivity Applications and Developments*, MD-Vol. 11, ASME, New York, 1988, pp. 29–34.

operate in high ambient magnetic fields. One may appropriately label some of these applications as "high power," (power transmission, for example) while others (involving various magnetic-field applications) are traditionally referred to as "high field."

Currently, high-field magnets constitute the most widespread application of bulk superconductors. Alloys of niobium and titanium, since they are ductile and therefore relatively easy to form into wires, are the most popular materials for construction of superconducting magnets. At liquid helium temperatures, these conductors can be used to generate fields up to about 9 T. Conventional copper magnets are limited to fields on the order of about 20% of this level (Foner & Orlando, 1988). Alloys of niobium and tin can be used to generate fields up to about 15 T, but this class of materials is relatively brittle and therefore not nearly so appropriate for fabrication of magnet wires as niobium-titanium. Even so, processing methods tailored to Nb_3Sn are yielding excellent magnets.

FIRST APPLICATIONS FOR HIGH-T_c MATERIALS: ELECTRONICS

Most of those who have worked in the field of superconductivity long before the advent of the high-temperature ceramics agree that initial application of the high-T_c materials will likely be in the area of electronics as opposed to high-field or high-power applications. One reason for this opinion is that most of the $YBa_2Cu_3O_x$ bulk ceramics are quite brittle, so that the production of useful wires for high-field applications may take some time to accomplish. Another reason, of course, is the limitation of critical current density in the bulk material.

On the other hand, high-T_c superconducting films, as would be required for the fabrication of a variety of electronic conductors, are already being produced in the laboratory. While early 1-2-3 films were rather unstable, improvements in processing techniques have led to much more satisfactory results. In addition, the thallium and especially the bismuth superconductors are much more stable than 1-2-3 ceramics, and very high critical currents have already been seen in the thallium compounds.

There are problems, however, even in the electronic applications. One of these is that the new high-T_c compounds must generally be processed at rather high temperatures, usually in excess of 900°C for 1-2-3 compounds. These temperatures are not a problem when a ceramic material ($SrTiO_3$ or MgO for example) is used for the substrate. In a wide variety of electronic applications the most advantageous combination would be the combination of superconductors with semiconductors. However, traditional semiconducting devices will not tolerate such high processing temperatures. How, then, does one mix the two technologies? Perhaps the superconducting portion could be processed first, on a ceramic substrate, with the semiconducting device processing performed afterwards.

Aside from the hybrid superconductor-semiconductor combinations considered above, what are the prospects for simply reproducing existing superconductor electronic systems by replacing the original materials (niobium, lead, etc.) with a high-T_c material? While the answer to this question must wait for the attempts to be made, John Clarke points out that, particularly in the case of superconducting detectors, the devices operating at liquid helium temperatures (~4.2 K) are successful, at least in part, because they have lower noise than competing devices operating at higher temperatures (Clarke, 1988). Merely replacing niobium with, for example, $Tl_2Ba_2Ca_2Cu_3O_{10}$ at 77 K, would not automatically lead one to expect improved operation of the detector device simply because of operation at a higher temperature. On the contrary, operation at a higher temperature would, in the absence of other factors, lead to an increase in detector noise levels. While noise is a critical factor in detector operation, the increased noise (induced by increasing temperature from 4.2 to 77 K) would not necessarily be a problem in digital circuitry and might be acceptable in many analog applications.

Consider the Josephson junction, the heart of superconductor computer technology. We refer to the layered junction, that is, superconductor-insulator-superconductor (SIS) configuration, as opposed to the weak link. While a variety of investigators have successfully made weak links with 1-2-3 material, the design of a layered (SIS) junction that operates as a high-quality Josephson device is more challenging than had been expected. (See Figure 4.1 for a comparison of weak-link and Josephson junctions.)

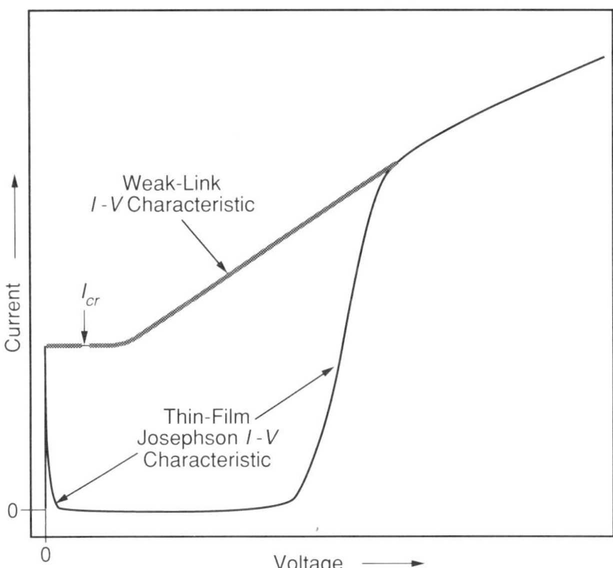

FIGURE 4.1 The current-voltage characteristic is compared for a weak-link and a Josephson junction. Since weak link characteristics are quite variable, many applications that use Josephson junctions cannot operate with weak links.

The problems have been outlined by John Clarke (Clarke, 1988). One problem is due to the high processing temperature for 1-2-3 compounds; the thin insulating barrier does not maintain its integrity at the processing temperatures required for the $YBa_2Cu_3O_x$. The difficulties are compounded by the fact that $YBa_2Cu_3O_x$ does not appear to superconduct on the surface. It remains to be seen whether the newer bismuth- or thallium-based compounds will provide solutions to these problems, but there is reason to be hopeful.

Recall that weak links, that is, regions that can sustain only a very small supercurrent without quenching, may also provide a somewhat degraded Josephson-like behavior. SQUID operations with 1-2-3 compounds to date have typically been realized either with "narrow-necked" weak links or weak links that arise when a sample is cracked. Even this approach is not straightforward, since the effective length of the junction must not exceed the coherence length ξ of the material. While ξ is on the order of 10,000 Å for niobium, ξ is only about 10 Å for $YBa_2Cu_3O_x$ and may be less, depending on the axis of the crystal along which ξ is determined. The reproducible construction of such short "bridges" will be difficult, particularly in mass production.

If these comments seem overly pessimistic about the future of high-T_c materials, it would certainly be premature to be discouraged. High-temperature superconductivity is, and will remain for some time, a very dynamic field of technology. Many, if not all the problems listed here, will probably be solved within one or two decades.

The attraction of these materials has been eloquently described by John Clarke, who uses the example of operation at a remote site: If a helium cryostat was designed to operate for a period of one week without attention, consider the fact that liquid nitrogen in the same cryostat would, due to its much higher latent heat of vaporization, last for approximately one year! In addition, the liquid nitrogen is much cheaper to produce than liquid helium (Clarke, 1988).

Consider the areas where economic advantages are significant. The applications of high-T_c superconductors are bound to increase in nuclear magnetic resonance (NMR) imaging devices (where refrigeration is a significant percent of capital outlay and operating costs), *IR* sensors, and a variety of electronic conductor and hybrid (superconductor/semiconductor) circuits. As values of critical current increase in materials that are easy to fabricate, applications will expand to a variety of other areas. Furthermore, if practical room-temperature superconductors are developed, all bets are off!

SQUIDS AND JOSEPHSON MAGNETOMETERS

Recall, in the earlier discussion of flux quantization, the flux quantum or fluxon (ϕ_o),

$$\phi_o = 2.07 \times 10^{-15} \text{ Wb (or } 2.07 \times 10^{-7} \text{ gauss-cm}^2\text{)} \qquad 4.1$$

The fluxoid $n\Phi_o$, where n is an integer, is the quantized amount of flux that may pass through a hole in a superconducting sheet (or thread a superconducting ring). When a Josephson (or weak-link) junction is placed in series with the current path around the hole containing the flux, the essential portion of a SQUID, that is, superconducting quantum interference device, is formed. The analysis of a superconducting ring with two weak links is analogous to a two-slit optical interference pattern, hence the origin of the term "interference." Even though rings with single weak links are not interference devices in this sense, the name SQUID is commonly applied to these as well (Webb, 1973). A list of SQUID applications follows:

1. Voltmeters (with sensitivities on the order of 10^{-15} V)
2. Sensitive magnetometers for biological applications
3. Magnetometers for geophysical applications
4. Physics instruments (detectors to search for magnetic monopoles and gravity waves)
5. Computer switching and memory applications

Josephson Junctions and Weak Links

The classic Josephson junction is typically formed by two narrow ribbons of superconducting material separated by a very thin insulating layer that is usually a metal oxide formed on one surface (see Figure 4.2). As mentioned earlier, the effective thickness of the SIS junction or "length" of the weak

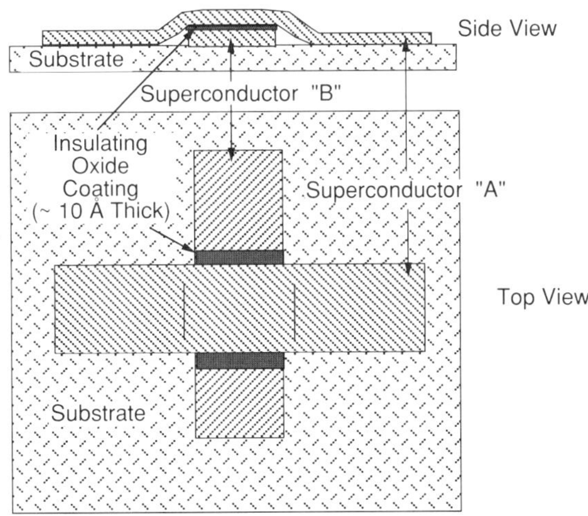

FIGURE 4.2 The classic Josephson junction is constructed by placing a very thin insulating layer (usually an oxide coating on one conductor) between two superconducting films.

link should not exceed the coherence length ξ of the superconducting material used to fabricate the device. This type of "sandwich" junction is currently extremely difficult to form with the high-T_c, copper-oxide superconductors, since the high processing temperature required to achieve appropriate oxygen stoichiometry tends to destroy the insulating oxide layer!

Other techniques are generally used to form SQUID junctions with the high-T_c copper oxides. It is possible to form a point-contact junction using the Y-Ba-Cu-O (Nakane, et al. 1987). DC SQUIDs have also been constructed with polycrystalline Y-Ba-Cu-O thin films; a pair of microbridges form the junctions (Kawabe, 1988) (see Figure 4.3).

The so-called weak links display similar behavior, that is, tunneling of Cooper pairs at zero applied voltage. A common form of weak link consists of a very narrow bridge of superconducting material that provides a conduction path between two larger volumes of superconductor. The bridge may be formed by placing a sharp point on a flat surface, or by forming a very narrow "neck" between two sections of a ribbon-shaped superconductor (see Figure 4.4). Another type of weak link is realized by placing a drop of solder on an oxidized niobium wire; this approach, known as either the solder-drop or the SLUG junction, was developed by John Clarke (Petley, 1971; Webb, 1973).

Regardless of whether a conventional Josephson junction or one of the varieties of weak links is used, the superconducting ring should generally have a relatively small inductance if operation at high frequencies is intended. It is common practice to achieve a self-inductance on the order of 10^{-9} H or less. The weak link is designed to reduce the value of critical current in the ring to a few μA, that is, the value that will force the weak link to quench. The quantized states in the ring will be fixed as previously indicated, since,

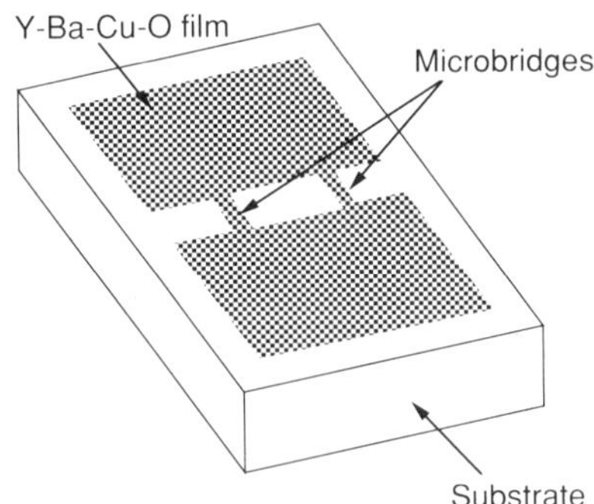

FIGURE 4.3 Techniques for the fabrication of conventional SQUIDs are not always appropriate for construction when using high-temperature, copper-oxide superconductors. While microbridge-type Y-Ba-Cu-O dc SQUIDs have been fabricated by photolithographic techniques, the relatively short coherence length makes precise control of parameters difficult. (After U. Kawabe, "Application of high-temperature superconductors to SQUID and other devices," Proc. of Int. Conf. on High Temperature Superconductors, Interlaken, Switzerland, *Physica C*, Vol. 153–155, Superconductivity, June 1988, pp. 1586–1591.)

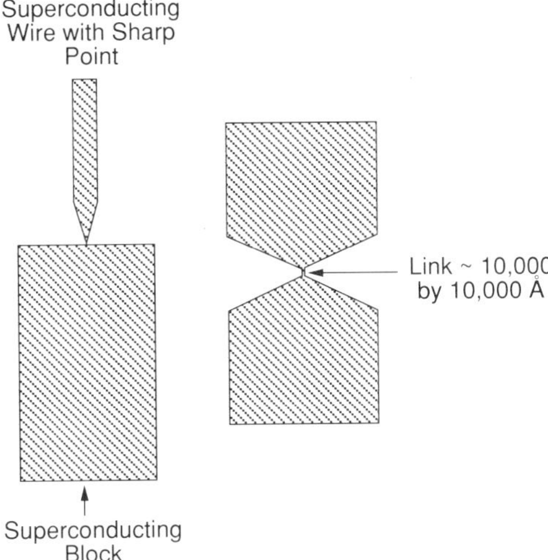

FIGURE 4.4 Weak links constructed by point-contact and microbridge are illustrated above.

with long-range quantum phase coherence, the superconducting state may be described by a single wave function:

$$\Psi = |\Psi| \epsilon^{-j\phi} \qquad 4.2$$

where the phase-angle ϕ changes by $2\pi n$ for a quantized change of $n(h/2e)$ in the flux threading the ring (see Figures 4.5 and 4.6).

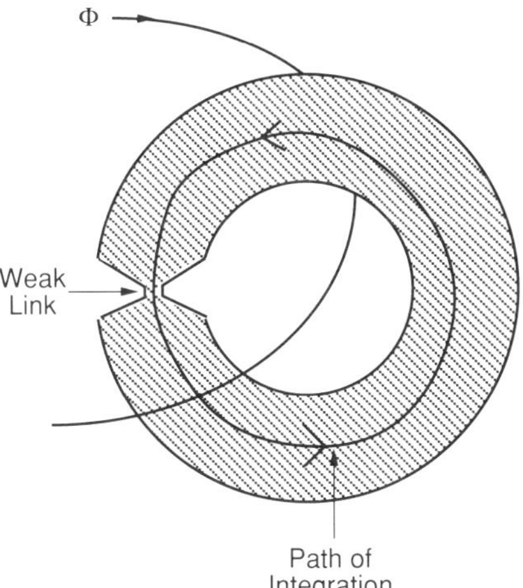

FIGURE 4.5 In a SQUID, such as the weak-link ac device shown above, the quantum wave function must be single-valued along any path that forms a loop around the open (sensing) region.

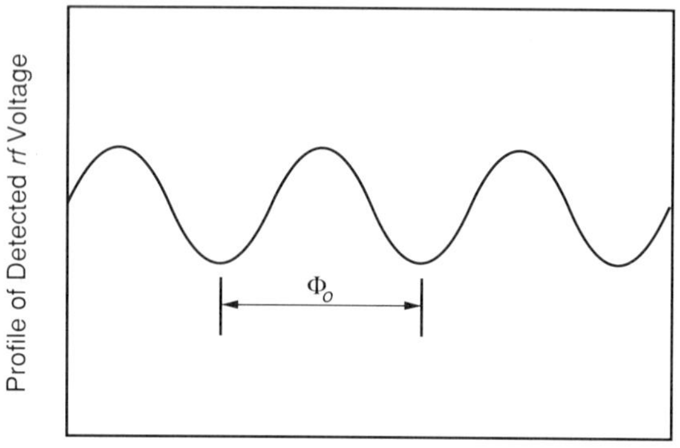

FIGURE 4.6 Phase-locked loops are used to detect minute changes of the magnetic field induced in a superconducting SQUID loop. A complete cycle represents the "fluxoid," $\Phi_o = h/2e$, but it is possible to resolve changes on the order of $\Phi_o/10,000$.

If a ring is interrupted by two weak links (as is the case for dc SQUIDs), the wave functions in each section need not be identical, since the weak links effectively isolate the two bulk regions. The wave-function phases of each section may be coupled via the weak links, however.

The effective length and cross-sectional area of the weak link may be defined as d and a respectively.

Recall that the penetration depth λ was defined earlier as:

$$\lambda = \sqrt{\frac{m}{4\pi n_s e^2}} \qquad 4.3$$

If i_s is the supercurrent flowing in the ring, the fluxoid quantization condition is:

$$4\pi\lambda^2 i_s\left(\frac{d}{a}\right) + \Phi = n\Phi_o \qquad 4.4$$

It is clear from this expression that the flux is quantized exactly as $n\Phi_o$ only in the case where i_s is precisely zero (Webb, 1973).

SQUID Operation

The essential feature of a SQUID device is that it must be biased so that small changes in flux will result in the critical current of the weak link being exceeded (which causes a voltage to be developed across the link). When the weak-link voltage is detected, external feedback circuitry adjusts flux (or current) to reverse the condition. It is the monitoring of the feedback circuitry that provides the useful output; the SQUID device itself serves as a nulling component. SQUID magnetometers are available in a variety of dc and *rf** feedback configurations, and are capable of resolving changes in the measured magnetic field on the order of $10^{-4}\Phi_o$.

DC and AC SQUIDs

Historically, SQUIDs have been divided into "dc" and "ac" devices. The names are somewhat misleading, since either type may measure relatively high-frequency magnetic fields. The primary difference is that the ac SQUID has a single weak link in the superconducting ring and receives its feedback from flux coupled via a resonant tank circuit, typically operating at a few MHz, but capable of operation at several GHz. The primary difference in the dc SQUID is that it receives a direct bias current and therefore requires two weak links so that the bias current is not shunted (see Figure 4.7). See Figure 4.8 for a dc SQUID electronics detector configuration.

DC SQUIDs have been constructed from $YBa_2Cu_3O_{7-\delta}$ using point-contact junctions, essentially a form of the weak link (Nakane, 1987). These devices have been operated successfully up to 70 K. Another group has formed bridges by cutting partial gaps in $YBa_2Cu_3O_7-\delta$; these bridges are typically much longer than the coherence length ($\xi \sim 10$Å) for the copper oxides, but a

* radio frequency

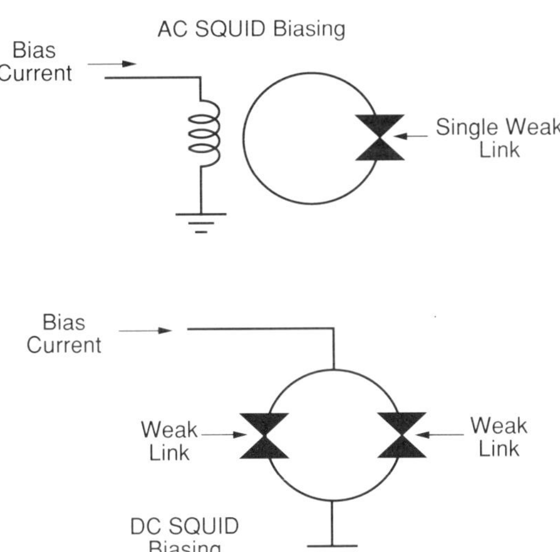

FIGURE 4.7 The terms "dc SQUID" and "ac SQUID" are somewhat misleading, since either device may be used to measure alternating magnetic fields. The terms merely specify whether the bias current is dc or ac coupled, as illustrated above. Note that the dc SQUID must have at least two weak links.

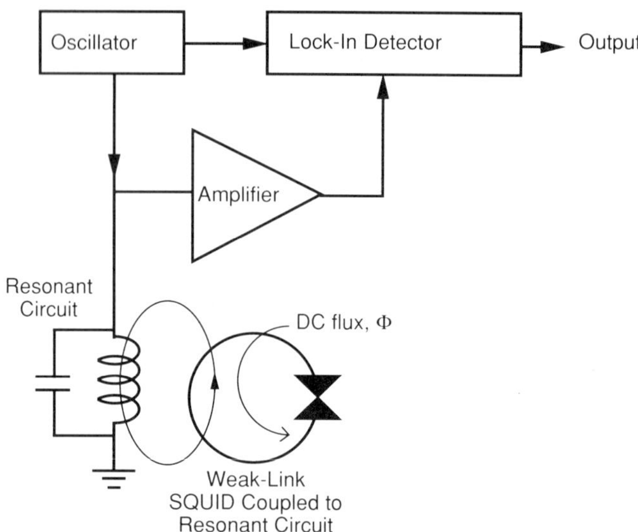

FIGURE 4.8 Simplified block-diagram of electronics used to detect changes in quantum-state ($\Psi = |\Psi|\epsilon^{j\phi}$) phase induced in a SQUID by changes in flux threading the weakly-superconducting ring. The system responds to a change in one flux quantum ($\Phi_o \sim 2 \times 10^{-15}$ Wb) with a change of 2π radians in the quantum wave-function phase (ϕ).

few appear to operate (Wu, et al., 1987). The authors believe that the grain structure of the $YBa_2Cu_3O_{7-\delta}$ may itself be providing a "complicated network of Josephson junctions." Frequency steps of both *nhf/2e* and *mhf/2e* are reported, where *n* is an integer but where *m* is an *even* integer. The presence of the even integers is believed to be related to standing waves (at 36 GHz) in the junction. Such behavior evidently has also been seen both in Pb-Pb and in Nb-Pb junctions.

Gradiometers, Flux Transformers

Aside from direct measurement of *H* fields, characterization of the gradient of such fields may be accomplished by the design of SQUIDs with multiple ports or loops. Such devices, called gradiometers, may be either first-derivative or second-derivative devices depending on the number and arrangement of detection loops. Alternately, separate flux transformers may drive a single SQUID (Falco & Schuller, 1981) (see Figure 4.9).

JOSEPHSON TUNNELING DEVICES AS SWITCHES

Recall that tunneling through insulating barriers with a thickness greater than about 50 Å is limited to quasiparticles (i.e., single electrons). In these

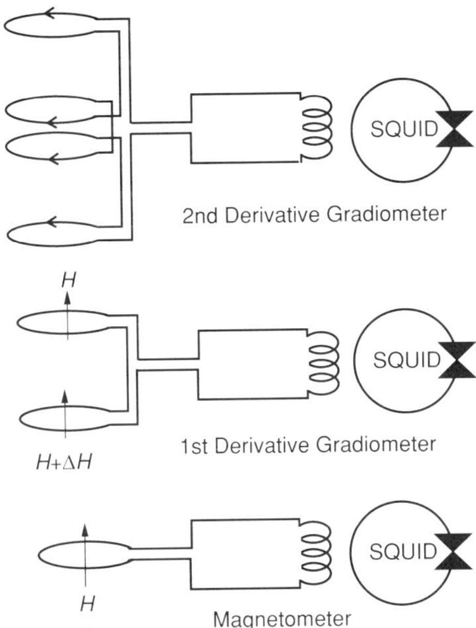

FIGURE 4.9 Use of flux transformer configurations with SQUIDs to achieve either a magnetometer or (first and second derivative) gradiometer. (After C. M. Falco and I. K. Shuller, "SQUIDs and their Sensitivity for Geophysical Apllications," in *Squid Applications to Geophysics*, H. Weinstock and W. C. Overton, Jr., eds., Soc. of Exploratory Geophysicists, Tulsa, 1981, pp. 13-18.)

"Giaever" junctions, the insulator is too thick to allow tunneling of superconducting electron pairs. When the insulating barrier is reduced to 10–20 Å, however, it is possible for paired electrons to tunnel. This is the Josephson effect; the superconducting pairs tunnel through the barrier even with zero potential across the insulating gap. As long as the current is below a critical level (I_{cr}), true zero-resistance tunneling is obtained. The 10–20 Å barrier in the classic Josephson junction has a negative resistance region; current is reduced as the barrier potential is increased. The weak-link or point-contact junctions usually do not exhibit the negative resistance region seen in the film junctions. The thin barrier, of course, also allows quasiparticle tunneling as well as Josephson tunneling, so that a combination of single electron and superconducting pair can be seen.

A thin-film Josephson junction has "off" resistance of only about 1Ω. This will sound rather low to the engineer who is familiar with CMOS switches, but since the "on" state has zero resistance, the resistance *ratio* of the "off" to the "on" state for the Josephson junction is infinite! In addition, this low "off" resistance (when combined with the junction capacitance, ~0.1 to 0.01 pF and a typical load resistance of about 10 Ω) allows the very fast switching times of these junctions (see Figure 4.11). Aside from junction capacitance and external resistance, the inherent delay (τ_d) in switching is limited by

$$\tau_d \geq \frac{\Phi_o}{V_g} = \frac{h}{2eV_g} \qquad 4.5$$

where Φ_o is the familiar flux quantum, $h/2e$.

RATIONALE FOR USING JOSEPHSON TECHNOLOGY IN COMPUTERS

Josephson junctions can serve as switches and therefore can be connected to behave as gates and memory elements for computer applications. The basic principle involved is that the junction voltage operating point can be switched from 0 to about 2–3 mV by a change either in junction current or local magnetic field. In most common configurations, gate current is interrupted to reset the junction to the zero-voltage state.

The motivation for using Josephson junctions in computer circuitry is to achieve very high speed and low dissipation. It is possible to achieve gate switching times under 10 ps on gates with areas on the order of 1.5–2.0 square microns. Switching power dissipation is typically on the order of a few microwatts per gate at maximum speed, about three orders of magnitude lower than that of the best semiconductor logic (Hayakawa, 1986). While liquid helium is not a very efficient coolant, providing about 600 mW/cm^2 heat transfer at 4.2 K, it should be possible to have a density on the order of 10^5 devices on a 5 mm^2 chip.

Josephson chips would have the additional advantage that the designer could employ superconducting ground planes as well as superconducting (stripline) transmission lines. The lines are virtually lossless at the frequencies to be transported, and distortion is undetectable when the lines are properly terminated.

BASIC JOSEPHSON JUNCTION CHARACTERISTIC

See Figure 4.10 for an illustration of the load line and stable states of the basic Josephson junction. The gate may be biased (by a gate current less than the critical current I_{cr}) in the "zero-voltage" state. The junction may be switched to the "non-zero" voltage state by one of two means:

1. Increasing the gate current (I_g) above the critical (I_{cr}) level.
2. Decreasing the I_{cr} level (by increasing the local magnetic field) below the existing gate current level.

Either operation will move the operating point from the zero-voltage level to the nonzero operating point as shown in Figure 4.10. The nonzero voltage is essentially the gap voltage, V_g, which is equal to $2\Delta/e$. Thus, the current steered to the load resistance R_L is approximately $2\Delta/(eR_L)$. Recall that $2\Delta/e$ is on the order of 2–3 mV for niobium junctions. There is a very small inherent delay in the Josephson junction switching, on the order of 1 ps. The voltage rise time on the junction is associated with parasitic junction capacitance and junction resistance, and may be as short as about 10–15 ps.

Once the junction is switched to the "resistive" state, it will remain in that mode until current is removed (or reduced to a very low value). This is typically accomplished by use of an alternating-polarity power supply (as opposed to the dc power supply one uses with semiconductor computer systems) that acts to reset every junction to the zero-voltage mode on each computer cycle (see Figure 4.17).

Generally, the superconductive state corresponds to a logic state of "0," the resistive state to a "1." The nonzero voltage level is essentially the threshold voltage V_g, which will range from about 1 to 3 mV for classical superconductors. Recall that, in an SIS junction utilizing identical supercon-

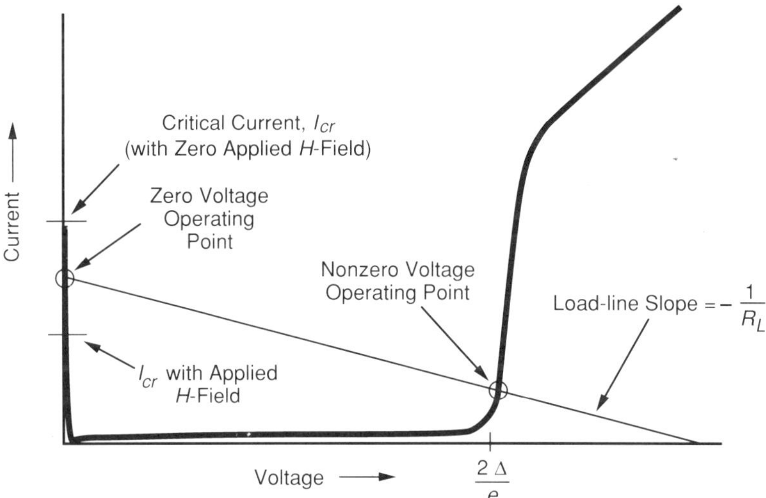

FIGURE 4.10 The basic Josephson junction has two operating points. The zero-voltage point is stable as long as the gate current is less than the critical level. Switching may be accomplished either by increasing the gate current (above the critical level) or by application of a magnetic field to lower the critical level below the existing gate current magnitude. Switching will occur in a few picoseconds to the nonzero voltage level illustrated above. This is the "latching mode"; gate current must be reduced to near zero before the junction will reset to the zero-voltage mode.

ductors on each side of the insulating film, V_g is related to the superconductor energy gap (Δ) by

$$V_g = \frac{2\Delta}{e} \tag{4.6}$$

and that the energy gap is related to the critical temperature (T_c) by

$$2\,\Delta \approx 3.53\,(k_B T_c) \tag{4.7}$$

where k_B is Boltzman's constant. Obviously,

$$V_g \approx \frac{3.53(k_B T_c)}{e} \tag{4.8}$$

so that selection of superconductors with higher T_c's results in larger values of V_g and, subsequently, larger signals developed by the switching event. See Figure 4.12 for the Josephson junction linearized *I-V* characteristic, schematic junction representation, and basic switching circuit.

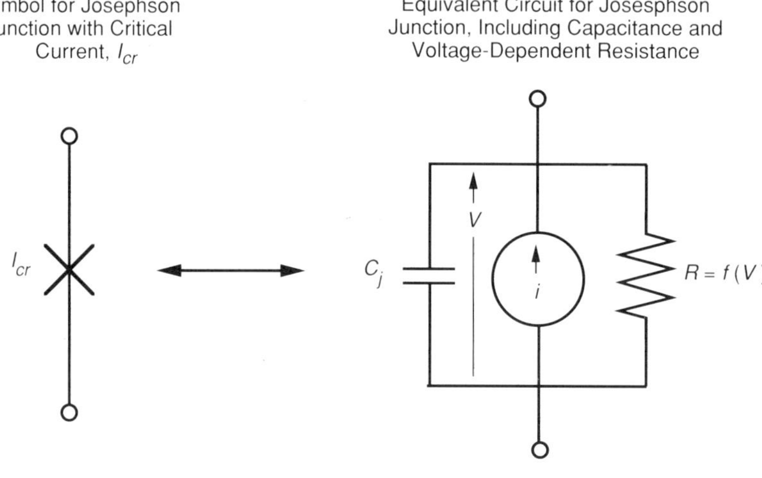

FIGURE 4.11 The symbol on the left is commonly used to represent the Josephson junction in circut diagrams. On the right is the Stewart-McCumber equivalent circut for a Josephson junction. Note the sinusoidal variation of current as a function of junction voltage *V*.

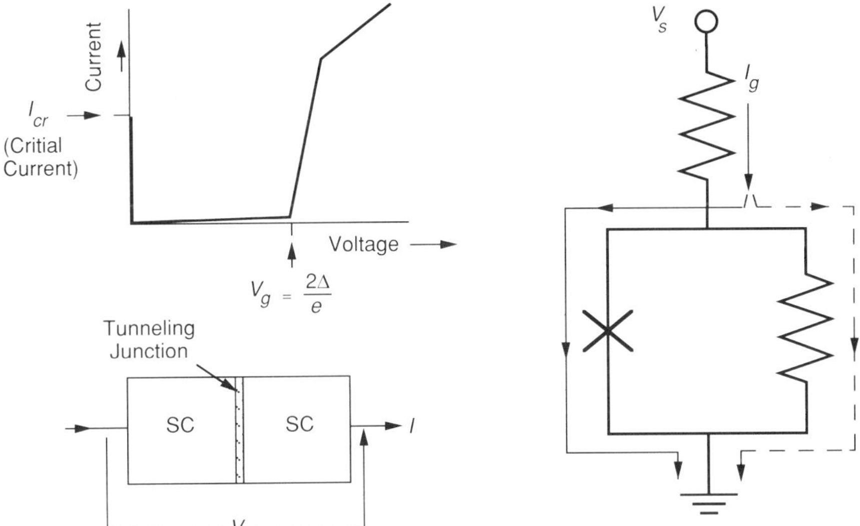

FIGURE 4.12 The Josephson junction is shown on the left along with its linearized *I-V* characteristic. The basic gate configuration, essentially a current-steering switch, is illustrated on the right. In the zero-voltage state, gate current (less than the critical level) flows through the junction (solid arrows). When current is increased above the critical level, the junction switches to the state where a few mV exist across the junction and steers current (dotted arrows) through the load resistor.

LOGIC GATES

Gates may be placed into two general categories by the way they are driven, as illustrated in Figures 4.13 and 4.14:

a. direct-coupled and
b. magnetic (induction) coupled.

In the direct-coupled mode, the junction is switched by the injection of an additional current into the gate. The combination of junction and load resistance can be constructed with very low inductance, resulting in fast switching. The induction coupling uses the junction in a SQUID configuration, where the introduction of a very small magnetic field is produced by a pulsed current in one or more control lines near the junction. A variety of standard AND and OR gates as well as other signal-processing logic circuits have been constructed using these basic techniques. Also see Figures 4.15 and 4.16.

Nonlatching Josephson Logic

Several proposals have been made to achieve nonlatching operation of Josephson logic, including the use of very low impedance transmission lines to a

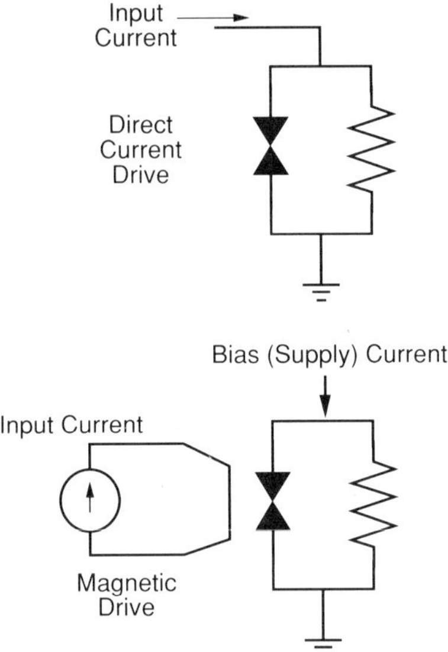

FIGURE 4.13 Two basic methods used to drive a Josephson junction for switching purposes are illustrated. While only single-input versions are shown here, multiple direct-drive inputs or additional magnetic coupling loops can be used. An input drives the device to its "off" mode for maximum voltage across the junction. Power supply current must be removed to reset the junction to its "on" or zero-voltage operating point.

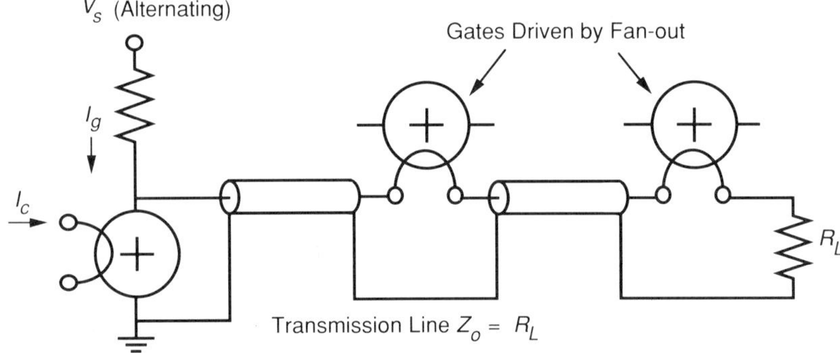

FIGURE 4.14 A superconducting transmission line, matched to a resistive load, serves as a means to connect the output of a single gate to the control lines of several gates. The inherent isolation and serial fan-out capabilities of SQUIDs in logic applications are considered to be significant advantages. Disadvantages include relatively large area per gate and a high sensitivity to stray magnetic fields. (After T. R. Gheewala, "Josephson-Logic Devices and Circuits," *IEEE Trans. Electron Devices*, Vol. ED-27, No. 10, October 1980, pp. 1857–1869.)

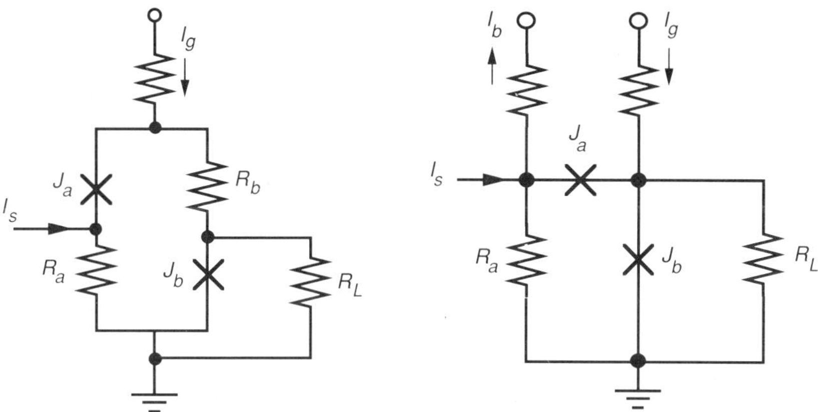

FIGURE 4.15 On the left, the basic "direct-coupled isolation" (DCI) element is illustrated. On the right, the Josephson Atto-Webber switch (JAWS) provides good isolation, but does require an additional bias current to achieve improved gain and noise margin. (After T. R. Gheewala, "Josephson-Logic Devices and Circuits," *IEEE Trans. Electron Devices*, Vol. ED-27, No. 10, October 1980, pp. 1857–1869.)

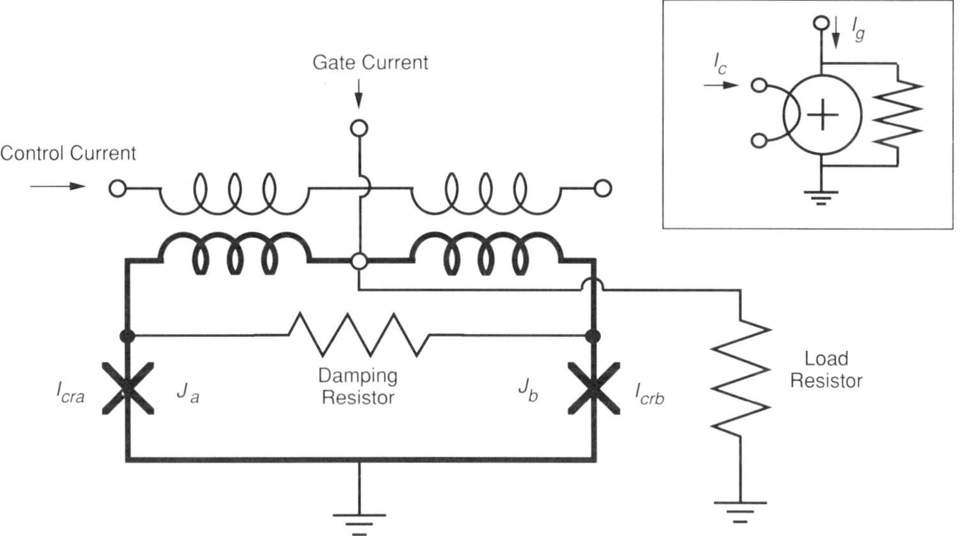

FIGURE 4.16 Input to a two-junction SQUID (heavy lines) may be isolated by means of magnetic coupling. Application of a current pulse to the transformer primary(s) steers the gate current to the load resistor as the Josephson junctions are switched to a nonzero voltage. The symbol for this configuration is shown in the inset.

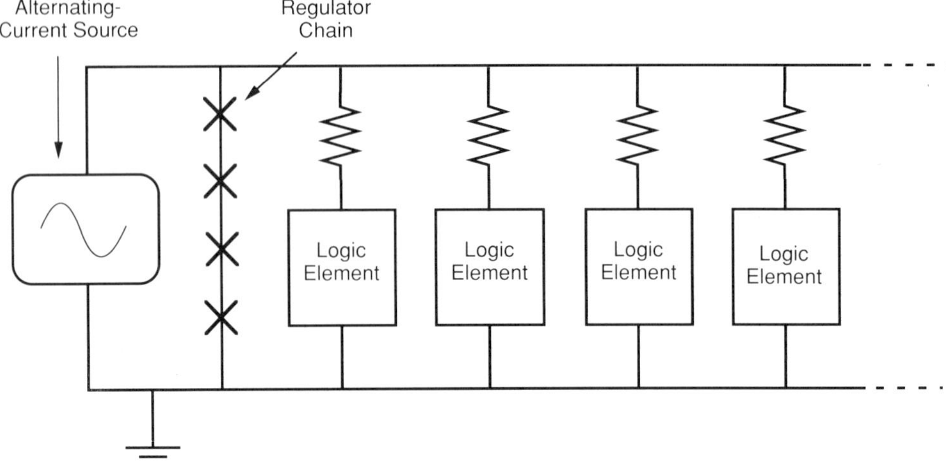

FIGURE 4.17 A major difference in Josephson logic gates and the more conventional semiconductor logic (i.e., TTL, DTL, etc.) is that the commonly used superconducting gates require an alternating current power supply to reset the gates after each operating cycle. In this design, several Josephson junctions are used to shape the power supply current pulse. (After T. Van Duzer, "Josephson Digital Devices and Circuits," *IEEE Trans. Microwave Theory & Tech.*, Vol. MTT-28, No. 5, May 1980, pp. 490–500.)

matched load (generally impractical given the impedance constraints on the lines that can be conveniently used in miniature circuits) or using a mismatch on the transmission line (Van Duzer, 1980). In the latter case, the inductive reflection would serve to reset the gate. A disadvantage with the mismatch-reflection reset is the relatively long delay in resetting the gate to the zero-voltage operating point.

Fabrication of Josephson Junctions; Problems with Lead

Josephson technology is especially attractive, promising a combination of smaller volume and increased speed for computers. As great as the promise is, there have been significant difficulties in getting the technology from the laboratory bench to applications in computers.

A basic problem results from the difference in coefficients of thermal expansion in the superconducting material and the insulating substrates and tunnel barriers. Lead films, for example, stretch inelastically when cooled below T_c; on warming to room temperature the lead film forms mounds that have sufficient height to rupture the oxide tunneling barrier. The problems associated with thermal cycling have been dealt with by the use of lead-gold-indium alloys (Van Duzer, 1980).

Some problems have been related to the use of alloys of lead, gold and indium. The Pb junctions tend to experience an alteration in their tunneling characteristics as well as an unacceptably high rate of initial failures in integrated circuit construction. The first problem is probably due to the diffusion of elements from other chip components into the junction. The failure rate may be due to the fact that the lead is too soft for the construction of highly reliable junctions.

Niobium Nitride

The more recent development of junctions using niobium nitride (T_c ~ 15 K) separated by a thin film of "artificial barriers" (as opposed to natural oxides derived from the superconductor material) have shown promise of solving some of these problems (Hayakawa, 1986). An example is NbN/MgO/NbN, which has a critical gap voltage in excess of 5 mV. In addition, the NbN is able to withstand the processing temperatures of semiconducting materials which are typically a few hundred °C. (The situation is somewhat reversed for $YBa_2Cu_3O_x$, where the *minimum* processing temperature generally required for the 1-2-3 materials is excessive for conventional semiconductor substrates.)

Niobium/Aluminum Oxide

The group headed by Shinya Hasuo at Fujitsu Laboratories has had excellent results in their development of 1.5 μm niobium junctions with aluminum oxide insulators, that is, Nb/AlO/Nb. Switching times of only ~2.5 ps are projected to drop to on the order of 1 ps when the same technology is scaled down to the 0.6 μm level. Characteristics are reliably reproduced, an essential feature for large-scale computer application. Average gate delays, using a three-phase sinusoidal power supply, are only 9.2 ps, approximately an order of magnitude faster than the best semiconductor gates. See Figure 4.18 for their "modified variable threshold logic" (MVTL) OR-gate.

Gates constructed from materials with higher T_c can be expected to increase the power dissipation of the logic gate, perhaps by as much as two orders of magnitude. Any significant increase in dissipation, of course, defeats one of the fundamental advantages of the Josephson junction over semiconductor logic. Whether the advantages of using liquid nitrogen cryogens over liquid helium will offset this disadvantage remains to be seen. Since high-T_c Josephson logic is not yet on the scene, the question is somewhat academic.

COMPUTER MEMORY

In conventional computers, the entire random-access memory is usually of one type. In Josephson superconducting computer systems, memory is typically segregated into two sections. The fastest section is also the smallest (in bytes of storage); this is referred to as cache memory and is used to work directly

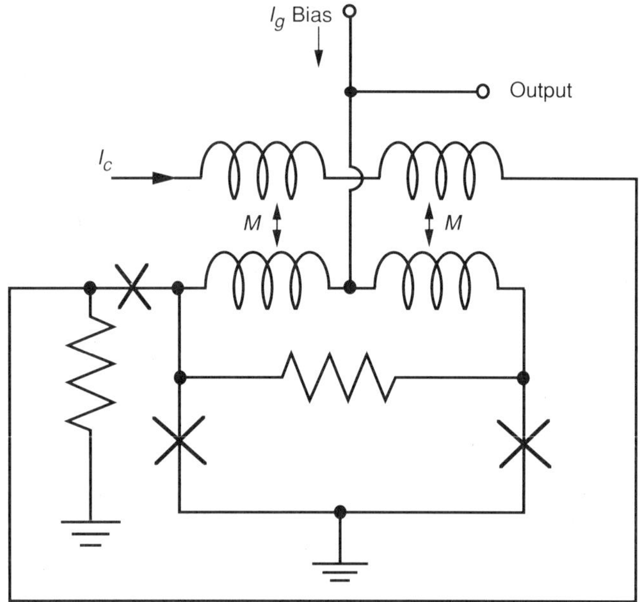

FIGURE 4.18 Fujitsu Laboratory's "modified variable threshold logic" (MVTL) gate shown here, uses 1.5 μm Nb/Al-O/Nb Josephson junctions. Gate delay is on the order of 9 ps, and is evidently quite reproducible. With 0.6 μm technology, junction switching times are expected to be reduced from 2.5 ps to $\sim$ 1 ps. Dissipation is approximately 11 μW per gate. (After Shinya Hasuo, "High-speed Josephson IC technology," presented at Applied Superconductivity Conference, P-1, San Francisco, August 1988.)

with the central processing unit (CPU). Cache memory may consist of 64 K of nondestruct read-only (NDRO) cells, with an access time of 4 ns (Van Duzer, 1980) (see Figure 4.19). With the cache memory present to handle the fast interactions with the CPU, the main random-access memory (RAM), which will usually consist of more than 10 megabytes, may be slower, trading off speed for lower power dissipation. There has been considerable interest in use of a two-junction, single-flux-quantum (SFQ) interferometer with control lines (see Figure 4.20). Such memory cells provide an access time on the order of 10 ns, and consume no power except during read and write cycles. Operation varies with the precise configuration; see the references for details (Van Duzer, 1980; Guéret, et al., 1980).

THE "PARAMETRON"; PARAMETRIC OSCILLATIONS AND DIGITAL APPLICATIONS

It has long been known that parametric oscillations occur in resonant systems when the energy-storing components are driven at twice the resonant frequency

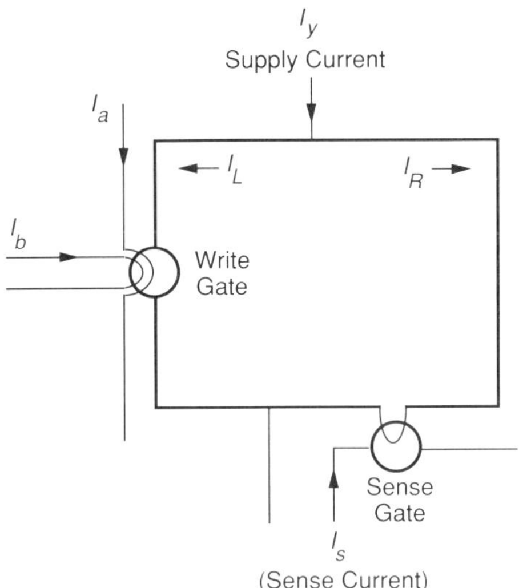

FIGURE 4.19 A random-access, nondestructive read-out (NDRO) memory cell suitable for fast cache applications in a Josephson computer. The major loop is superconducting; a logical "zero" exists when there is no circulating current; a "one" exists when a clockwise persistent current flows. Supply current is split into right and left components, with levels depending on the inductance of each branch. After current is established in the "write" gate, currents I_a and I_b are applied to force the left branch current to the right side, establishing a persistent current in the loop. The cell is read by applying the supply current (I_y) simultaneously with the sense current (I_s). Other NDRO schemes use circulating currents of opposite direction to represent the zero and one logical states. (For details see T. Van Duzer, "Josephson Digital Devices and Circuits," *IEEE Trans. Microwave Theory & Tech.*, Vol. MTT-28, No. 5, May 1980, pp. 490–500. or: W. H. Henkels and H. H. Zappe, "An Experimental 64-Bit Decoded Josephson NDRO Random Access Memory," *IEEE J. Solid-State Circuits*," Vol. SC-13, No. 5, October 1978, pp. 591–600.)

of the system. For electronic applications, the modulated element is either the inductance or the capacitance. The remarkable (and useful) property of such parametric oscillations is that the resultant oscillation has two stable states that differ only by a 180° phase difference. During the 1950s, Eiichi Goto suggested using these two phase states as "digital" states, where one would be designated as a logical "0," the other as a logical "1" (Goto, 1959). In early circuitry, logic functions were realized using either nonlinear inductors or capacitors. The most straightforward circuits used ferrite-core inductors; inductance was varied periodically at twice the resonant frequency ω by driving isolated loops through the inductor toroid at 2ω, so that the inductance L varied as

$$L_{2\omega} = L_o(1 + 2\Gamma \sin[2\omega t]) \qquad 4.9$$

where L_o is the unperturbed inductance and Γ is an amplitude parameter used by Goto. If the resonant circuit is tuned to $f(\omega/2\pi)$, one may consider a current

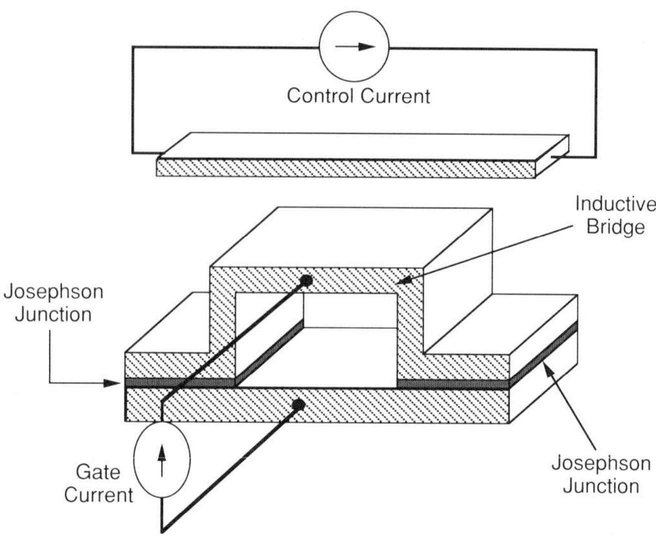

FIGURE 4.20 A single-flux-quantum [SFQ] memory cell is ideal for main memory applications due to low power consumption, even though its speed is somewhat less than memory elements used for cache applications. There are two quite distinct switching transitions, depending on whether a vortex mode is crossed above or below a critical value of the gate current. For details of operation, see the following reference. (After P. Guéret, A. Moser, and P. Wolf, "Investigations for a Josephson Computer Main Memory with Single-Flux-Quantum Cells," *IBM J. Res. Develop.*, Vol. 24, No. 2, March 1980, pp. 155–166.)

I_f, which has sine and cosine components:

$$I_f = I_s \sin(\omega t) + I_c \cos(\omega t) \qquad 4.10$$

If the amplitudes I_s and I_c are small compared to the induced voltages from the inductance variation, the tank voltage V will be approximately

$$L_{2\omega}\frac{dI_f}{dt} = \omega L_o(I_s \sin \omega t + I_c \cos \omega t) + \Gamma\omega L_o(I_c \cos \omega t - I_s \sin \omega t) \qquad 4.11$$

A term involving $3\omega t$ has been ignored in this expansion, since it is rather far off resonance. Important points to note in expression 4.11 include the significance of the negative sign on the $\sin \omega t$ term; this implies negative resistance and results in an exponentially growing term, while the $\cos \omega t$ term tends to decay. The negative resistance amplitude is $(-)\Gamma\omega L_o$. The resultant phase of the final oscillation depends on the sign of the sine component of the small initial oscillations; that is, the initial state of the signal that perturbs the reaction.

Early versions of parametron logic circuits generally utilized ferrite-core inductors as the nonlinear inductive element. Drive signals at 2ω were coupled

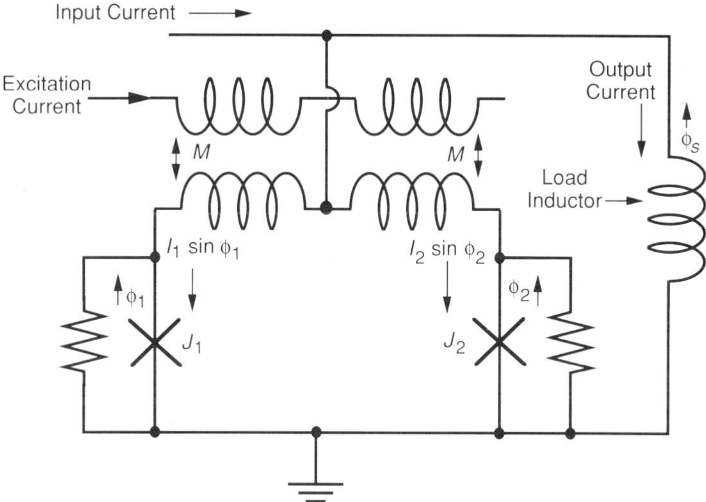

FIGURE 4.21 Schematic representation of a quantum flux parametron [QFP] logic element. Parametric oscillations with opposite phases represent the "1" and "0" logical states. The load inductor is on the order of 5–6 pH; loop inductors are ~ 2 pH. Resistors across the Josephson junctions are provided to damp ringing induced by switching. (After Y. Harada, H. Nakane, N. Miyamoto, U. Kawabe, E. Goto, and T. Soma, "Basic operations of the quantum flux parametron," *IEEE Trans. Magn.*, Vol. MAG-23, No. 5, September 1987, pp. 3801–3807.)

through turns that were separate from the turns for the inductance section. These circuits had the advantage of being passive; gain was available in the superregenerative sense. Even so, parametric logic circuits using ferrite transformers are not noted for speed.

Following the development of practical Josephson junctions in the 1950s, the parametric-oscillation technique was applied to superconducting logic applications (Loe & Goto, 1985). An example is the Quantum Flux Parametron, or QFP, which is illustrated in Figure 4.21 (Harada, et al., 1987).

Essentially, the QFP consists of a superconducting loop containing two Josephson junctions, a pair of loop inductors coupled to an excitation line, and a load inductor. A minute signal current is injected directly into the superconducting loop to be amplified by the excitation current which is coupled through the loop inductors. The amplified output current flows to the load inductor. The output current may flow in either of two directions; these directions indicate the "0" or "1" logical state of the system. Simulation results indicate that the dissipation may be on the order of about 1 nW at a 10 GHz clock rate. This amounts to three orders of magnitude less than the power requirement for a conventional dc SQUID logic element! While simulations predict stable operation to frequencies as high as 10 GHz, tests to date have been restricted to frequencies of a few kHz.

The parametron concept has also been applied (with Josephson junctions) to flash-type A/D conversion (Shimizu, et al., 1988). This device reportedly converts the input analog signal to periodic "digital" signals of a single quantum flux (Φ_o) per cycle. While confirmation of operation at 100 kHz has been proven experimentally, computer simulations of the circuit suggest that operation at frequencies up to 20 GHz is possible.

LOW-ENERGY ANALOG AND *rf* APPLICATIONS

The nonlinearity of a Josephson junction, as with conventional diodes, is a useful characteristic for *rf* signal detection as well as for heterodyne mixing applications, particularly at mm and sub-mm wavelengths.

Photon-Assisted Tunneling

When a Josephson junction is exposed to an *rf* field, its voltage-current characteristic acquires steps. Recall that the normal threshold voltage is $2\Delta/e$; this level is offset toward lower voltages as a function of the applied frequency (f) by voltage steps of hf/e (Tinkham, 1985). This amounts to an absorption of one (or multiple) photons of energy (hf) in the tunneling process, so the effect is referred to as photon-assisted tunneling (Richards, 1986). For this effect to be obvious, the "knee" on the *I-V* curve must be sharp compared to the interval hf/e; observations are typically at frequencies in the range of 30–100 GHz (see Figure 4.22).

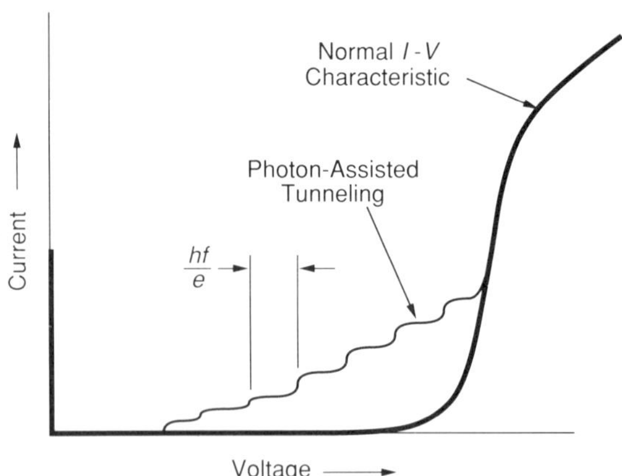

FIGURE 4.22 The heavy curve shows the normal current-voltage characteristic of a Josephson junction. The light curve illustrates the steps induced by the presence of photon-assisted tunneling. Such steps can only be observed in high-quality junctions that exhibit rather sharp transitions. (Photon-assisted tunneling is not to be confused with Shapiro steps, which begin at zero junction voltage and have half the width.)

SIS Mixer Applications

Since 1979, the SIS tunnel-junction mixer has been considered to be virtually the ideal frequency converter for millimeter-band receivers. The SIS mixer requires very little local-oscillator power, has conversion gain and, most importantly, exhibits extremely low shot noise. The SIS junction has been used to design mixer circuits with gain and noise figures that approach the quantum limit, that is, each incident microwave photon produces one tunneling electron. Despite early expectations of wide use as high-sensitivity mm-band mixers, SIS mixers have not yet become as widespread as initially predicted. While about ten radio telescopes currently use SIS mixers, most exhibit sensitivities that are not much better than Schottky-diode receivers.

The problems that arise in practice are largely a result of providing appropriate impedances at signal, harmonic sideband, and image frequencies. Harmonic sidebands may be shorted by using SIS junctions with relatively large capacitance, but this capacitance must be tuned out over the signal and image bands. This is typically accomplished by use of an inductance in parallel with the SIS junction. Since a short-circuit at dc (or the intermediate frequency) across the junction cannot be tolerated, a blocking capacitance must also be included. (See Figure 4.23.)

S. K. Pan and his colleagues have an improved integrated-circuit SIS mixer design that demonstrates that theoretical sensitivity can be approached (Pan, et al., 1988). Significantly, this circuitry, using Nb-Pb technology, can be constructed using reproducible fabrication techniques. In a test receiver for the 85–116 GHz range, this group used their integrated-circuit SIS mixer in

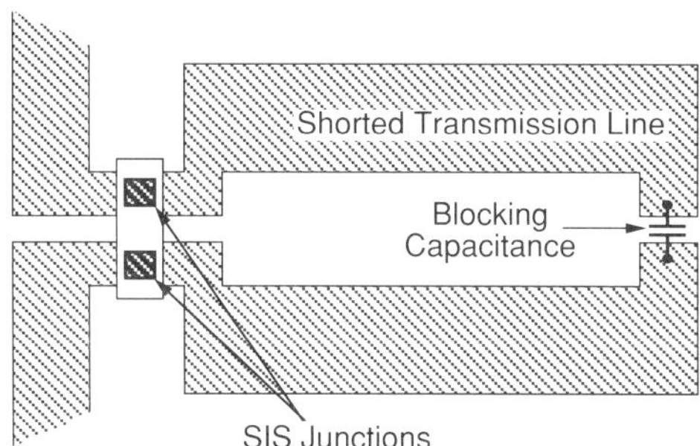

FIGURE 4.23 This simplified drawing of a two-junction SIS mixer illustrates the use of a shorted transmission line to tune the capacitance of the SIS junctions. A dc blocking capacitor is shown at the short location. In practice, the short position is adjustable for variable tuning. Such mixers are typically used in the millimeter wavelength range. While it is possible to approach the quantum limit with SIS junctions, they do exhibit a rather limited dynamic range and are therefore saturated at relatively low power levels. The most common application of SIS mixers is in radio astronomy. (After S. K. Pan, A. R. Kerr, M. J. Feldman, and A. W. Kleinsasser, "SIS technology boosts sensitivity of mm-wave receivers," *Microwaves & RF*, Vol. 27, No. 9, September 1988, pp. 139–146.)

combination with a low-noise HEMT 1.4 GHz IF amplifier to achieve a total noise figure of 0.14 db. This is within a factor of two of the photon noise temperature, hf/k.

While SIS mixers are extremely sensitive at mm wavelengths, they do suffer from a lack of dynamic range, that is, they are easily saturated at very low input power. When saturation begins to occur, operation becomes nonlinear and the usefulness of the mixer is compromised. While considerable effort has been invested to understand this problem where the input has been assumed to be monochromatic (single frequency), it has recently been suggested that out-of-band signals may play a significant role in SIS mixer saturation (D'Addario, 1988). If this is the case, bandpass filtering at the input may be useful.

As opposed to this superconductor-insulator-superconductor (SIS) junction, other junctions using a superconductor separated from a normal material (SIN) have also proved to be capable of very low-noise mixing and detection of microwave signals.

Shapiro Steps; Voltage Standards

When the SIS junction is biased at a dc voltage and pumped with an external microwave field, there are beat frequencies with the Josephson current that oscillates naturally at $2eV/h$. As a result, sharp current steps occur in the current-voltage curve where the beat frequency is zero. These distinct steps in current are seen at voltages of $nhf/2e$, where f is the applied frequency and n is an integer. These steps were first detected by S. Shapiro, and are commonly referred to as "Shapiro steps" (Shapiro, 1973) (see Figure 4.24).

Since the applied frequency can be measured with great precision, it is possible to use this phenomenon to measure the applied junction voltage

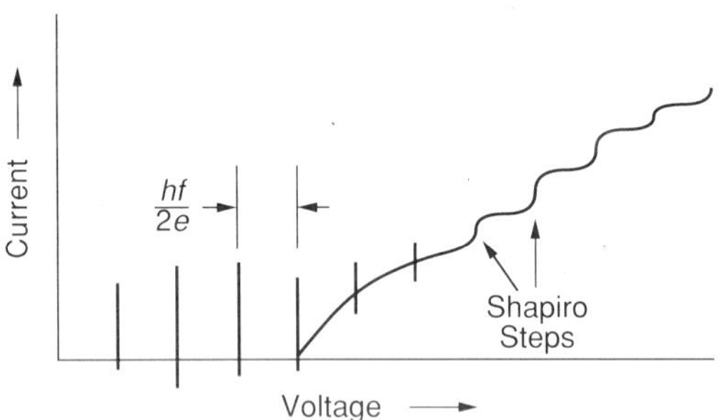

FIGURE 4.24 "Shapiro" tunneling in a high capacitance results from the beat of an applied microwave field and the natural Josephson oscillation frequency. At low voltages, the only stable operating points are located at the voltage intervals separated by $hf/2e$. This configuration, in large series arrays, is used as a voltage standard. (After P. L. Richards, "Analog superconducting electronics," *Physics Today*, March 1986, pp. 54–62. Also see: J. Clarke, "Small-scale analog applications of high-transition-temperature superconductors," *Nature*, Vol. 333, No. 6168, May 5, 1988, pp. 29–35 and M. Tinkham, *Introduction to Superconductivity*, Robert E. Krieger, Malabar, FL, 1985, p. 195.)

to great accuracy. SIS junctions are connected in series for use as voltage standards; since 1972, the U.S. legal volt has been defined on this basis (Richards & Shen, 1980). More recent systems use on the order of 20,000 junctions and operate at 10 V; these may be used to calibrate Zener diodes for secondary standards. More specifically, a standard with the following specifications has been developed at the U.S. National Bureau of Standards (Hamilton & Lloyd, 1988):

Junctions: 18,992 in dc series (16 parallel *rf* sections of 1187 each)

Junction material: $Nb/Nb_2O_5/PbInAu$

Critical current density: 70 A/cm^2

Critical junction current: 200 μA

Junction length/width: 12 μm/24 μm

Plasma frequency: 22 GHz

Lowest resonant mode: 175 GHz

Rf drive: 87 GHz at 5 mW

The Shapiro step is similar in appearance to the steps that result from photon-assisted quasiparticle tunneling, so the reader is cautioned not to confuse the two different effects. The following differences are worth noting:

Photon-assisted tunneling is characterized by less distinct steps of hf/e, which begin at the critical gap voltage, $2\Delta/e$.

Shapiro steps are very sharp, begin at zero gap voltage, and are separated by integral voltages of $hf/2e$.

OTHER ANALOG AND HYBRID APPLICATIONS

Josephson devices have also been used for amplification of *rf* signals, either in SQUID circuitry or in a long-junction configuration. In the latter case, analog amplification is accomplished by modulation of the junction's voltage-current characteristic. This is done by applying an alternating magnetic field to the junction, the result is a magnetic-to-voltage converter or "flux-flow amplifier" (FFA) (Yoshida, et al., 1987).

In the superconducting current-injection transistor (Super-CIT) configuration, a control current is injected directly into one of the electrodes adjacent to the Josephson junction (see Figure 4.25). The magnetic field created by this control current flow reduces the maximum Josephson junction current. As this magnetic field is increased, current flow in the circuit is eventually reduced to V/R, where V is the junction voltage and R is the external damping resistance placed across the junction. Such devices can, at least theoretically, operate at very high frequencies, but typically suffer from high feed-through

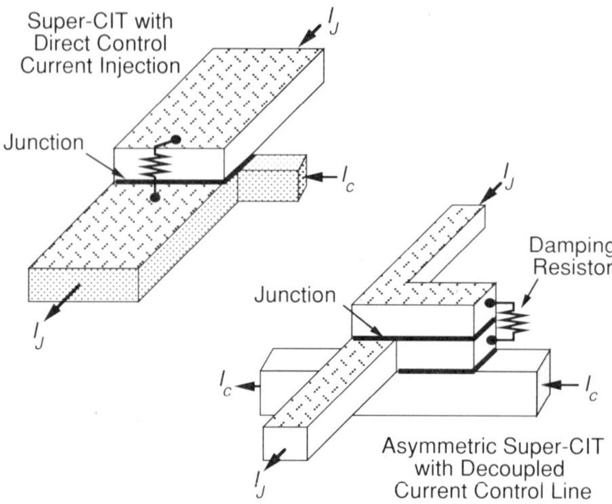

FIGURE 4.25 A superconducting current-injection transistor (Super-CIT) can be realized as illustrated above-left, by the formation of a relatively long Josephson junction. In this basic design, the magnetic field developed by a control current (injected directly into the lower electrode) controls the maximum Josephson current flow. When the current control line is decoupled, the CIT may be used in distributed amplifier configurations. In the decoupled-control line Super-CIT shown on the lower right, small-signal gain is increased by the asymmetric bias configuration. In either case, damping resistors are placed across the junction to eliminate hysteresis. (After D. P. McGinnis, J. B. Beyer, and J. E. Nordman, "A modified superconducting current injection transistor and distributed amplifier design," presented at 1988 Applied Superconductivity Conference, San Francisco, August 21–25, 1988, (EJ-11), Abstracts, p. 52.)

capacitance. Since individual current-controlled elements have very low characteristic impedances as well as low gain, it has been suggested that these limitations might be overcome by connecting a large number of elements in a parametric amplifier configuration.

Use of these elements in a parametric amplifier requires isolation of the current-control line so that it may be associated with each CIT element. Therefore, the three-terminal CIT device becomes a four-terminal component. A modified Super-CIT has been modeled that appears to provide higher gain, and reduced effects from feedthrough capacitance and low characteristic impedance (McGinnis, et al., 1988) (see Figure 4.26). This device, using an isolated control-current line and asymmetric bias configuration, is illustrated in Figure 4.25. It appears that gains of 15 db over bandwidths of 90 GHz are possible with parametric amplifiers constructed with this modified Super-CIT.

Other representative applications include analog signal correlators and superconducting delay lines (Green, et al., 1987; Delaney, et al., 1987). Such applications are typically combined with semiconductors and, as such, may be considered hybrid circuits.

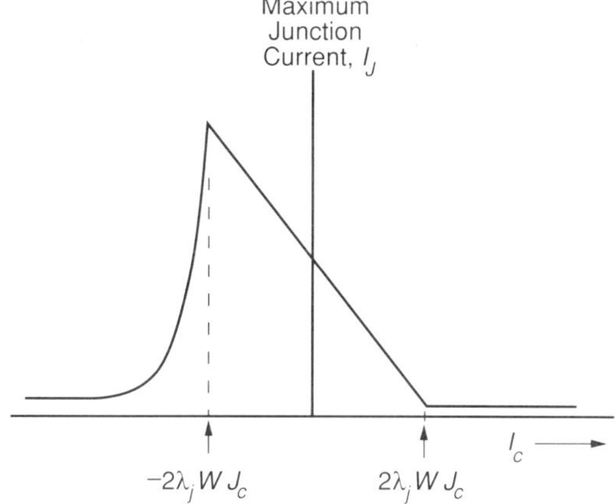

FIGURE 4.26 Modulation of maximum Josephson junction current (I_j) by control current (I_c) in a Super-CIT gain element (with asymmetric bias configuration) on a decoupled control-current line. λ_j is the Josephson penetration depth, W is the width of the isolated control line, and J_c is the critical current density for the junction. (After D. P. McGinnis, J. B. Beyer, and J. E. Nordman, "A modified superconducting current injection transistor and distributed amplifier design," presented at 1988 Applied Superconductivity Conference, San Francisco, August 21–25, 1988, (EJ-11), Abstracts, p. 52.)

Josephson devices have also been constructed using a variety of barrier materials other than the usual oxide insulators. Semiconductor barriers, in particular, can support relatively large densities of Josephson current flow (Kroger, 1980). The fact that semiconductor barriers have a lower dielectric constant and may be used with greater thickness than the more conventional oxide barriers could provide speed advantages in digital applications. Amorphous, single-crystal and polycrystalline barriers have also been investigated, as have light-sensitive CdS barriers. Each of these approaches may yield novel (and perhaps useful) current-voltage characteristics for specific applications.

The mixture of high-temperature superconductor material with conventional semiconductor components appears to improve operation in certain combinations. There are several advantages over semiconductors operated at room temperature to be gained by such a hybrid configuration at 77 K:

a. Decreased noise in amplifiers
b. Decreased power dissipation
c. Improved thermal conduction

A first concern is how the various semiconductor types operate at liquid nitrogen temperature. The mobility of holes and electrons in semiconductors increases as temperature is lowered, so the performance of majority-carrier devices will improve. This includes all of the field-effect transistor technologies

(MOSFET, JFET, MESFET, etc.) but does not include bipolar transistors that utilize minority-carrier injection. MOSFET drain-to-source "on" resistance, a critical parameter in many applications, decreases by approximately a factor of four when the device is cooled from 300 K to 77 K. Ferrites generally operate poorly at 77 K, but nickle-iron powdered magnetic materials perform reasonably well at liquid nitrogen temperature (Mueller, 1989).

CMOS and Superconducting Interconnects

It has already been pointed out that, due to limitations in processing semiconductors, it will not be straightforward to mix semiconductor and high-T_c ceramic superconductor processing techniques to produce intimate mixtures of these two technologies. However, the use of copper-oxide interconnects combined with semiconductor technology does hold some promise. A case in point is the use of superconducting interconnects, perhaps based on the bismuth or thallium compounds, with CMOS. Operation at 77 K, now possible with certain CMOS technologies, would be appropriate for such a combination (Malozemoff, 1988).

The possibilities include using high-T_c superconductors as interconnects both between chips, and as interconnects on the CMOS chip itself. With reduced interconnect resistance, delay times could be reduced, but the reduction would arise more from reduced capacitance (due to the possibility of using a smaller line cross-section) than directly due to the decrease in resistance. This is because much of the delay is a function of length. In Malozemoff's notation, the total delay τ_{TOT} is a function of device output resistance R_D, line resistance per unit length R_L, line length L, line capacitance per unit length C_L, and load capacitance C_D:

$$\tau_{TOT} = (R_D + LR_L)(LC_L + C_D) \qquad 4.12$$

which reduces to:

$$\tau_{TOT} = R_D C_D + L(R_D C_L + R_L C_D) + R_L C_L L^2 \qquad 4.13$$

This expression illustrates how there is a first term that is constant, depending on the device alone. The second and third terms are linear and quadratic functions of line length, which tends to dominate the balance of the total time delay.

It is also useful to consider the intrinsic limits on superconducting transmission lines using the London estimate for conductivity as a function of reduced temperature ($t = T/T_c$):

$$\sigma = \sigma_1 - j\,\sigma_2 \approx \sigma_{n,s} t^4 - j\,\frac{(1 - t^4)}{\omega \mu_o \lambda_o} \qquad 4.14$$

where $\sigma_{n,s}$ is the normal-state conductivity for temperatures near T_c (i.e., $t \sim 1$), ω is the angular frequency ($2\pi f$), μ_o is the permittivity of free space and λ_o is the London penetration depth. This estimate is reasonably accurate for temperatures not too far below T_c, and for frequencies below the "energy gap" frequency of $2\Delta/h$. If the Y-Ba-Cu-O superconductor, with a nominal T_c of 90 K, behaves in BCS fashion, the value of 2Δ would be 27.4×10^{-3} eV, that is,

$$2\Delta = 3.53 k_B T_c = (3.53)(8.625 \times 10^{-5}\ \text{eV/K})(90\ \text{K}) = 27.4 \times 10^{-3}\ \text{eV}$$

If this value for 2Δ is used to infer the gap frequency $2\Delta/h$,

$$f_{\text{gap}} = \frac{2\Delta}{h} = \frac{(27.4 \times 10^{-3}\ \text{eV})}{(4.14 \times 10^{-15}\ \text{eV-sec})} = 6.6 \times 10^{12}\ \text{Hz}$$

or 6.6 THz. In fact, gap frequencies up to 16 THz have been reported for Y-Ba-Cu-O (Malozemoff, 1988).

In any case, the London estimate for conductivity implies the following values for phase velocity v_{ps} and attenuation constant α_s in superconducting transmission strip-lines:

Phase velocity:

$$v_{ps} = \frac{c}{\epsilon_r \sqrt{1 + (2\lambda/d)}} \qquad 4.15$$

Attenuation constant:

$$\alpha_s = \frac{\omega^2 \sigma_n \epsilon_r^{1/2} \epsilon_o^{1/2} \mu_o^{3/2} \lambda^3 t^4}{d(1 + t^2)(1 + t)^2} \qquad 4.16$$

where d is the thickness of the dielectric substrate, and ϵ_r and ϵ_o are the relative and normal permittivities respectively, and

$$\lambda = \frac{\lambda_o}{\sqrt{1 - t}} \qquad 4.17$$

It is instructive to compare these expressions for the superconductive state to equivalent expressions for phase velocity v_{pn} and attenuation constant σ_n for normal conductors:

$$v_{pn} = \frac{c}{\epsilon_r \sqrt{1 + (\delta/d)}} \qquad 4.18$$

$$\sigma_n = \frac{1}{d}\sqrt{\frac{\omega \epsilon_r \epsilon_o \rho}{2(1 + (\delta/d))}} \tag{4.19}$$

where ρ is the normal-state resistivity, and δ is the normal skin depth defined by

$$\delta = \sqrt{\frac{2\rho}{\omega \mu_o}} \tag{4.20}$$

For common metallic resistivities, the δ term becomes significant at frequencies as low as 1 GHz, leading to a marked dispersion in the phase velocity. Again, for normal metallic conductors, the characteristic ($1/\epsilon$) attenuation distance is on the order of 10 cm at 1 GHz. This implies that high-performance (i.e., low attenuation and low phase dispersion) operation of normal strip lines at 1 GHz is limited to packages on the order of 10 cm or less.

By contrast, a superconducting strip line at 77 K is expected to have a characteristic propagation length of 5000 cm at 10 GHz, or alternately, a 70 GHz bandwidth for characteristics lengths of 100 cm (Malozemoff, 1988). Phase propagation in these instances would be virtually dispersionless.

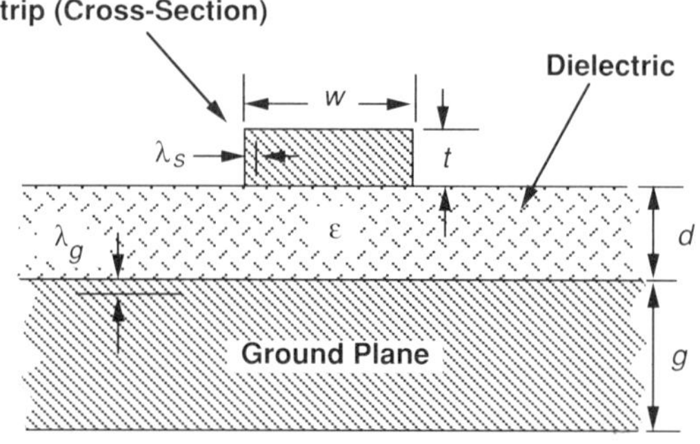

$$z_o = \frac{d}{W}\sqrt{\frac{\mu_o}{\varepsilon}\left[1 + \frac{\lambda_g}{d}\coth\left(\frac{g}{\lambda_g}\right) + \frac{\lambda_s}{d}\coth\left(\frac{t}{\lambda_s}\right)\right]}$$

$$\tau_d = \sqrt{\mu_o \varepsilon\left[1 + \frac{\lambda_g}{d}\coth\left(\frac{g}{\lambda_g}\right) + \frac{\lambda_s}{d}\coth\left(\frac{t}{\lambda_s}\right)\right]}$$

FIGURE 4.27 A basic superconducting strip-line. Both the "strip" and the ground plane are superconducting, with penetration depths as indicated. Expressions for the characteristic impedance and time delay are shown in the illustration. (After H. H. Zappe, "Josephson Computer Technology," in *Advances in Superconductivity*, B. Deaver and J. Ruvalds, eds., Plenum Press, New York, pp. 66 and 67.)

RF-magnetron sputtering has been used to form films at 650°C on single-crystal MgO, which was then patterned by Ar ion beam etching into 2 μm-wide strip lines. These microstrip lines of 1-2-3 material (Gd-Ba-Cu-O) exhibited a critical current of 1×10^6 A/cm^2 (Enokihara, et al., 1988) (see Figure 4.27 for a basic superconducting strip line).

THE HYPRES® INSTRUMENTATION

Perhaps the most impressive application of Josephson junctions to commercial technology (as opposed to one-of-a-kind research devices) is represented by the sampling digital oscilloscope and time domain reflectometry (TDR) instrumentation produced by HYPRES, Inc., of Elmsford, NY (see Figures 4.28 and 4.29). Based on niobium alloy Josephson-junction technology, the HYPRES

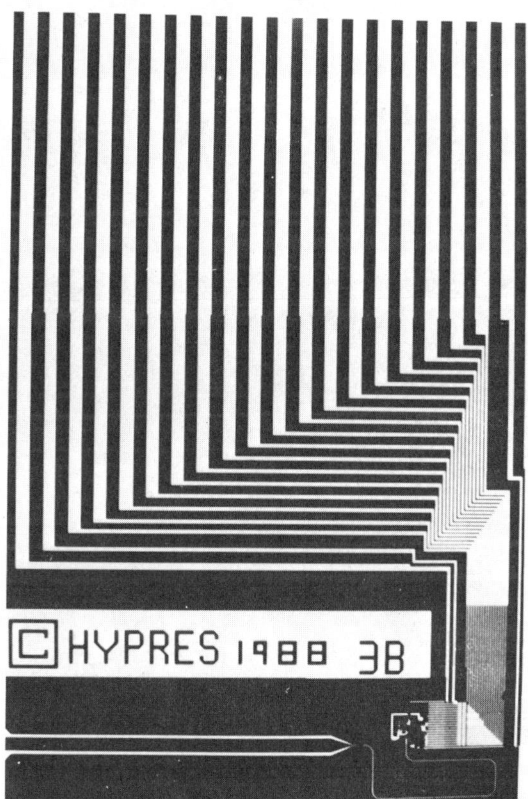

FIGURE 4.28 In this example of hybrid technology, the superconducting integrated circuit on the lower right is spray-cooled with liquid helium. Normal interconnects, shown at the upper portion of the figure, are at room temperature. The input transmission line is seen at the lower left. (Photograph provided by HYPRES, Inc.)

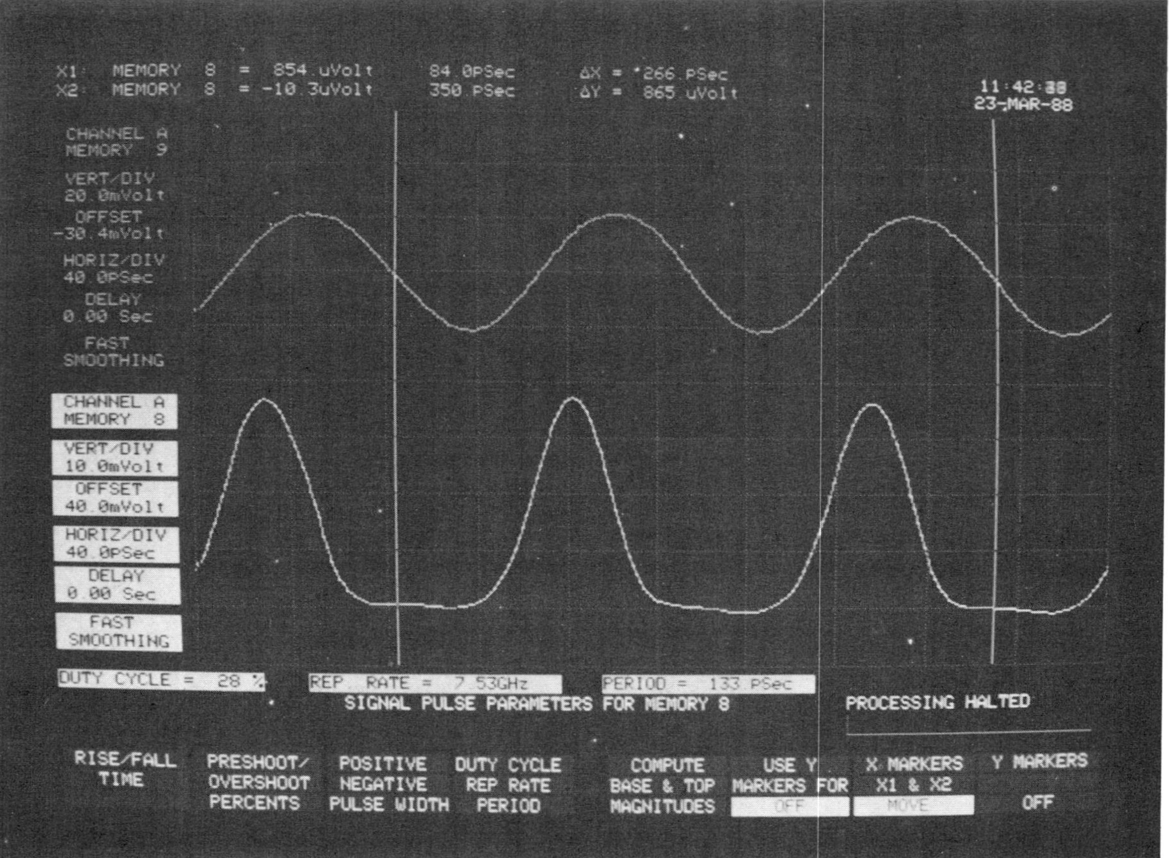

FIGURE 4.29 This close-up view of the HYPRES sampling oscilloscope display clearly reveals the nonlinear effects of an active microwave device operating at 7.5 GHz. Sensitivity of this instrument is $\sim 50\ \mu V$. (Photograph provided by HYPRES, Inc.)

instrumentation achieves almost dispersionless transmission lines and reliable Josephson interferometers to produce impressive specifications for their PSP-1000 workstation:

System rise time	5 ps
Sensitivity	50 μV
Bandwidth	70 GHz

The design is conservative as well as novel. The conservative element: a 3 μm geometry on a fuzed-quartz substrate make it possible to achieve high yields (Weber, 1987). A novel feature: The conventional cryogenics system is replaced by a technique where liquid helium is sprayed onto the superconducting electronics while the interface to the normal circuitry (only about 1 cm away) is held at room temperature.

Future plans to improve these already successful systems include submicrometer technology for even greater speed, and niobium-nitride junctions (operating at a higher temperature, i.e., ~16 K) which will allow use of a closed-cycle refrigerator and eliminate the need to refill the helium dewar. (See Figure 4.30 for a HYPRES time-domain analysis system.)

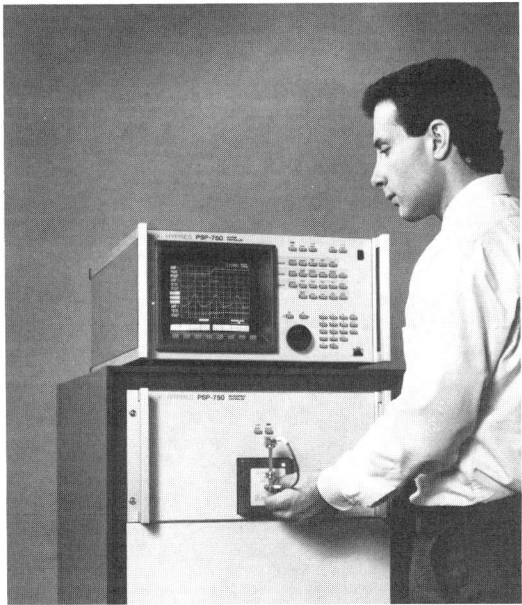

FIGURE 4.30 Time-domain analysis at the picosecond level is possible with 3 μm niobium Josephson junction technology. This sampling oscilloscope provides a bandwidth of 70 GHz. With internal triggering, virtually jitter-free measurements with 5 ps resolution can be achieved. (Photograph provided by HYPRES, Inc.)

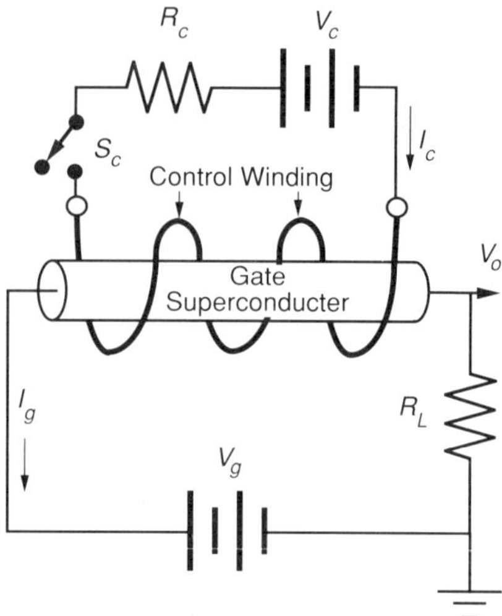

FIGURE 4.31 In the basic cryotron, the gate element is constructed from a superconducting material with a lower critical field than is used in the superconducting control element. The gate remains superconducting until a critical control current (I_{cc}) flows which is capable of producing a critical magnetic field at the surface of the gate element.

CRYOTRONS: FOR SLOW SWITCHING

The term cryotron applies generally to a class of low-temperature superconductive switches. The switching is usually accomplished by raising one element (the gate) above its critical magnetic field by passing a sufficient electric current through an adjacent superconductor (the control element). This quenching of the gate element results in normal electrical resistivity in the gate (see Figures 4.31 and 4.32). In cases where the cryotron must operate

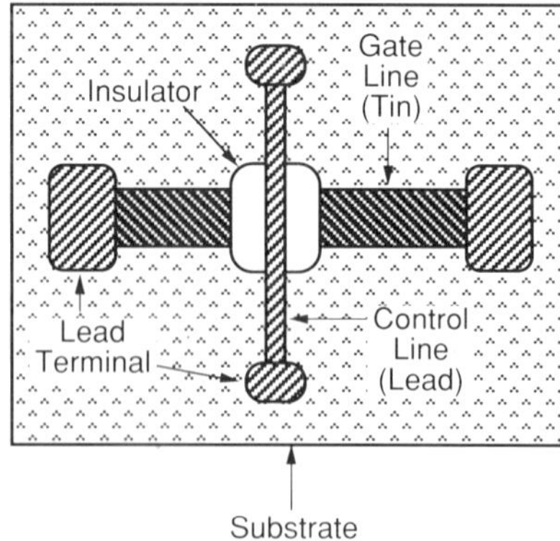

FIGURE 4.32 Large-scale cryotron systems were generally formed from single-element control lines rather than control solenoids, and these were typically films deposited on a substrate such as glass.

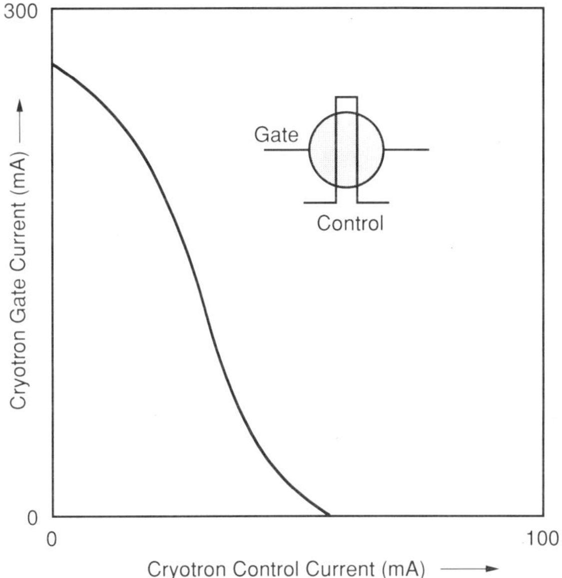

FIGURE 4.33 A cryotron is a very low-impedance device, even when compared a power MOSFET. The curve above is for a constant voltage ($\sim 0.01\ \mu V$) across the gate terminals. After a control current of ~50 mA reduces the gate current from >200 mA to a few μA, d.c. gate resistance will have increased to only a few hundred $\mu\Omega$.

in relatively high ambient magnetic fields the gate element may be heated to cause quenching.

Cryotrons are used for switching currents as well as for amplification. While multiple cryotrons may be used to form logic or memory elements, this type of switch is so slow compared to Josephson junction technology that it has been abandoned as far as fast computer applications are concerned. Nevertheless, cryotron switches still find use in relatively slow applications, especially where high powers or high magnetic fields are involved. See Figures 4.33 and 4.34.

Effects in Inductive Branches

Prior to discussing the use of cryotrons, it is helpful to consider how currents are divided between branches where each has zero dc resistance. Consider two branches a and b connected in parallel. The resistance of each branch (R_a and R_b) is zero. It is, perhaps, not clear how a dc current (I_T) applied to this parallel circuit will divide between the two branches, since the normal current-divider equation for the current in branch a is

$$I_a = I_T\left(\frac{R_b}{R_a + R_b}\right) \qquad 4.21$$

This approach is not useful, since it results in a division by zero. The answer lies in the fact that each branch must have a nonzero inductance (L_a and L_b)

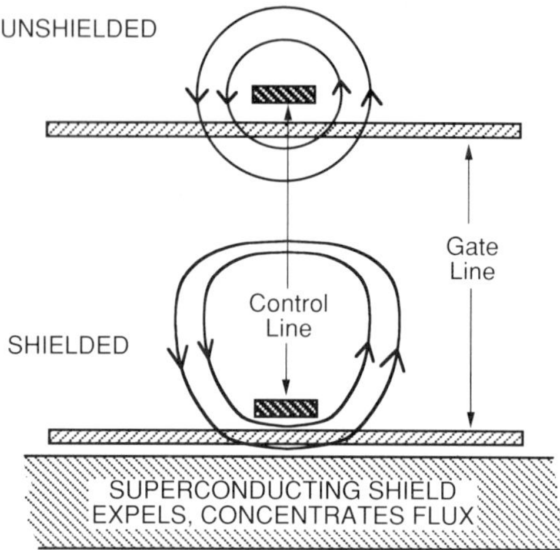

FIGURE 4.34 The addition of a superconducting shield under a cryotron gate segment tends to increase sensitivity to the control current by concentrating the magnetic field in the vicinity of the gate. In this film cryotron, the gate film is assumed to be thin compared to the London penetration depth (λ) so that the magnetic field lines penetrate the entire structure.

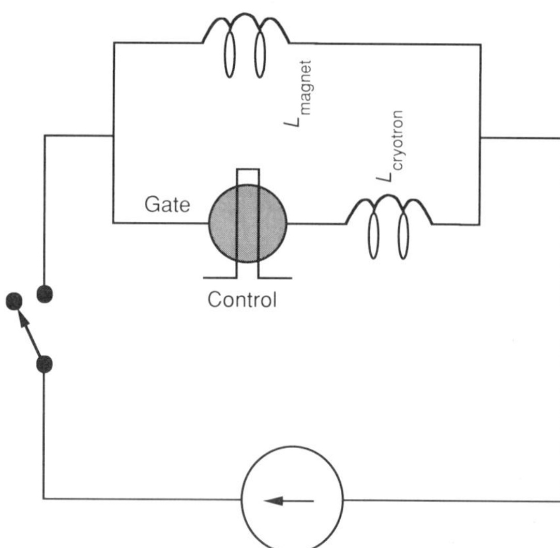

FIGURE 4.35 A cryotron switch (either current- or temperature-activated) may be used in parallel with an inductive winding to generate a persistent current. After external current is applied, the cryotron is switched first to the resistive state and then back to the superconductive state. When the external current source is removed, a persistent current flows through the cryotron-inductor loop. Details appear in the following figure.

and that the current must change from zero to some finite value. The current in each inductive branch $i_a(t)$ and $i_b(t)$, subjected to a voltage varying with time $v(t)$ is

$$i_a(t) = \left(\frac{1}{L_a}\right)\int_0^t v(t)dt \qquad 4.22$$

and

$$i_b(t) = \left(\frac{1}{L_b}\right)\int_0^t v(t)dt \qquad 4.23$$

Since the integral is identical in each instance,

$$\frac{i_a(t)}{i_b(t)} = \frac{L_b}{L_a} = \frac{I_a}{I_b} \qquad 4.24$$

(a) Current initially divides according to inductance in each parallel branch.

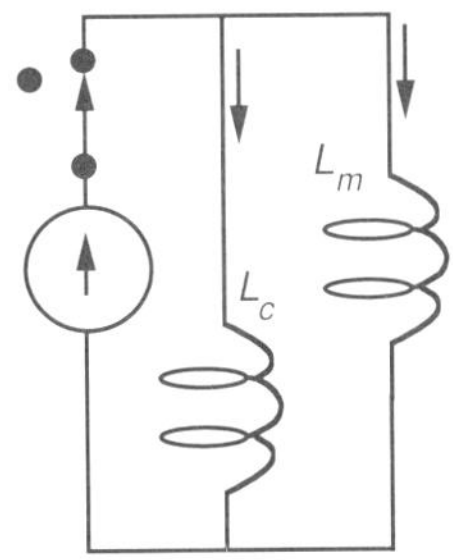

(b) Appearance of resistance in one branch shifts all current to the zero-resistance branch.

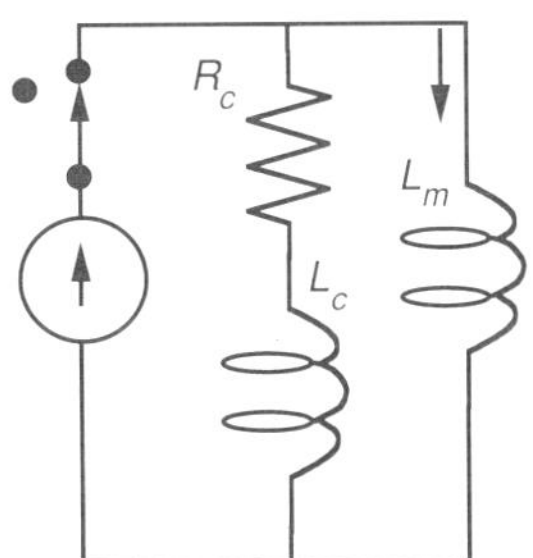

(c) Disappearance of resistance does not affect current flow since there is no potential across the superconducting branch.

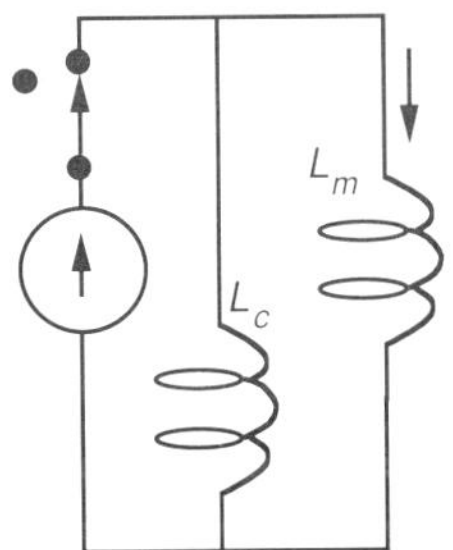

(d) Interruption of external current source results in persistent current through internal loop.

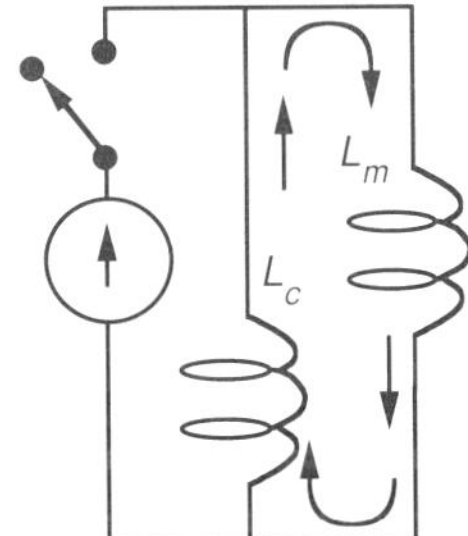

FIGURE 4.36 Steps for producing persistent current flow in magnet inductance L_m and cryotron branch L_c. The resistance R_c appears when the cryotron is quenched.

The point worth remembering is that the relative magnitude of a branch current (in the absence of any resistance) is inversely related to the relative inductance of the branch, and, of course, the sum of the branch currents must equal the applied current. An application of these principles is illustrated in the accompanying Figure 4.36 which illustrates how external current may be injected into a "persistent loop" by use of a cryotron switch. Inductances are labeled L_c and L_m to represent the cryotron and magnet branches respectively. Assume that each branch is superconducting when initial current is applied, so that currents will divide through each branch according to inductance. If the cryotron is then switched to a resistive state, all of the current will be diverted to the magnet winding where resistance remains zero. If the cryotron is now switched back to the superconducting state, cryotron current *remains* at the zero level, since the voltage across the magnet winding is zero. If, however, the external current source is now removed, a counter-emf will be developed across the parallel circuit by the magnet inductance (to keep the magnet current constant) and current will now flow in a "frozen" or "persistent" mode through the magnet-cryotron loop.

Crossed-film cryotron amplifiers have proven useful in the investigation of thermal noise analysis as it applies to the input of electronic amplifiers (Radhakrishnan and Newhouse, 1971). More recently, cryotron-like switching has proven to be effective in the design of superconducting rectifiers operating

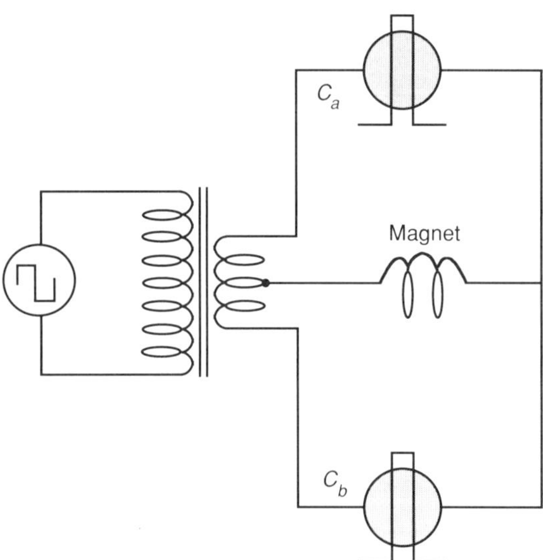

FIGURE 4.37 In this flux-pump circut, the superconducting transformer is driven into saturation by the virtually square-wave voltage on the primary. The cryotons C_a and C_b are switched at appropriate intervals by timing circutry (not shown) to provide a direct current for the magnet winding. Note the similarity to a conventional full-wave SCR rectifier. Persistent direct currents flow through the cryotrons and the magnet winding, via the transformer secondary.

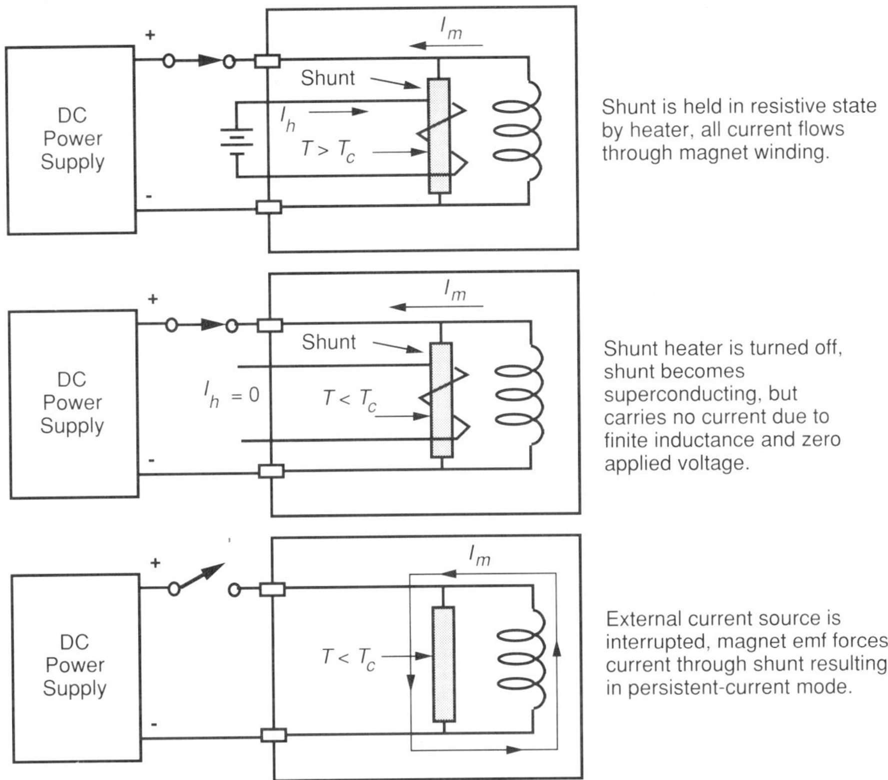

FIGURE 4.38 In this application, a thermally-activated cryotron switch is used during the turn-on cycle of a persistent-current mode magnet. (This applications described to the author by James C. Sturrock.)

at 30 A and approximately 4 Hz (Mulder, et al., 1988) (see Figure 4.37). Others have used a pulse-mode superconducting switch that is opened (quenched) to divert current from an energy storage inductor to operate a rail gun (Leung, et al., 1988). A magnet switching application of a cryotron is illustrated in Figure 4.38.

HIGH-ENERGY APPLICATIONS

High-Field Inductors

Superconducting inductors that must transport relatively high current densities in high magnetic fields have a variety of applications:

1. Coils for windings in motors and generators (Utility, automotive, marine propulsion applications)
2. High-field magnets for research applications (Particle accelerators, materials research)

3. Magnetic Levitating (MAGLEV) coils for high-speed ground transportation
4. Superconducting Magnetic Energy Storage (SMES) (Electric utilities, Military applications)
5. Magnetic containment fields for thermonuclear fusion research, exemplified by the 340 ton Nb-Ti "Yin-Yang" coils at Lawrence Livermore National Laboratory in California, which have produced peak fields of 8 T (Huang, 1988)
6. MHD (magnetohydrodynamic) EMT (electromagnetic thrust) systems for marine propulsion applications
7. MRI (Magnetic Resonance Imaging) which requires extremely uniform magnetic fields at the 10–20 kgauss level (formerly known as NMR, nuclear magnetic resonance).

While the inductors, that is, coils, in these applications will vary enormously in size and configuration, each has the requirement that high current densities must be transported without significant electrical losses in the presence of a high magnetic field, which is usually generated by the current flow. Some of the criteria of design will be considered in the following section.

Superconductors for High Field Applications

Type II superconductors are required in high-field applications, due to their ability to transport relatively high current densities in the presence of high magnetic fields (see Table 4.2). Typical conventional materials used include Nb_3Sn, V_3Ga, Nb_3Al, $Nb_3(Ge_{0.2}Al_{0.8})$ and Nb-48% Ti (Stekly, 1973). The

TABLE 4.2 Properties of selected Type II superconductors. It is instructive to compare these materials to Pb (a Type I material), where T_c is 7.22 K, but operation is limited to fields less than 0.1 T.

Compound or Material	T_c (K)	H_c @ 4.2 K (T)
Nb_3Al	16–20	20–35
$Nb_3(Ge_{0.2}Al_{0.8})$	20.7	41.0
NbN	16–18	20–35
Nb_3Sn	17–18	19–25
Nb(45–50 wt%)-Ti	8.8–9.3	10.5–11
$PbMo_6S_8$	14	40–55
V_3Ga	16.8	21.0

Compiled from:
Larbalestier, D., G. Fisk, B. Montgomery, and D. Hawksworth, "High-Field Superconductivity," *Physics Today*, March 1986, pp. 24–33.
and:
Stekly, Z. J. J., "Superconducting Coils," in *The Science and Technology of Superconductivity*, Gregory, W. D., W. N. Mathews, Jr., and E. A. Edelsack, eds., Vol. 2., Plenum Press, New York, 1973, p. 499.

ductile Nb-Ti alloys may be formed into wires by ordinary techniques and are commonly used for magnet applications where fields are under 100 kgauss (10 T). At 4.2 K and 50 kgauss (5 T), Nb-Ti alloys can transport current densities on the order of 10^5 A/cm^2.

While Nb_3Sn is relatively brittle and far more difficult to fabricate into useful conductors (e.g., vapor deposition onto a steel substrate), it is capable of operation without quenching in fields well over 200 kgauss (20 T). At 4.2 K and 100 kgauss, Nb_3Sn can transport current densities in excess of 2×10^5 A/cm^2.

Since the 1-2-3 materials ($YBa_2Cu_3O_x$ and its rare earth counterparts RE-$Ba_2Cu_3O_x$) may have critical fields in excess of 200 T at 0 K, it is natural to consider using these copper oxides to construct magnets of unprecedented field strength. Even if these materials (or others on the horizon) are eventually practical in terms of ductility and J_c, the limitations on field strength by the enormous forces involved at higher fields may be the primary factor in limiting the construction of magnets, rather than H_{c2} of the material.

The high-T_c ($YBa_2Cu_3O_{7-\delta}$) copper oxides are potentially interesting due to the possibility of operating at the liquid nitrogen temperature of 77 K, where refrigeration systems are smaller, more reliable and far less expensive than liquid helium systems. However, it must be pointed out that many experienced investigators believe that operation at 77 K may not be practical for 90 K superconductors; estimates for practical operating temperatures vary from $0.5T_c$ to $0.75T_c$, with the former "rule of thumb" being the most widely accepted.

The much higher H_{c2} of the new ceramics would also be attractive for construction of very high field magnets but, as mentioned previously, this is not the case since the Lorentz forces (which tend to destroy the structure) for fields above ~100 T are the limiting factor in magnet construction. Aside from these limitations of extremely high field operation, bulk $YBa_2Cu_3O_x$ superconductors are not able to operate in a field approaching the H_{c2} of the individual grains, due to the poor connections at grain boundaries. At present, it seems that insufficient current density in bulk materials as well as difficulty of fabrication will delay the widespread introduction of the 1-2-3 copper oxides into this range of applications. Nevertheless, work is already underway to design novel types of 1-2-3 wires and tapes for high-field, high current density use.

Small steps in the direction of magnet construction with the new materials are already underway. In one instance, an experimental high-T_c superconducting magnet using several discs of $YBa_2Cu_3O_{7-x}$ has been tested. In this small-scale experiment the field produced was a scant 0.02 T with a current flow of 65 A (Kirschner, et al., 1988). To avoid the problem of making electrical contact with the ceramic superconductor, current was induced into the stack of 1-2-3 toroids.

Since the recently discovered compounds of Bi-Sr-Ca-Cu-O and Al-Bi-Sr-Ca-Cu-O are reportedly somewhat more ductile than Y-Ba-Cu-O, perhaps some variant of these materials will find use in the construction of high-temperature magnet wire. There is some indication that untextured films of the thallium superconductor can support rather high critical current density in the presence of moderate **H** fields.

A fundamental limitation on use of the high-T_c ceramics for magnet windings is related to the fact that all of the materials are quite brittle when compared to their conventional counterparts such as niobium-tin. The enormous stress exerted on the winding structure by high magnetic fields is already a limiting factor with the relatively ductile conventional superconductors; little imagination is required to visualize the effect of these forces on the brittle high-T_c ceramics! More will be said about the nature of these forces in the following section.

SUPERCONDUCTING MAGNETS; DESIGN CONSIDERATIONS

The use of superconductors in a variety of alluring applications is of more immediate interest, but discussions of these subjects should be preceded by a consideration of an element common to all: the construction and safe operation of superconducting inductors for the production of high magnetic fields.

Superficially, the design of a superconducting coil does not seem to be particularly complex: maintain the current flow below the critical level and keep the coolant flowing. As long as the critical field H_{c2} is not exceeded and the cryogen supply is sufficient to hold temperature well below T_c (for the particular magnetic field environment), operation should be routine. In fact, well-designed systems do operate much in this fashion, but design is far from simple. Reliable operation requires careful attention to a number of details.

It has been pointed out in the chapter on superconducting phenomena that superconductors used in practice are not "perfect," as are the idealized materials that serve as models for the development of theories and calculations to estimate maximum performance criteria. The materials that are manufactured for use in magnet or generator field windings may not be prepared with as much care as those small samples from the research laboratory. (If they were, perhaps no one would be able to afford the cost!)

In addition, construction techniques are critical factors in the actual performance of a superconducting coil. Some of these factors will be discussed in the following section.

Thermal Conductivity

Typical Type II superconductors, which do have excellent critical magnetic field (H_c) and critical current density (J_c) characteristics, also have a serious shortcoming. As indicated in an earlier chapter, thermal conductivity tends to become relatively low at superconducting temperatures. For example, the ratio of the thermal resistivity of copper (k-$Cu_{4.2K}$ = 7000 mW/cm-K) and Cu thermal diffusivity (7900 cm/s^2) to the same parameters for several Type II superconductors is listed for T = 4.2 K (Stekly, 1973).

TABLE 4.3 Thermal properties of Nb_3Sn compared to copper and liquid helium.

Parameter	Symb.	dim.	Nb_3Sn	Cu	Liq. Helium
Thermal Conductivity	k	mW/cm-K	0.4	7000	2.72
Specific Heat	C	mJ/gm-K	0.21	0.099	4480
Resistivity (Normal)	ρ_n	$\mu\Omega$-cm	26	0.03	—
Thermal Diffusivity	$*k/\gamma C$	cm^2/s	0.35	7900	0.005

* (γ is density, g/cm^3)
(After Stekly, Z. J. J., "Superconducting Coils," in *The Science and Technology of Superconductivity*, Gregory, W. D., W. N. Mathews, Jr., and E. A. Edelsack, eds., Vol. 2., Plenum Press, New York, 1973, p. 500.)

Superconductor	*k-Cu/k-SC*	*Th.Df-Cu/Th.Df-SC*
Nb_3Sn	17,500	22,570
Nb-Zr	8750	1463
Nb-Ti	5833	6695

These examples indicate rather clearly that copper is a far better material for conducting heat than typical Type II superconductors. For this reason, copper is a favorite material for use as a thermal and mechanical support for superconducting wires used in magnets and similar applications. See Figures 4.39 and 4.40.

Regardless of the construction method used, it is necessary to maintain a sufficiently low temperature in the superconductor to avoid quenching. This requires sufficient heat flux removal per unit conductor area exposed to cryogen flow. For liquid helium, the maximum value is on the order of 0.3 W/cm^2, but this figure varies with the exact configuration of the cooling passage (Stekly, 1973).

Minimizing Strand Diameter

The diameter of a cylindrical superconductor, given the low thermal conductivity of these materials, must obviously not be too great, regardless of the coolant properties at the surface of the wire. Stekly provides the following approximate criteria for maximum diameter (d) of a round strand (Stekly, 1973):

$$d < \frac{2}{J_c}\sqrt{\left\{\frac{A_{sb}}{A_{sc}}\right\}\frac{k(T_{ch} - T_b)}{\rho}} \qquad 4.25$$

where k is the superconductor's thermal conductivity, J_c is the superconductor's critical current density in the maximum ambient magnetic field, A_{sb}/A_{sc} is the ratio of substrate to superconductor area, T_b is the cryogen (bath) tem-

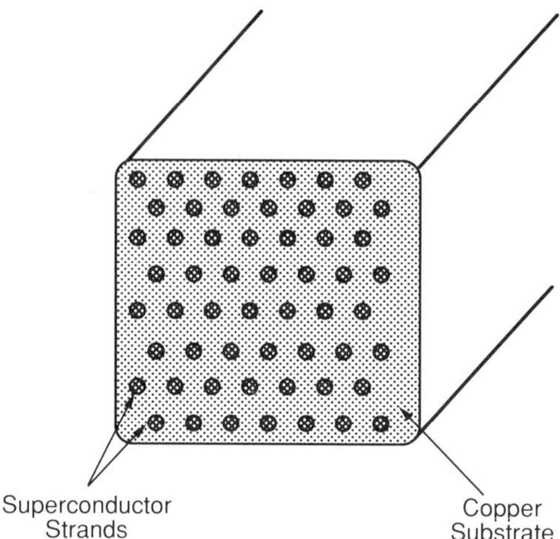

FIGURE 4.39 Representative configuration of a superconducting magnet "wire" consisting of multiple superconducting strands in a substrate of copper. The strands are often twisted to increase stability by reducing induced currents in the presence of time-varying magnetic fields.

perature, T_{ch} is the critical superconductor temperature in the local magnetic field (at zero current flow), and ρ is the resistivity of the substrate material.

The same author (Stekly, 1973) also presents a maximum diameter condition for "intrinsic" stability for limiting the heat generation by magnetic flux changes, where heat transfer to a substrate is not taken into account. This condition is:

$$d < \frac{\pi}{2J_c}\sqrt{\frac{\gamma C(T_{ch} - T_b)}{\mu_o}} \tag{4.26}$$

where γ is the density in kg/m^3 of the superconductor, μ_o is the permeability of free space ($4\pi \times 10^{-7}$ H/m), and C is the superconductor's specific heat in J/kg-K.

Aside from the obvious advantages of effective heat transfer to the cryogen, it is also clear that stability is enhanced by using a small diameter for the superconducting wire and by providing a highly conductive (electrically and thermally) substrate such as copper. It is common practice to use a number of small diameter superconducting wires in close communication with a copper substrate.

Strand diameter in dipole magnets for the 6.6 T Superconducting Super Collider (SSC) accelerator will be 3–5 μm (Larbalesteir, et al., 1986).

Twisting of Superconducting Strands

An eddy current may be induced in adjacent strands of superconductor that are embedded in a normally-conducting substrate. This current, if substantial,

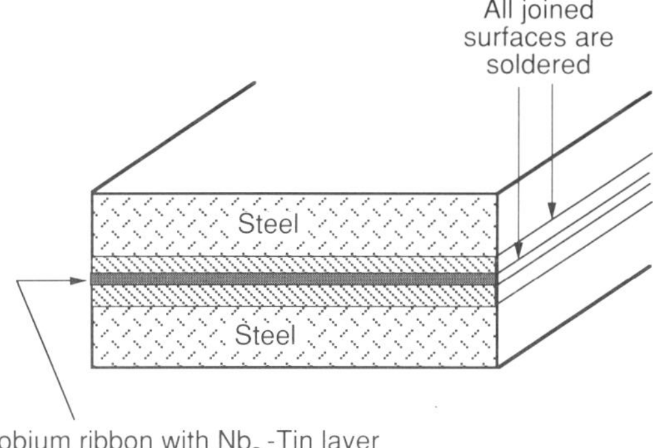

FIGURE 4.40 While niobium-tin is too brittle for use where more ductile superconductors are effective, Nb_3Sn is used where fields in excess of 10 T are encountered, due to its high critical field strength. Vapor-deposition onto stainless steel may be used, or, as illustrated above, a thin layer of Nb_3Sn on a niobium ribbon may be soldered in a sandwich of copper (for thermal conductivity) and steel (for strength). (After Z. J. J. Stekly, "Superconducting Coils," in *The Science and Technology of Superconductivity*, W. D. Gregory, W. N. Mathews, Jr., and E. A. Edelsack, eds., Vol. 2, Plenum Press, New York, pp. 497–538.)

can lead to quenching of the superconducting state. The eddy current is reduced by a practice familiar to electrical engineers who wish to reduce external magnetic fields around lines carrying alternating currents, that is, twisting of the individual pairs of superconductors. The shorter the twist length L, of course, the more effective the technique. An approximate formula for the induced eddy current J_{ec} is (Stekly, 1973):

$$J_{ec} = \frac{L^2}{\pi^2 \rho d_s}\left(\frac{\partial B}{\partial t}\right) \qquad 4.27$$

where ρ is the resistivity of the substrate (Cu?), B is the magnitude of the magnetic flux density in the substrate, and d_s is the diameter of a single superconducting strand. The characteristic decay time τ of the eddy currents after the source (B) is removed is also dependent on L^2:

$$\tau = \frac{\mu_o L^2}{\pi^2 \rho} \qquad 4.28$$

Even though twisting is very effective for relatively slow changes in B, it is clear from the expression above that the characteristic decay time is limited to some minimum value by how short the twist length L can be in practice. Viewed another way, the twisting technique looses effectiveness as frequency is increased. (See Figure 4.41 for an illustration of twisting of superconducting strands.)

Brittle materials like Niobium-Tin (Nb_3Sn) are not generally constructed into fine filaments, but instead may be vapor-deposited on other metals (cop-

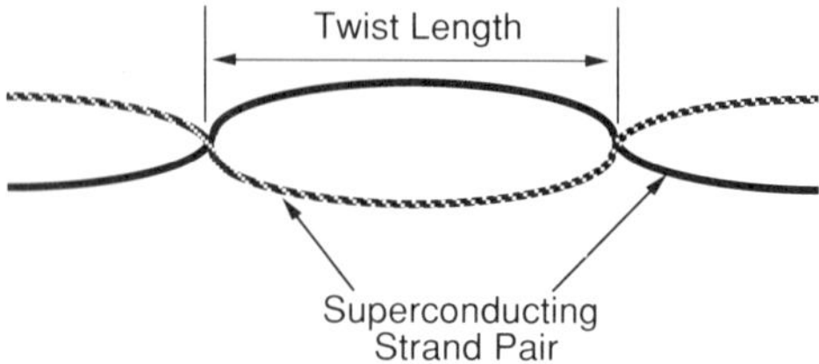

FIGURE 4.41 Strands of superconductor for high-field magnet applications are usually twisted to reduce the effect of eddy currents induced by time-varying fields.

per or silver-plated stainless steel for example). Conversely, Nb_3Sn reaction-formed onto a ribbon of niobium may be sandwiched between layers of copper and stainless steel, the former for thermal conduction and the latter for strength. Nb_3Sn is usually used in fields above 100 kgauss, where other Type II superconductors will quench regardless of current density. For fields well below 100 kgauss, other materials (usually niobium-titanium) are used since construction is relatively straightforward compared to the methods required for Nb_3Sn.

In practice, strands are spaced rather closely together within the normal matrix material. In the multifilimentary Nb-Ti superconductors used in the design of magnets for the superconducting supercollider (SSC) particle accelerator, it was found that critical current (J_c) could be increased significantly by reducing the spacing-to-diameter (S/D) ratio from 0.35 to 0.13. Filament diameter was approximately 9 μm. The highest values of J_c achieved were 265,000 A/cm^2 and 109,100 A/cm^2 at 5 T and 8 T, respectively (Kreilick & Gregory, 1987).

Protecting against Instabilities

Electrical engineers are perhaps somewhat more familiar with the dangers involved in the storage of large amounts of energy in capacitor banks rather than in large magnets. Recall that for a capacitance C charged to a voltage V, energy storage W_c is:

$$W_c = \frac{CV^2}{2} \tag{4.29}$$

while for an inductance L, with a constant current I flowing, stored energy W_L is

$$W_L = \frac{LI^2}{2} \tag{4.30}$$

Since the entire purpose of using superconducting magnets is to achieve very high fields by use of high currents, energy storage can be quite significant.

Since superconducting magnets (due to small wire cross-section) are relatively small compared to their copper-conductor counterparts, energy density ($\mu H^2/2$) may be high. With high energy storage (or simply high energy density), care must be taken to avoid situations where this energy is suddenly diverted from the field to create heat in the conductor or havoc in the workplace.

With the most careful design, a small region of a magnet superconductor may shift from the superconducting to the normal phase; this is commonly referred to as "quenching." This may occur due to a defect in the superconductor, a local failure in cooling, or a sudden increase in local flux density which causes unexpected local losses in the superconductor. In either of these scenarios, it is possible for the situation to be regenerative. For instance, a local normal region will tend to heat rapidly from current flow, causing a localized boiling of the cryogen. The loss of liquid cryogen at a point reduces heat transfer from the conductor, which heats further, creating more loss of cryogen. The conductor may actually melt. In large superconducting magnets, the stored energy, when dissipated in a sudden manner, rapidly boils away virtually all the liquid helium. The resultant impulsive increase in pressure may rupture the enclosure vessel. It is possible to have catastrophic results including destruction of an expensive system or even loss of life.

Clearly, the mechanical design of the coolant system must provide for a possibility like the one described above. Redundancy is desirable, that is, there should be a variety of precautions to assure that even a sudden overpressure in the cryogen system may be safely relieved. It is desirable, however, to provide as many means as possible to assure that such failures do not occur. The problems may be addressed by two approaches, stabilization and protection. (See Figure 4.42 for a superconducting persistent-mode magnet unprotected, with internal "transformer" protection, and with external resistive dissipation.)

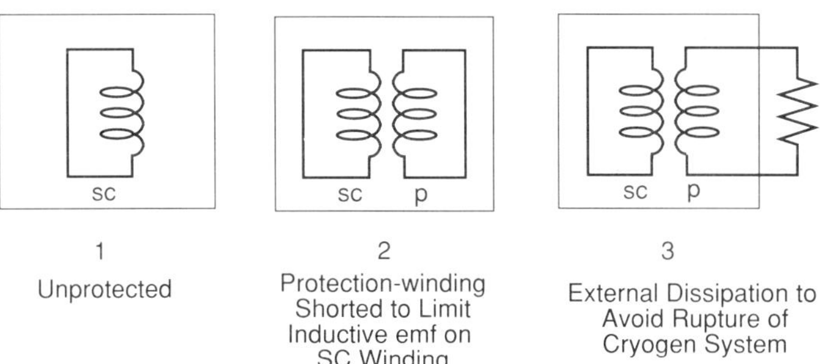

FIGURE 4.42 An additional winding may be added for protection of a superconducting coil. This "transformer" technique may be either a shorted secondary inside the cryogen enclosure (2) or the secondary may be connected to an external dissipative element (3). Scheme (2) is preferable where it is only necessary to limit the voltage produced across the winding by the quenching event, but (3) is used where it is necessary to dissipate the stored energy outside the cryogen container. (After V. A. Al'tov, V. B. Zenkevich, M. G. Kremlev, and V. V. Sychev, *Stabilization of Superconducting Magnetic Systems*, Trans. by G. D. Archard, Plenum Press, New York, 1977.)

The first, stabilization, encompasses those techniques used in magnet design that are intended to provide a system in which a localized, abrupt change to normal in a section of superconductor is extremely unlikely. The most common technique involves the use of a copper substrate to provide good thermal conduction paths as well as alternate current paths in the case where a section of superconductor fails. While effective, this method has the disadvantage that the volume of copper is typically many times greater than the volume of superconductor so that the size, weight, and cost of the system are increased.

Electrical protection schemes may be divided into two general types, those used when an external supply is permanently connected to the magnet and others where the winding is operated in the "persistent current" or "frozen" mode, where current flows through a closed loop that is isolated from direct external connections. (See Figure 4.43 for external storage versus external dissipation of fault energy for a superconducting magnet.)

In the first instance, there are a variety of means to divert or dissipate the energy stored in the system through the connections made at the power supply terminals, or at least to limit the voltage developed at those terminals if a fault occurs. Common methods include connections across the power supply terminals of a resistive element (for dissipation) or of a series diode-capacitor combination for limiting the counter-emf and providing a means for storage of energy formerly in the magnetic field. The diode is connected so that the capacitor will not be charged by the power supply, but will provide a current path for a counter-emf (opposite in polarity from the power supply) generated by the magnetic circuit. The diode also prevents oscillation of the capacitor-inductor combination. It is worth noting that the diode specifications (maximum current flow, maximum reverse voltage) must be considered carefully for this application. Parallel-connected diodes may be used to achieve the current-flow capability required. Clearly, the capacitor must be capable of storing the energy to be diverted from the magnet.

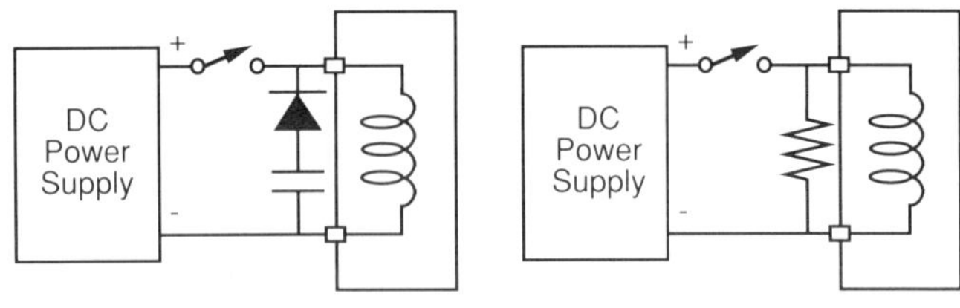

FIGURE 4.43 During a fault condition, energy from the magnetic field may be diverted to an external dissipative element (resistor) or to a capacitive storage element. In either case, caution must be taken to ensure that the element is capable of dissipating or storing the amount of energy to be diverted from the H-field. When the magnet is operated in the persistent-current mode and effectively isolated from external terminals energy may be diverted through inductive (transformer) techniques.

In the second case, where the magnetic winding is isolated in the persistent-current mode, "transformer" protection may be used to limit voltage or to divert energy to an external load (Al'tov, et al., 1977). A winding that is coupled to the superconducting winding will have no effect on the magnet under dc current conditions, but will tend to sense an abrupt change in flux caused by any rapid change in current in the superconducting winding (as would be caused by a sudden fault). See Figure 4.42.

The protective winding can be shorted inside the cryostat if it is only necessary to limit the voltage (to prevent an arc-over) that might build up on the superconducting winding. This approach has the disadvantage that the stored energy is dissipated inside the cryostat as joule losses in the resistance of the protective winding and its "short." If it is essential to dissipate a large proportion of the stored energy outside the magnet enclosure (to prevent boil-off or even rupture of the vessel), the protective winding terminals may be placed outside the cryostat on a resistive load capable of dissipating the total energy stored in the magnet, or at least the proportion to be diverted.

The value of external load resistance should be selected with at least two considerations in mind, that is, the L/R time constant of the discharge and the proportion of the energy to be externalized from the cryostat. Using Al'tov's symbols, R_{pc} and R_{ext} specify the resistance of the protective coil and the external resistance, respectively. If L is the effective inductance of the protective coil during a fault, the characteristic time constant of energy discharge is simply

$$\frac{L}{(R_{pc} + R_{ext})} \tag{4.31}$$

A zero value for R_{ext} provides the longest time constant and the best protection for overvoltage on the superconducting winding, but obviously does not provide for any externalization (from the cryostat) of energy. An extremely large value for R_{ext} provides *neither* overvoltage protection nor externalization of energy from the cryostat. If R_{ext} is smaller than R_{pc}, most of the stored energy in the magnetic field will be dissipated inside the cryostat. There is no method for specifying Rext that will suffice for all applications, since both externalization of dissipated energy and overvoltage on the superconducting winding must be considered (Al'tov, 1977).

However it may be terminated, the protective winding is often arranged to be bifilar with the superconducting coil for optimum coupling. The protective winding is generally constructed from copper.

Magnetic Forces on Coils

While there are an enormous variety of coils used in superconducting applications, only the solenoid will be considered here for illustrative purposes. Consider a solenoid whose length is much greater than its diameter (d), and where winding thickness is small compared to diameter, that is, a "long, thin solenoid." The solenoid is affected by two magnetic forces that should be considered during the design process, radial (outward) and compressive along the axis. The radial force (F_r) per unit area (a) is usually referred to as the magnetic energy density or "pressure" P_m:

$$P_m = \frac{F_r}{a} = \frac{B^2}{2\mu_o} \tag{4.32}$$

where the magnetic field B is a linear function of winding current density J, winding thickness τ, and permeability μ_o:

$$B = \tau\mu_o J \tag{4.33}$$

The axial force F_a acting on the ends of the solenoid (compressing the windings) has the form

$$F_a = \left(\frac{B^2}{2\mu_o}\right)\left(\frac{\pi d^2}{4}\right) = P_m\left(\frac{\pi d^2}{4}\right) \tag{4.34}$$

Note again the presence of the $B^2/2\mu_o$ magnetic pressure term, acting compressively on $\pi d^2/4$, the area of the bore of the long solenoid. (See Figure 4.44 for magnetic field-induced forces on a long solenoid.)

A range of values of B versus magnetic pressure are given in Table 4.4. If a cylinder of thickness t is used to support the solenoid, the hoop stress (S_H) and axial stress (S_A) will be (Stekly, 1973)

$$S_H = P_m\left(\frac{d}{2t}\right) \tag{4.35}$$

and

$$S_A = P_m\left(\frac{d}{4t}\right) \tag{4.36}$$

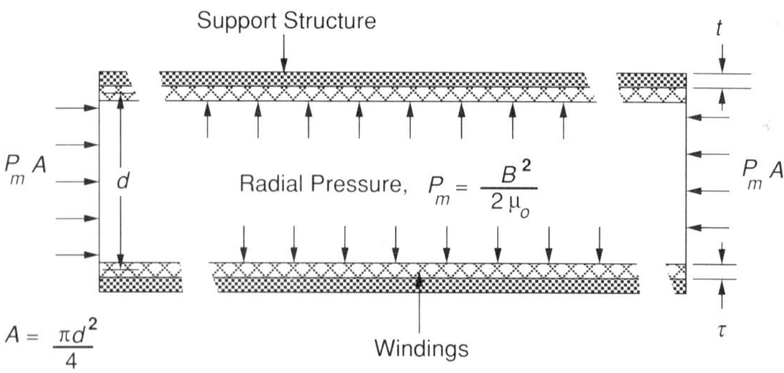

FIGURE 4.44 A long solenoid behaves as if there were an internal radial pressure and an axial compression with a magnitude proportional to the square of the internal flux density **B**. The support structure required to maintain the magnet under these pressures is the major limiting factor in the construction of high-field magnets.

TABLE 4.4 Magnetic pressure on a long, thin solenoid ($B^2/2\mu_o$) as a function of magnetic flux B.

Magnetic Flux Density (T),(Wb/m^2)	Magnetic Pressure on long, Thin Solenoid (N/m^2)	(lb/in^2)
0.1	3.96×10^3	0.6
0.2	1.59×10^4	2.4
0.5	9.9×10^4	1.5×10^2
1.0	3.96×10^5	6.0×10^2
2.0	1.59×10^6	2.4×10^4
5.0	9.9×10^6	1.5×10^3
10.0	3.96×10^7	6.0×10^3
20.0	1.59×10^8	2.4×10^4

(After Stekly, Z. J. J., "Superconducting Coils," in *The Science and Technology of Superconductivity*, Gregory, W. D., W. N. Mathews, Jr., and E. A. Edelsack, eds., Vol. 2., Plenum Press, New York, 1973, p. 533.)

NMR/MRI SUPERCONDUCTING MAGNET APPLICATIONS

Nuclear magnetic resonance (NMR), is now more commonly known as magnetic resonance imaging (MRI). This name change is evidently a public relations move to avoid mentioning the dreaded "N" word in the presence of a patient who does not realize that his body mass is primarily comprised of atomic nuclei.

MRI is used for noninvasive imaging of soft tissues without the need for exposure to ionizing radiation. This application of superconductivity is particularly significant aside from the important medical benefits, since MRI is currently the major market for superconducting magnets. Sales of MRI magnets are on the order of 500 per year at a cost of approximately $350,000 per unit. Cost of a typical commerical MRI system is ~ $1.5 million, generating revenues on the order of $750 million per year.

The most interesting technical aspect of superconducting MRI magnets is the requirement for very uniform *H* fields over relatively large volumes. MRI requires fields at the 10–20 T level, with a field uniformity of ~10 ppm over distances of approximately 0.5–1 m.

The MRI process depends on the existence of nuclear magnetic moments associated with quantized nuclear spins. When a magnetic field is applied, these moments give rise to distinct Zeeman nuclear energy levels. Applied *rf* electromagnetic fields can induce measurable spectroscopic transitions between these nuclear energy levels, leading to identification of specific nuclei.

The MRI technique involves the application of simultaneous *rf* and magnetic fields to study the transitions between nuclear spin states. The sample to be studied is surrounded by an *rf* induction coil and placed in a magnetic field on the order of several tesla. The magnetic field is swept through a small range with a small auxilliary coil. A separate *rf* pickup coil (orthogonal to the *rf* induction coil) is used to sense resonances as the magnetic field is varied. At fields of 10 T, resonance frequencies are on the order of a few tens of

MHz. For example, the H^1 nucleus has a resonance of 42.6 MHz in a field of 10 T. MRI units are primarily used to identify and localize malignant tumors.

MAGNETIC SEPARATION

Magnetic separation of ores is a time-honored technique that has been used with success for decades in the mining industry. Aside from the applications in the processing of iron ores, magnetic separation is also used in the processing of kaolin clay to remove dark stains (Iwasa & Montgomery, 1975). It is even possible to purify water by a process involving magnetic seeding prior to exposure to a high-field magnetic separator.

In such separation, one depends on the different magnetic characteristics of the materials to be mutually isolated. If the material to be separated is ferromagnetic, that is, iron, cobalt or nickel, or minerals such as magnetite, ilmenite, or pyrrhotite, relatively weak magnetic fields may be used and there is no need for superconducting magnets. Indeed, permanent magnets are usually sufficient. On the other hand, weakly magnetic minerals such as hematite, garnet, or limonite require the use of very intense fields for separation from a (nonmagnetic) matrix. Since surface tension may be comparable to the magnetic forces, such orepowders must also be quite dry for magnetic separation to be effective.

In purification of kaolin clay and coal desulfization high fields are required, as well as reasonably large separation volumes. For examination of this type of application, an experimental device has been assembled at the Institute of Electrical Engineering in Beijing. Parameters for this Nb-Ti superconducting magnet include peak fields of 5 T, and a separation chamber that is 40 cm long by 8 cm in diameter (Yan, et al., 1988). A pair of compensating coils are used to trim the field to a homogeneity of $\pm 5\%$ over the active volume. In tests of this apparatus with high-sulphur coal, reduction of total sulphur ranged between 30 and 47.5%, while reduction of sulphur of pyrite ranged from 70 to 83.3%. Ash was reduced on the order of 25–30%. In other tests, the investigators report that the increase in Kaolin clay brightness has reached about 4%.

In short, effectiveness for this class of superconducting-magnet separation is acceptable for both weakly and strongly magnetic minerals, with particle sizes ranging from microns to millimeters. As certain types of resources on the planet are depleted (leading to exploitation of lower-grade ores) and pollution increases, the use of magnetic separators can be expected to increase.

MAGNETIC LEVITATION (MAGLEV)

There are a variety of potential applications of magnetic levitation that are currently the topic of speculation, such as levitating runways for aircraft and evacuated levitation tubes for travel between cities at velocities on the order of 2000 mph. While such concepts will probably be relegated to the realm of science fiction for the next 100 years or so, there are fascinating applications of transport levitation currently under development.

Magnetic levitation or suspension of trains for speeds above about 250 mph appears attractive for relatively short distances between large cities such as Boston and New York or even Los Angeles and San Francisco (where the time spent at airports is comparable to the flying time).

Currently, the major players in the development of magnetically levitated trains are Japan and West Germany. While the world's fastest conventional train, France's TGV *(train à grande vitesse)* can push 190 mph at full throttle, both the normal and superconducting versions of MAGLEV are expected to coast along at an exhilarating 300 mph. Indeed, West German TR-06 has already reached 256 mph on a straightaway, and its successor (the TR-07) is expected to operate at 300 mph.

This German Transrapid (TR) development uses conventional magnet technology. The German group abandoned the superconducting approach in 1979, due to technical difficulty. The Japan Railways (JR) Group, somewhat behind the Germans in production plans, have spent nearly $400 million on their superconducting design that will operate at liquid helium temperature. The JR group has reached a speed of more than 320 mph on a 4.4 mi straight track.

The German group may reap the early commercial rewards of MAGLEV; there is serious talk of a 95 mi line between Hanover and Hamburg, at a cost of one billion dollars (Benjamin, 1988). MAGLEV routes are also being considered between Moscow and Leningrad, Montréal and Ottawa, as well as a line connecting Medina and Mecca in Saudia Arabia (Neffe, 1988). In contrast to this latter chariot for the prayerful are plans for a MAGLEV line betwixt Los Angeles and Las Vegas. This high-tech transit will presumably be devoted to propelling consignments of eager gamblers to that supreme expression of artificiality in the desert, where they will be skillfully separated from their greenbacks. One may safely assume that most of these newly-impoverished souls will return to the City of the Angels by less expensive, if more conventional, means of transport.

There are a variety of ways that magnetic levitation of trains can be achieved. One generic approach uses permanent magnets placed along the track which repel other permanent magnets mounted on the moving vehicle. A second scheme uses electromagnets on the vehicle to generate an attractive force via a continuous rail of laminated iron that is above the on-board electromagnet. This attraction lifts the vehicle above the railbed. Obviously, the electromagnets on the vehicle could be superconducting to provide high fields, which would provide additional lift force and greater bottom clearance. A third technique uses on-board superconducting rectangular loops to create eddy currents on a continuous conductive sheet on the railbed. As the vehicle moves, eddy currents are generated in the conductive sheet (see Figure 4.45). The resultant repulsion lifts the vehicle. Lift is a function of velocity (zero at zero velocity) so

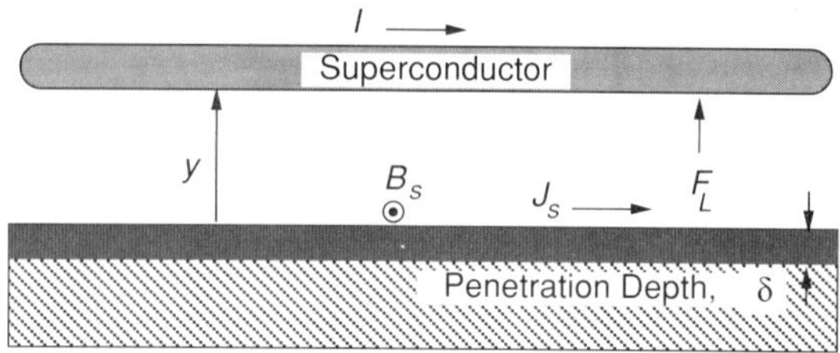

FIGURE 4.45 Relationship between rectangular-loop superconductor and sheet-conductor for induced eddy currents under one type of magnetically levitated train. (After T. A. Buchhold, "Superconductive Machinery," in *Applied Superconductivity*, V. L. Newhouse, ed., 2 vols., John Wiley & Sons, New York, 1975.)

that wheels must be provided for low-speed operation. The higher the velocity, the smaller the penetration depth of the eddy currents. Buchhold provides several illuminating expressions for this type of levitated vehicle, which will be discussed (Buchhold, 1975). (See Figures 4.46 and 4.47.)

If the superconducting loop (conducting a current of I A) is a distance y cm above the conducting sheet, the force F (per cm) on the superconductor will be

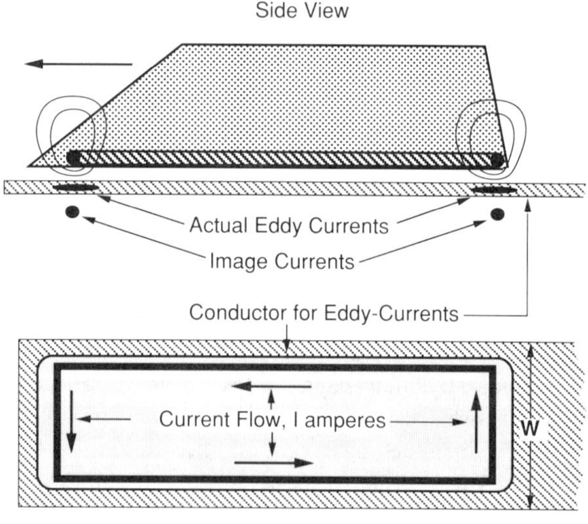

FIGURE 4.46 Magnetic levitation (MAGLEV) results from current flowing through internal coils generating H fields (when the vehicle is moving) that interact with eddy currents produced in a conductive sheet under the vehicle. In this illustration, the "image" currents are also indicated. (After T. A. Buchhold, "Superconductive Machinery," in *Applied Superconductivity*, V. L. Newhouse, ed., 2 vols., John Wiley & Sons, New York, 1975.)

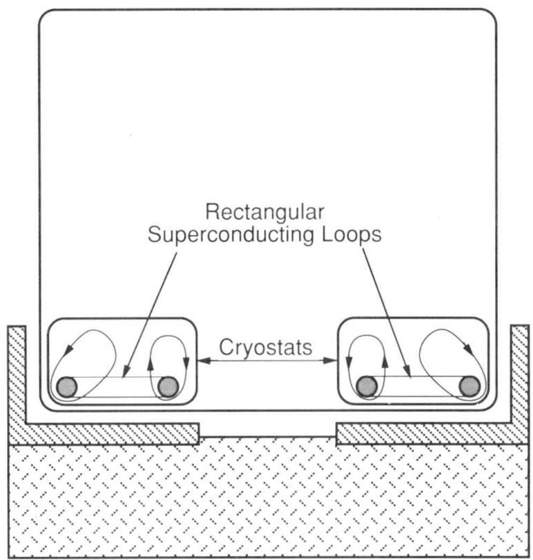

FIGURE 4.47 Stability can be achieved by using shaped conductors on the "railbed" to generate eddy-current fields that oppose sideways motion of the vehicle. (After T. A. Buchhold, "Superconductive Machinery," in *Applied Superconductivity*, V. L. Newhouse, ed., 2 vols., John Wiley & Sons, New York, 1975.)

$$F = \left(\frac{I^2}{2y}\right) 2.04 \times 10^{-8} \ \frac{\text{kg}}{\text{cm}} \qquad 4.37$$

If the current driving the system was sinusoidal, the skin depth into the sheet of electrical conductivity σ and magnetic permeability μ at a single frequency f would be the usual

$$\delta = \sqrt{\frac{1}{\pi f \mu \sigma}} \qquad 4.38$$

Since the current will not generally be sinusoidal, an empirical representation for skin depth is suggested (Buchhold, 1975):

$$\delta_{emp} = \frac{K_1}{\sqrt{v \sigma}} \qquad 4.39$$

In this form, the velocity (v) of the vehicle replaces the frequency term and K_1 is a value that is empirically determined for the system under consideration. The dissipation per cm^2 (P_s) of conductor surface is a function of the current density J_s and the surface resistance (R_s) of the eddy-current conductor. The surface resistance depends on skin depth (δ), however, and δ_{emp} is a function of vehicle velocity v:

$$P_s = J_s^2 R_s = J_s^2 \left(\frac{1}{\sigma \delta} \right) = J_s^2 \left(\frac{\sqrt{\sigma v}}{\sigma K_1} \right) = \left(\frac{J_s^2}{K_1} \right) \sqrt{\frac{v}{\sigma}} \qquad 4.40$$

A relevant factor here is that the loss in the conductive sheet under the vehicle is proportional to $v^{1/2}$. The flux density B_s at the surface of the conductive sheet is, of course, a function of the current density J_s:

$$B_s = J_s \left(\frac{4\pi}{10} \right) \qquad 4.41$$

The pressure P exerted by this flux density on the vehicle can be approximated (for small separation between sheet and vehicle, compared to sheet width) as

$$P = \left(\frac{B_s}{5000} \right)^2 \approx \left(\frac{J_s}{4000} \right)^2 \frac{\text{kg}}{\text{cm}^2} \qquad 4.42$$

Since the current density can be expressed in terms of the surface power dissipation and the empirical form for skin depth,

$$P = \left(\frac{1}{4000} \right)^2 P_s K_1 \left(\sqrt{\frac{\sigma}{v}} \right) \qquad 4.43$$

the "lift" pressure P is proportional to the dissipation per unit area of conductor P_s. If the loss per unit area is replaced with the total dissipation P_t, the pressure term becomes the total lifting force F_L:

$$F_L = \left(\frac{1}{4000} \right)^2 P \ K \left(\sqrt{\frac{\sigma}{v}} \right) \qquad 4.44$$

The total dissipation P_T causes a drag force proportional to this dissipation. The ratio of drag force (F_D) to lifting force (F_L) is

$$\frac{F_D}{F_L} = \frac{K_2}{\sqrt{\sigma v}} \qquad 4.45$$

where K_2 is a constant for a particular system. The significant point of this expression is that the drag force becomes excessive as v decreases; at velocities in excess of 200 mph, drag ratios on the order of 10–15% may be expected. (See Figure 4.48 for lifting force versus velocity.)

Magnetic levitation need not, of course, be relegated exclusively to the field of transportation. Engineers in Japan at Nippon Steel have studied the feasibility of continuous-strand steel casting where superconducting magnets would be used to generate fields both for confinement and floating of molten

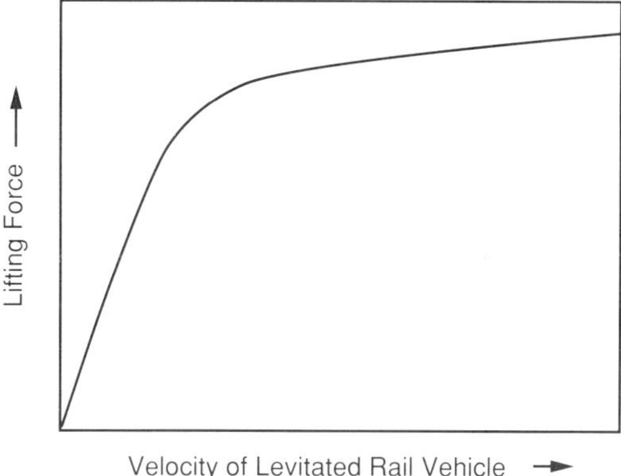

FIGURE 4.48 Since the force holding the magnetically levitated vehicle away from the eddy-current surface is proportional to velocity in the version of MAGLEV discussed previously, wheels must be supplied for starting and stopping. (After T. A. Buchhold, "Superconductive Machinery," in *Applied Superconductivity*, V. L. Newhouse, ed., 2 vols., John Wiley & Sons, New York, 1975.)

steel. This magnetic management of the molten product would be followed by further magnetic levitation of the cold steel for various in-plant handling requirements (Robyn, et al., 1988–1989).

MARINE PROPULSION: MOTORS AND MAGNETOHYDRODYNAMICS

It is hardly surprising that an interest in the use of superconducting electric motors developed in the 1960s, almost as soon as high-field superconductors became practical. The U.S. Navy supported a small program to investigate these applications. There was the usual problem: while the superconducting motors would be smaller and much more efficient than their normal counterparts, there would be the need for rather bulky liquid helium refrigeration apparatus. The U. S. Navy was less than enthusiastic. Japan has a very active development program in this same area; superconducting motors will turn the propellers on otherwise conventional ships.

MHD

The use of magnetohydrodynamic (MHD) forces is a far more novel notion for the propulsion of ships. While surface ships are being considered, a more interesting idea involves the use of MHD to drive large cargo-carrying submarines at relatively shallow depths. The basic approach is a magnetofluid pump; orthogonal **E** and **H** fields are applied to seawater; thrust is developed by the Lorentz force as the electrically conductive saltwater is pumped by the combined fields.

There are two basic designs. In the first, the orthogonal **E**-**H** fields are applied at the outside surface of the vessel. This has the advantage of efficiency; large volumes of fluid can be pumped. A drawback, particularly for military submarines, is that the enormous **H** fields (and perhaps even the **E** fields) could be easily detected. Indeed, metallic junk could even be attracted from the ocean floor!

Stewart Way and C. Devlin constructed and tested a model MHD submersible in the late 1960s (Cramer & Pai, 1973) (see Figure 4.49). This external-field submarine model had much the appearance of a torpedo, and carried five 217 A-h, 6V batteries for the energy source. While the center section was an 18-in diameter steel pipe, reinforced epoxy fiberglass was used to construct the tail and nose section. Current drain for the (nonsuperconducting) magnets and the seawater drain on the E-field electrodes was ~ 115 A. Some of the specifications are listed

Length	300 cm
Diameter	45 cm

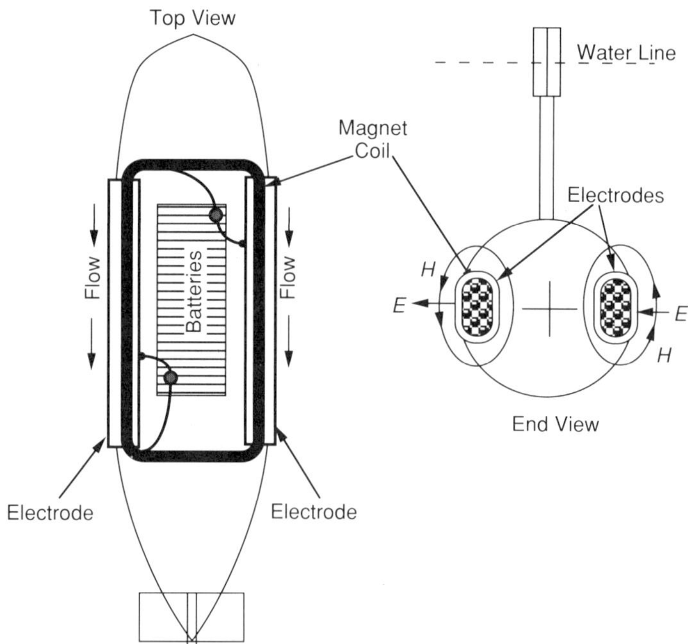

FIGURE 4.49 This sketch illustrates construction of the Way-Devlin MHD-propelled submarine. Length of the model was ~ 10 ft; power for the magnet winding and the electrodes was supplied by five 6 V lead-acid cells. Seawater served as the MHD fluid; the orthogonal **E** and **H** fields develop a Lorentz force that forces the water at the hull toward the rear of the submersible. While such designs do provide propulsion, superconducting magnets combined with an internal tube would be more practical for a variety of reasons, including nondetectability. (After K. R. Cramer and Shih-I Pai, *Magnetofluid Dynamics for Engineers and Applied Physicists*, McGraw-Hill (Scripta), New York, 1973.)

Displacement	408 kg
Lead-Acid Battery Weight	345 lb (156.5 kg)
Active Field Length	100 cm
Characteristic B Field	0.015 Wb/m^2
Design Speed	0.4 m/s
Thrust Power	0.472 W
Magnet	16,900 A/turns
Electrode Current	91.3 A

The model performed as designed; actual speed was up to 0.5 m/s; maximum magnetic field strength was 0.0155 Wb/m^2.

In the second approach, the electrodes are housed in a longitudinal tube that is contained within the vessel. While somewhat less efficient, this approach has the advantage that it is possible to largely confine the E and H fields within the vessel.

In Japan, MHD designs are currently being pursued by a combination of Kobe Steel Limited, Mitsubishi, and Toshiba. The seawater flowing through a long tube (or set of tubes) on the ship's hull serves as the fluidic electrical conductor. The currents forced through the conductor are subjected to a strong magnetic field, generating a force on the current-carrying water that pushes it through the tube. Anyone who has read Tom Clancy's *The Hunt for Red October* will be familiar with the notion of a water-jet propelled submarine. Clearly, eliminating the noise from propellers could lead to very quiet underwater vessels. Indeed, the importance of the successful development of such propulsion systems for ships has been compared to the improvements of aircraft performance which followed the development of the jet engine (Yoder, 1988).

UTILITY AND RELATED APPLICATIONS

Generators

Perhaps the most impressive elements in an electric utility installation are the massive electric generators, which may be driven by water power or steam. For the purposes of comparison, we will consider a rather simplified version of the conventional ac generator. As illustrated in Figure 4.50, this generator consists essentially of a magnetic-steel rotor that is enclosed in coaxial fashion by the stator that supports the armature windings. Copper field windings are in rotor slots; copper armature windings are in stator slots. As the rotor is turned, the rotation of the field flux cuts the armature windings to produce an alternating current output at the armature terminals. With an appropriate load and a sufficient level of field current, the output voltage is sinusoidal. Depending on local standards, frequency of operation is generally either 50 Hz or 60 Hz. For operation at 60 Hz, revolution is typically 3600 rpm.

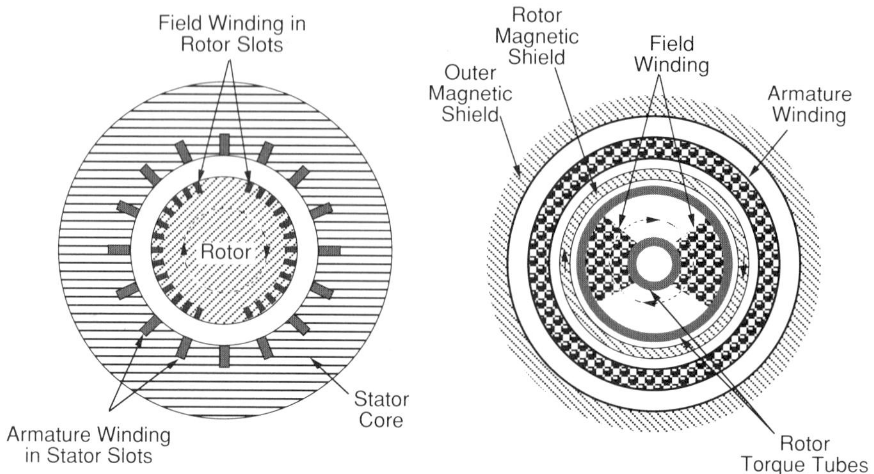

FIGURE 4.50 In a conventional generator (left), the field windings are carried in rotor slots. The rotor is stainless steel and the stator is comprised of sheets of magnetic steel. In the air-core superconducting version (right), the field winding is supported (and restrained) by coaxial torque tubes. The external magnetic shield confines the rotating magnetic flux to the generator, partly for safety reasons. The internal (rotor) magnetic shield prevents eddy-current induction in the rotor. (After J. L. Kirtley, Jr., "Application of superconductors to electric power generators," presented at 1988 Conf. on Electrical Applications of Superconductivity, Orlando, FL, September 21-23, 1988.)

The stator core, constructed from thin sheets of magnetic steel, is held at ground potential, thus limiting the voltage on the armature windings to approximately 26 kV (Kirtley, 1988). With the required insulation, armature-slot current densities are limited to approximately 300 A/cm^2 with the use of copper conductors. For such a machine to have appropriate transient characteristics and voltage regulation, the reactive impedance of the armature winding (i.e., "synchronous reactance") must be held below a certain level. This leads to the requirement that the air-gap between rotor and stator be larger than some minimum level, since synchronous reactance is inversely proportional to the air-gap. A larger air gap leads to the need for an increased excitation current in the field winding. In a conventional generator, the increase in excitation current leads to increased I^2-R losses in the field winding, which reduces overall efficiency of operation. A superconducting field winding would naturally be of some advantage in improving efficiency, but this is not the primary reason that superconducting generator configurations are being considered for utility applications. The reduction in size and weight, along with the capability of higher current densities, is of greater importance. Refer again to Figure 4.50, which illustrates an air-core generator design using superconducting windings.

The difference in design (compared to the conventional generator) is remarkable (Kirtley, 1988). The rotor now supports the field winding on coaxial "torque tubes;" the massive magnetic steel stator core has been eliminated. The torque tubes serve as torsional supports, but also restrain the field windings from the centrifugal forces and the enormous magnetic forces resulting from the intense field currents. The armature winding (placed in stator-slots on the conventional generator) is now an air-core winding coaxial to the torque tubes. The elimination of magnetic iron provides a number of advantages:

1. The inherent inductive reactance of the armature is greatly reduced, resulting in improved dynamic machine performance and voltage regulation.
2. Space for the armature winding is increased, which increases potential power density and generator efficiency.
3. By elimination of the interleaved stator iron (at ground potential), insulation requirements on the armature are reduced and/or much higher voltage may be delivered at the armature terminals. Higher terminal voltage can eliminate the need for a step-up transformer.

A pair of magnetic shields are used on the superconducting generator. The external shield serves the conventional purpose of shielding nearby metallic objects from eddy-current induction as well as preventing forces from being exerted on magnetic objects. The rotor shield (which turns with the rotor) serves to prevent alternating voltage induction in the rotor structural components and in the superconducting field winding.

It is important to note that superconducting generators, while benefitting enormously from the higher current densities that may be achieved with conventional superconductors, do not require the current densities (~100,000–200,000 A/cm^2) that are commonly necessary for a variety of high-field electromagnet applications. For operation of niobium-based superconductors at liquid helium temperature (4.2 K), current densities on the order of 16,000 A/cm^2 are quite adequate. Kirtley also reports that a similar current density (14,000 A/cm^2) would be required for liquid-nitrogen operation to be attractive for large generators (Kirtley, 1988). While this level of current density may be within striking distance for the high-temperature materials, the use of high-T_c ceramics is still made uncertain by a variety of factors, including the need to fashion reliable bulk conductors that will provide these current densities in high magnetic fields. Even so, Kirtley points out some other potential advantages of operating superconducting generators at 77 K in contrast to 4.2 K operation:

1. A liquid nitrogen cryogen system is more tolerant of small leaks and generally easier to maintain than a liquid helium system.
2. Machine reliability is improved because virtually any contaminant is a solid at liquid helium temperatures, potentially destroying gas bearing surfaces and/or clogging cryogen flow paths.
3. Since incremental heat capacity of solid materials is higher at 77 K than at 4.2 K, conductor stability is probably better at liquid nitrogen temperatures.

These advantages of 77 K liquid nitrogen systems over liquid helium operation clearly apply not only to large generators, but also to other high-field systems.

One is obviously not limited to the air-core choice when replacing conventional iron-core machines with superconducting versions. The air-core version has been presented here since it involves a rather complete departure from conventional machinery, illustrating the advantage of the reduction of inductive reactance with associated improvements in dynamic performance. While one may simply replace the copper windings in conventional machines with superconducting windings to achieve an increase in efficiency due to reduced I^2-R losses, this approach does not lead to an increase in the maximum flux density, since that is a function of the saturation characteristics of the iron. The presence of iron actually limits the performance of superconducting generators, since air-core superconductors are capable of producing flux densities much higher than the saturation density of iron. Nevertheless, smaller-scale applications where achievement of zero field loss would be significant, could benefit by the replacement of copper windings with superconducting windings.

It is generally accepted that the use of superconducting generators would primarily be advantageous for relatively large-scale utility applications, that is, ~ 500 MVA or greater. Machine design might be of a type where only the rotor was superconducting. While reducing losses would be moderately important, the major advantages would be reduced size and weight, and perhaps higher voltage output to eliminate the need for generator step-up transformers. Immediate market potential is not great, but the market by the year 2000 could become significant as it becomes necessary to replace older equipment and meet increased energy demands.

Experimental generators are being constructed and tested in a variety of locations. Since protection of these machines from short-circuits across the output is a serious problem, many machines are operated open circuit until design and experimental testing is complete.

In the United States, fundamental research on superconducting generators has been conducted at the Massachusets Institute of Technology. Experimental superconducting generators have been developed by the General Electric and Westinghouse corporations, with ratings of 20 MVA and 5 MVA, respectively, but these were not intended for deployment in the utilities industry. With funding from the U.S. Department of Energy, MIT has constructed a 10 MVA superconducting generator which is believed to be sufficiently robust for operation in a utility environment. Other work has been funded by the Electric Power Research Institute (EPRI).

In Japan, Hitachi has operated a superconducting 50 MVA generator under no-load conditions, and this machine has been installed into the power grid. Mitsubishi and Fuji Electric have collaborated to produce two experimental generators, rated at 6 and 30 MVA, and both of these have been operated under no-load conditions. Likewise, Toshiba Electric has tested a 6 MVA superconducting generator.

In Europe, several interesting small experimental machines have been tested, and several of these have novel configurations. At the Technical University of Munich, a 300 kVA "inside-out" machine has been constructed. In Grenoble, a freely spinning field-winding version of a 500 kVA generator has been tested. At the Technical University of Gratz, a full-sized rotor (i.e., large enough for a full-sized generator) has been cooled with supercritical helium. The Soviet Union has long had an active program for the development of superconducting machinery, for power levels up to 300 MVA (Kirtley, 1988).

To date, such generators have not yet demonstrated a capability to produce power output at the 100 MW level and have not been operated in the conventional manner to put power into an electric grid. In addition, even open-circuit tests are of short duration, usually limited to a few hours. In short, this is a relatively unexploited application of superconductors which requires considerable further development and holds great promise for improvement of dynamic performance and efficiency.

Superconducting Motors

While many of the problems of high-temperature superconducting motors are expected to be quite similar to those encountered with superconducting generators, the former subject nevertheless deserves some special attention. There are such a wide variety of electric motor applications, that these remarks will be primarily concerned with large-scale applications such as those found in the electric utility and oil refining industry. This would include, for example, motors for pump drives and high pressure fans with typical ratings of 10,000 to 20,000 hp. A significant factor in the future of large-scale motor applications is the requirement for variable-speed drives to provide increased efficiency at the lower speeds for fan and pump applications. This, in turn, will lead to increased use of adjustable-frequency power converters to drive such motors. Another factor that is expected to influence the design of large-scale motors is the trend toward factory-assembled power plants with outputs on the order of 200 MW (Jordan, 1988).

A variety of motor types might be considered for high-T_c superconducting designs (Lipo, 1988):

Induction
Induction/Synchronous Hybrid
Reluctance
Homopolar Inductor
Homopolar dc
Synchronous ac

While all of the motor types on this list have been considered for superconducting designs, the last two have actually been demonstrated in superconducting applications. The homopolar dc motor is generally considered to be an excellent architecture for superconducting applications since the coil is stationary and therefore easy to cool by submersion in the liquid cryogen. On the other hand, power requirements tend toward low voltage and high current. While this combination may cause only minor problems in providing an appropriate energy supply, high current flow through the contacts on the rotating portion of the motor could be a serious drawback for the homopolar dc configuration.

H. E. Jordan believes that the synchronous ac motor is quite appropriate for use with high-temperature superconducting ceramics (Jordan, 1988). He points out the following advantages:

1. Ease of adaption to air-core design.
2. The Cu armature may be adapted for a voltage rating appropriate for horsepower requirements.
3. Increased space available for the armature winding due to the absence of stator iron teeth reduces primary I^2-R loss for a given armature winding current density.

While these are significant advantages, there are drawbacks associated with the necessity to cool the rotating member. If high-temperature ceramics are used in the design, these must be able to withstand the mechanical forces to which the coil is normally subjected. Nevertheless, a design study that assumed a field-coil current-density of 10^5 A/cm^2 in an H-field environment of 5 T (a projection of future high-T_c ceramic performance) for a 10,000 hp motor design led to the conclusion that losses would be approximately 40% of those in a synchronous motor of conventional design. In addition, the plot of efficiency versus load for the superconducting motor is much flatter than that for the conventional design (Jordan, 1988). This is of considerable importance in applications where operation under reduced load conditions represents a large proportion of the motor service. In addition to these advantages, the volume of the superconducting motor was only 30% of the conventional design.

While the cooling requirements generally make large-scale motor applications more favorable than small ones, there are airborne and space applications where either the particular requirements of the application or the availability of cryogens lead to the consideration of small-scale superconducting motors. In a specific example, a commercial stepper motor has been modified with niobium-tin superconductors (Formvar-insulated and clad with 0.025 mm copper) for use in a piezoelectric fine-focusing mechanism for a cryogenic acoustic microscope (Moulthrop & Muha, 1988).

Bearings

The Meissner effect naturally leads to the consideration of superconductors for bearing applications. Consider a superconducting rod surrounded by a hollow magnetic shaft. The superconductor will repel the field from the magnetic shaft and thus repel the shaft in the same manner that magnets are levitated above superconductors. Friction is virtually eliminated, and high rotation velocity can be supported. Potential applications include gyroscopes and high-speed computer storage disk drives.

Superconducting Magnetic Energy Storage (SMES)

The public utilities have long needed reliable, large-capacity energy storage for load leveling, that is, the storage of energy during minimum-use periods, with discharge during peak use periods. Power-generation systems tend to be more economical and trouble-free when power can be delivered at a constant rate. Obviously, investors are not making income on steam-driven systems while they are shut down for the night. In addition, thermal cycling has a tendency to increase maintenance costs. Nuclear plants generally cannot be shut down on 24-hour cycles; they are usually operated continuously, at full output. If a hydroelectric plant is not run around the clock, energy is simply wasted.

A variety of energy storage techniques have been considered, ranging from flywheel systems to elevation and discharge of modest-sized lakes. Approximately 2.5% of the electric power used in the United States is stored for peak use, and all of this is pumped hydropower. As indicated above, water is pumped to an elevated location at night, and drained through turbines to generate power the next day. In West Germany and Japan, this proportion of stored energy is closer to 10%. Limitations on pumped hydropower include finding an appropriate site, relatively low (75%) efficiency, and environmental impacts.

It may well be that the largest impact of superconductivity technology during the first few years of the twenty-first century will be in SMES applications. The

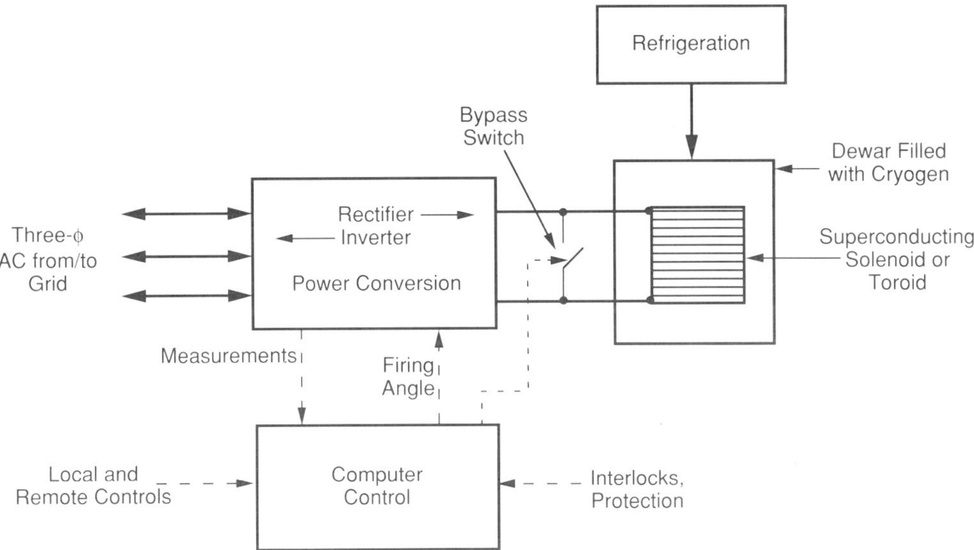

FIGURE 4.51 Schematic representation of a "Superconducting Magnetic Energy Storage" (SMES) system. The large superconducting solenoid (or toroid) would be buried underground and bathed in an appropriate cryogen. Such systems are under consideration by electric utilities for slack-period storage of energy that would be discharged during peak-requirement periods. There are also potential military applications for SMES systems, including short-duration discharge for operation of high peak power lasers.

SMES concept is simple; the energy is stored in the magnetic field of a rather large, underground superconducting solenoid (see Figure 4.51). For a persistent current level I in the SMES inductor L, stored energy is $LI^2/2$. The solenoid is reversibly connected to the external power system via a rectifier/inverter power conversion system. When demand is low, the ac line is rectified to supply dc current to the solenoid. During peak demand periods, the dc current in the solenoid is inverted to 60 Hz and fed back into the power grid.

The SMES is inherently low-loss, with ~95% efficiency expected. Most of the 5% loss (about 3%) is expected to be dissipated during the rectification of ac to dc and the inversion back to ac. The balance of the loss is primarily due to energy used in refrigeration of the cryogen.

SMES units are also very rapid in response, being able to switch from the storage to the discharge mode in a time on the order of 1 s. Such fast response could, in some situations, eliminate brownouts or even total blackouts.

To get an idea of the scope of SMES facilities, consider the representative specifications for coils used in a 100 MW-hr and a 5000 MW-hr installation using NbTi superconductors in a copper matrix (Huang, 1988):

100 MW-hr SMES:	
Coil diameter	150 m
Coil height	8 m
Coil weight	80,000 kg
Hoop Force	8×10^9 N
5000 MW-hr SMES:	
Coil diameter	630 m
Coil height	20 m
Coil weight	1,200,000 kg
Hoop Force	79×10^9 N

The force on the coils, particularly in the latter instance, is enormous. The restraining structure for the coils is one of the most expensive considerations in the construction of large-scale SMES installations. The cost scaling for an SMES installation is dominated by the fact that the H field is produced within a volume defined by a superconductor surface. For example, doubling the diameter and height of the coil increases stored energy by a factor of eight, while the cryogenic surface and amount of superconducting material is increased by a factor of four. In short, the cost of SMES installations (per unit energy stored) decreases with increasing size of the installation.

While a relatively small (30 MJ) SMES unit was built for the Bonneville Power Administration to stabilize the voltage on transmission lines fed by its Tacoma power plant, this installation was shut down after intractable problems were encountered with refrigeration. Construction of larger units is, at present, still in the planning stages.

In the United States, there are plans to fund development of SMES technology for military applications. The major Strategic Defense Initiative (SDI) application is related to a ground-based laser that would direct its beam toward an orbiting reflector. One estimate of the energy required is on the order of 1300–1600 MW for several minutes (Loyd, 1988). A typical nuclear power plant would be hard pressed to supply energy at this rate! The SMES approach seems to be particularly appropriate for this type of requirement, and it appears that funding will become available for some level of R&D. Initial planning is for design of a 20 MW-h inductor that would be approximately 100 m in diameter. Contracts for studies have been awarded to Ebasco Services and to Bechtel corporation.

If successful, such research and development could lead to large-scale SMES applicable to public utility energy storage requirements. Bechtel has designs for a 5000 MW-h unit that would be capable of charging or discharging at a rate of 1000 MW for a period of five hours. This design is capable of absorbing the night-time excess load from a nuclear power plant. Cost would be on the order of $900 million; this may seem like a large price tag, but it is less than the cost of a pumped hydropower plant with similar capacity.

The Electric Power Research Institute (EPRI) studies on the subject suggest that the market for large-scale SMES units could range between 10% and 30% of total U.S. generating capacity; obviously the demand increases as the price drops. Presumably, a similar market exists in other countries as well.

Energy Transmission

The interest in using superconductors to transmit energy has two basic driving forces. The first is that of technical advantage; it is worthwhile to consider any move that would significantly improve the efficiency of energy transmission. The second is aesthetic; the use of superconductive lines is one of the methods for putting the power lines underground. There have been a variety of complaints about high-voltage overhead lines aside from aesthetics, including the suspicion that the relatively high electric fields at ground level might have negative effects on the health of animals and humans.

The use of superconducting lines to transmit power for utility applications, is, at least in concept, simple. One must design the superconducting cables in a form where it is possible to provide cryogenic flow without significant thermal losses. See Figure 4.52 for a schematic illustration of a three-phase ac superconducting line. Simpler configurations can also be used for superconductive dc transmission at lower voltages.

Clearly, it is necessary to take considerable care to protect such lines from accidental rupture. The approach most generally considered is to bury the line so that the thermal environment is relatively stable, hopefully at a sufficient depth and well-marked location that is free from the depredations of those well-meaning road-builders and city plumbers who are seemingly drawn to subterranean gas and telephone lines as moths to flame. It would seem prudent to use rather substantial armoring to protect such transmission lines as well as the unsuspecting operator of the back-hoe! In addition to the necessity to protect these lines from accidental destruction, there is the question of

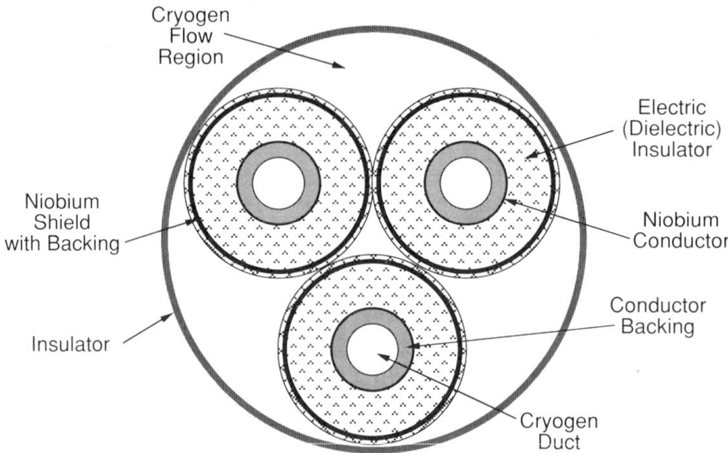

FIGURE 4.52 Three-phase superconductors for operation at 50–60 Hz could be placed underground for protection and reduction of heat absorption in the cryogen. (After H. M. Long and J. Notaro, "Design features of ac superconducting cables," *J. Appl. Phys.*, Vol. 42, No. 1, January 1971, pp. 155–162.)

maintenance, such as repair of small leaks. Mechanical engineers who have had the responsibility for maintaining localized cryogenic systems may approach the thought of managing such large, distributed systems with some trepidation.

The proposed use of superconducting power transmission lines also serves as an example of the economic dilemma that involves decisions to use superconductors in many other high-energy applications. It is estimated that losses of 60 Hz power in the United States account for approximately 5% of the energy transmitted on high-voltage lines. This is significant; the cost is on the order of $8–10 billion annually. Even so, the initial capital cost of constructing underground refrigerated lines does not make superconducting transmission economically attractive.

It has been pointed out that the only way to make such refrigerated transmission cost effective would involve the transfer of very large amounts of energy over a single line (Foner & Orlando, 1988). This approach suggests the image of a large city depending on a single transmission line for all its electrical energy. Clearly, the question of whether to use existing superconductor technology for a large project like power transmission raises many issues other than technological ones.

While there have been a variety of studies of the properties of superconducting power lines, two outdoor facilities have been constructed to study the problems that might be associated with realistic electric utility applications. One of these was in Austria where a relatively short superconducting cable was energized as a portion of an operational power transmission cable. The other was a 1000 MVA test site at the Brookhaven National Laboratory on Long Island. While each of these efforts was successful, both have been terminated. The Brookhaven experiment, which used a niobium-tin superconductor, was shut down by the U.S. Department of Energy during 1986, probably because the market for this rather expensive technology in the public utility industry was not at hand. Three paper studies of superconducting power transmission lines (including ac and dc approaches) indicate that the superconducting technology is on the order of three times as expensive as an overhead 500 kV line with comparable capabilities.

It is interesting that the critical current requirement for superconducting power transmission lines is not nearly as stringent as for a variety of other applications. In a list of "suggested characteristics" for a high-T_c power transmission application, Forsyth suggests that an operating current density of 2000 A/mm^2 (20 A/cm^2) and a critical current density of 15,000 A/mm^2 (1500 A/cm^2) are adequate (Forsyth, 1988). Bulk high-T_c ceramics can already perform this well. Additionally, there would be considerable advantage to using liquid nitrogen cryogen over liquid helium for this application.

Unfortunately, the brittleness and tendency to break under moderate stress are serious drawbacks for the high-temperature ceramics in this type of application. While laminates (i.e., normal conductor plated with superconductor) of the ceramic superconductors have been considered, the coefficients of thermal expansion of the superconductor and the normal substrate must match reasonably well if superconductor cracking is to be avoided during thermal cycling.

PARTICLE ACCELERATORS

Resonant Cavities and Superconducting Magnets

Superconductivity is applied to particle accelerator technology in two primary areas. To save energy and produce high H fields, superconducting magnets may be used. A prime example of such use is seen at Fermilab's Tevatron. The second area for accelerator application is the use of superconductors (usually niobium, occasionally lead) in the construction of very high-Q superconducting accelerator cavities. The primary purpose in this instance is to achieve high accelerating field gradients and to reduce the radio frequency power requirements. The CEBAF facility in Virginia is an example of this use of superconductors.

Accelerating structures are commonly categorized as "high β" or "low β," depending on whether low-mass particles (electrons, positrons) or heavier particles (protons, oxygen ions) are to be accelerated. Machines for acceleration of low-mass particles to nearly the velocity of light are referred to as $\beta = 1$ structures (β is defined here as v/c, where v is particle velocity and c is the velocity of light). For medium-energy proton accelerators, β is on the order of ~0.8–0.9, and for heavy-ion accelerators, β is considerably less than 1.

Superconducting Microwave Cavities. While superconducting microwave cavities find use in a variety of amplifier, oscillator, and filtering applications, perhaps the most dramatic resonant-cavity applications are those for particle acceleration. Superconducting magnets are commonly used in particle accelerators as well; the proposed superconducting super-collider [SSC] would use a large number of superconducting electromagnets.

The major purpose in using superconducting materials for accelerator cavities is that of reducing the cavity loss, that is, increasing the quality factor (Q):

$$Q \approx \frac{\Gamma}{R_s} \tag{4.46}$$

where Γ is a geometry factor that must have dimensions of ohms, with typical values of a few hundred Ω (Cohn, et al., 1988). R_s, for copper at room temperature (at 3 GHz), is on the order of 0.01 Ω, yielding a Q of 20,000 for a geometry factor of 200.

It is not uncommon, when using niobium cavities at 4.2 K, to reduce R_s to values on the order of 10^{-5} that of room-temperature copper. This reduction, in combination with a similar geometry factor Γ, will yield values for Q in excess of 10^9!

Power loss in the cavity wall is proportional to the square of cavity electric field E, frequency ω, and inversely proportional to Q, that is,

$$P \approx \frac{\omega E^2}{Q} \tag{4.47}$$

To achieve something approaching the ideal behavior of theoretical superconducting cavities, it is necessary to avoid any normal-conducting (or moderately lossy) impurities on the cavities' inner surface. In practice, this requires considerable preparation of the niobium surface by acid etching and cleaning with very pure distilled water. Niobium accelerator cavities are normally prepared and assembled in a clean room to minimize the possibility of contaminating the surface.

A reduction of surface resistance (R_s) by up to five orders of magnitude is realized by replacing room-temperature copper accelerator structures with niobium at 4.2 K. This is an enormous benefit in reducing losses and increasing attainable accelerating gradients. Early problems with multipacting, a high vacuum resonant electron avalanche process, have been largely dealt with by redesign of the conventional cavity configuration. When designing superconducting accelerator structures, one prefers cavities with a spherical or elliptical inner surface to avoid the resonances in electron field emission commonly seen with more conventional pillbox structures.

In the United States, the use of superconducting cavity structures began in earnest in 1964, with the construction by Stanford's High Energy Physics Laboratory of its "Superconducting Recyclotron," a recirculating electron linear accelerator (linac) for nuclear physics experiments. While there had been hope of achieving unprecedented high accelerating gradients (on the order of 20

FIGURE 4.53 Five-cell superconducting Nb cavity designed for use in the CEBAF particle accelerator. (Photo provided by Babcock & Wilcox, Inc.)

MV/m) in this pioneering effort, multipacting in the resonant cavities limited fields to approximately 10% of this ambitious goal. Beam current was limited to a few microamperes in multipass acceleration; this was largely due to beam breakup caused by undamped higher order cavity resonances or "modes." In subsequent designs, considerable effort was made to damp these higher-order modes. The Q at these higher-order modes (HOM) is denoted Q_{HOM}.

During the same time period, the development of low-β ($\beta \ll 1$) resonators for the acceleration of heavier particles began at Argonne National Laboratory (ANL), Caltech, and Karlsruhe. The ANL heavy-ion linac began operation in 1978, serving as a post-accelerator for the Argonne tandem Van de Graff.

Concerted efforts are underway at a number of locations in the United States (Cornell, Stanford), Europe (CERN, DESY, Karlsruhe, Orsay, Wuppertal), and Japan (KEK) to develop improved $\beta = 1$ superconducting rf structures. As a result of advances achieved by these research groups, it is not uncommon to achieve accelerating gradients of 10 MV/m in multicell cavities and gradients of 20 MV/m in single-cell cavities (Hartline, 1988).

A variety of particle accelerators and storage rings have been constructed during the past 25 years, and more are planned or under construction. An example is the U.S. Department of Energy's Continuous Electron Beam Accelerator Facility (CEBAF) machine in Newport News, Virginia.

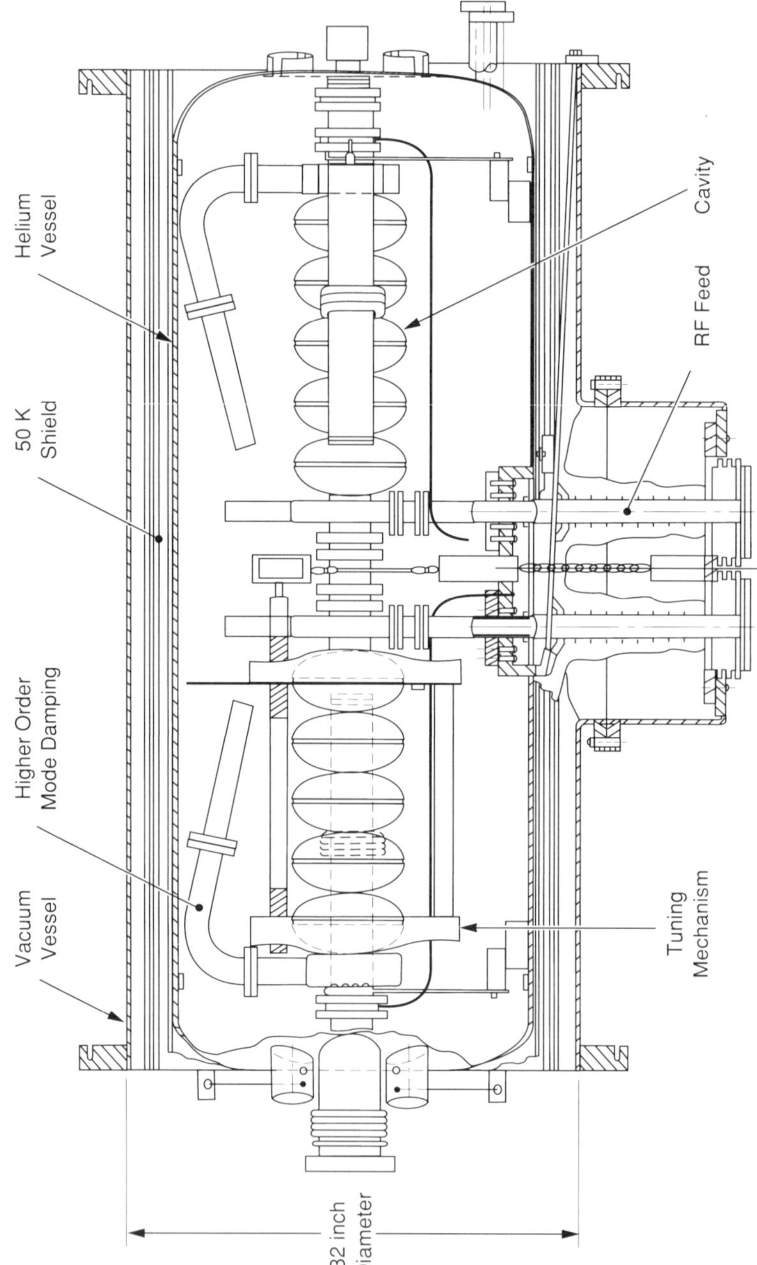

FIGURE 4.54 Detail of CEBAF Nb cavity pair in cryostat. (Illustration provided by CEBAF, Newport News, VA)

CEBAF, planned for completion in 1994 at a cost of $265 million, is a superconducting, recirculating, linear electron accelerator. Performance objectives are:

Energy (E)	0.5 GeV $\leq E \leq$ 4.0 GeV
Beam Current (I)	$I \leq 200\ \mu$A
Duty Factor	Continuous

The primary purpose of this machine is the study of the nuclear many-body problem and quark substructure, as well as the strong and electroweak interactions that determine the behavior of this state of matter. CEBAF's superconducting niobium cavity structure was developed by a group at Cornell University, and has been successfully prototyped by industry. The industry prototypes have met or exceeded CEBAF's specifications:

Frequency	1497 MHz
Accelerating Gradient	$\geq$ 5 MV/m
Unloaded Q (Q_o)	$\geq 2.4 \times 10^9$ @ 2 K
Loaded Q (fundamental mode)	6.6×10^6
External HOM Q (damped)	$10^3 \leq Q_{\text{HOM}} \leq 10^5$
Shunt Impedance	960 Ω/m

The goal of future developments in niobium accelerator cavities will be lower surface resistance ($R_s \sim 10^{-7}\ \Omega$) and higher accelerating gradients (~15–20 MV/meter). (See Figure 4.55 for the assembled Nb accelerator cavity pair.)

FIGURE 4.55 Assembled Nb accelerator cavity pair. (Photo provided by CEBAF, Newport News, VA)

How and when will the new high-temperature ceramics be applied to particle accelerator cavities? There is, quite naturally, considerable interest in exploiting these new materials. The primary reason is the usual one: simplicity and lower cost of liquid-nitrogen refrigeration compared to liquid helium refrigeration. The higher heat capacity of liquid nitrogen is another important factor, since most particle accelerator systems generate considerable heat. Recall that, even with the best niobium superconductor, there is a finite loss in the presence of alternating fields, and that these losses increase as ω^2.

Unfortunately, the bulk copper oxide materials tend to have a value of R_s that is orders of magnitude greater than niobium, and are rather limited in transporting sufficient current for developing electric magnetic fields required for acceleration of particles. While textured films of Y-Ba-Cu-O have relatively higher values of critical current density (J_c) and lower values of R_s than bulk polycrystalline samples, plating the inside of a copper cavity with textured films is hardly an appealing task.

Aside from the technical difficulty of such a procedure, the frequent arcs experienced in accelerator cavities could be expected to play havoc with such a thin, delicate structure. Nevertheless, a careful investigation of the application of high-temperature superconductors in particle accelerator cavities is justified when one considers the advantages that could accrue with success. If early

FIGURE 4.56 Fully assembled CEBAF Nb cavity pair in cryostat. Waveguides for *rf* power can be seen at the center of the assembly. (Photo provided by CEBAF, Newport News, VA)

experiments yielding very high J_c in untextured thallium superconducting films are replicated, this superconductor might be far superior to Y-Ba-Cu-O for accelerator-cavity applications.

In any case, the requirements of superconducting microwave cavities in filters, oscillators, and so forth are much less stringent and these will undoubtedly be among the first to benefit from the high-T_c materials.

Superconducting Supercollider (SSC) Magnets

The proposal for a superconducting supercollider particle accelerator using 10,000 niobium-titanium magnets operating at 6.6 T and cooled by liquid helium has raised questions in the U.S. Congress about delaying the development until the new high-temperature materials could be used in the magnets. Superficially, this argument has some appeal. If the new materials could be used to produce reliable magnet conductors, one could consider cooling with liquid nitrogen and realize significant savings in refrigeration. Unfortunately, there are two major problems.

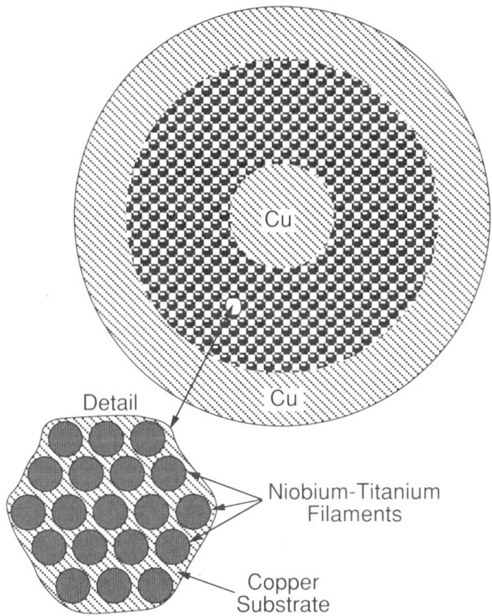

FIGURE 4.57 Cross-section of a filamentary superconducting cable similar to those used in high-field magnets for particle-accelerator applications. Filaments are closely spaced in the copper matrix to maximize critical current. Diameter of superconducting filaments may be as small as a few μm, with critical current density in excess of 200,000 A/cm^2 for fields of several T. (After T. S. Kreilick and E. Gregory, "Further improvements in current density by reduction of filament spacing in multifilamentary Nb-Ti superconductors," *Cryogenics*, Vol. 27, July 1987, pp. 401–403.)

The first is technological: the ceramic superconductors appear to be a long way from being capable of being used as magnet wires. As has been pointed out earlier, the limitations for application of bulk material is due to the relatively low critical current densities and quenching in magnetic fields that are extremely low compared to the intrinsic levels of H_{c2}. Each of these limitations is apparently related to problems at the grain-boundary, where current flow is mediated by Josephson-like effects via weak links.

The second limitation is related to the overall significance of the actual savings that could be predicted. The cost of magnets (using current technology) on the SSC is only about 5% of the cost of the $5–6 billion machine. It is probably not advisable, since answers to the fundamental questions in high-energy physics are desired promptly, to delay the development and construction of the SSC based on potential savings that would probably be on the order of 1–2% of total costs, even if the high-temperature superconductor technology was ready to use. The cross-section of a superconducting magnet "wire" is illustrated in Figure 4.57.

ANTENNAS, WAVEGUIDES

The new high-temperature superconductors may have advantages in antenna applications. It is important in many applications to have antennas that are small compared to wavelength of the energy to be transmitted or received. This is not a particularly efficient approach with metals in the normal state, since the ohmic losses in the antenna may be large compared to the effective radiation resistance. Clearly, any means of reducing this inherent resistance would tend to improve radiation efficiency, and this observation leads quite naturally to the consideration of superconducting antennas where such would be practical. (See Figure 4.58 for a superconducting Y-Ba-Cu-O dipole antenna.)

Low-temperature superconductors have been used to construct fractional-wavelength antennas, leading to a significant improvement in radiation efficiency. Obviously, the use of liquid helium as a cryogen tends to limit the application of such antennas. The possibility of operating superconductors at liquid nitrogen temperature (77 K) or even higher might open up a variety of applications for electrically short antennas. Applications in space are one apparent potential for such antennas. A group in the United Kingdom has reported progress in this area (Gough, et al., 1988). A short dipole was operated at 550 MHz in liquid nitrogen. The experimental antenna consisted of a matching network and short dipole constructed from Y-Ba-Cu-O wire, mounted on a Tufnul substrate. The assembly was placed in liquid nitrogen, inside an unsilvered glass dewar. Radiation was detected by an external antenna, while the internal assembly was driven at 550 MHz. Results were compared to those using an identical copper antenna assembly, which was operated both at 77 K and at 300 K.

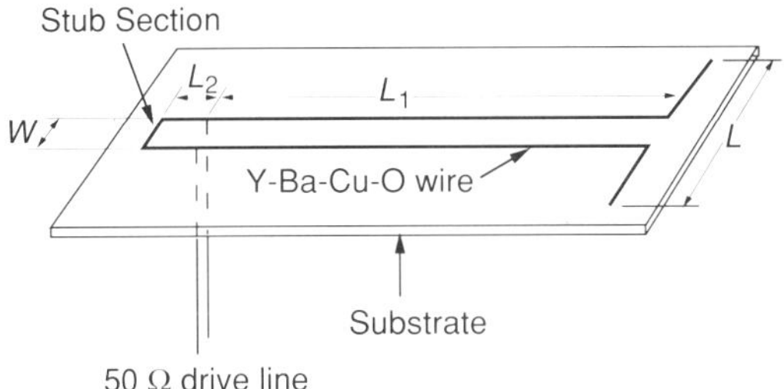

FIGURE 4.58 Superconducting antennas have significant advantages over normal-state metal antennas when dimensions are short compared to signal wavelength. In this case, an electrically short Y-Ba-Cu-O dipole is illustrated that performed much better than a copper antenna (of the same size) at 550 MHz. (After C. E. Gough, S. K. Khamas, T. S. M. Maclean, M. J. Mehler, N. McN. Alford, and M. M. Harmer, "Critical currents in a high-T_c superconducting short dipole antenna," presented at 1988 Applied Superconductivity Conference, San Francisco, August 21–25, 1988, (EK-6), Abstracts, p. 56.)

The superconducting Y-Ba-Cu-O antenna exhibited a gain enhancement of 12 db above the room-temperature copper antenna, and 6 db enhancement over the same copper antenna at 77 K. With the expected improvements in ceramic superconductor critical current, it is likely that these encouraging results will be improved on. With high temperature ceramics, antennas on the order of 5% of the size of normal antennas could be operated in liquid nitrogen.

Another potential application for superconductors is in the construction of electromagnetic waveguides. The advantage over conventional metal waveguides (usually aluminum or brass) would be at the higher frequencies. In the case of mm-sized waveguides, attenuation becomes prohibitive except for applications where the guide length is very short, that is, usually less than a meter. At mm wavelengths, conventional metal guides have attenuations on the order of 10 db/m due to the high value of surface resistance (R_S) of the metal walls at ~200 GHz.

The use of conventional low-T_c superconductors would go a long way in alleviating this problem, but the cost of liquid helium for kilometer-long waveguide runs would be impractical for most applications. For the usual economic reasons involving replacement of liquid helium by liquid nitrogen, there is interest in using the high-T_c ceramics for mm-waveguide applications (Winters & Rose, 1988). Unfortunately, the R_S value for the ceramics is currently too high for mm-waveguide applications. It is expected that advances in thin-film ceramic superconductor technology will eventually make such applications practical.

DETECTION OF VISIBLE LIGHT AND INFRA-RED

An ideal bulk superconductor, if operated at a temperature well below T_c, is expected to be a perfect reflector of electromagnetic radiation at low-to-moderate frequencies. If, however, the energy associated with a particular photon frequency ($h\nu$) is greater than the superconductor's energy gap the material will absorb the electromagnetic radiation much like a normal metal. Several of the earliest investigations of the energy gap relied on measurements of *IR* and microwave absorption and reflection. Although such measurements are now well understood for "conventional" superconductors, behavior of the newer copper oxide ceramics is quite complex. Nevertheless, there are important applications for photon detection.

Superconducting films are used for detection of microwave and *IR* radiation. Superconducting photon detectors have the following advantages over their semiconducting counterparts:

1. Less power is required.
2. Response is faster.
3. Bandwidth is greater, extending to long-wave *IR*.

To be more specific on (3), the sensitivity of conventional superconducting detectors extends to wavelengths of several hundred μm, while typical semiconductor detectors are useful to no more than $\sim$ 20 μm.

The detection techniques are generally classified as thermal (i.e., bolometric) or nonthermal. In the bolometric case, the temperature of the sensor must be increased by the absorption of radiation. In the latter, one depends on a nonlinearity in the voltage-current characteristic of the film (Rose, et al., 1975). In the nonthermal mechanism, one depends on the creation of quasiparticle pairs induced by the incident photon. The actual measurement is of a reduced gap energy, that is, a modification of the Josephson junction current-voltage curves (Talvacchio, et al., 1989).

Consider a superconductor biased with a bias current I_b; for relatively small changes in temperature (ΔT) or current (ΔI), the change in voltage (ΔV) across the film is approximately:

$$\Delta V = I_b\left(\frac{\partial R}{\partial T}\Delta T\right) + \Delta I\left(\frac{\partial V}{\partial I}\right) \qquad 4.48$$

where the partial derivative of resistance (R) with respect to temperature is often replaced by the symbol γ. The second term accounts for the nonlinearities just mentioned, while the first term represents the thermal (bolometer) effect. While either the bolometric or nonlinear responses may be used in photon detection, the bolometric response is relatively slow. The response time for the nonlinear or "nonequilibrium" response is determined by the quasiparticle recombination time, which may be on the order of 100 ps for a strong-coupling superconductor operated at a temperature near T_c (Talvacchio, et al., 1989). The bolometric response will be considered first. (See Figure 4.59.)

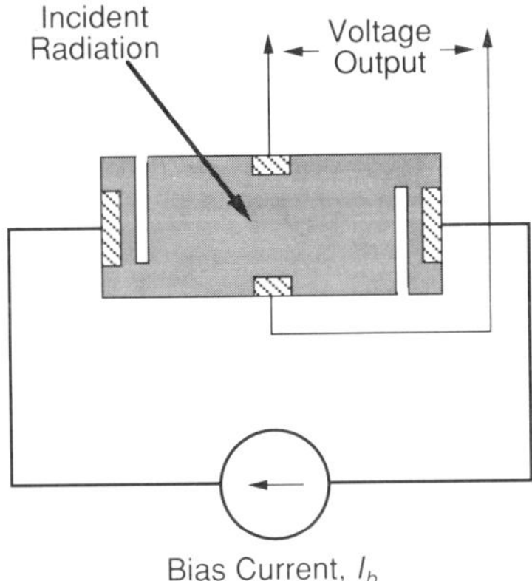

FIGURE 4.59 Schematic diagram of a superconducting thin-film bolometer for long-wave *IR* or microwave detection. Detectors of this type are limited by Johnson noise as well as by noise in the preamplifier. (After K. Rose, C. L. Bertin, and R. M. Katz, "Radiation Detectors," in *Applied Superconductivity*, Vol. 1, V. L. Newhouse, ed., Academic Press, 1975, pp. 267–308.)

The thermal term represents one of the more useful characteristics of high-quality superconductors; that is, the rapid increase of resistivity with very small changes in temperature when operated near T_C. Rather sensitive far-infrared detectors (using niobium nitride, for example) may be realized simply by biasing a superconductor near T_C with a combination of external temperature control and the level of applied current.

If the thickness of a superconducting film is d and the real part of conductivity is σ (which is generally complex at the frequencies of interest here), the resistance (per square) is

$$R_S = \frac{1}{\sigma d} \qquad 4.49$$

The fraction of incident power (normal to the film's surface) absorbed is related to the ratio of the characteristic impedance of free space ($Z_O = 377\ \Omega$) and the resistance of the film (R_S). A radiation coupling coefficient may be represented by α where the ratio of Z_O to R_S is given by g (Bertin & Rose, 1968, 1971a, 1971b)

$$\alpha = \frac{2g}{(g + 1)^2} \qquad 4.50$$

This expression is applicable only when the film thickness (d) is not large compared to skin depth (δ):

$$\delta = \sqrt{\frac{2}{\omega\mu\sigma}} \tag{4.51}$$

Such films are typically biased at a current level (I_b) that is approximately 10% of the critical (quenching) level (I_c). The sample is typically placed in one leg of a bridge circuit for detection of the small voltage. Very small *IR* signals ($\sim 10^{-12}$ W) create sufficient changes in resistance to be detectable. It is even possible, using this technique, to detect single alpha particles in thin tin films. Individual alpha's generate a detectable increase in temperature at the collision location.

A useful parameter for specifying the performance of a thermal detector (bolometer) is the responsivity, (r) which is merely the output (rms) voltage divided by the input power (W):

$$r = \frac{V_o}{P_{in}}\left(\frac{\text{volts}}{\text{watt}}\right) \tag{4.52}$$

A more specific formula for responsivity is (Rose, et al., 1975)

$$r = \frac{\alpha F \gamma I_b}{K_e\sqrt{1 + \omega^2\tau^2}} \tag{4.53}$$

where K_e is an "effective" thermal conductance,

$$K_e = K - I_b^2\gamma \tag{4.54}$$

K is the thermal conductance (W/K) of the bolometer material, and τ (s) is the bolometer's thermal time constant

$$\tau = \frac{C}{K_e} \tag{4.55}$$

where C is the effective heat capacity (J/K) of the bolometer at the temperature of operation. Response times are generally on the order of a few nanoseconds. The bias current must be limited so that K_e does not become negative; thermal runaway will result if

$$I_b \geq \sqrt{\frac{K}{\gamma}} \tag{4.56}$$

Some investigators who measure the response of small-area, weak-link detectors find it more appropriate to define responsivity in terms of volts/(watt/cm^2) (Osterman, et al., 1988).

As in all detectors, one must first overcome the system noise before considering the gain of the system. The Johnson (i.e., thermal) noise within a specified bandwidth (f_H-f_L) across the terminals of a resistance (R) operating at an absolute temperature (T, Kelvin) is given by

$$v_n(\mathrm{rms}) = \sqrt{4k_B T R(f_H - f_L)} \qquad 4.57$$

where k_B is Boltzman's constant. Due to the presence of Johnson noise, there is no advantage to be gained by increasing the resistance of the film for the purpose of increasing the signal-to-noise ratio. The fact that thermal noise is an important factor in the design of superconducting *IR* detectors leads quite naturally to a discussion of high-temperature superconductors in these applications.

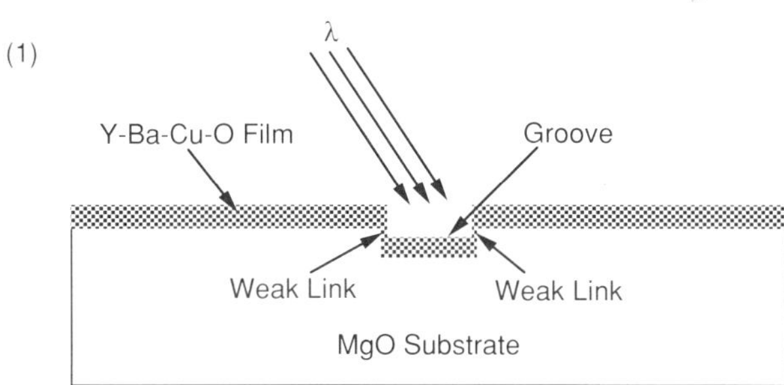

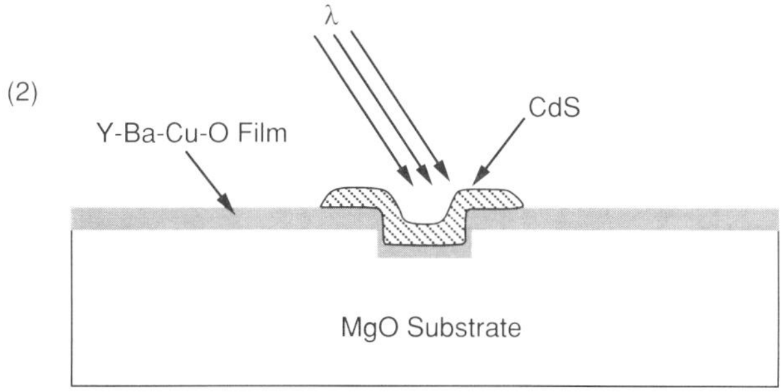

FIGURE 4.60 Weak links in Y-Ba-Cu-O films are used as photon detectors. Current in the weak link decreases with exposure to the photons, probably due to quasiparticle excitation. In the lower illustration, a film of CdS photoconductive semiconductor has been vacuum-deposited over the weak link, greatly enhancing detection sensitivity. (After U. Kawabe, "Application of high-temperature superconductors to SQUID and other devices," Proc. of Int. Conf. on High Temperature Superconductors, Interlaken, Switzerland, *Physica C*, Vol. 153–155, Superconductivity, June 1988, pp. 1586–1591.)

Weak-Link IR Detection. Small changes in temperature may also be detected by single-particle (quasiparticle) tunneling devices, utilizing the phenomenon referred to as Giaever tunneling. The basic sensor is a pair of thin superconducting films separated by an insulating oxide layer, that is, an SIS junction. Virtually no current flows until the threshold voltage exceeds $2\Delta/e$, where 2Δ is the superconductor energy gap and e is the electron charge. Tunneling of electrons will occur at lower voltages (than $2\Delta/e$) in the presence of photons of frequency ν if the incident photon energy is at least equal to the difference in the energy gap and eV :

$$h\nu \geq 2\Delta - eV \qquad 4.58$$

where V is the gap bias voltage. See Figures 4.60 and 4.61.

Because such detectors do not require high values of J_c, and since 1-2-3 polycrystalline material comes equipped with multiple grain boundaries, that is, "built-in" weak links, *IR* detection may be one of the earliest applications to exploit $YBa_2Cu_3O_x$ copper oxide superconductors.

A group from NTT Opto-Electronics Laboratories in Japan has tested a polycrystalline film of $BaPb_{0.7}Bi_{0.2}O_3$ for photon detection (Enomoto, et al.,

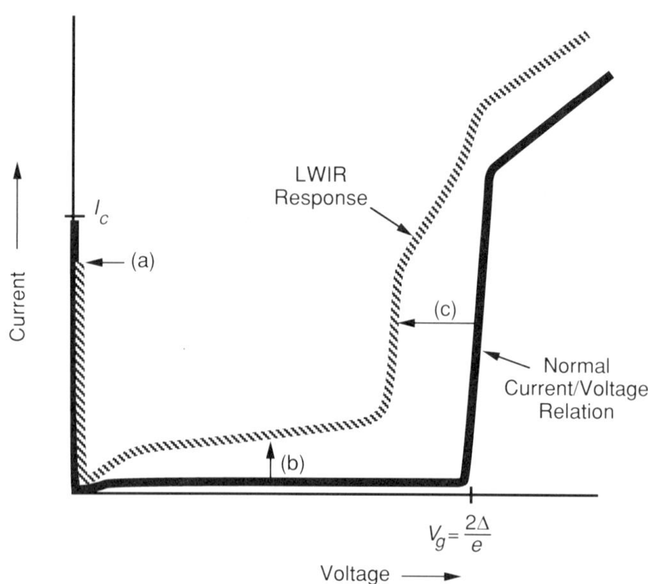

FIGURE 4.61 In Josephson junction long-wave infrared (LWIR) detection, the general departures induced in the more standard Josephson junction *I-V* curve include (a) decreased critical current, (b) increased current below the voltage transition, and (c) shift of the voltage transition below the usual $2\Delta/e$ level.

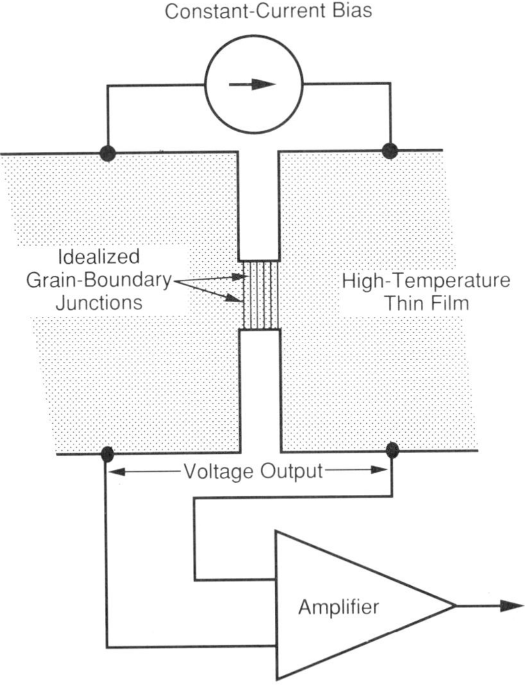

FIGURE 4.62 Bias arrangement for a high-temperature ceramic weak-link LWIR detector. In this application, the Josephson-like weak links at the boundaries of the Y-Ba-Cu-O grain junctions, rather than representing a limitation as in most other applications, function as the IR detection mechanism.

1988). Quasiparticles, created by incident photons as they break Cooper pairs, are detected by Josephson junctions at the grain boundaries. At wavelengths above 1 μm, sensitivity was 10^4 V/W. The authors point out that high-T_c 1-2-3 oxide films are also capable of operating in this fashion, but assert that response times are three orders of magnitude slower than the $BaPb_{0.7}Bi_{0.2}O_3$ film, which is capable of operation up to 1 GHz at a 1 μm wavelength. See Figure 4.62 for the bias arrangement.

As pointed out earlier in this chapter, an increase in operating temperature (from liquid helium 4.2 K to liquid nitrogen 77 K, for example) can be expected to increase noise, assuming all other factors are constant. Nevertheless, there is considerable interest in detectors that could operate at higher temperatures. Some applications are limited by the cost of refrigeration; deep space detectors using high-T_c materials might operate without any artificial refrigeration at all.

The U.S. Naval Research Laboratory has reported the successful testing of *IR* detectors constructed of evaporated $YBa_2Cu_3O_{7-\delta}$ film grains in a nonsuperconducting matrix. A pulsed *IR* laser on the current-biased film produced a response time (τ) of approximately 10 ns (U.S. Navy, 1987).

Even though most of the early experiments with high-T_c materials have produced only bolomeric response, it has been demonstrated that fast, non-bolometric response can be seen with sufficiently thin ($\sim$ 80 nm) films (Kwok, et al., 1989).

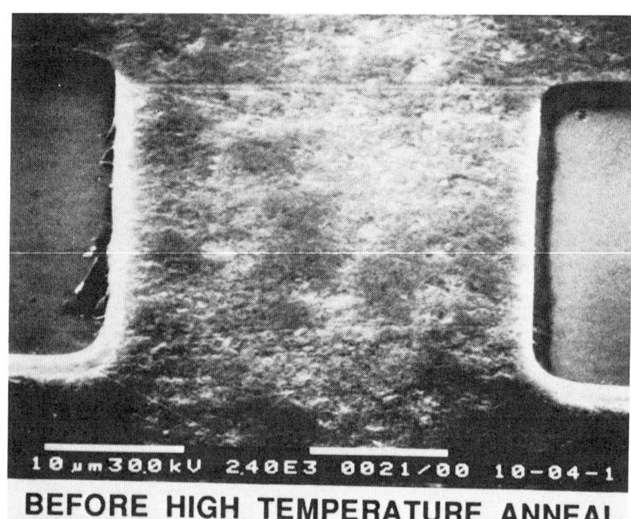

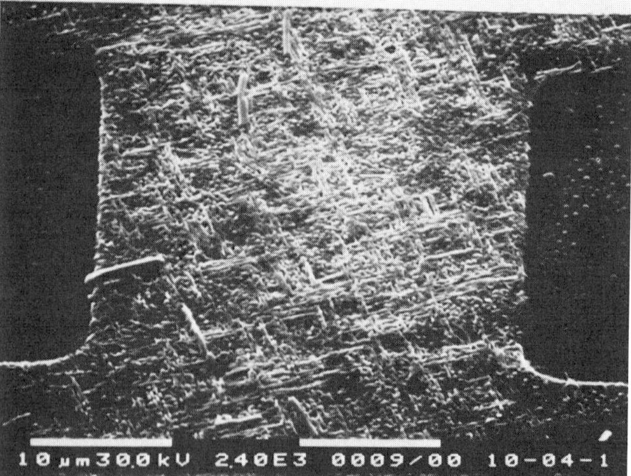

FIGURE 4.63 Y-Ba-Cu-O LWIR detector bridge structure before and after anneal. Note smooth texture prior to anneal and appearance of crystalline structure following anneal. For weak-link structure, anneal temperature is 850°C, compared to 900°C when the weak links are not desired. (Provided by Group MEE-11, Electronics Research and Exploratory Development, Los Alamos National Laboratory)

FIGURE 4.64 Array of experimental Y-Ba-Cu-O weak-links for use as LWIR detectors. Substrate is strontium-titanate. (Provided by Group MEE-11, Electronics Research and Exploratory Development, Los Alamos National Laboratory)

MAGNETIC SHIELDING

For those engineers who have had to deal with the problem of shielding sensitive circuitry and electronic sensors from external electromagnetic fields, there is little need to make a case for the importance of any improvements that might be gained from the electromagnetic properties of superconductors. With appropriate geometries, it is possible to provide lightweight, effective shielding that would be economically impractical with ordinary materials due to the large mass of (normal) shielding that would be required. There is, of course, the requirement for cryogenic systems to cool the superconductor well below T_c, and this factor limits the use of superconductive shielding to applications where the expense and additional complexity of the system can be justified by the need to improve over conventional shielding.

Shielding from high-frequency electromagnetic fields and low-frequency electric fields is relatively straightforward with conventional materials. Alternating magnetic flux in the incident wave induces voltages in the conductor; the induced voltages create currents that generate magnetic fields to cancel the incident **H** fields. This is, of course, a statement of Lenz's law. As long as the thickness and radius of curvature of the conducting surface is large compared to skin depth, shielding using ordinary conductors such as copper or aluminum is relatively straightforward and inexpensive. Generally, ~20 skin depths is considered sufficient for effective shielding. At *rf* frequencies, therefore, shielding applications do not usually require the use of superconductors.

In contrast, shielding of low-frequency (or especially dc) magnetic fields normally involves the use of relatively expensive (and sometimes very thick) magnetic materials. One may consider these materials to be "short-circuiting" the applied magnetic flux. The reduction of these low-frequency or dc fields to arbitrarily low values becomes essentially a problem of cost and the mass of shielding that can be tolerated for a particular application.

In space applications, the weight of the shielding is the limiting factor. If superconducting shields are to replace normal magnetic shields for such applications, the added weight of any cryogenics systems must be considered. In cases where cryogenics supplies are already on board, this may not be a significant factor. If the high-T_c ceramics prove to be useful, some space applications will not require cryogens for the superconducting shield, that is, ambient temperature will suffice.

The concepts of zero (dc) resistivity and the Meissner effect, which have already been introduced are, of course, of critical importance in the use of superconductors for electromagnetic shielding. Limiting factors in the use of superconductors for shielding applications involve the critical field (H_c) above which the material "quenches" and behaves like a normal conductor and the fact that superconductor resistivity increases with frequency of the applied electromagnetic field. The subjects of flux pinning and flux creep are also critical to shielding for low-frequency or dc magnetic fields.

Recall Ohm's law and Maxwell's equation based on Faraday's induction law:

$$\mathbf{J} = \sigma \mathbf{E} \tag{4.59}$$

and

$$\nabla \times \mathbf{E} = -\frac{\partial \mathbf{B}}{\partial t} \tag{4.60}$$

E must be zero since dc conductivity (σ) is infinite, but if **E** is zero then **B** cannot change with time in the superconductor. If external magnetic fields (at the superconductor surface) are changing in time, and **B** cannot change in the superconductor, Lenz's law requires that screening currents be developed on the surface of the material. The screening currents must develop superficial fields to cancel the external **H**-field effects inside the superconductor. Since the conductivity is infinite, the currents are persistent. The arguments so far only take infinite conductivity into account and this is quite sufficient for shielding at high frequencies. For dc magnetic shielding the Meissner effect and flux quantization must also be considered.

If one had a perfect Type I superconductor shield in small **H** fields, all magnetic flux would be expelled as the shield is cooled below T_c. Shielding would be effective until surface fields exceed H_c. In practice, even rather pure Type I materials will contain a small concentration of sites where flux will be pinned and therefore not expelled in the superconducting state. While the number of pinning sites can be reduced by annealing the shield following fabrication, there are always some limiting number of pinning sites at grain boundaries and dislocations. The point is, the Meissner effect is never perfect. It is therefore important to reduce the amount of flux that must be expelled during the transition from the normal to the superconducting state, and the most common technique is to reduce the ambient field as low as possible during cooling. Once the transition is complete, the shielding should be effective until the critical field (H_c) is exceeded.

Flux pinning may even be reduced by spinning the shield while it is cooled through the phase transition (Cabrera & Hamilton, 1973). Eddy-currents induced by spinning the shield in the ambient magnetic field tend to screen the time-varying field (that results from the spinning). This rotation effectively excludes the ambient flux as the spinning shield temperature passes below T_c.

AC Shielding

Most of the screening currents on a Type I superconductor will flow between the surface and the London penetration depth (λ_L), typically a few hundred Å.

In the normal state, surface resistance is a function of the skin depth (δ) and conductivity (σ) of the conductor, that is,

$$R_s = \frac{1}{\sigma\delta} \tag{4.61}$$

the surface resistivity (R_s) tends to increase with frequency, since skin depth is inversely proportional to the square root of frequency:

$$\delta = \sqrt{\frac{1}{\pi f \mu\sigma}} \tag{4.62}$$

If one is primarily concerned with the required thickness of the shield, superconductors are superior to normal metals until a frequency is reached where skin depth for a normal metal becomes comparable to London penetration depth for the superconductor (λ). For example, one might wish to know the frequency where copper would have a penetration depth comparable to a superconductor with $\lambda \sim 500$ Å, (5×10^{-6} cm). From the previous expression,

$$f = \frac{1}{\pi\mu\sigma\delta^2} \tag{4.63}$$

For copper at room temperature, σ is approximately 5.65×10^5 mhos, $\mu = \mu_o = 4\pi \times 10^{-9}$ H/cm, the relevant frequency for a skin depth of 5×10^{-6} cm is approximately 1.8×10^{12} Hz, or 1800 GHz.

This approach is oversimplified, however. A comparison of skin depth and London penetration depth is not the dominating consideration. The superconductor will become normal at a cutoff frequency (f_{co}) where the "photon" energy hf_{co} is equal to the full energy gap (2Δ):

$$f_{co} = \frac{2\Delta}{h} \approx \frac{3.53 k_B T_c}{h} \approx 72.8\text{GHz}(T_c) \tag{4.64}$$

The approximation sign appears in this expression since the $2\Delta = 3.53k_BT_c$ relationship is approximate and, in any case, is not yet firmly established for the variety of high-temperature superconductors now available. If we consider a superconductor such as niobium, with $T_c = 9.46$ K, the corresponding cutoff frequency is $\sim$ 688 GHz.

For conventional superconductors, f_{co} typically falls in the range of 100–1000 GHz. To the extent that the high-T_c superconductors have a higher energy gap (i.e., if $2\Delta \sim 3.53kT_c$), the cutoff frequency will also be higher, perhaps on the order of 10^{13} Hz.

There is interest in using the high-temperature ceramic superconductors in shielding applications for the usual reasons: the availability and low cost of liquid nitrogen.

DC and Low-Frequency Shielding

Consider a shield of cylindrical configuration with radius R, its axis aligned with the z-axis (see Figure 4.65). If the external field is H_e, (also along the z-axis) the axial magnetic field in a cylinder where no flux could penetrate the material would be

$$H_z = H_e\epsilon^{-3.8z/r} \tag{4.65}$$

In practice, values for the exponent constant of 3.4 have been achieved by a spinning technique (Cabrera & Hamilton, 1973). The transverse component should also decrease exponentially:

$$H_t = H_e\epsilon^{-1.8z/r} \tag{4.66}$$

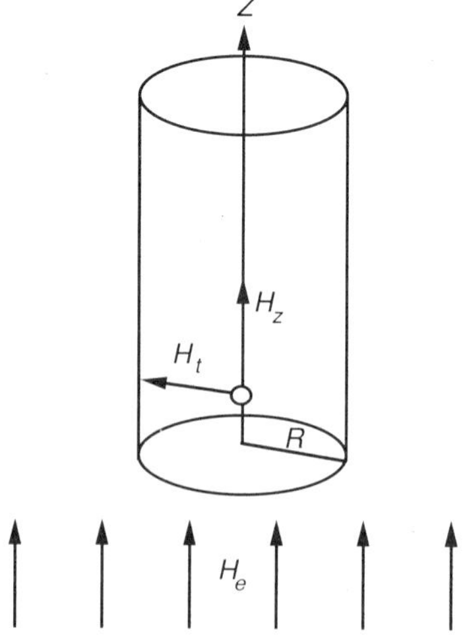

FIGURE 4.65 Radial H-field along the axis of a cylindrical shield. See the text for discussion.

An interesting technique for further decreasing the trapped field during the cooling process involves use of a collapsed or "folded" shield that is then "inflated" or expanded when $T < T_c$ (see Figure 4.66). A fixed amount of flux is trapped by the deflated shield, but when the shield is expanded (below T_c), the internal flux density is reduced due to the greater effective area of the *expanded* shield. This method can (theoretically) be repeated, that is, expanding shields within expanding shields, to eventually provide a net shield where no trapped lines will exist in the innermost shielded volume. By this method, fields have been decreased on the order of 500 per folded enclosure (Cabrera & Hamilton, 1973). Such configurations are also referred to in the literature as "bladders" and "balloons" (see Figure 4.67).

There is, of course, great interest in the possibility of using the new high-T_c superconductors in shielding applications. Shielding measurements on Y-Ba-Cu-O samples have been made in the frequency range between 1 and 100 kHz (Hattori, et al., 1988). In these experiments, the sample under investigation was placed between a pair of induction coils; the effect of shielding in modifying the coupling coefficient (k) and phase of the voltage induced in the secondary coil was measured to determine shielding effects. The configuration was not optimum for the development of effective shielding.

Using discs of Y-Ba-Cu-O, shielding ratios of 125 (at 12 Hz) to 6000 (at 15 kHz) were measured (Macfarlane, et al., 1988). These particular samples

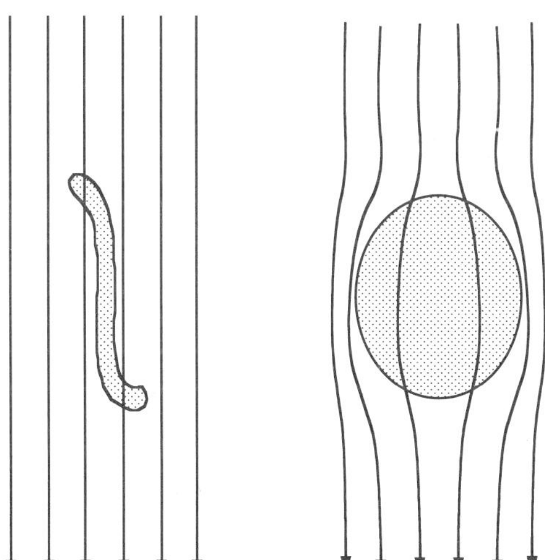

FIGURE 4.66 If a collapsed Type II superconducting shield is unfolded or "inflated" after temperature is reduced well below T_c, only the original number of pinned fluxoids remain and the internal flux density will be reduced. This technique can be improved on by nesting multiple collapsed shields, all inflated after cooling. (After B. Cabrera and W. O. Hamilton, "Electric and Magnetic Shielding with Superconductors," in *The Science and Technology of Superconductivity*, W. D. Gregory, W. N. Mathews, Jr., and E. A. Edelsack, eds., Vol. 2, Plenum Press, New York, 1973, pp. 587–606.)

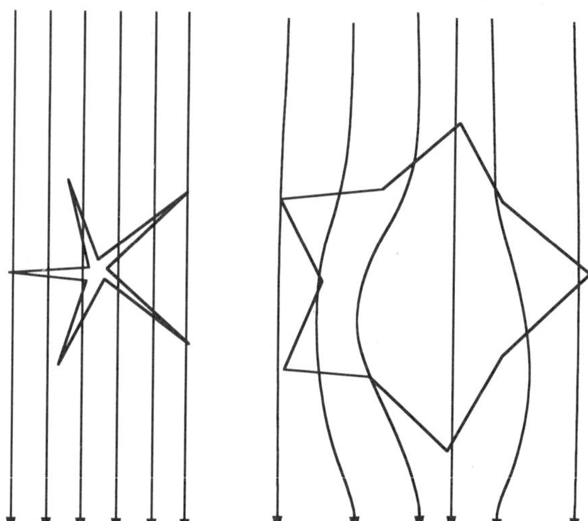

FIGURE 4.67 Another version of the collapsed shield shown in the previous figure. (After B. Cabrera and W. O. Hamilton, "Electric and Magnetic Shielding with Superconductors," in *The Science and Technology of Superconductivity*, W. D. Gregory, W. N. Mathews, Jr., and E. A. Edelsack, eds., Vol. 2., Plenum Press, New York, 1973, pp. 587–606.)

were stable, even following brief exposure to water. Others have measured the dc magnetic shielding characteristics of hollow Y-Ba-Cu-O tubes, and have determined that this particular high-T_c ceramic exhibits a maximum shielding capability that is in general agreement with calculations based on the Bean critical state model (Willis, et al., 1988). The effectiveness of 1-2-3 materials in shielding applications is limited by imperfect flux pinning, flux creep, and weak links; shielding properties will improve as these problems are solved. (See Figure 4.68 for apparatus for shielding measurments.)

One group of investigators has reported "intensification" or "focusing" of magnetic fields passing through a $YBa_2Cu_3O_7$ disc (Marshall, et al., 1988). These results were seen with the disc cooled both before and after being placed over a magnet. Such an effect (not replicated or understood at the time of this writing) would have obvious implications for shielding.

SUMMARY

The major advantage of high-T_c superconductors at present is implied in the name, a high critical temperature (~90–125 K) which will make it possible to operate certain superconducting applications at the liquid nitrogen temperature of 77 K, while others should be cooled to $T_c/2$ or less. Cooling with liquid nitrogen is quite attractive in applications where the refrigeration component is a major part of the entire system, since lower cost, higher reliability, and

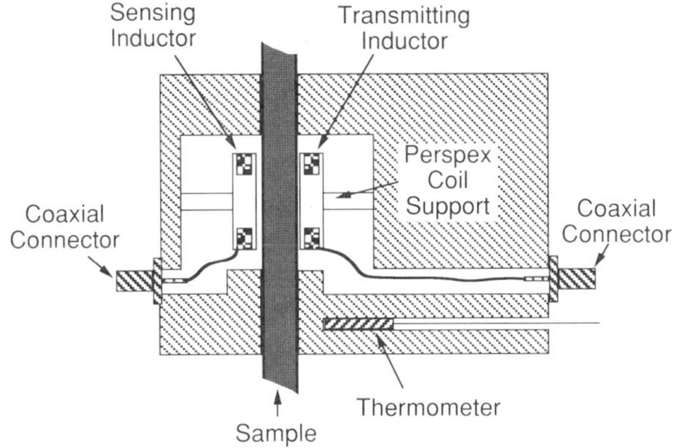

FIGURE 4.68 Apparatus for shielding measurements on Y-Ba-Cu-O superconducting samples. In this experiment, signals were applied to the transmitting inductor. The output signal at the receiving inductor was an indication of the shielding capability of the sample. Data was taken above and below T_c to determine the effect of the superconducting state in improving shielding. Shielding ratios increased from 125 to 6000 as frequency was changed from 12 Hz to 15 KHz. (After J. C. Macfarlane, R. Driver, R. B. Roberts, E. C. Horrigan, and C. Andrikidis, "Electromagnetic shielding properties of yttrium barium cuprate superconductor," Proc. of Int. Conf. on High Temperature Superconductors, Interlaken, Switzerland, *Physica C*, Vol. 153–155, Superconductivity, June 1988, pp. 1423–1424.)

smaller size of such refrigeration systems can be expected. Examples include NMR imagers, use of Josephson junctions in applications such as the SQUID magnetometer, far-infrared, or microwave detectors as well as low-frequency electromagnetic shields and coatings for high-Q resonators.

While rapid progress is being made in the design of bulk conductors, the widespread use of high-T_c superconductors in high current density applications may require considerable more development. Note that high current applications include not only the obvious uses such as high-field magnets. Computer-chip applications (primarily to gain speed and reduce size) also require current densities on the order of 10^6 A/cm^2 in some instances. Even though the peak currents in a conductor carrying a digital signal may be on the order of only tens of milliamperes, conductor cross-sections in some instances are sufficiently small to produce current densities that will press the capabilities of even the highest-quality high-T_c films. If history is any guide, designers of the newest systems will specify a current level and conductor cross-section that will exceed the capabilities of the best superconducting material!

The notion of operating conventional semiconductors at 77 K (in association with $YBa_2Cu_3O_x$, Bi-Sr-Ca-Cu-O, or thallium compounds) is attractive because this is almost an ideal temperature for semiconductor operation. (At far lower temperatures the charge carriers introduced by doping become ineffective, while at higher temperatures semiconductor lattice vibrations increase losses and reduce speed of operation.) Currently, however, the high processing temperatures required for the superconducting ceramics would require separate assembly of the semiconducting and superconducting portions of the system or circuit.

Progress is being made in the area of manufacturing hybrid combinations of high-temperature superconductors and semiconductors. Fujitsu is reportedly using a 600°C process to manufacture superconductor and semiconductor components on a single ceramic film. Hitachi, meanwhile, is using a 450°C microwave-assisted plasma reactive evaporation system to form a high-T_c superconducting thin film over a silicon substrate (Sakurai, 1989).

While it is quite natural to consider which of the existing applications of "conventional" superconductors might be exploited with high-T_c ceramics, one should also take care to be on the lookout for new applications that take advantage of some of the unusual characteristics of the copper oxide materials, which include a much smaller coherence length (ξ, 10–15 Å), a larger energy gap (Δ), higher upper critical field (H_{c2}, 250–300 T?), and, of course, the higher temperature of operation.

While even well-informed scientists are properly cautious about prospects, some observers do not hide their cynicism about the future of high-temperature superconductors:

> . . . research continued as physicists looked for superconductors that would work at higher temperatures. It was a long shot at best, but they somehow knew that a room-temperature superconductor would have applications in space travel, nuclear power, and the refrigerator-magnet industry.
>
> (Runté, 1988)

REFERENCES

Al'tov, V. A., V. B. Zenkevich, M. G. Kremlev, and V. V. Sychev, (1977). *Stabilization of Superconducting Magnetic Systems*, Trans. by G. D. Archard, Plenum Press, New York, pp. 48–54.

Benjamin, D., (1988). "Floating trains: what a way to go!" TIME, August 22, pp. 58–59.

Bertin, C. L., and K. Rose, (1968). "Radiant-Energy Detection by Superconducting Films" *J. Appl. Phys.*, Vol. 39, No. 6, May, pp. 2561–2568.

Bertin, C. L., and K. Rose, (1971a). "Comparison of Superconducting and Semiconducting Bolometers," *J. Appl. Phys.*, Vol. 42, No. 1, January, pp. 163–166.

Bertin, C. L., and K. Rose, (1971b). "Enhanced-Mode Radiation Detection by Superconducting Films," *J. Appl. Phys.*, Vol. 42, No. 2, February, pp. 631–642.

Buchhold, T. A., (1975). "Superconductive Machinery," in *Applied Superconductivity*, Newhouse, V. L., ed., 2 vols., John Wiley & Sons, New York.

Cabrera, B., and W. O. Hamilton, (1973). "Electric and Magnetic Shielding with Superconductors," in *The Science and Technology of Superconductivity*, W. D. Gregory, W. N. Mathews, Jr., and E. A. Edelsack, eds., Vol 2., Plenum Press, New York, pp. 587–606.

Clarke, J., (1988). "Small-scale analog applications of high-transition-temperature superconductors," *Nature*, Vol. 333, No. 6168, May 5, pp. 29–35.

Cohn, D. R., L. Bromberg, W. Halverson, B. Lax, P. P. Woskov, (1987). "Possible high-frequency cavity and waveguide applications of high temperature superconductors," Massachusets Institute of Technology, PFC/JA/87–49, October 1.

Cramer, K. R., and Shih-I Pai, (1973). *Magnetofluid Dynamics for Engineers and Applied Physicists*, McGraw-Hill (Scripta), New York, p. 223.

D'Addario, L., (1988). "Saturation of the SIS mixer by out-of-band signals," *IEEE Trans. Microwave Theory and Tech.*, Vol. 36, No. 6, June, pp. 1103–1105.

Delaney, M. A., R. S. Withers, A. C. Anderson, J. B. Green, and Robert W. Mountain, (1987). "Superconductive Delay Line with Integral MOSFET Taps," *IEEE Trans. Magn.*, Vol. MAG–23, No. 2, March, pp. 791–795.

Enokihara, A., H. Higashino, K. Setsune, T. Mitsuyu, and K. Wasa, (1988). "Superconductivity in 2-μm wide strip line of Gd-Ba-Cu-O thin film fabricated by low-temperature process," *Jap. J. Appl. Phys*, Vol. 27, No. 8, August, pp. L1521–1523.

Enomoto, Y., T. Murakami, and M. Suzuki, (1988). "Infrared optical detector using superconducting oxide thin film," Proc. of Int. Conf. on High Temperature Superconductors, Interlaken, Switzerland, *Physica C*, Vol. 153-155, Superconductivity, June, pp. 1592–1597.

Falco, C. M., and I. K. Schuller, (1981). "SQUIDs And Their Sensitivity for Geophysical Applications," in *Squid Applications to Geophysics*, Weinstock, H., and W. C. Overton, Jr., eds., Soc. of Exploratory Geophysicists, Tulsa, pp. 13–18.

Foner, S., and T. P. Orlando, (1981). "Superconductors: The Long Road Ahead," *Technology Review*, Vol. 91, No. 2, February-March, pp. 36–47.

Forsyth, E. B., (1988). "Underground superconducting power transmission systems and the outlook for this technology," presented at 1988 Conf. on Electrical Applications of Superconductivity, Orlando, FL, September 21–23.

Goto, E., (1959). "The parametron, a digitial computing element which utilizes parametric oscillation," Proc. IRE, Vol. 47, August, pp. 1309–1316.

Gough, C. E., S. K. Khamas, T. S. M. Maclean, M. J. Mehler, N. McN. Alford, and M. M. Harmer, (1988). "Critical currents in a high-T_c superconducting short dipole antenna," presented at 1988 Applied Superconductivity Conference, San Francisco, August 21–25, (EK-6), Abstracts, p. 56.

Green, J. B., L. N. Smith, A. C. Anderson, S. A. Reible, and R. S. Withers, (1987). "Analog Signal Correlator using Superconductive Integrated Components," *IEEE Trans. Magn.*, Vol. MAG–23, No. 2, March, pp. 895–898.

Guéret, P., A. Moser, P. Wolf, (1980). "Investigations for a Jospehson Computer Main Memory with Single-Flux-Quantum Cells," *IBM J. Res. Develop.*, Vol. 24, No. 2, March, pp. 155–166.

Hamilton, C. A., and F. L. Lloyd, (1988). "A 10-V Josephson voltage standard," presented at 1988 Applied Superconductivity Conference, EA-2, San Francisco, August 21–25, Abstracts, pp. 1–2.

Harada, Y., H. Nakane, N. Miyamoto, U. Kawabe, E. Goto, and T. Soma, (1987). "Basic operations of the quantum flux parametron," *IEEE Trans. Magn.*, Vol. MAG-23, No. 5, September, pp. 3801–3807.

Hartline, B. K., (1988). "CEBAF: A superconducting radio frequency accelerator for nuclear physics," CEBAF publication, Newport News, VA.

Hattori, T., N. Higemoto, S. Kanazawa, and M. Kobayashi, (1988). "Magnetic shielding using high-T_c superconductor," *Jap. J. Appl. Phys.*, Vol. 27, No. 6, June, pp. L1120–L1122.

Hayakawa, H., (1986). "Josephson computer technology," *Physics Today*, March, pp. 46–52.

Huang, C. Y., (1988). "Applications of Superconductivity," Lockheed Missiles & Space Co. Publication.

Iwasa, Y., and D. B. Montgomery, (1975). "High-field superconducting magnets," in *Applied Superconductivity*, Vol. 1, Newhouse, V. L., ed., Academic Press, pp. 387–487.

Jordan, H. E., (1988). "Feasibility study of electric motors constructed with high temperature superconducting materials," presented at 1988 Conf. on Electrical Applications of Superconductivity, Orlando, FL, September 21–23.

Kawabe, U., (1988). "Application of high-temperature superconductors to SQUID and other devices," Proc. of Int. Conf. on High Temperature Superconductors, Interlaken, Switzerland, *Physica C*, Vol. 153–155, Superconductivity, June, pp. 1586–1591.

Kirschner, I., M. Lamm, T. Porjesz, J. Matrai, G. Y. Kovacs, T. Trager, T. Karman, J. György, and G. Zsolt, (1988). "First experiments with a high–T_c superconducting magnet," Proc. of Int. Conf. on High Temperature Superconductors, Interlaken, Switzerland, *Physica C*, Vol. 153–155, Superconductivity, June, pp. 1417–1418.

Kirtley, J. L. Jr., (1988). "Application of superconductors to electric power generators," presented at 1988 Conf. on Electrical Applications of Superconductivity, Orlando, FL, September 21–23.

Kreilick, T. S., and E. Gregory, (1987). "Further improvements in current density by reduction of filament spacing in multifilamentary Nb-Ti superconductors," *Cryogenics*, Vol. 27, July, pp. 401–403.

Kroger, H. (1980). "Josephson Devices Coupled by Semiconductor Links," *IEEE Trans. Electron Devices*, Vol. ED–27, No. 10, October, pp 2016–2026.

Kwok, H. S., J. P. Zheng, Q. Y. Ying, and R. Rao, (1989). "Nonthermal optical response of Y-Ba-Cu-O thin films," Reprint provided to author, 1989.

Larbalestier, D., G. Fisk, B. Montgomery, and D. Hawksworth, (1986). "High-Field Superconductivity," *Physics Today*, March, pp. 24–33.

Leung, E. M. W., R. E. Baily, and P. H. Michels, (1988). "Using a small hybrid pulse power transformer unit as a component of a high–current opening switch for a railgun," presented at 1988 Applied Superconductivity Conference, San Francisco, August 21–25, (LL–2), Abstracts, p. 62.

Lipo, T. A., (1989). "The potential for high temperature superconducting AC and DC motors," *Electric Machines and Power Systems*, Vol. 13, No. 6.

Little, W. A., (1965). "Superconductivity at room temperature," *Scientific American*, Vol. 212, No. 2, February, pp. 21–27.

Loe, K. F., and E. Goto, (1985). "Analysis of flux input and output Josephson pair device," *IEEE Trans. Magn.*, Vol. MAG–21, No. 2, March, pp. 884–887.

Loyd, R. J., (1988). "Bechtel's program in superconducting magnetic energy storage," *Supercurrents*, Vol. 2, February.

Macfarlane, J. C., R. Driver, R. B. Roberts, E. C. Horrigan, and C. Andrikidis, (1988). "Electromagnetic shielding properties of yttrium barium cuprate superconductor," Proc.

of Int. Conf. on High Temperature Superconductors, Interlaken, Switzerland, *Physica C*, Vol. 153–155, Superconductivity, June, pp. 1423–1424.

Malozemoff, A. P., (1988). "Computer applications of high temperature superconductivity," Proc. of Int. Conf. on High Temperature Superconductors, Interlaken, Switzerland, *Physica C*, Vol. 153–155, Superconductivity, June, pp. 1049–1054.

Marshall, D. B., R. E. DeWames, P. E. D. Morgan, and J. J. Ratto, (1988). "Flux penetration in high-T_c superconductors: Implications for magnetic suspension and shielding," Preprint: submitted to *Phys. Rev. Lett.*

McGinnis, D. P., J. B. Beyer, and J. E. Nordman, (1988). "A modified superconducting current injection transistor and distributed amplifier design," presented at 1988 Applied Superconductivity Conference, San Francisco, August 21–25, (EJ–11), Abstracts, p. 52.

Moulthrop, A., and M. S. Muha, (1988). "Superconducting stepper motors," *Rev. Sci. Instrum.*, Vol. 59, No. 4, April, pp. 649–650.

Mueller, O., (1989). "RF components at low temperatures," *RF Design*, January, pp. 29–39.

Mulder, G. B. J., H. H. J. ten Kate, H. J. G. Krooshoop, and L. J. M. van de Klundert, (1988). "Thermally and magnetically controlled superconducting rectifiers," presented at 1988 Applied Superconductivity Conference, San Francisco, August 21–25, (LL–13), Abstracts, p. 63.

Nakane, H., T. Nishino, M. Hirano, K. Takagi, and U. Kawabe, (1987). "DC-SQUID Using High-Critical-Temperature Oxide Superconductors," *Jap. J. Appl. Phys.*, Vol. 26, No. 10, October, pp. L1581–L1582

Neffe, J., (1988). "German maglev trains way ahead of rivals," *Nature*, Vol. 335, No. 6189, September 29, pp. 388.

Osterman, D. P., R. Drake, R. Patt, E. K. Track, M. Radparvar, and S. M. Faris, (1988). "Optical response of YBCO thin films and weak links," *IEEE Trans. Magn.*, Vol. 25, No. 2, March, pp. 1323–1326.

Pan, S. K., A. R. Kerr, M. J. Feldman, and A. W. Kleinsasser, (1988). "SIS technology boosts sensitivity of mm-wave receivers," *Microwaves & RF*, Vol. 27, No. 9, September, pp. 139–146.

Petley, B. W., (1971). *An Introduction to the Josephson Effects*, Mills & Boon, London, pp. 1–66.

Radhakrishnan, V., and V. L. Newhouse, (1971). "Noise Analysis for Amplifiers with Superconducting Input," *J. Appl. Phys.*, Vol. 42, No. 1, January, pp. 129–132.

Richards, P. L., (1986). "Analog superconducting electronics," *Physics Today*, March, pp. 54–62.

Richards, P. L., and Tek–Ming Shen, (1980). "Superconductive Devices for Millimeter Wave Detection, Mixing, and Amplification," *IEEE Trans. Electron Devices*, Vol. ED-27, No. 10, October, pp. 1909–1920.

Robyn, D., W. W. Fletcher, and J. A. Alic, (1988–1989). "Bringing superconductivity to market," *Issues in Science & Technology*, Vol. V, No. 2, Winter, pp. 38–45.

Rose, K., C. L. Bertin, and R. M. Katz, (1975). "Radiation Detectors," in *Applied Superconductivity*, Vol. 1, Newhouse, V. L., ed., Academic Press, pp. 267–308.

Runté, T., (1988). "Last Word," *Omni*, Vol. 11, No. 1, October, p. 230.

Sakurai, M., (1989). "Superconductors gain," Electronic Engineering Times, January, pp. 65–66.

Shapiro, S., (1973). "Millimeter and Submillimeter Detectors and Devices," in *The Science and Technology of Superconductivity,* Gregory, W. D., W. N. Mathews, Jr., and E. A. Edelsack, eds., Vol 2., Plenum Press, New York, pp. 631–652.

Shimizu, N., Y. Harada, N. Miyamoto, and E. Goto, (1988). "A new A/D converter with quantum flux parametron," Vol 25, No. 2, *IEEE Trans. Magn.,* March, 1989, pp. 865–868.

Stekly, Z. J. J., (1973). "Superconducting Coils," in *The Science and Technology of Superconductivity,* Gregory, W. D., W. N. Mathews, Jr., and E. A. Edelsack, eds., Vol 2., Plenum Press, New York, pp. 497–538.

Talvacchio, J., M. G. Forrester, and A. I. Braginski, (1989). "Photodetection with high–T_c superconducting films," in *Science and Technology of Thin–Film Superconductors,* McConel, R., and S. A. Wolf, eds., Plenum Press, New York.

Tinkham, M., (1988). *Introduction to Superconductivity,* Robert E. Krieger, Malabar, FL.

Van Duzer, T., (1980). "Josephson Digital Devices and Circuits," *IEEE Trans. Microwave Theory & Tech.,* Vol. MTT-28, No. 5, May, pp. 490–500.

Webb, W. W., (1973). "Magnetometers and Interference Devices," in *The Science and Technology of Superconductivity,* Gregory, W. D., W. N. Mathews, Jr., and E. A. Edelsack, eds., Vol 2., Plenum Press, New York, pp. 653–677.

Weber, S., (1987). "Found! A practical way to turn out Josephson junction chips," *Electronics,* February 19, pp. 49–52.

Willis, J. O., M. E. McHenry, M. P. Maley, and H. Sheinberg, (1988). "Magnetic shielding by superconducting Y-Ba-Cu-O hollow cylinders," presentation at 1988 Applied Superconductivity Conference, San Francisco, August 21–25, (MT–3), Abstracts, p. 76.

Winters, J. H., and C. Rose, (1988). "High–T_c superconductor waveguides: Theory and applications," submitted to *IEEE Trans. Microwave Theory and Tech.,*

Wu, P. H., Q. H. Cheng, S. Z. Yang, J. Chen, Y. Li, Z. M. Ji, J. M. Song, H. X. Lu, X. K. Gao, J. Wu, and X. Y. Zhang, (1987) "The Josephson Effect in a Ceramic Bridge at Liquid Nitrogen Temperature," *Jap. J. Appl. Phys.,* Vol. 26, No. 10, October, pp. L1579–L1580.

Yan, L. G., Y. J. Yu, Z. K. Wang, Z. Y. Kao, Z. X. Ye, C. L. Xue, P. Ye, Y. L. Cheng, X. M. Li, Q. W. Kong, S. S. Song, H. L. Nan, Y. M. Dai, H. T. Tang, (1988). "A laboratory superconducting high gradient magnetic separator," *IEEE Trans. Magn.,* March, pp. 1873–1876.

Yoder, S. K., (1988). "Japan plans speedy superconductor ships," *Wall Street Journal,* August 17.

Yoshida, K., T. Hashimoto, T. Nagatsuma, and K. Enpuku, (1987). "Josephson Analog Amplifier with High Current Gain," *IEEE Trans. Magn.,* Vol. MAG–23, No. 2, March, pp. 723–726.

FURTHER READING

Adkins, C. J., "Two-particle Tunneling between Superconductors," *Phil. Mag.,* Vol. 8, No. 90, June 1963, pp. 1051–1061.

Anacker, W., "Josephson Computer Technology: An IBM Research Project," *IBM J. Res. Develop.,* Vol. 24, No. 2, March 1980, pp. 107–112.

Arnett, P. A., and D. J. Herrel, "Power Design for Gigabit Josephson Logic Systems," *IEEE Trans. on Microwave Theory & Tech.*, Vol. MTT–28, No. 5, May 1980, pp. 500–508.

Bégin, G., P. R. Chritchlow, and R. Roberge, "Technological impact study of superconductivity at liquid nitrogen temperature for an electric utililty," Pres. at 1988 Conf. on Electrical Applications of Superconductivity, Orlando, FL, September 21–23, 1988.

Bloch, F., "Josephson Effect in a Superconducting Ring," in *The Science and Technology of Superconductivity*, Gregory, W. D., W. N. Mathews, Jr., and E. A. Edelsack, eds., Vol. 1., Plenum Press, New York, 1973, pp. 149–162.

Braginski, A. I., M. G. Forrester, J. Talvacchio, and G. R. Wagner, "Prospects for thin-film electronic devices of high–T_c superconductors," Proc. 5th Int. Workshop on Future Electron Devices—High Temperature Superconducting Devices, June 2–4, 1988, pp. 171–179.

Braginski, A. I., "Material constraints on electronic applications of oxide superconductors," Proc. of Int. Conf. on High Temperature Superconductors, Interlaken, Switzerland, *Physica C*, Vol. 153–155, Superconductivity, June 1988, pp. 1598–1603.

Bremer, John W., *Superconductive Devices*, McGraw–Hill, New York, 1962.

Chan, H. W. K., and T. Van Duzer, "Josephson Nonlatching Logic Circuits," *IEEE J. Solid–State Circuits*, Vol. SC–12, No. 1, February 1977, pp. 73–79.

Clarke, J., "Advances in SQUID Magnetometers," *IEEE Trans. Electron Devices*, Vol. ED–27, No. 10, October 1980, pp. 1896–1908.

Clarke, J., "SQUIDs, brains, and gravity waves," *Physics Today*, March 1986, pp. 36–44.

Deutscher, G., "Superconducting devices," in *McGraw Hill Encyclopedia of Science & Technology*, Vol. 13, 5th Ed., McGraw–Hill, New York, 1982, pp. 339–348.

Dhong, S. H., and T. Van Duzer, "Minimum-Width Control-Current Pulse for Josephson Logic Gates," *IEEE Trans. Electron Devices*, Vol. ED–27, No. 10, October 1980, pp. 1965–1973.

Finnegan, T. F., "Superconductivity in DC Voltage Metrology," in *The Science and Technology of Superconductivity*, Gregory, W. D., W. N. Mathews, Jr., and E. A. Edelsack, eds., Vol. 2., Plenum Press, New York, 1973, pp. 565–586.

Geballe, T. H., and J. K. Hulm, "Superconductivity—The state that came in from the cold," *Science*, Vol. 239, No. 4838, January 22, 1988, pp. 367–375.

Gheewala, T. R., "Design of 2.5-Micrometer Josephson Current Injection Logic (CIL)," *IBM J. Res. Develop.*, Vol. 24, No. 2, March 1980, pp. 130–154.

Gheewala, T. R., "Josephson-Logic Devices and Circuits," *IEEE Trans. Electron Devices*, Vol. ED–27, No. 10, October 1980, pp. 1857–1869.

Greiner, J. H., C. J. Kircher, S. P. Klepner, S. K. Lahri, A. J. Warnecke, S. Basavaiah, E. T. Yen, J. M. Baker, P. R. Brosious, H.-C. W. Huang, M. Murakami, and I. Ames, "Fabrication Process for Josephson Integrated Circuits," *IBM J. Res. Develop.*, Vol. 24, No. 2, March 1980, pp. 195–205.

Hasuo, S., "High-speed Josephson IC technology," presented at Appl. Superconductivity Conference, P-1, San Francisco, August 1988.

Henkels, W. H., and H. H. Zappe, "An Experimental 64-Bit Decoded Josephson NDRO Random Access Memory," *IEEE J. Solid-State Circuits*, Vol. SC-13, No. 5, October 1978, pp. 591–600.

Kotani, S., N. Fujimaki, S. Morohashi, S. Ohara, T. Imamura, and S. Hasuo, "High-Speed Unit-Cell for Josephson LSI Circuits," *IEEE Trans. Magn.*, Vol. MAG-23, No. 2, March 1987, pp. 869–874.

Laundrie, A., "Superconducting ICs generate and detect signals to 100 GHz," *Microwaves & RF*, Vol. 27, No. 9, September 1988, pp. 163–164.

Levi, B. G., and B. Schwarzschild, "Super Collider Magnet Program Pushes Toward Prototype," *Physics Today*, Vol. 41, No. 4, April 1988, pp. 17–21.

Long, H. M., and J. Notaro, "Design features of ac superconducting Cables," *J. Appl. Phys.*, Vol. 42, No. 1, January 1971, pp. 155–162.

Martens, J. S., and J. S. Beyer, "Measure RF losses of superconductors in planar circuits," *Microwaves & RF*, Vol. 27, No. 9, September 1988, pp. 150–156.

Martinelli, A. P., "Application of Superconductivity in Thermonuclear Fusion Research," in *The Science and Technology of Superconductivity*, Gregory, W. D., W. N. Mathews, Jr., and E. A. Edelsack, eds., Vol. 2, Plenum Press, New York, 1973, pp. 459–482.

Matisoo, J., "Overview of Josephson Technology Logic and Memory," *IBM J. Res. Develop.*, Vol. 24, No. 2, March 1980, pp. 113–129.

Matisoo, J., "Superconductive Computer Devices," in *The Science and Technology of Superconductivity*, Gregory, W. D., W. N. Mathews, Jr., and E. A. Edelsack, eds., Vol. 2, Plenum Press, New York, 1973, pp. 607–624.

Matisoo, J., "The Tunneling Cryotron—A Superconductive Logic Element Based on Electron Tunneling," *Proc. IEEE*, Vol. 55, No. 2, February 1967, pp. 172–180.

McDonald, D. G., R. L. Peterson, C. A. Hamilton, R. E. Harris, and R. L. Kautz, "P:icosecond Applications of Josephson Junctions," *IEEE Trans. Electron Devices*, Vol. ED-27, No. 10, October 1980, pp. 1945–1965.

Meyerhoff, R. W., and W. T. Beall, Jr., "High-Current ac Losses in Large Superconducting Niobium Tubes," *J. Appl. Phys.*, Vol. 42, No. 1, January 1971, pp. 147–153.

Meyerhoff, R. W., "Superconducting Power Transmission," in *The Science and Technology of Superconductivity*, Gregory, W. D., W. N. Mathews, Jr., and E. A. Edelsack, eds., Vol. 2, Plenum Press, New York, 1973, pp. 433–458

Miyahara, K., M. Mukaida, M. Tokumitsu, S. Kubo, and K. Hohkawa, "Abrikosov Vortex Memory with Improved Sensitivity and Reduced Write Levels," *IEEE Trans. Magn.*, Vol. MAG-23, No. 2, March 1987, pp. 875–878.

Morisue, M., M. Kaneko, and H. Hosoya, "A Content Addressable Memory Circuit using Josephson Junctions," *IEEE Trans. Magn.*, Vol. MAG-23, No. 2, March 1987, pp. 743–746.

Mukhanov, O. A., V. K. Semenov, and K. K. Likharev, "Ultimate Performance of the RSFQ Logic Circuits," *IEEE Trans. Magn.*, Vol. MAG-23, No. 2, March 1987, pp. 759–762.

Myers, W. E., C. W. Taylor, J. J. Pachot, D. P. Hartmann, N. E. Fuller, J. F. Hauer, J. J. Vithayathil, D. R. Matheson, E. C. Starr, and C. L. Schaufelberger, "Superconductivity applications on an electric power system: A system engineering study," Bonneville Power System publication, August 1988.

Nakagawa, H., I. Kurosawa, S. Takada, and H. Hayakawa, "Jospehson 4-bit Digital Counter Circuit Made by Nb/Al-oxide/Nb Junctions," *IEEE Trans. Magn.*, Vol. MAG-23, No. 2, March 1987, pp. 739–742.

Newhouse, V. L., "Superconducting devices," in *McGraw Hill Encyclopedia of Science & Technology*, Vol. 13, 3rd Ed., McGraw-Hill, New York, 1982, pp. 304–308.

Newhouse, Vernon L., *Applied Superconductivity*, 2 vols., John Wiley & Sons, New York, 1975.

Padamsee, H., "High-T_c superconductors and their impact on niobium based accelerator applications," Cornell University publication CLNS 87/112, Pres. at 1987 Electron Beam Melting and Refining Conference, Reno, NV.

Padamsee, H., K. Green, J. Gruschus, J. Kirchgessner, D. Moffat, D. L. Rubin, J. Sears, and Q. S. Shu, "Microwave Superconductivity for Particle Accelerators—How the High-T_c Superconductors Measure Up?," Presented at the Conference on Superconductivity and Applications, Institute on Superconductivity, Buffalo, NY, April 18–20, 1988.

Padamsee, H., "Superconducting structures for electron accelerators," Presented at MIT DOE/EPRI Workshop on Novel Applications of High-Temperature Superconductors, Salem, MA, June 1988.

Petersen, D. A., H. Ko, and T. Van Duzer, "Dynamic Behavior of a Josephson Junction Latching Comparator for use in a High-Speed Analog-to-Digital Converter," *IEEE Trans. Magn.*, Vol. MAG-23, No. 2, March 1987, pp. 891–894.

Piel, H., M. Hein, N. Klein, U. Klein, A. Michalke, G. Müeller, and L. Ponto, "Superconducting perovskites in microwave fields," Proc. of Int. Conf. on High Temperature Superconductors, Interlaken, Switzerland, *Physica C*, Vol. 153–155, Superconductivity, June 1988, pp. 1604–1609.

Pond, J. M., J. H. Classen, and W. L. Carter, "Kinetic Inductance Microstrip Delay Lines," *IEEE Trans. Magn.*, Vol. MAG-23, No. 2, March 1987, pp. 903–906.

Przbysz, J., and R. D. Blaugher, "Josephson Data Latch for Frequency Agile Shift Registers," *IEEE Trans. Magn.*, Vol. MAG-23, No. 2, March 1987, pp. 777–780.

Schooley, J. F., "Superconductors in Thermometry," in *The Science and Technology of Superconductivity*, Gregory, W. D., W. N. Mathews, Jr., and E. A. Edelsack, eds., Vol. 2, Plenum Press, New York, 1973, pp. 625–630.

Smith, J. L., Jr., "Applications of Superconductors to Motors and Generators," in *The Science and Technology of Superconductivity*, Gregory, W. D., W. N. Mathews, Jr., and E. A. Edelsack, eds., Vol. 2, Plenum Press, New York, 1973, pp. 483–496.

Soderman, D. A., and K. Rose, "Microwave Studies of Thin Superconducting Films," *J. Appl. Phys.*, Vol. 39, No. 6, May 1968, pp. 2610–2617.

Terman, F. E., R. A. Helliwell, J. M. Pettit, D. A. Watkins, W. R. Rambo, *Electronic and Radio Engineering*, 4th Ed., McGraw–Hill, New York 1955, pp. 35–40.

Wilson, M. S., *Superconducting Magnets*, (Monographs on Cryogenics series), Clarendon Press, Oxford, 1986.

Wolsky, A. M., R. F. Giese, and E. J. Daniels, "The new superconductors: Prospects for applications," *Scientific American*, Vol. 260, No. 2, February 1989, pp. 60–69.

Yntema, G. B., "Niobium Superconducting Magnets," *IEEE Trans. Magn.*, Vol. MAG-23, No. 2, March 1987, pp. 390–395.

Zappe, H. H., and B. S. Landman, "Experimental investigation of resonances in low-Q Josephson interferometer devices," *J. Appl. Phys.*, Vol. 49, No. 7, July 1978, pp. 4149–4154.

Zappe, H. H., "Josephson Computer Technology," in *Advances in Superconductivity*, Deaver, B., and J. Ruvalds, eds., Plenum Press, New York, 1983, pp. 51–128.

Zappe, H. H., "Memory-Cell Design in Josephson Technology," *IEEE Trans. Electron Devices*, Vol. ED-27, No. 10, October 1980, pp. 1870–1882.

CHAPTER 5

PROCESSING OF HIGH-TEMPERATURE SUPERCONDUCTORS

INTRODUCTION

While the reader may not be directly involved in the fabrication of superconducting materials, it is nevertheless useful to familiarize oneself with some of the interesting processes involved. This chapter will describe representative techniques for the fabrication of bulk samples, thin films, single crystals, and other useful superconducting forms.

SAMPLE MATERIAL PREPARATION

Two general approaches are commonly used in the preparation of high-T_c samples: the solid state reaction technique and the coprecipitation method.

Solid State Reaction Method

In the former case, one begins with either the carbonate or oxide forms of the individual constituents; several stages of grinding and mixing are involved. The solid state reaction method has the advantage of being relatively rapid; samples may be prepared in just a few hours. A disadvantage of this approach is that the final sample invariably has additional undesirable phases. In the case of Y-Ba-Cu-O, typical undesirable phases are $BaCuO_2$ and CuO (Chen, et al., 1987) (See Figure 5.1).

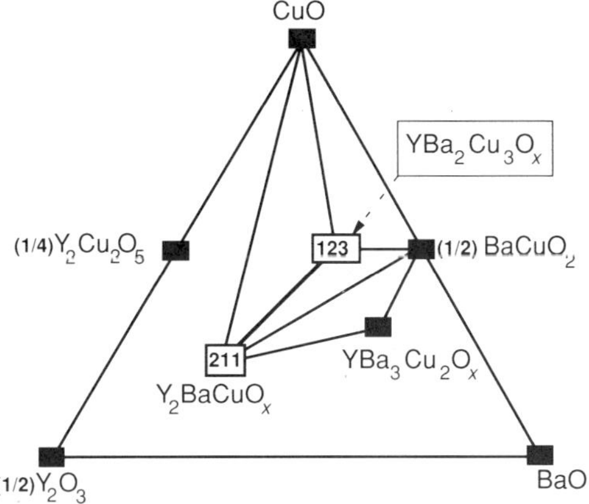

FIGURE 5.1 Ternary phase diagram for thin-film Y-Ba-Cu-O. The constant multipliers associated with the molecular formulas are for normalization. (After K. Oka, K. Nakane, M. Ito, M. Saito, and H. Unoki, "Phase-equilibrium diagram in the ternary system Y_2O_3-BaO-Cu-O," *Jap. J. Appl. Phys.*, Vol. 27, No. 6, June 1988, pp. L1065–L1067.)

Coprecipitation Method

In contrast, when using the coprecipitation method, one begins with the nitrate forms of the constituents; these are precipitated out of solution to provide their corresponding carbonate forms. This material is heated to remove H_2O. The coprecipitation method has the advantage (over the solid state reaction technique) of more reliable production of a desired phase; mixing on the atomic scale is more uniform so that unwanted phases are less likely to be produced in significant quantities.

Other coprecipitation methods include the oxalate and citrate techniques, where the nitrates of the powders are dissolved in solution. The formation of double salts is avoided in the oxalate method by careful monitoring of pH.

Sintering, Annealing

For detailed instructions using either the coprecipitation or the solid state reaction methods, the reader is referred to the excellent exposition by Xiao-Dong Chen et al., which is the source of much of the following information concerning sintering and oxygen annealing.

Once the powdered oxides are prepared and thoroughly mixed (typically in a vortex mixer or similar device where mechanical vibrations are used for mixing), they may be ground by mortar and pestle until the color is uniform. The powders are then placed in an appropriate furnace and heated to approximately 900°C. During this process, water and CO_2 are released. Following this initial baking process for 8–10 hours, the furnace is cooled to

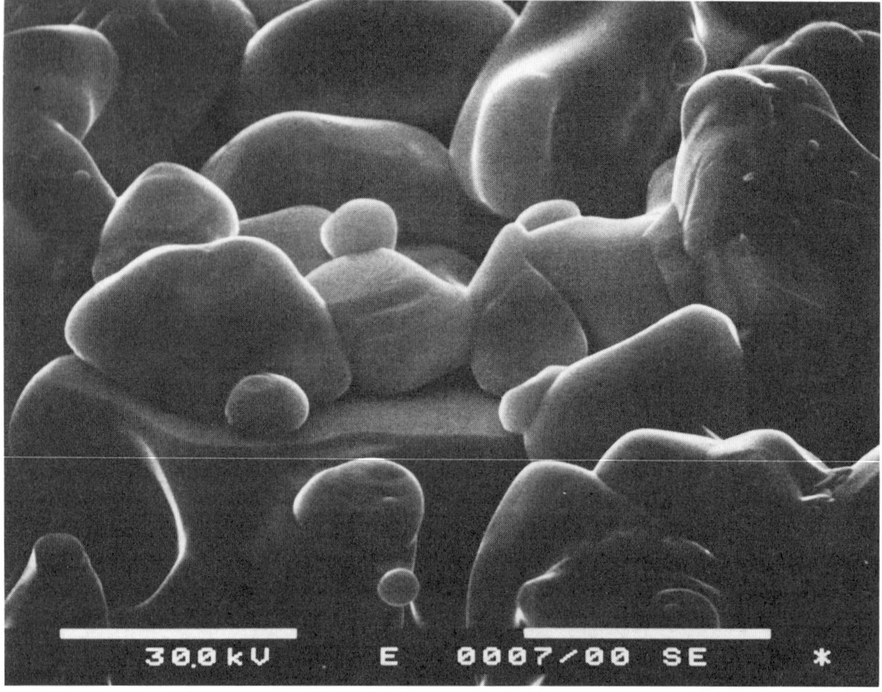

FIGURE 5.2 Typical granular configuration of polycrystalline Y-Ba-Cu-O. Calibration bars are 10 μm. (Provided by Group MEE-11, Electronics Research and Exploratory Development, Los Alamos National Laboratory)

approximately 400°C just prior to removing the powder. The solid material is then ground again and baked. This procedure is followed about three times. When the prebaking procedures are considered sufficient, the powder may be pressed into disks or pellets, using 1400–1500 psi. While the best sintering temperature varies, the disks are typically heated to about 900°C. The most general rule is that the disk should be sintered at as high a temperature as possible as long as melting is avoided. For 1-2-3 materials, the furnace is cooled to 500°C; at this temperature the oxygen flow may begin. Oxygen is typically provided through an appropriate flow monitor connected to a quartz tube. The oxygen annealing process time varies depending on the particular operator's preference, but 10–14 hours is fairly typical.

Analysis of Specimen During Sintering

While the development of most processing techniques amounts to an educated cut-and-try approach, it would be very helpful to monitor relevant material parameters during the process for optimization. Recently, an effort has

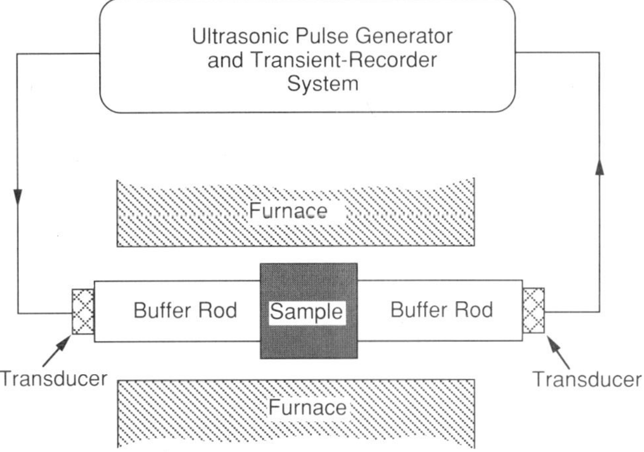

FIGURE 5.3 Information (in real time) on the condition of samples undergoing the sintering process has been determined by measurement of transmitted and reflected ultrasound pulses. This approach has already proven to be valuable in investigations of superconductor processing techniques, and may have industrial applications such as closed-loop feedback control of processing parameters to yield optimum results. (After J. H. Gieske and H. M. Frost, "Sound velocity measurements in green-body ceramics as a function of sintering temperature," Proceedings: Review of Progress in QNDE, La Joya, CA, August 1–5, 1988.)

been made to gather information about the sintering phenomena, in real time. Investigators have transmitted 5 MHz ultrasound pulses through a specimen as it undergoes the sintering process. Sound velocity typically varies from approximately 1 to 3 mm/μs during the processing schedule, indicating densification of the sample. Analysis of a variety of data from the sample *during* processing could eventually provide the real-time feedback required for optimum processing (Gieske & Frost, 1988) (see Figure 5.3).

Melt-Textured Growth

The current transport of $YBa_2Cu_3O_x$ has been significantly improved by a processing technique referred to as "melt-textured growth" (Jin, et al., 1988). This process begins with $YBa_2Cu_3O_x$ material produced by conventional sintering techniques. These precursor samples are melt-processed by heating above the levels typical for sintering (900–950°C). Temperatures used for melt processing range from 1030°C to 1180°C for production of a "liquid plus solid" phase, and as high as 1320°C for a single-phase liquid. The melt is held at these temperature levels for periods ranging up to two hours, then cooled in a thermal gradient of approximately 50°C/cm.

During the cooling process, needle-like crystals are formed, with lengths ranging typically from 40 to 600 μm. Minor dimensions of the needles are on

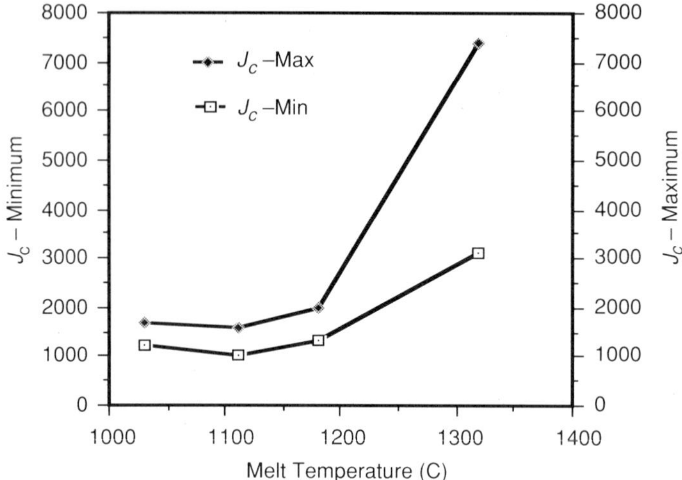

FIGURE 5.4 In melt-textured growth procedures, the Y-Ba-Cu-O material is melted at temperatures above 1000°C, then cooled in a temperature gradient on the order of 50°C/cm. This process generates a locally textured array of long, needle-like structures with diameters of ~ 2–5 μm and lengths up to several hundred μm. Additional heat treatments provide stress relief, increase oxygen content, and produce a more homogeneous structure. Above, the relation between melt processing temperatures (during the melt) and the range of current densities is illustrated. By contrast, the critical current following a conventional sintering process is on the order of 1000 A/cm^2 and decreases much more rapidly with the application of a magnetic field. (After S. Jin, T. H. Tiefel, R. C. Sherwood, R. B. van Dover, M. E. Davis, G. W. Kammlott, and R. A. Fastnacht, "Melt-textured growth of polycrystalline $YBa_2Cu_3O_{7-\delta}$ with high transport J_c at 77 K," *Phys. Rev. B*, Vol. 37, No. 13, May 1, 1988, pp. 7850–7853.)

the order of 2–5 μm. Zero-field current density in melt-textured samples of Y-Ba-Cu-O are as high as 7400 A/cm^2, compared to values generally much less than 1000 A/cm^2 for sintered bulk samples (see Figure 5.4).

SUPERCONDUCTING FILMS

While the production of bulk samples of high-temperature superconductors is important for the construction of wire and ribbon conductors, the production of high-quality films is essential for a wide variety of electronic applications ranging from strip-lines and hybrid interconnects to antennas. For this reason, it is useful to consider a few of the representative methods for production of superconducting films, along with their individual merits and shortcomings.

The process of depositing films can, for the purposes of this discussion, be broken down into two main categories. These are

a. Processes involving droplet transfer:
   i. Arc Spraying
   ii. Plasma Spraying
b. Processes involving transfer of atoms:
   i. Sputtering
   ii. Evaporation
   iii. Ion Plating
   iv. Chemical Vapor Deposition
   v. Electrodeposition

Droplet-transfer mechanisms, like arc spray and plasma spray methods, are relatively straightforward and capable of applying fairly thick films, that is, up to tens of μm. These methods also have the disadvantage that the final coating is "grainy" and porous. Such methods may prove to be useful for construction of films for applications where critical current density (J_c) need not be large. Examples of such applications include SQUIDs and infrared detectors.

In one example, Y-Ba-Cu-O films that were plasma-sprayed on stainless steel and then annealed in oxygen at 950°C for approximately one hour were superconducting when thickness was greater than 100 μm (Karthikeyan, et al., 1988) (see Figure 5.5).

Many other applications, ranging from high-speed computer circuitry to coatings for microwave cavities, require a film with a relatively low surface resistance and a high critical current density. Such applications will probably require texturing of the film. In the instance of the 1-2-3 materials, texturing implies that the long (c) axis of the crystal is perpendicular to the plane of the film since current flow direction is in the plane, that is, along the a and b crystal axes where resistance is minimal.

Texturing is normally accomplished, at least in part, by selection of a single-crystal substrate (as opposed to polycrystalline) whose crystal lattice dimensions are similar to the dimensions of the superconductor crystal. For 1-2-3 materials, strontium-titanate ($SrTiO_3$) has an excellent lattice match but has a high dielectric constant ($\sim$ 270 at 9.3 GHz) and is lossy at high frequencies. Lanthanum gallate ($LaGaO_3$) is a perovskite-like material which, in its single-crystal configuration, may have applications as a substrate for several of the ceramic superconductors. In contrast to $SrTiO_3$, $LaGaO_3$ has a relative dielectric constant that is an order of magnitude lower at 9.3 GHz, that is, 27 versus 270.

Evaporation Processes, Electron Beam Heating

In principle at least, evaporation techniques are simple. The materials to be evaporated are placed in a vacuum chamber near the substrate where plating is to be accomplished. A variety of methods may be used to heat the material,

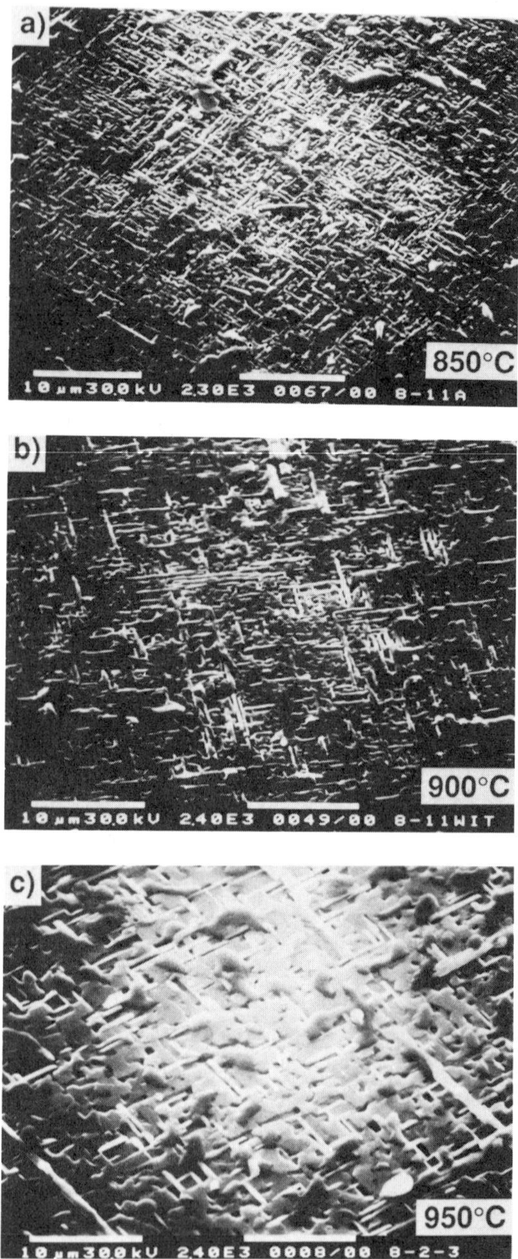

FIGURE 5.5 Effects of annealing temperature on Y-Ba-Cu-O film. While 900°C is near optimum for high-current applications, 850°C results in a more granular film with weak links that are useful in *IR* detector applications. Annealing at 950°C leads to melting and destruction of the film's superconducting characteristics. (Provided by Group MEE-11, Electronics Research and Exploratory Development, Los Alamos National Laboratory)

but bombardment by electron beams, laser beams or electrical discharge are common methods. The vacuum is generally in the range of 10^{-5} to 10^{-6} torr, allowing the evaporated atoms to have a line-of-sight trajectory to the target substrate without significant probability of collision en route. (See Figure 5.6 for the basic electron-beam evaporation process.)

See Figure 5.7 for a more detailed schematic view of a typical electron-beam evaporation system. Since the ejected material will not be deposited uniformly on the target, the target may be moved during the process to improve uniformity of coating.

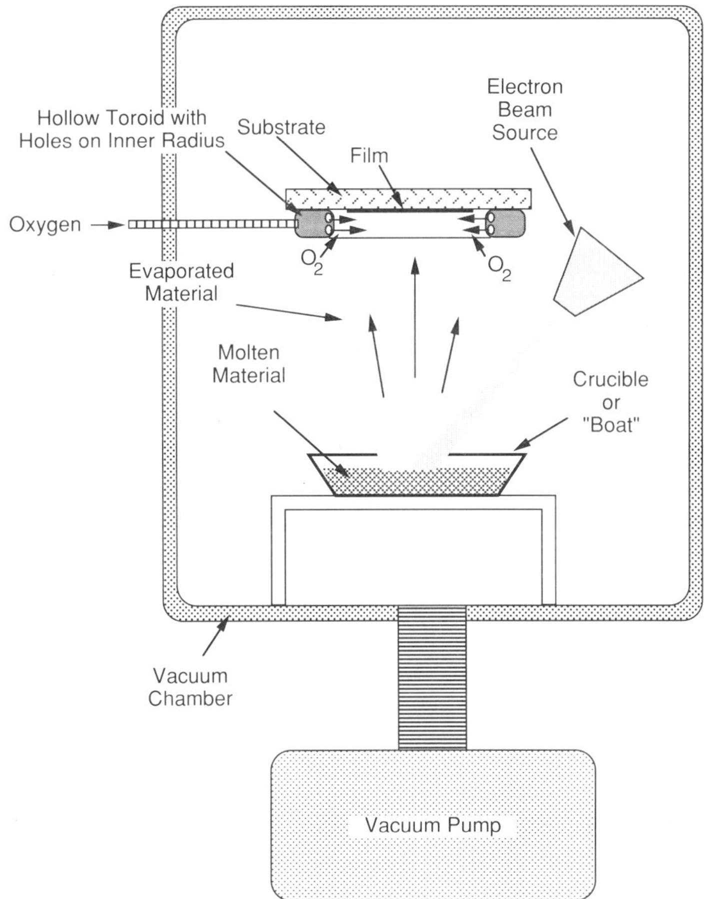

FIGURE 5.6 Highly schematic view of basic electron-beam evaporation process for production of superconducting Y-Ba-Cu-O films. The hollow metal toroid surrounding the substrate is used to supply oxygen during the deposition process. A laser could be substituted (for the e-beam) as the heat source; multiple crucibles are generally used instead of the single unit shown here for simplicity.

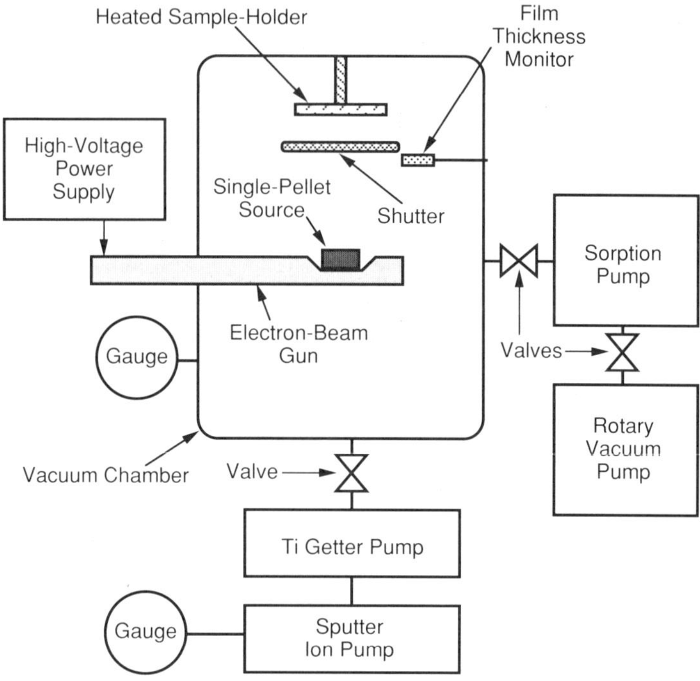

FIGURE 5.7 This more detailed view of an electron-beam evaporation system for a single-pellet source illustrates vacuum pumps, valves, vacuum gauges, a shutter, and the film thickness monitor. (After I. Terasaki, Y. Nakayama, K. Uchinokura, A. Maeda, T. Hasegawa, and S. Tanaka, "Superconducting films of $YBa_2Cu_3O_x$ and Bi-Sr-Ca-Cu-O fabricated by electron-beam deposition with a single source," *Jap. J. Appl. Phys.*, Vol. 27, No. 8, August 1988, pp. L1480–L1483.)

Diode Ion-Coating

The ion-coating process is quite similar in that an electron beam is used to evaporate the material(s). The difference is that the evaporated material passes through a glow discharge on its trajectory to the substrate target, ionizing a proportion of the evaporated atoms. The substrate is biased to a few kV (negative) to ionize argon that is admitted to the chamber for this purpose. This results in the substrate being bombarded by energetic gas ions that constantly clean the substrate surface. This effect is desirable due to the better adhesion properties and reduction of surface impurities, but it obviously also reduces the deposition rate of the intended plating material. This technique is known as "diode ion plating."

Sputtering Processes

Sputtering techniques rely on positive gas ions (such as argon) that are produced in a glow discharge; these energetic particles bombard a cathode or "target."

The cathode is the material that is to be deposited on the substrate. See Figures 5.8 and 5.9, which illustrate the basic sputtering process. Sputtering is not a particularly efficient process, yielding approximately one atom on the substrate for each 500 eV argon ion that impinges on the target (Bunshah, 1967).

Single-target sputtering shows considerable promise in the production of thin 1-2-3 films. The target may be either stoichiometric (Terada, et al., 1988) or a "single, slightly composition-compensated target" (Sandstrom, et al., 1988).

Thin Film Deposition by Multiple-Target Sputtering

One is not, of course, limited to a single sputtering target. A dual-target method has been reported where one target of yttrium was supplemented with another of Ba_2Cu_3; oxygen was added in the post-annealing process (Makous, et al., 1987). This technique, utilizing sequential passes of the substrate past the two targets to deposit 10 Å layers, produced multiple-phase polycrystalline films with T_c's on the order of 89 K. For most purposes, of course, only the single 1-2-3 phase is desired. Relative amounts of specific elements in thin films may be determined by Rutherford back-scattering (RBS). See Figure 5.11.

Once the film is deposited on the substrate, typically some type of ceramic, it is usually necessary to place the sample in an oven for a heat-cycling and annealing procedure that is generally not the same schedule used in the processing of bulk samples. Nevertheless, *in situ* processes where no post-anneal is required are gradually becoming more common.

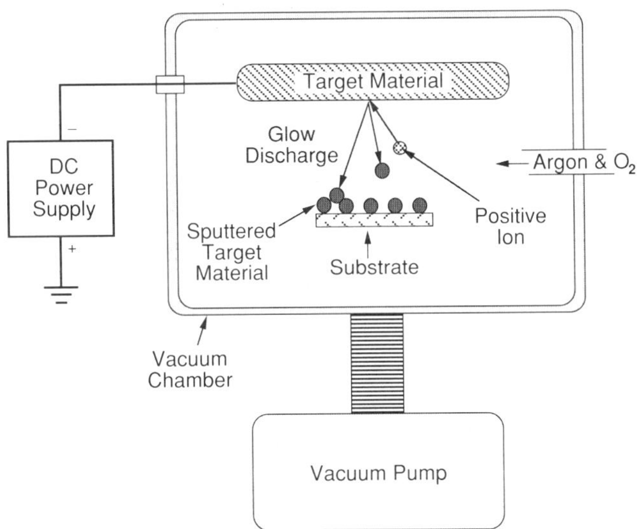

FIGURE 5.8 In the basic dc sputtering process shown above, positive argon atoms in the plasma are accelerated towards the negative target where they dislodge groups of atoms. Some of these atoms are deposited onto the substrate to form the film.

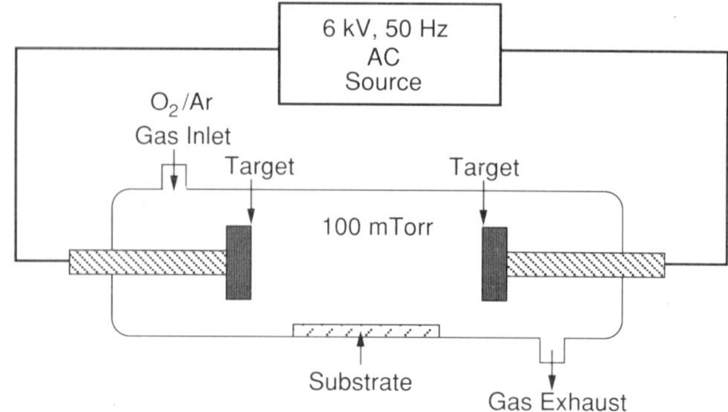

FIGURE 5.9 An ac sputtering arrangement, used in this instance to plate the bismuth superconductor system on substrates of polycrystalline yttria-stabilized zirconia (YSZ). Targets are composed of various Bi-Sr-Ca-Cu-O compounds; see text for details. Annealing times ranged from 20 to 300 min at temperatures ranging from 750 to 900°C. X-ray diffraction revealed highly oriented periodic superstructures. One specimen showed a T_c-onset of 115 K and zero dc resistivity at 73 K, typical of the dual-phase character of the bismuth system. (After H. Koinuma, M. Kawasaki, S. Nagata, K. Takeuchi, and K. Fueki, "Preparation of high-T_c Bi-Sr-Ca-Cu-O superconducting thin films by ac sputtering," *Jap. J. Appl. Phys.*, Vol. 27, No. 3, March 1988, p. L376.)

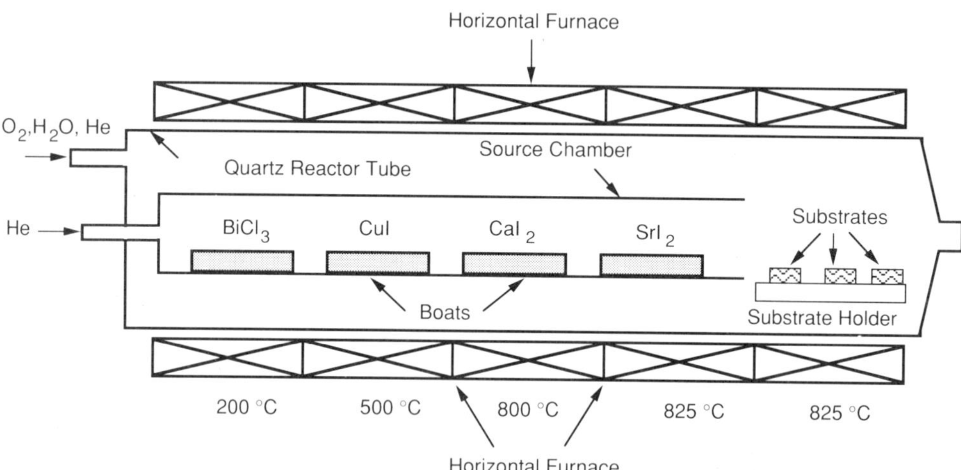

FIGURE 5.10 An open-tube chemical vapor deposition [CVD] system for Bi-Sr-Ca-Cu-O, using a horizontal furnace and a quartz reactor tube. Note the different temperatures for the individual sources, ranging from 200°C to 825°C. (After M. Ihara, T. Kimura, H. Yamawaki, and K. Ikeda, "High-Tc BiSrCaCuO superconductor grown by CVD technique," *IEEE Trans. on Magnetics*, Vol. 25, No. 3, March 1989, pp. 2470–2473.)

Thin Films by Electron-Beam Coevaporation

Thin high-T_c films of excellent quality are now commonly produced by electron beam coevaporation, that is, multiple electron beams on specific constituents of the superconductor which are to be evaporated onto the substrate. In one report, several substrates were investigated following e-beam deposition and a course of heating as follows: 650°C in oxygen flow for three to six hours, followed by one-hour steps at 750 and 850°C, then gradual cooling (Oh, et al., 1987). The worst substrate evaluated was Al_2O_3 and the best was $SrTiO_3$. In the case of the latter substrate, films of $YBa_2Cu_3O_{7-\delta}$ showed evidence of epitaxial growth and relatively high current densities. The values reported for J_c were 90,000 A/cm^2 at 78 K and 2×10^6 A/cm^2 at 4.2 K.

High-T_c Film Production by Laser Coevaporation

Laser coevaporation is essentially quite similar to electron-beam coevaporation; the primary difference is that lasers are substituted for the electron sources. With lasers, of course, it is relatively easy to focus on a very small target area, and this can be an advantage when small pellets are used as the primary source of high-T_c material. In one report thin films of 1-2-3 material were deposited

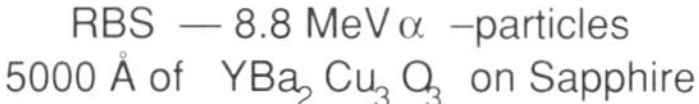

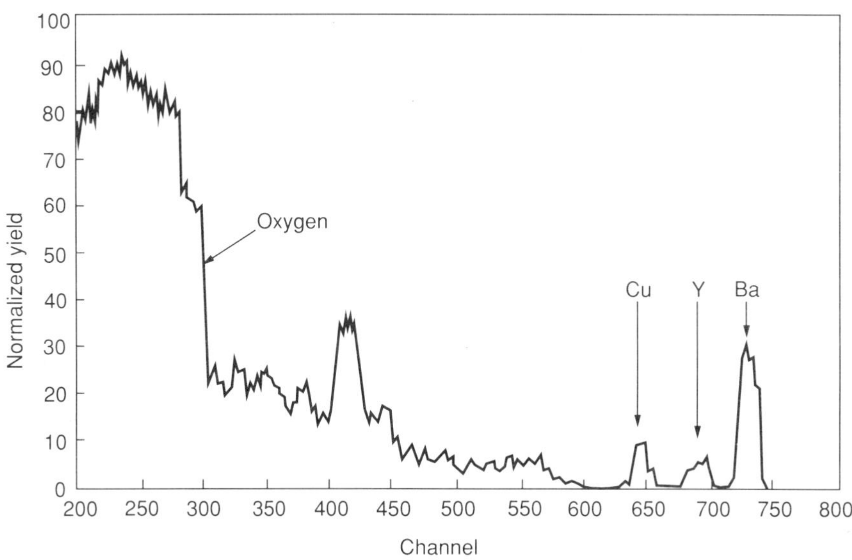

FIGURE 5.11 Rutherford back-scattering [RBS] is commonly used to determine the relative amounts of the elements in thin films. (Provided by Group MEE-11, Electronics Research and Exploratory Development, Los Alamos National Laboratory)

by illuminating a bulk sample of $YBa_2Cu_3O_{7-\delta}$ with a pulsed excimer laser (Wu, et al., 1987). Vacuum was kept at 5×10^{-7} torr and the target sample was held at 450°C. These films were annealed in oxygen at 900°C for only one hour, with subsequent slow cooling to room temperature. As in other studies, the films deposited on $SrTiO_3$ showed indications (via X-ray diffraction) of a preferential alignment (texturing) with the 1-2-3 crystal c-axis normal to the plane of the film. At the present state of the art, there are limitations on the size of the area where deposition is uniform; it is difficult to produce uniform films with areas greater than a few square cm.

See Figure 5.12 for a pulsed excimer laser evaporation technique.

Electrophoretic Deposition of Films

Investigators at the University of Wuppertal have had success with electrophoretic plating of $YBa_2Cu_3O_x$ onto silver substrates. A colloidal suspen-

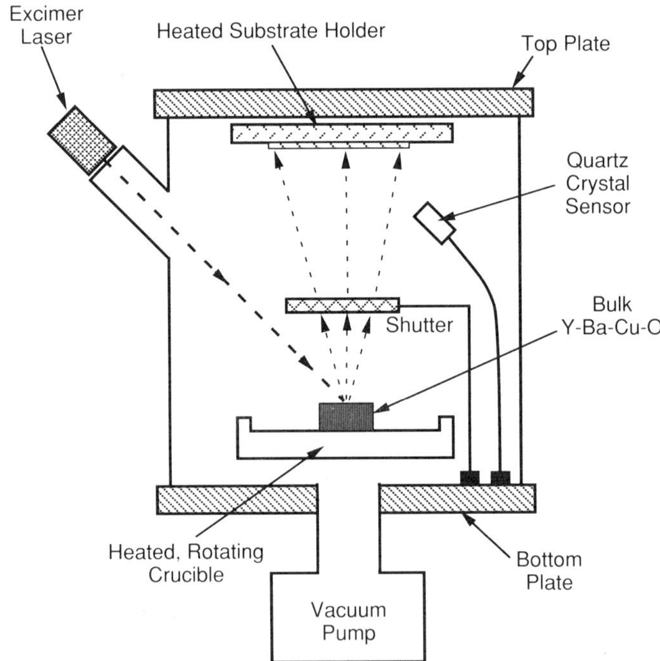

FIGURE 5.12 Investigators* from Rutgers University and Bell Communications Research Laboratory have had success with pulsed excimer laser evaporation of bulk Y-Ba-Cu-O pellets on substrates heated to 450°C. Laser-induced evaporant from the pellet is deposited on the substrate, which is typically strontium titanate. Films are post-annealed for one hour at 900°C, following a gradual cooling to room temperature. (*After D. Dijkkamp, and T. Venkatesan, "Preparation of Y-Ba-Cu oxide superconductor thin films using pulsed laser evaporation from high-T_c bulk material," *Appl. Phys. Lett.*, Vol. 51, No. 8, August 24, 1987, pp. 619–621.)

sion of $YBa_2Cu_3O_x$ powder was deposited in a stepwise fashion onto a silver disk. After each deposition, which amounted to approximately 10 μm thickness, the disk was heat treated in an O_2 atmosphere at 930°C. Final film thickness was ~ 50 μm (Piel, et al., 1988; Hein, et al., 1988).

Laser Etching of Thin Films

Lasers are commonly used for etching, with particular success in the preparation of miniature electronic devices. Similar processes will certainly be used in the preparation of films with specific shapes. See, for example, the discussion of *IR* detectors in the chapter on applications. In one report, a pulsed KrF excimer laser with a 248 nm wavelength and 30 ns pulse width was used to etch 1-2-3 films (Inam, et al., 1987). The etch threshold energy density was reported to be 0.11 J/cm^2, and etch depth as a function of the number of pulses applied was linear over a wide range of energy densities. The etch rate followed a Beer-Lambert logarithmic relation as a function of incident laser energy, that is,

$$\frac{\text{etch depth}}{\text{pulse}} = \frac{1}{\alpha} \ln \left(\frac{E_{\text{inc}}}{E_{\text{th}}} \right)$$

where E_{inc} is the incident laser energy, E_{th} is the absorption threshold energy, and $1/\alpha = 44$ nm. At a laser energy density of 2 J/cm^2, the etch rate was ~ 1000 Å/pulse.

Chemical Etching of Thin Films

Chemical etching has been investigated, using thin films of Y-Ba-Cu-O, deposited by *rf* sputtering of a stoichioimetric target. Substrates used were glass slides, alumina, zirconia, and Y_2O_3-coated silicon (Shih & Qiu, 1988). Etching solutions that yielded reproducible results included H_3PO_4/H_2O, HNO_3/H_2O, and HCl/H_2O. Volumetric proportions of acid in water ranged from 1% to 10%. Etch rate was, as would be expected, greatly affected by solution temperature. It was determined that H_3PO_4/H_2O in volumetric ratios between 1/10 and 1/60 was effective in varying the etch rate from 1.4 to 0.2 μm/min. The investigators found the 0.2 μm/min rate to be appropriate for etching fine features; high-quality lines of Y-Ba-Cu-O as narrow as 3 μm were produced on the silicon substrate. The properties of the Y-Ba-Cu-O film were not significantly affected by the positive photoresist or by the chemical etching process. Others have not seen useful results from similar chemical etching processes; severe undercutting has been a drawback.

Other techniques used for controlled removal of high-T_c film include ion milling and sputter etching. These two approaches are quite similar, except that ion milling is much more directional than sputter etching. The latter technique is used for depositing film as well as etching.

TABLE 5.1 Results of experiments comparing four-terminal dc resistivity versus temperature response for coated and uncoated films.

Coating, Thickness	T_{co} (K)	T_z (K)	J_c (A/cm^2 at 4.2 K)
None	85	65	4000
Nb_2O_5, 600 Å	85	60	2800
Al_2O_3, 600 Å	85	60	2000
Nb_2O_5, 1350 Å	85	~ 10	1000

Dielectric Coating of Thin Films

In a variety of applications where thin superconducting films are used, it is useful to be able to coat the superconductor with dielectric films. Such coatings not only protect from accidental short-circuits when using circuits constructed from the superconductors, but are often essential in protecting the systems from environmental contaminants. Given the sensitivity of Y-Ba-Cu-O to moisture, such coatings may be of some importance.

Investigators at Matsushita Electric have tested the effects of coating Y-Ba-Cu-O films with thin layers of Nb_2O_5 and Al_2O_3 (Ichikawa, et al., 1988). The superconducting films were first deposited onto sapphire substrates by sputtering from a sintered target of stoichiometric $YBa_2Cu_3O_{7-\delta}$. This was followed by post-annealing in O_2 for five hours at 850°C. Gold electrodes were then deposited on the film, which was approximately 5000 Å thick. Either the Nb_2O_5 or the Al_2O_3 films were then deposited onto the Y-Ba-Cu-O by *rf*-magnetron sputtering; the gold electrodes were shielded by a metal shadow-masking technique.

These experiments compared the four-terminal dc resistivity versus temperature response for coated and uncoated films. Uncoated films exhibited a T_c-onset (T_{co}) of 85 K and "zero" dc resistivity at 65 K. While the coated films had approximately the same T_{co}, the zero-resistance temperature (T_z) was lower and critical current density was also decreased. Details appear in the Table 5.1. While the uncoated superconductor showed the traditional positive resistance coefficient that typifies metallic behavior, all of the coated systems had a negative temperature coefficient that is suggestive of semiconductive behavior. The investigators conclude that overcoating these dielectric films by sputtering causes degradation in the Y-Ba-Cu-O films due to interdiffusion between the dielectric overcoating and the superconducting film. In particular, there is some evidence of an increase in insulating grain boundaries caused by the overcoating process.

PREPARATION OF SINGLE CRYSTAL HIGH-T_c SAMPLES

Single crystals of high-T_c superconductors are grown much in the same manner as are single crystals of other materials. Essentially, the process involves melting a mixture of proper stoichiometry, reducing the temperature below the melting level, and holding the mixture at the new temperature for several days. Generally, the longer the process, the larger the crystals will grow. In one report a mixture of $YBa_2Cu_3O_{7-x}$ was melted with some additional CuO at 1150°C (Damento, et al., 1987). Following the melting process, the platinum boat containing the molten material was rapidly cooled to room temperature to avoid "excessive segregation and nucleation" of the material. Following this procedure, the temperature was increased to 900°C and held at this level for four days. Final cooling to room temperature was at the rate of 1°C/min. Crystals were reported in the form of thin "plates," typically with the crystal's c-axis normal to the face of the plate. Crystal size was generally approximately 1 mm diameter and 0.2 mm thick; T_c was about 89 K. The authors state their belief that crystal growth was aided by a liquid phase that remained at the holding temperature of 900°C.

In a similar report a mixture of Y_2O_3, $BaCO_3$, and $Cu_2(CO_3)(OH)_2$ were first preheated at 960°C for two hours in Pt crucibles (Kaiser, et al., 1987). Samples were then heated to 1000–1075°C prior to cooling. Cooling rates varied from 4°C/hr to 20°C/hr, concluding at either 800 or 950°C. The result was solidified melts containing cavities; platelets of single-crystal $YBa_2Cu_3O_{7-x}$ were far less common than crystals of off-phase materials and were difficult to remove. Values of critical temperature up to 85 K were reported.

See Figures 5.13 and 5.14 for single crystal growth on a zirconia tube.

There is considerable interest in the production of large, uniform single crystals of all of the new high-T_c materials. The single crystals, in contrast to

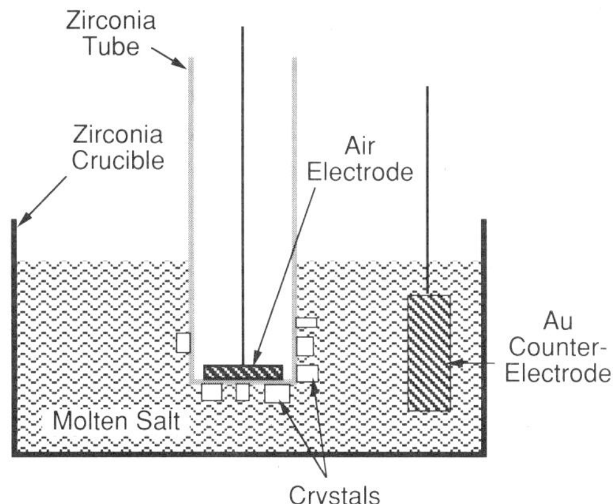

FIGURE 5.13 Single crystals of Y-Ba-Cu-O may be grown as illustrated above, by application of a potential between an electrode (inside the zirconia tube) and a gold counter-electrode in the molten salt. A locally high oxygen partial pressure is maintained at the melt electrolyte interface; nonequilibrium crystal growth occurs at this location. Oxygen stoichiometry of the single crystal is a function of applied voltage. (Provided by Group MEE-11, Electronics Research and Exploratory Development, Los Alamos National Laboratory)

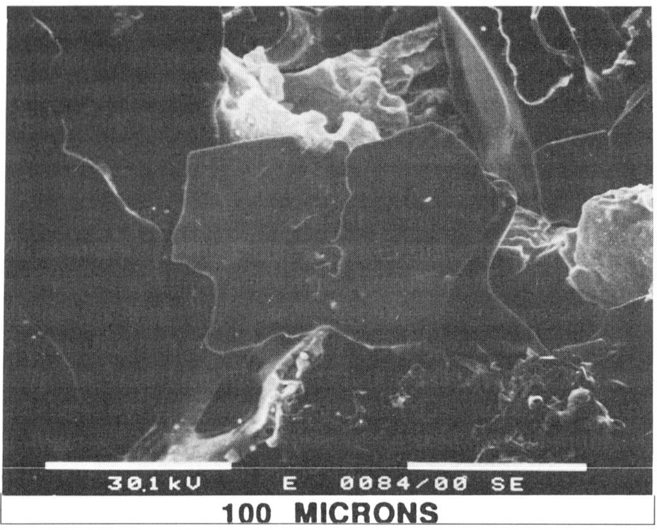

FIGURE 5.14 Y-Ba-Cu-O single crystal grown on zirconia tube. (Provided by Group MEE-11, Electronics Research and Exploratory Development, Los Alamos National Laboratory)

polycrystalline samples or even moderately textured films, are more likely to approach the pure form of the particular superconductor.

This statement deserves some clarification. A single crystal, by the strict definition, is *always* a pure form of the material. In practice, however, the ceramic superconductor specimens that are casually referred to as "single crystals" tend to be contaminated, and virtually always exhibit twinning and incomplete oxygenation. Our statements generally refer to the *real* crystals rather than the theoretically pure form.

Nevertheless, it is by careful measurement of the properties of the best available "single crystals" that one can determine the intrinsic values of important parameters, with current density being perhaps of greatest interest. In other words, whatever value of J_c is measured in a single crystal, it cannot be greater (and will usually be much less) in a bulk polycrystalline sample.

USE OF FLUORINE IN 1-2-3 PROCESSING

It is not uncommon to use small amounts of fluorine in the processing of Y-Ba-Cu-O compounds to improve the electromagnetic properties of the superconductor. It has been reported that addition of the limited amount of fluorine does not appear to change the structure of the 1-2-3 unit crystal, but is suspected of strengthening the weak-link connections between the superconducting grains (Hakuraku, et al., 1988). In addition, the use of fluorine makes the processing somewhat easier and reduces the sensitivity of 1-2-3 ceramics to water.

SUBSTRATES

Thermal Expansion, Mismatch to Substrates

It is a common (and frustrating) problem to observe cracks in the ceramic superconductors. This problem is especially noticeable in thin films deposited on a variety of substrates. A very useful study has shown a close relation between crack formation and the difference in thermal expansion coefficient between the superconducting material and the particular substrate (Hashimoto, et al., 1988a). Of particular interest in this report were temperatures in the processing range, that is, 30 to 900°C. Significantly, the investigators found a sudden change in the thermal expansion coefficient at 350°C and 650°C, corresponding to the change in concentration of oxygen and the phase transition, respectively. See Table 5.2 for information on the average coefficient of thermal expansion for representative ceramic superconductors and common substrates.

Reaction between Y-Ba-Cu-O and Substrates

There is, of course, great interest in the reactions between the new superconducting ceramics and a variety of substrates. Rutherford backscattering (RBS) investigations of interface reactions and interdiffusion of $YBa_2Cu_3O_x$ films

TABLE 5.2 The difference in thermal coefficient of expansion for a superconducting film and its substrate can result in cracking during high-temperature processing. The figures below represent average coefficients over rather wide temperature ranges; rapid changes in expansion are seen in 1-2-3 materials at approximately 350°C and 650°C.

Material	Average Coefficient of Thermal Expansion $\{\alpha = \Delta L/(L\Delta T)°C^{-1}\}$ (30→500°C)	(30→900°C)
Superconductor:		
$YBa_2Cu_3O_{7-\delta}$	1.44×10^{-5}	1.69×10^{-5}
$YbBa_2Cu_3O_{7-\delta}$	1.27×10^{-5}	1.50×10^{-5}
$(La_{0.9}Sr_{0.1})_2CuO_{4-d}$	1.44×10^{-5}	1.38×10^{-5}
Substrate:		
MgO	1.24×10^{-5}	1.30×10^{-5}
$SrTiO_3$	1.02×10^{-5}	1.11×10^{-5}
YSZ	1.00×10^{-5}	1.03×10^{-5}
sapphire	0.68×10^{-5}	0.75×10^{-5}
alumina	0.62×10^{-5}	0.70×10^{-5}
glass (Corning #7059)	0.50×10^{-5}	—
quartz	less than 10^{-6}	less than 10^{-6}
$LaGaO_3$	*9×10^{-6}	—

(After Hashimoto, et al., 1988a)

*Published data from Litton-Airtron, Morris Plains, NJ.

on single-crystal MgO, sapphire, and fuzed quartz indicate that MgO is the most stable substrate following annealing procedures (Nakajima, et al., 1988). These same experiments show that the copper from the 1-2-3 superconductor diffuses much more rapidly into the various substrates than yttrium or barium; Cu diffusion was greatest in MgO and least in fuzed quartz. When sapphire substrates were tested, aluminum atoms exhibited a significant diffusion into the $YBa_2Cu_3O_x$ film.

A representative list of common substrates that are used for deposition of superconducting $YBa_2Cu_3O_{7-\delta}$ films follows:

Strontium Titanate: $SrTiO_3$
Magnesium Oxide: MgO
Yttria-Stabilized Zirconia (YSZ): ZrO_2-Y_2O_3
Spinel: $MgO \cdot Al_2O_3$
Sapphire: Al_2O_3
YAG: $Y_3Al_5O_{12}$
Rutile: TlO_2
Lanthanum gallate: $LaGaO_3$
Lanthanum aluminate: $LaAlO_3$
$KTaO_3$
$BaZrO_3$
Ba_2SiO_4
BaF_2

Single-crystal $KTaO_3$ is a cubic perovskite that may be used as a substrate for Y-Ba-Cu-O epitaxial thin films, with minimal film-substrate interaction. Interesting applications are possible, since appropriate doping will convert $KTaO_3$ to either a ferroelectric or an n-type semiconductor (Feenstra, et al., 1988).

Single-phase samples of $BaZrO_3$, Ba_2SiO_4, and BaF_2 do not interact strongly with quenched films of $YBa_2Cu_3O_x$, and are therefore considered to be good candidates as substrates for thin 1-2-3 films. In the same set of experiments, $BaAl_2O_4$ was not a suitable substrate for 1-2-3 films; no superconductivity was seen above 77 K. It is suspected that contamination from the Al may have shifted the 1-2-3 unit crystal from the orthorhombic to tetragonal state (Komatsu, et al., 1988b).

USEFUL CONFIGURATIONS FOR HIGH-T_c SUPERCONDUCTORS

The extensive use of any of the new high-temperature superconductors will not be forthcoming unless the materials can be formed into useful shapes, that is, wires, ribbons, and the like. In spite of the brittle nature of these ceramics,

TABLE 5.3 Current density achieved by several firms in the manufacture of high-temperature superconducting wire.

Company	Process	J_c (A/cm^2)
American SC	MS	1000
AT&T	PT	1100
AT&T	MTG	17,000
Furakawa	PT	1300
Fujikura	PTM	11,000
Hitachi	PRM	2000
Mitsubishi	PM	5680
Toshiba	PT	725

Processing code:

PT = Powder in tube*
PTM = Modified powder in tube*
MTG = Melt-textured growth
PM = Powder with ~1000 C processing
MS = Melt spin metal, then oxidize

* (Usually a silver tube)

Compiled from: Schlabach, T. D., "High-T_c superconductor wire fabrication and prospects for use in large scale applications," ASME Ann. Mtng., November 30, 1988, pub. in *Superconductivity Applications and Developments*, MD-Vol. 11, ASME, New York, 1988, pp. 29–34.

there has been some progress in this direction (see Table 5.3). In one instance the construction of wire formed by molten oxide processing was demonstrated by the production of 1.2 mm diameter, 10 mm long samples (Jin, et al., 1987).

In another instance, tapes with thickness ranging from 0.05 to 0.5 mm were formed by cold rolling $YBa_2Cu_3O_{7-\delta}$ "wires" that had 2.8 mm diameter (Okada, et al., 1988). The tapes were reheated to 910°C for 5 to 150 hours in flowing oxygen, then cooled at a rate of approximately 200°C/hr. Critical current was measured in a field that ranged from 0 to 7 T, applied perpendicular to the tape surface. Levels of J_c up to 3330 A/cm^2 were seen in a tape with a thickness of 0.06 mm, and the magnitude of J_c varied inversely with tape thickness. (Also see the discussion of melt-quenching production of glass-like Bi-Ca-Sr-Cu-O superconductors later in this chapter.)

BISMUTH-SYSTEM PROCESSING

These systems are barely a year old as this is being written, so the information on the processing of Bi-Ca-Sr-Cu-O is necessarily sketchy. Even so, it is a starting place for the reader. There was some initial difficulty in separation of the high (~110 K) and low (~80 K) phases of Bi-Sr-Ca-Cu-O. More recently, there have been reports of processing techniques for isolating the high-T_c phase. In one instance, Bi-Sr-Ca-Cu-O thin films are produced by *rf*-magnetron sputtering. Critical temperatures above 100 K are seen when substrate temperature control is used to maintain a partially melted underlayer (Fukutomi, et al., 1988). In another approach, co-decomposition of mixed nitrates of Bi, Pb, Sr, Ca, and Cu at 830°C in low oxygen pressure accomplishes the same goal with a zero-resistance T_c of 107.5 K (Endo, et al., 1988).

Preparation of Bismuth Films

AC Sputtering. Thin films of Bi-Ca-Sr-Cu-O have been prepared by ac sputtering techniques (Koinuma, et al., 1988). Target A was comprised of $Bi_1Sr_1Ca_1Cu_1O_x$, while the second target (B) was one of

$Bi_3Sr_5Ca_1Cu_6O_x$
$Bi_1Sr_1Ca_1Cu_4O_x$

The individual targets (connected along the quartz cylinder axis) were prepared by mixing the appropriate amounts of CuO, Bi_2O_3, $SrCO_3$, and $CaCO_3$ powders in ethanol and calcining the powder mixture for two hours at 800°C. The calcined powders were then pressed into pellets and sintered for two more hours at 820°C. Film substrates, placed horizontally inside the quartz cylinder, were sintered polycrystalline 2.5% yttria-stabilized zirconia (YSZ). The gas mixture used during the sputtering process was a 1/1 mixture of O_2/Ar at a pressure of 100 mtorr. Sputtering/annealing conditions of 750°C for 240 min did not produce a superconductor. Annealing temperatures of 800–900°C, for periods ranging from 50 to 300 min produced T_c-onsets in the range of 92–112 K, and zero dc resistivity typically in the range of 54–76 K, but as low as 18 K for the sample annealed at 900°C. A plot of dc resistivity versus temperature in good samples typically showed a small "bump" indicating the presence of a phase near 115 K, while the major transition was close to 80 K.

Screen Printing. Thick films of Bi-Ca-Sr-Cu-O have been deposited on either YSZ or $SrTiO_3$ substrates by a screen-printing method (Hashimoto, et al., 1988b). The investigators began with the same ingredients listed previously (i.e., CuO, Bi_2O_3, $SrCO_3$, and $CaCO_3$ powder mixture), which were processed for 12 hours at 800°C. The resultant ceramic powders were $Bi_1Sr_1Ca_1Cu_2O_x$ and $Bi_2Sr_2Ca_1Cu_3O_x$.

A paste was made by mixing 1 g of powder with 0.05 ml of octyl alcohol; mixing was in an agate mortar. Substrates used were alumina, quartz, YSZ, and single-crystal $SrTiO_3$. Paste was applied to the various substrates through a "150 mesh" screen. The films were first dried (under vacuum) for three hours at 120°C, then sintered at temperatures between 830 and 800°C for one hour. Thickness of the films ranged from 8 to 12 μm.

Electrical characteristics of the films were found to be strongly dependent on substrate material and sintering temperature. (This is also the case for screen-printed films of La-Sr-Cu-O and Y-Ba-Cu-O.) Generally, the YSZ substrates were preferable to either $SrTiO_3$ or alumina; films formed on quartz tended to be insulating.

Magnetron Sputtering. Films of Bi-Ca-Sr-Cu-O have been formed by magnetron sputtering onto MgO, $SrTiO_3$ and Al_2O_3 substrates (Nakao, et al., 1988) (see Figure 5.15). The compound target was sintered $Bi_1Ca_1Sr_1Cu_2O_x$. Sputtering was done at room temperature in ultrapure argon with pressures ranging from 3 to 30 mtorr, with a target voltage of 2300 V and a target-substrate separation of 4.5 cm. Films had a thickness (prior to annealing) of 1–2 μm. After sputtering, the films were annealed for one hour at 870 to 890°C in an oxygen atmosphere. Indium contacts were used for four-probe measurement of dc resistance as a function of temperature.

The films deposited onto Al_2O_3 developed a gray appearance after annealing and showed no evidence of superconductivity. The investigators report that Bi-Ca-Sr-Cu-O films deposited on $SrTiO_3$, while superconducting, were inferior to those deposited onto MgO. Annealing time, however was a critical parameter, with a T_c of ~ 85 K for an annealing temperature of 880°C for one hour. When the annealing temperature was changed in either direction by only 10°C (i.e., 870 or 890°C), T_c dropped to approximately 65 K. See Figure 5.16.

Melt-Quenching. Sintering, the common method for preparation of bulk high-T_c superconductor samples, implies heating without melting. It has been

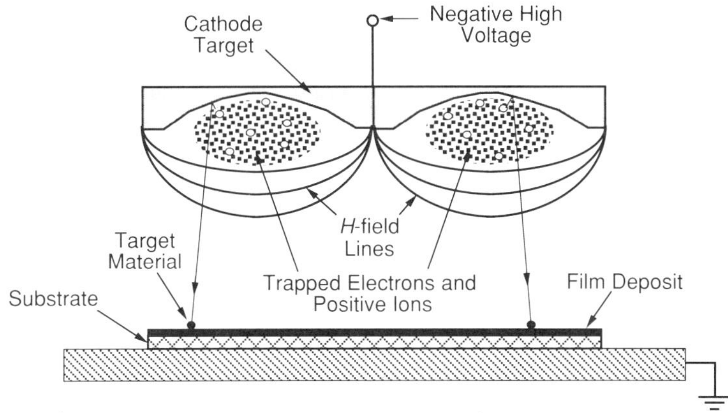

FIGURE 5.15 In magnetron sputtering, a shaped magnetic field traps electrons near the negative target-cathode. The high density of electrons ionizes the sputtering (argon) gas close to the target surface. The positive argon ions (represented here by empty circles) are attracted to the negative target, where their impact dislodges target-material atoms (represented by filled circles). Some of these target atoms are deposited on the substrate. (Based partially on information in literature provided by U.S., Inc., Campbell, CA)

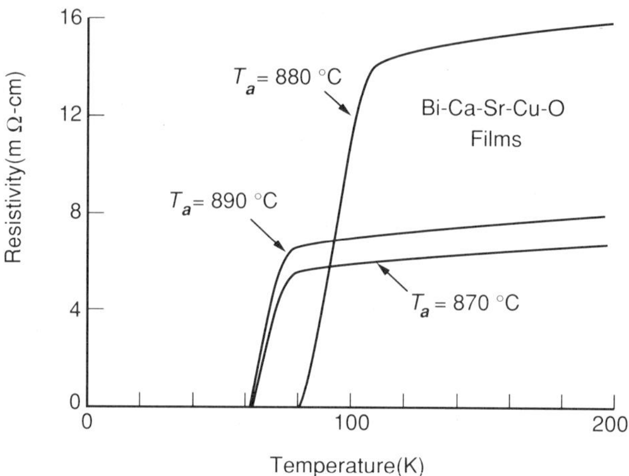

FIGURE 5.16 This set of plots, for Bi-Ca-Sr-Cu-O magnetron-sputtered thin films, is the result of changing the anneal temperature from 870°C to 890°C in 10°C steps. Anneal time was fixed at 1 hour. Clearly, anneal temperature was a critical parameter in the preparation of these superconducting films. The substrate in this instance was MgO. (After M. Nakao, H. Kuwahara, R. Yuasa, H. Mukaida, and A. Mizukami, "Magnetron sputtering of Bi-Ca-Sr-Cu-O thin films with superconductivity above 80 K," *Jap. J. Appl. Phys.*, Vol. 27, No. 3, March 1988, pp. L378–L380.)

shown that heating above the melting temperature, followed by cooling at a modest rate, does not generally produce a sample that superconducts.

Researchers at Japan's Technological University of Nagaoka have demonstrated that processing at temperature levels above the melting point produces a glass-like superconductor, if the melt is *rapidly* quenched (Komatsu, et al., 1988a).

In the work with Bi-Ca-Sr-Cu-O, powders of Bi_2O_3, $CaCO_3$, $SrCO_3$, and CuO were mixed and then melted in a platinum crucible at a temperature of 1150°C for 30 min. The melted compound, of $BiCaSrCu_2O_x$, was poured onto a room-temperature iron plate and immediately pressed to a thickness of ~ 3 mm. Some workers describe this part of the process graphically, if somewhat inelegantly, as "splat" processing. No evidence of superconductivity was seen at this point; annealing in oxygen was necessary.

In the specific case at hand, the samples were then annealed for 24 hours in an oxygen atmosphere at temperatures ranging from 800 to 850°C. Samples were then cooled slowly or "rapidly" (by immersion in liquid nitrogen!). While the T_c-onset was approximately 85 K with a very broad transition for the slowly cooled samples, T_c was 95 K and the transition was quite sharp for the rapidly cooled samples. The investigators state that, as opposed to some earlier melt-quenched superconductors that were a mixture of crystals and did not show the amorphous structure of glass, these samples show combination of a true glass transition effect (i.e., an endothermic peak) at 427°C (Komatsu, et al., 1988a).

All of the phases are superconducting, with their stoichiometry and T_c's listed

Phase	T_c
$Bi_{1.5}CaSrCu_2O_x$	65 K
$Bi_{1.5}Ca_{1.5}SrCu_2O_x$	63 K
$Bi_{0.5}Pb_{0.5}CaSrCu_2O_x$	71 K
$BiAl_{0.1}CaSrCu_2O_x$	73 K
$BiAl_{0.3}CaSrCu_2O_x$	78 K

The principle advantage of these "glass ceramics" is that they can readily be formed into a variety of useful shapes. The investigators expect it will be straightforward to draw the glass into small-diameter wires or tapes, using the same techniques that have already been developed for the manufacture of optical fibers. Similar melt-quenching techniques, when applied to the processing of Y-Ba-Cu-O, produce samples that are microcrystalline rather than glass-like (Swinbanks, 1988).

Thick Films by the Sol-Gel Method

Most of us are familiar with the concept of a gel or "jelly," which is a two-phase colloidal system comprised of a solid and a liquid. The primary difference in a sol and a gel is that the latter has a tendency to retain its characteristic shape, while a sol (i.e., a colloidal dispersion) assumes the shape of its container. Setting of a sol to the gel phase is accompanied by very modest heat effects; such a transformation is known as the isothermal sol-gel transformation (Milligan, 1971).

The sol-gel technique for processing of superconducting films has the advantage that pure, homogeneous ceramics can be prepared at relatively low sintering temperatures compared to other conventional techniques. The successful preparation of thick Bi-Ca-Sr-Cu-O films by the sol-gel technique has been reported as follows:

Commercially available ethoxide solutions of Bi, Sr, Ca, and Cu were mixed to prepare appropriate molar ratios of ethanol solutions of Bi:Sr:Ca:Cu:O. The final concentration of this solution was 3–6 mmol Bi-Sr-Ca-Cu-O/100 ml EtOH. This ethanol solution was "dropped" on a MgO<100> surface, where the oxide film was quickly pyrolyzed at 600–700°C in an oxygen atmosphere. This combination dropping and thermal decomposition of the ethoxide was repeated dozens of times. The resultant films were then sintered for two hours at several test temperatures ranging from 850–900°C in O_2. Cooling rates were on the order of 1°C/min. Final film thickness ranged between 30 and 50 μm (Kobayashi, et al., 1988).

Spray Pyrolysis

This technique has the advantages of speed, simplicity and low cost. In addition, thick Bi(Pb)-Sr-Ca-Cu-O films have been produced that show evidence of orientation. In one example, an aqueous solution of Pb, Bi, Ca, Sr, and Cu nitrates were prepared from precursors of $Bi(NO_3)$, $5H_2O_3$, $Pb(NO_3)_2$, $Sr(NO_3)_2$, $Ca(NO_3)_2$ $4H_2O$, and $Cu(NO_3)_2$ $3H_2O$, which were dissolved into triply distilled water at a ratio of 25 mmol/100 cc. The aqueous solution was sprayed onto a single-crystal MgO<100> substrate that was maintained at 400°C by a hot plate. Carrier gas for the spray was O_2 compressed to approximately 1 kg/cm^2; distance from the spray nozzle to the MgO substrate was ~ 20 cm. The hot plate was turned off after the spray was complete; cooling required about 2 hours. The variety of samples made were stacked and maintained at 845°C in air for periods ranging from 2 to 80 hours; a minimum of 15 hours was considered sufficient. The authors report that stacking, which limited exposure to air and made the processing oxygen-deficient, yielded better results than trials with unstacked samples. Final film thickness was 40–50 μm. The zero-resistance T_c was as high as 110 K (Nobumasa, et al., 1988).

Bi-Ca-Sr-Cu-O Fibers by Laser-Heated Pedestal Growth

Another interesting technique for the production of "wires" of the new ceramics, is the use of CO_2, highly-focused (~ 50 μm), laser-beam heating of a superconducting source material. An oriented "seed" is placed into contact with the molten zone and gradually pulled away to form the superconducting fiber. The method has the advantage that it is "containerless;" the process does not require an oven since the heat is generated by the laser in a highly localized fashion (see Figure 5.17).

Superconducting fibers of $Bi_2CaSrCu_2O_8$ have been produced in this fashion (Feigelson, et al., 1988). The *a-b* planes of the crystal may be generated parallel to the fiber axis, providing maximum current transport capability along the resultant fiber. Recall that the copper-oxygen planes are parallel to the *a-b* axis, and that current transport is optimum along these planes. Typical fiber diameter ranged from 0.25 to 1.0 mm, with lengths up to 40 mm. Current density, measured by a pulse technique, was as high as 60,000 A/cm^2 at 68 K. T_c was 90–95 K.

Unstable Phases in Bismuth Systems

While the major phases of Bi-Ca-Sr-Cu-O that have been investigated are quite stable, there is evidence that most of these preparations may also contain relatively small amounts of unstable phases. Decomposition of transient phases has been reported as a result of temperature cycling, where low temperatures destroy the unstable phases (Inoue, et al., 1988). The initial formation of these unstable phases was largely a function of the sintering atmosphere.

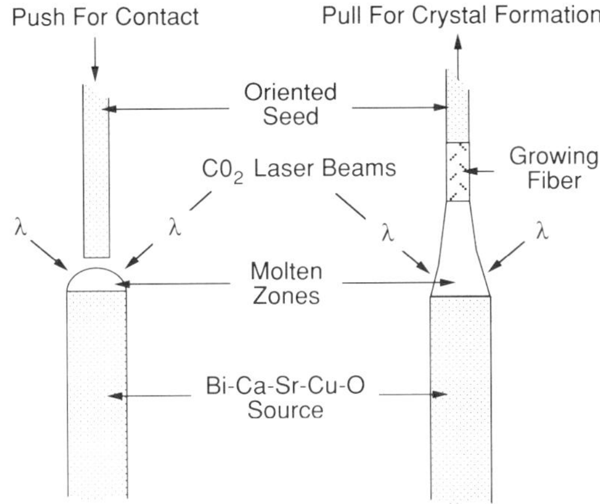

FIGURE 5.17 Laser-heated pedestal growth has been used to grow Bi-Ca-Sr-Cu-O fibers with diameters ranging from 0.25 to 1.0 mm, with lengths ranging from about 1 to 4 cm. Growth rate of the crystal fiber was typically about 1.5 to 24 mm/hr. Critical current density, measured by a pulse technique, was 60,000 A/cm^2 at 68 K. (After R. S. Feigelson, D. Gazit, D. K. Fork, T. H. Geballe, "Superconducting Bi-Ca-Sr-Cu-O fibers grown by the laser-heated pedestal growth method," *Science*, Vol. 240, No. 4859, June 17, 1988, pp. 1642–1645.)

IN COMBINATION WITH SEMICONDUCTORS

Superconducting Films on Silicon

One of the problems facing those who wish to exploit the characteristics of the various new high-T_c ceramics in electronic applications is the fact the new superconductors are easily "poisoned" by use of an inappropriate substrate. Silicon, in particular, diffuses into the 1-2-3 ceramic superconductor and destroys its properties. Investigators at the Hiroshima University have had some success by epitaxially growing a layer of tetragonal ZrO_2 as a buffer layer on <100>Si wafers (Nasu, et al., 1988). They measure an onset of superconductive behavior for bismuth films at 97 K, with zero resistivity at 50 K. The same investigators have also used MgO, $SrTiO_3$, and platinum as a buffer between the silicon and bismuth superconductor, but prefer ZrO_2 since it effectively blocks diffusion of silicon and is inexpensive.

Other investigators have used a sol-gel technique to make ZrO_2-coated silicon, and have been successful in screen-printing Y-Ba-Cu-O on this buffer layer (Kitagawa, et al., 1988). They report that the ZrO_2 buffer layer protects the 1-2-3 film from the silicon during the firing process. The resultant temperature for zero dc resistance was 85 K for this approach (see Figure 5.18). A group at Itami Research Laboratories has used magnetron sputtering to deposit Y-Ba-Cu-O films on silicon, with a 50 nm buffer layer of either ZrO_2 or MgO. While the

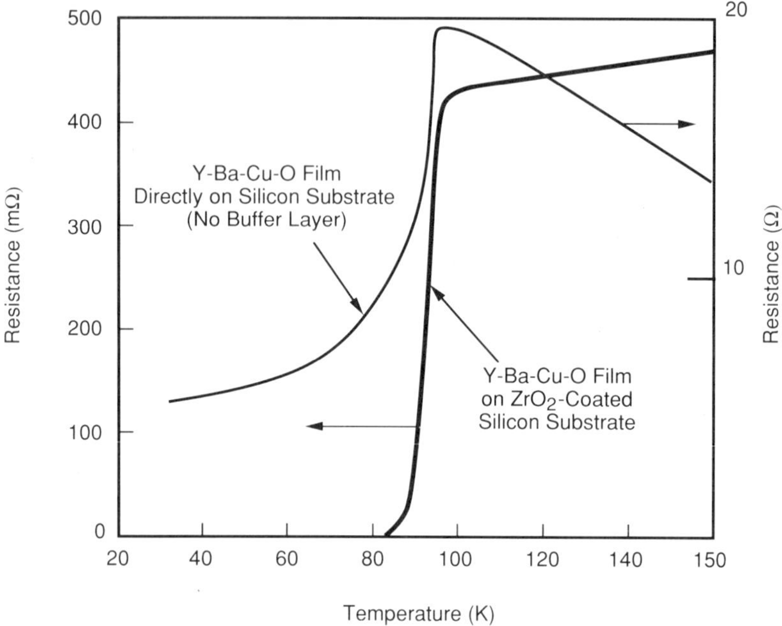

FIGURE 5.18 When Y-Ba-Cu-O superconducting films are plated directly onto silicon (important for hybrid applications with semiconductors) penetration of silicon into the superconducting film during heat-processing dramatically degrades film performance. While a transition can still be seen at T_c, films resistance remains high below the transition. When ZrO_2 is used as a buffer layer between the Y-Ba-CuO-O and the silicon, film integrity is largely preserved. (After T. Kitagawa, S. Shibata, H. Okazaki, T. Kimura, and Y. Enomoto, "High-T_C superconducting film on silicon substrate with ZrO_2 buffer layer," *Jap. J. Appl. Phys.*, Vol. 27, No. 6, June, pp. L1113–L1115.)

ZrO_2 buffer successfully prevented interdiffusions between the silicon and the 1-2-3 film, the MgO buffer did not perform as well in this regard. The critical transition for the film on ZrO_2-buffered silicon had an onset at 93 K with zero dc resistivity at 82 K. The same 1-2-3 film on silicon/MgO exhibited a very broad transition, finally reaching zero dc resistance at 45 K (Harada, et al., 1988).

Y-Ba-Cu-O and Germanium

Investigators at Osaka University have studied the effects of growing germanium thin films on $YBa_2Cu_3O_x$ films, since Ge is of considerable interest in semiconductor applications, and because <100>Ge is a good lattice match to <001>$YBa_2Cu_3O_x$ (Tonouchi, et al., 1988). The substrate for the 1-2-3 film was magnesium oxide. Crystalized germanium was prepared on the polycrystalline $YBa_2Cu_3O_x$ films with the 1-2-3/MgO substrate temperature ranging from 100°C to 300°C. A temperature above 200°C was necessary for crystallization of the Ge onto the $YBa_2Cu_3O_x$ film. The Ge films were evidently formed without an intense reaction with the $YBa_2Cu_3O_x$.

THALLIUM PROCESSING—PROCEED WITH CAUTION!

Thallium and many of its compounds (including oxides, e.g., Tl_2O_3) are extremely toxic. This toxicity leads to restrictions that must be considered in the planning of any process that uses thallium, particularly since the toxin will pass through unbroken skin.

While solid state reactions with short heating times have been used, as well as an approach where samples are heated in a sealed quartz tube containing oxygen, there is a continual search for simpler, safer methods (Zhou, et al., 1988).

In an effort to solve some of the problems associated with processing these toxic materials, the group at the University of Arkansas that discovered the thallium superconducting compounds has experimented with a technique where gasified thallium oxide (Tl_2O_3) is combined with a precursor made of barium, calcium, and a copper oxide in a hermetically sealed reaction chamber. Tl_2O_3 melts at 717°C and evaporates readily above this temperature. The Tl_2O_3 vapor reacts with a Ba-Ca-Cu-O pellet precursor to form the superconductor. Since the thallium and the precursor materials are contained in the sealed chamber, risks of contamination of either the processing equipment or personnel during processing is minimized. The thallium evidently penetrates 2–3 mm into the precursor material, quite sufficient for most applications (Sheng, et al., 1988; Carlson, 1988).

While the problems with processing thallium have been mentioned briefly here, it is well beyond the scope or intent of this reference to provide detailed descriptions of methods for the safe handling of thallium. The reader is urged to consult the literature on this subject and discuss proposed procedures with safety experts before processing any material that is highly toxic. There is not only the obvious danger to those who operate the instrumentation, but also the very real possibility of contaminating processing apparatus.

REFERENCES

Bunshah, R. F., (1967). "Physical vapor deposition of metals, alloys, and ceramics," in *New Trends in Materials Processing*, American Society for Metals, pp. 200–267.

Carlson, L., (1988). "The Arkansas Method," *Superconductor Industry*, Vol. 1, No. 2, Winter, p. 18.

Chen, Xiao-Dong, Sang Young Leeh, J. P. Golbeng, Sung-Ik Lee, R. D. McMichael, Yi Song, Tae W. Noh, and J. R. Gaines, (1987). "Practical preparation of copper oxide superconductors," *Rev. Sci. Instrum.* 58(9), September, pp. 1565–1571.

Damento, M. A., K. A. Gschneidner, Jr., and R. W. McCallum, (1987). "Preparation of single crystals of superconducting $YBa_2Cu_3O_{y-x}$ from CuO melts," *Appl. Phys. Lett.*, Vol. 51, No. 9, August 31, pp. 690–691.

Endo, U., S. Koyama, and T. Kawai, (1988). "Preparation of the high-T_c phase of the Bi-Sr-Ca-Cu-O superconductor," *Jap. J. Appl. Phys.*, Vol. 27, No. 8, August, pp. L1476–L1479.

Feenstra, R., L. A. Boatner, J. D. Budal, D. K. Christen, M. D. Galloway, and D. B. Poker, (1988). "Epitaxial superconducting thin films of $YBa_2Cu_3O_{7-x}$ on $KTaO_3$ single crystals," Submitted to *Appl. Phys. Lett.*

Feigelson, R. S., D. Gazit, D. K. Fork, and T. H. Geballe, (1988). "Superconducting Bi-Ca-Sr-Cu-O fibers grown by the laser-heated pedestal growth method," *Science*, Vol. 240, No. 4859, June 17, pp. 1642–1645.

Fukutomi, M., J. Machida, Y. Tanaka, T. Asano, T. Yamamoto, and H. Maeda, (1988). "New technique for preparation of BiSrCaCuO thin films with T_c of 100 K and above," *Jap. J. Appl. Phys.*, Vol. 27, No. 8, August, pp. L1484–L1486.

Gieske, J. H., and Frost, H. M., (1988). "Sound velocity measurements in green-body ceramics as a function of sintering temperature," Proceedings: Review of Progress in QNDE, La Joya, CA, August 1–5.

Hakuraku, Y., F. Sumiyoshi, and T. Ogushi, (1988). "Role of added fluorine to enhance the electromagnetic properties of superconducting Y-Ba-Cu-O compounds," *Appl. Phys. Lett.*, Vol. 52, No. 18, May 2, pp. 1528–1530.

Harada, K., N. Fujimori, and S. Yazu, (1988). "Y-Ba-Cu-O thin films on Si substrate," Vol. 27, No. 8, *Jap. J. Appl. Phys*, August, pp. L1524–L1526.

Hashimoto, T., K. Fueki, A. Kishi, T. Azumi, and H. Koinuma, (1988a). "Thermal Expansion Coefficients of High-T_c Superconductors," *Jap. J. Appl. Phys.*, Vol. 27, No. 2, February, pp. L214–L216.

Hashimoto, T., T. Kosaka, Y. Yoshida, K. Fueki, and H. Koinuma, (1988b). "Superconductivity and substrate interaction of screen-printed Bi-Sr-Ca-Cu-O films," *Jap. J. Appl. Phys.*, Vol. 27, No. 3, March, pp. L384–L386.

Hein, M., G. Müeller, H. Piel, and L. Ponto, (1988). University of Wuppertal Report No. WUB-88-1.

Ichikawa, Y., H. Adachi, T. Mitsuyu, and K. Wasa, (1988). "Effect of overcoating with dielectric films on the superconductive properties of the high-T_c Y-Ba-Cu-O films," *Jap. J. Appl. Phys.*, Vol 27, No. 3, March, pp. L381–L383.

Inam, A., X. D. Wu, T. Venkatesan, S. B. Ogale, C. C. Chang, and D. Dijkkamp, (1987). "Pulsed laser etching of high-T_c superconducting films," *Appl. Phys. Lett.*, Vol. 51, No. 14, October 5, pp. 1112–1114.

Inoue, O., S. Adachi, and S. Kawsashima, (1988). "Formation and deposition of superconducting $BiSrCaCu_2O_x$ ceramics," *Jap. J. Appl. Phys.*, Vol. 27, No. 3, March, pp. L347–L349.

Jin, S., T. H. Tiefel, R. C. Sherwood, G. W. Kammlott, and S. M. Zahurak, (1987). "Fabrication of dense $Ba_2YCu_3O_{y-\delta}$ superconductor wire by molten oxide processing," *Appl. Phys. Lett.*, Vol. 51, No. 12, September 21, pp. 943–945.

Jin, S., T. H. Tiefel, R. C. Sherwood, R. B. van Dover, M. E. Davis, G. W. Kammlott, and R. A. Fastnacht, (1988). "Melt-textured growth of polycrystalline $YBa_2Cu_3O_{7-\delta}$

with high transport J_c at 77 K," *Phys. Rev. B.*, Vol. 37, No. 13, May 1, pp. 7850–7853.

Kaiser, D. L., F. Holtzberg, B. A. Scott, and T. R. McGuire, (1987). "Growth of $YBa_2Cu_3O_x$ single crystals," *Appl. Phys. Lett.*, Vol. 51, No. 13, Sept. 28, pp. 1040–1042.

Karthikeyan, J., K. P. Sreekumar, M. B. Kurup, D. S. Patil, P. V. Anantapadmanabhan, N. Venkatramani, and V. K. Rohatgi, (1988). "Plasma sprayed superconducting $Y_1Ba_2Cu_3O_{7-x}$ coatings," *J. Phys. D*, Vol. 21, No. 7, July, pp. 1246–1249.

Kitagawa, T., S. Shibata, H. Okazaki, T. Kimura, and Y. Enomoto, (1988). "High-T_c superconducting film on silicon substrate with ZrO_2 buffer layer," *Jap. J. Appl. Phys.*, Vol. 27, No. 6, June, pp. L1113–L1115.

Kobayashi, T., K. Nomura, F. Uchikawa, T. Masumi, and Y. Uehara, (1988). "Superconducting Bi-Sr-Ca-Cu-O thick films by the Sol-Gel method," *Jap. J. Appl. Phys.*, Vol. 27, No. 10, October, pp. L1880–L1882.

Koinuma, H., M. Kawasaki, S. Nagata, K. Takeuchi, and K. Fueki, (1988). "Preparation of high-T_c Bi-Sr-Ca-Cu-O superconducting thin films by ac sputtering," *Jap. J. Appl. Phys.*, Vol. 27, No. 3, March, pp. L376–L377.

Komatsu, T., K. Imai, R. Sato, K. Matusita, and T. Yamashita, (1989a). "Preparation of high-T_c superconducting Bi-Ca-Sr-Cu-O ceramics by the melt quenching method," *Jap. J. Appl. Phys.*, Vol. 27, No. 4, April, pp. L533–L535.

Komatsu, T., O. Tanaka, K. Matusita, and T. Yamashita, (1988b). "On the new substrate materials for high-T_c superconducting Ba-Y-Cu-O thin films," *Jap. J. Appl. Phys.*, Vol. 27, No. 9, September, pp. L1686–L1689.

Makous, J. L, L. Maritato, C. M. Falco, J. P. Cronin, G. P. Rajendran, E. V. Uhlmann, and D. R. Uhlmann, (1987). "Superconducting and structural properties of sputtered thin films of $YBa_2Cu_3O_{7-x}$," *Appl. Phys. Lett.*, Vol. 51, No. 25, December 21, pp. 2164–2166.

Milligan, W. O., (1971). "Gel," in *McGraw-Hill Encyclopedia of Science and Technology*, Vol. 6, McGraw-Hill, New York, p. 97.

Nakajima, H., S. Yamaguchi, K. Iwasaki, H. Morita, H. Fujimori, and Y., Fujino, (1988). "Interdiffusion and interfacial reaction between $YBa_2Cu_3O_x$ thin film and substrate," to be published in *Appl. Phys. Lett.*, October 10.

Nakao, M., H. Kuwahara, R. Yuasa, H. Mukaida, and A. Mizukami, (1988). "Magnetron sputtering of Bi-Ca-Sr-Cu-O thin films with superconductivity above 80 K," *Jap. J. Appl. Phys.*, Vol. 27, No. 3, March, pp. L378–L380.

Nasu, H., H. Myoren, Y. Ibara, S. Makida, Y. Nishiyama, T. Kato, T. Imura, and Y. Osaka, (1988). "Formation of high-T_c superconducting $BiSrCaCu_2O_x$ films on ZrO_2/Si(100)," *Jap. J. Appl. Phys.*, Vol. 27, No. 4, April, pp. L634–L635.

Nobumasa, H., K. Shimizuh, Y. Kitano, and T. Kawai, (1988). "Formation of a 100 K superconducting Bi(Pb)-Sr-Ca-Cu-O film by spray pyrolysis," *Jap. J. Appl. Phys.*, Vol. 27, No. 9, September, pp. L1669–L1671.

Oh, B., M. Naito, S. Arnason, P. Rosenthal, R. Barton, M. R. Beasley, T. H. Geballe, R. H. Hammond, and A. Kapitulnik, (1987). "Critical current densities and transport in superconducting $YBa_2Cu_3O_{7-\delta}$ films made by electron beam coevaporation," *Appl. Phys. Lett.*, Vol. 51, No. 11, September 14, pp. 852–854.

Okada, M., A. Okayama, T. Morimoto, T. Matsumoto, K. Aihara, and S. Matsuda, (1988). "Fabrication of Ag-Sheathed Ba-Y-Cu Oxide Superconductor Tape," *Jap. J. Appl. Phys.*, Vol. 27, No. 2, February, pp. L185–L187.

Piel, H., M. Hein, N. Klein, U. Klein, A. Michalke, G. Müeller, and L. Ponto, (1988). "Superconducting perovskites in microwave fields," Proc. of Int. Conf. on High Temperature Superconductors, Interlaken, Switzerland, *Physica C*, Vol. 153-155, Superconductivity, June, pp. 1604–1609.

Sandstrom, R. L., W. J. Gallagher, T. R. Dinger, R. H. Koch, R. B. Laibowitz, A. W. Kleinsasser, R. J. Gambino, B. Bumble, and M. F. Chisholm, (1988). "Reliable single-target sputtering process for high-temperature superconducting films and devices," *Appl. Phys. Lett.*, Vol. 53, No. 5, August 1, pp. 444–446.

Sheng, Z. Z., L. Sheng, H. M. Su, and A. M. Hermann, (1988). "Tl_2O_3-vapor-process of making Tl-Ba-Ca-Cu-O superconductors," submitted to *Appl. Phys. Lett.*

Shih, I., and C. X. Qiu, (1988). "Chemical etching of Y-Cu-Ba-O thin films," *Appl. Phys. Lett.*, Vol. 52, No. 18, May 2, pp. 1523–1524.

Swinbanks, D., (1988). "High-critical-temperature superconductor made of glass," *Nature*, Vol. 332, No. 6165, April 28, pp. 575.

Terada, N., H. Ihara, M. Jo, M. Hirabayashi, Y. Kimura, K. Matsutani, K. Hirata, E. Ohno, R. Sugise, and F. Kawashima, (1988). "Sputter synthesis of $Ba_2YCu_3O_y$ as deposited superconducting thin films from stoichiometric target—A mechanism for compositional deviation and its control," *Jap. J. Appl. Phys.*, Vol. 27, No. 4, April, pp. L639–L642.

Tonouchi, M., Y. Sakaguchi, K. Hashimoto, Y. Yoshizako, and T. Kobayashi, (1988). "Germanium thin film growth onto high-T_c superconducting films," *IEEE Trans. Magn.*, Vol. 25, No. 2, March, pp. 957–960.

Wu, X. D., D. Dijkkamp, S. B. Ogale, A. Inam, E. W. Chase, P. F. Miceli, C. C. Chang, J. M. Tarascon, and T. Venkatesan, (1987). "Epitaxial ordering of oxide superconductor thin films on (100)$SrTiO_3$ prepared by pulsed laser evaporation," *Appl. Phys. Lett.*, Vol. 51, No. 11, September 14, pp. 861–863.

Zhou, X. Z., A. H. Morrish, Y. L. Luo, M. Raudsepp, and I. Maartense, (1988). "Bulk high-temperature superconductivity in the Tl-Ca-Ba-Cu-O system," *J. Phys. D*, Vol. 21, No. 7, July, pp. 1243–1245.

SUGGESTED READING ON CERAMIC SUPERCONDUCTOR PROCESSING

Adachi, H., Y. Ichikawa, K. Setsune, S-I Hatta, K. Hirochi, and K. Wasa, "Preparation and properties of superconducting Bi-Sr-Ca-Cu-O thin films," *Jap. J. Appl. Phys.*, Vol. 27, No. 4, April, 1988, pp. L643–L645.

Aizaki, N., K. Terashima, J-I Fujita, and S. Matsui, "$YBa_2Cu_3O_y$ Superconducting Thin Film Obtained by Laser Annealing," *Jap. J. Appl. Phys.*, Vol. 27, No. 2, February 1988, pp. L231–L233.

Beasly, M. R., "Advanced Superconducting Materials for Electronic Applications," *IEEE Trans. Electron Devices*, Vol. ED-27, No. 10, October 1980, pp. 2009–2015.

Dijkkamp, D., and T. Venkatesan, "Preparation of Y-Ba-Cu oxide superconductor thin films using pulsed laser evaporation from high T_c bulk material," *Appl. Phys. Lett.*, Vol. 51, No. 8, August 24, 1987, pp. 619–621.

Donaldson, G. B., and H. Faghihi-Nejad, "An Ellipsometric Study of RF Sputter Oxidation of Lead-Indium Alloys," *IEEE Trans. Electron Devices*, Vol. ED-27, No. 10, October 1980, pp. 1988–1997.

Fujii, H., H. Kawanaka, W. Ye, S. Orimo, and H. Fukuba, "Effect of hydrogen absorption on superconductivity in $YBa_2Cu_3O_{6.91}$ and $GdBa_2Cu_3O_{6.89}$," *Jap. J. Appl. Phys.*, Vol. 27, No. 4, April 1988, pp. L525–L528.

Fukutomi, M., J. Machida, Y. Tanaka, T. Asano, H. Maeda, and K. Hoshino, "Sputter deposition of BiSrCaCuO thin films," *Jap. J. Appl. Phys.*, Vol. 27, No. 4, April 1988, pp. L632–633.

Harada, K., N. Fujimori, and S. Yazu, "Y-Ba-Cu-O thin film on Si substrate," *Jap. J. Appl. Phys.*, Vol. 27, No. 8, August 1988, pp. L1524–L1526.

Honda, T., T. Wada, M. Sakai, M. Miyajima, N. Nishikawa, S-I Uchida, K. Uchinokura, and S. Tanaka, "Preparation of high-T_c (105 K) superconducting phase in the Bi-Sr-Ca-K-Cu-O system," *Jap. J. Appl. Phys.*, Vol. 27, No. 4, April 1988, pp. L545–L547.

Hyland, G. J., "On the fluorination dependence of T_c in $YBa_2Cu_3O_{7-\delta}$," *Jap. J. Appl. Phys.*, Vol. 27, No. 4, April 1988, pp. L598–L599.

Ihara, M., T. Kimura, H. Yamawaki, and K. Ikeda, "High-T_c BiSrCaCuO superconductor grown by CVD technique," *IEEE Trans. Magn.*, Vol. 25, No. 2, March 1989, pp. 2470–2473.

Kato, T., K. Aihara, J. Kuniya, T. Kamo, and S.-P. Matsuda, "Phase transition in $YBa_2Cu_3O_{7-x}$ through hydrogen ion implantation," *Jap. J. Appl. Phys.*, Vol. 27, No. 4, April 1988, pp. L564–L566.

Koch, R. H., and P. Chaudhari, "Improving thin-film techniques," *Nature*, Vol. 332, No. 6166, April 21, 1988, pp. 682–683.

Komatsu, T., R. Sato, K. Imai, K. Matusita, and T. Yamashita, "High-T_c superconducting glass ceramics based on the Bi-Ca-Sr-Cu-O system," *Jap. J. Appl. Phys.*, Vol. 27, No. 4, April 1988, pp. L550–L552.

Liu, R. S., Y. T. Huang, P. T. Wu, and J. J. Chu, "Epitaxial growth of high-T_c Bi-Ca-Sr-Cu-O superconducting layer by LPE process," *Jap. J. Appl. Phys.*, Vol. 27, No. 8, August 1988, pp. L1470–1472.

Mukaida, M., M. Yamamoto, Y. Tazoh, K. Kuroda, and K. Hohkawa, "Synthesis of Superconducting $YBa_2Cu_3O_{7-\delta}$ Thin Films by Electron Beam Deposition," *Jap. J. Appl. Phys.*, Vol. 27, No. 2, February 1988, pp. L211–L213.

Murphy, D. W., D. W. Johnson, Jr., S. Jin, R. E. Howard, "Processing techniques for the 93 K superconductor $Ba_2YCu_3O_7$," *Science*, Vol. 241, No. 4868, August 19, 1988, pp. 922–930.

Nakamori, T., H. Abe, Y. Takahashi, T. Kanamori, and S. Shibata, "Preparation of superconducting Bi-Ca-Sr-Cu-O printed thick films using a coprecipitation of oxalates," *Jap. J. Appl. Phys.*, Vol. 27, No. 4, April 1988, pp. L649–L651.

Nasu, H., S. Makida, Y. Ibara, T. Kato, T. Imura, and Y. Osaka, "Preparation of $BiSrCaCu_2O_x$ films with $T_c > 77$ K by pyrolysis of organic acid salts," *Jap. J. Appl. Phys.*, Vol. 27, No. 4, April 1988, pp. L536–L537.

Oka, K., K. Nakane, M. Ito, M. Saito, and H. Unoki, "Phase-equilibrium diagram in the ternary system Y_2O_3-BaO-Cu-O," *Jap. J. Appl. Phys.*, Vol. 27, No. 6, June 1988, pp. L1065–L1067.

Schneemeyer, L. F., R. B. van Dover, S. H. Glarum, S. A. Sunshine, R. M. Fleming, B. Batlogg, T. Siegrist, J. H. Marshall, J. V. Waszcsak, and L. W. Rupp, "Growth of superconducting single crystals in the Bi-Sr-Ca-Cu-O system from alkali chloride fluxes," *Nature*, Vol. 332, No. 6163, March 31, 1988, pp. 422–424.

Shinohara, K., F. Munakata, and M. Yamanaka, "Preparation of Y-Ba-Cu-O superconducting thin films by chemical vapor deposition," *Jap. J. Appl. Phys.*, Vol. 27, No. 9, September 1988, pp. L1683–L1685.

Takahashi, K., M. Nakao, D. R. Dietderich, H. Kumakura, and K. Togano, "Preparation and properties of Tl-Ca-Ba-Cu-O," *Jap. J. Appl. Phys.*, Vol. 27, No. 8, August 1988, pp. L1457–L1459.

Tanaka, Y., M. Fukutomi, T. Asano, and H. Maeda, "Effects of synthesis conditions on the properties of a superconducting Bi-Sr-Ca-Cu-O system," *Jap. J. Appl. Phys.*, Vol. 27, No. 4, April 1988, pp. L548–L549.

Terasaki, I., Y. Nakayama, K. Uchinokura, A. Maeda, T. Hasegawa, and S. Tanaka, "Superconducting films of $YBa_2Cu_3O_x$ and Bi-Sr-Ca-Cu-O fabricated by electron-beam deposition with a single source," *Jap. J. Appl. Phys.*, Vol. 27, No. 8, August 1988, pp. L1480–L1483.

Togano, K., H. Kumkura, H. Maeda, K. Takahashi, and M. Kakao, "Preparation of high-T_c Bi-Sr-Ca-Cu-O superconductors," *Jap. J. Appl. Phys.*, Vol. 27, No. 3, March 1988, pp. L323–L324.

Tomita, M. T. Hayashi, H. Takaoka, Y. Ishii, Y. Enomoto, and T. Murakami, "Cross-sectional TEM observation of $Ba_2YCu_3O_{7-x}$ film on $SrTiO_3$," *Jap. J. Appl. Phys.*, Vol. 27, No. 4, April 1988, pp. L636–L638.

Yang, K. Y., H. Homma, R. Lee, R. Bhadra, M. Grimsditch, S. Bader, "Phase diagram and oxygen stoichiometry of Y-Ba-Cu-O thin films," *Appl. Phys. Lett.*, September 7, 1988.

CHAPTER 6

MEASUREMENTS

"To Comprehend Through Measurement"
Kamerlingh Onnes' Motto
(de Bruyn Ouboter, 1987)

INTRODUCTION

The characterization of superconductors requires a familiarity with a variety of measurement techniques, but the determination of resistivity and magnetic susceptibility are rather fundamental. This is not to say that there are not other important characteristics to be determined, including the various thermoelectric parameters, thermal conductivity, heat capacity, ultrasound attenuation, *IR* absorption, microwave surface resistance, and the like. Since all of these parameters will usually be measured as a function of temperature, the experimentalist should have a working knowledge of thermometry at cryogenic temperatures.

TEMPERATURE MEASUREMENT

The measurement and control of temperature is, of course, usually central to experiments or applications involving superconductivity. Exquisite control may be required for an infrared sensor which must be operated very close to T_c, while a rather more straightforward monitor will suffice for a system that is simply operated under a liquid helium bath. This chapter will discuss a variety of sensors that are commonly used in cryogenic applications. For a

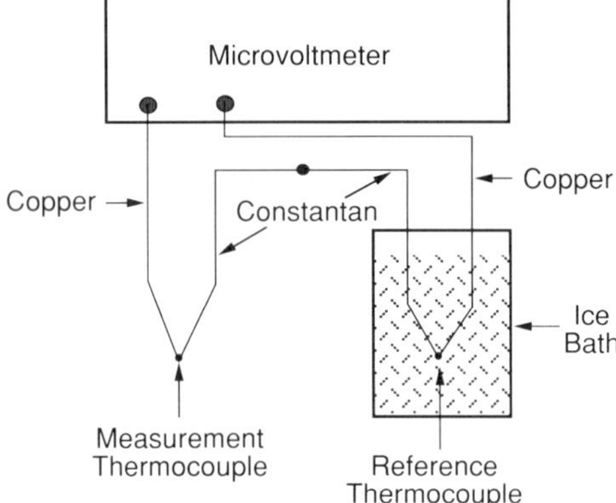

FIGURE 6.1 Differential thermometry with thermocouples is a straightforward technique for temperature measurements, even if the method has limitations in cryogenic applications. Electronic "reference junctions" that replace the ice bath shown above are available for several thermocouple types.

list of the relations between different temperature scales (Fahrenheit, Celsius, Kelvin, Rankine) see the appendix on that subject.

Temperature Sensors

Our attention will be restricted to a few types of sensors that are particularly useful at temperatures well below 100 K.

Thermocouples. Thermocouples are particularly useful for applications where a very small sensor is required for high spatial resolution or where low thermal capacity of the sensor is important. When high accuracy is required, thermocouples may be calibrated against a temperature standard over the range of interest, but other sensors may be preferable. Thermocouples also have the advantage that the circuitry required for monitoring is simple, that is, a microvoltmeter and a water-ice bath for the reference junction (see

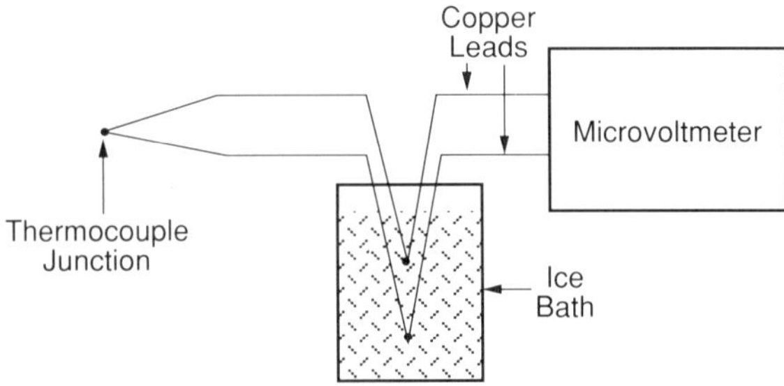

FIGURE 6.2 An alternate method for connecting thermocouple reference junctions.

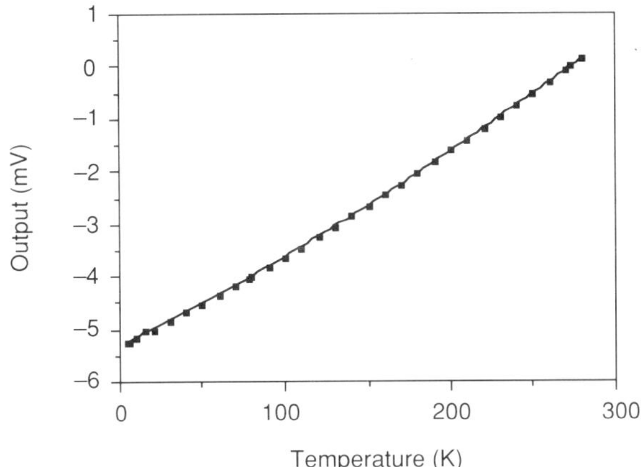

FIGURE 6.3 A typical calibration curve for the Au + 0.07%Fe vs. chromel thermocouple. A desirable feature of this particular thermocouple is the relatively constant sensitivity as temperature is decreased into the cryogenic region. Unfortunately, the precise calibration is quite sensitive to the iron content in the gold, rendering the response rather batch-dependent.

Figures 6.1 and 6.2). A thermoelectric refrigerator may replace the ice bath, and cold-junction "compensation" circuits are also available.

Copper-constantan (Type T), for example, may be used down to about 20 K with reasonable accuracy (Hoge, 1955). Type N thermocouples, comprised of $Ni_{84.5}Cr_{14.2}Si_{1.4}$ versus $Ni_{95.5}Si_{4.4}Mg_{0.1}$, are calibrated below 10 K, but sensitivity drops noticeably as temperature decreases below about 30 K (McGee, 1988). A preferred thermocouple at low temperatures is gold doped with iron versus chromel, (i.e., Au[0.07% Fe] versus chromel). The sensitivity of this (0.07% Fe) thermocouple is relatively constant at low temperatures.

At an iron doping of 0.02%, Au-Fe has been proposed as an absolute thermometer down to 0.4 K, due to the anomalously high thermopower (Rosenbaum, 1968). Rosenbaum suggests that individual lengths of thermocouple be calibrated against a known standard if "accuracies of better than ± 0.05 K in temperature are desired below 1 K." See Figures 6.3 and 6.4 for comparative output voltage from the Au-Fe/chromel thermocouple and Type N.

Consideration of the effects of electromagnetic "noise" in the vicinity of the measurement is required, since thermocouple systems tend to be sensitive to high-frequency electromagnetic fields. In the presence of electromagnetic fields, it is usually necessary to insulate the thermocouple and may be advisable to keep the thermocouple leads orthogonal to the direction of the electric field component. The induction of currents in the thermocouple leads can be reduced by increasing the resistance of the leads. Other approaches for noise reduction include twisting the leads to reduce common-mode capacitive pickup, eliminating ground loops, providing shielding for leads and amplifier circuitry, and using a battery (not referenced to local grounds) as a floating power supply rather than using 50 or 60 Hz ac power to operate the thermocouple electronics.

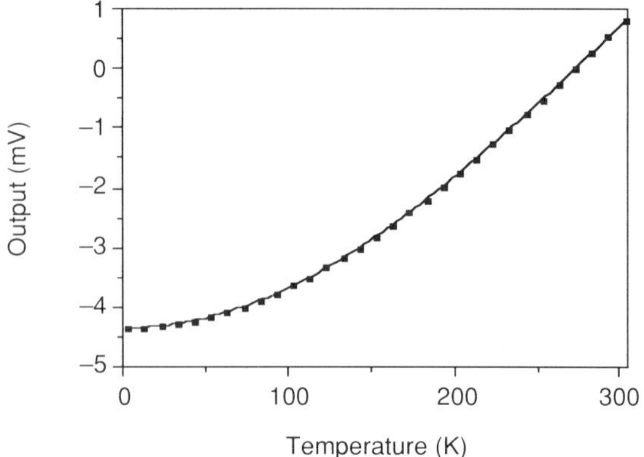

FIGURE 6.4 Thermocouples that are quite effective at higher temperatures, such as the Type N represented by this calibration curve, are not particularly useful in cryogenic applications due to the gradually decreasing sensitivity as temperature is decreased below ~100 K. Compare this plot to that of Figure 6.3.

Carbon and Carbon-Glass Resistors (CGRs). Favorite temperature sensors for cryogenic applications are the 1/8- or 1/4-watt carbon-composition (Allen-Bradley®) resistor and the Airco-Speer type SR-1/2, Grade 1002 1/2-watt carbon resistor (Sample, et al., 1974). These devices are quite rugged and show a very high sensitivity to temperature in the range below approximately 20 K, where resistance increases very rapidly as temperature decreases. See Figure 6.5 for a typical temperature-resistance curve. As with other resistive sensors where voltage-drop on the current-supplying wires interfere with accurate measurement, carbon resistors are also used in a four-lead measurement configuration, that is, separate pairs of leads for constant-current flow and voltage measurement.

Unfortunately, neither the Allen-Bradley nor Airco-Speer resistor types mentioned is currently available commercially—if you have access to some that were produced years ago, count yourself lucky. An alternative is the Matsushita 1/8-watt resistor, grades ERC-18GK, ERC-18SG, and ERC-18SCJ. These have been tested to 0.01 K and found to have suitable performance. The Matsushita resistors, manufactured in Japan, are now available in several countries (Rubin, et al., 1982).

As an alternative to using conventional resistors for cryogenic thermometry, highly stable carbon resistors designed especially for thermometry, that is, carbon-glass resistors (CGRs), are available. These hermetically-sealed devices may be purchased with individual temperature calibration sheets over the useful range of about 1.4 to 325 K (Lake Shore, 1988). The fact that CGRs, like ordinary carbon resistors, require individual calibration is a drawback, but one that is often tolerated in light of the excellent stability and commercial availability of these sensors.

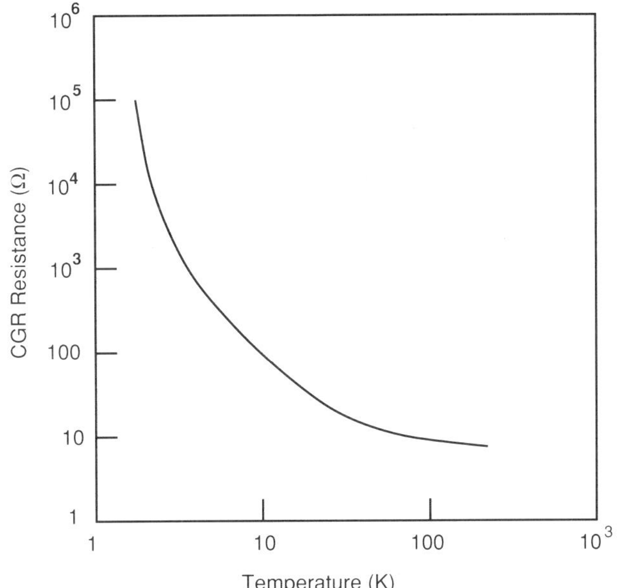

FIGURE 6.5 The resistance vs. temperature plot for a typical carbon composition resistor exhibits a high sensitivity at cryogenic temperatures. Commercial "carbon-glass" resistor (CGR) thermometers may be provided with calibration data in the range from about 1.4 K to 300 K. (Compiled from data provided by Lake Shore Cryotronics, Inc., Westerville, Ohio)

Since superconductor measurements must often be made in high magnetic fields, it is important to note that CGRs are relatively insensitive to such fields up to about 10 T, and are typically specified up to 19 T (Couach, et al., 1982) (see Figure 6.6). It is not unusual for CGR thermometers to show a slight orientation dependence of their magnetoresistance, although the effect is generally more noticeable at 77 K than at 4 K (Sample, et al., 1982).

The manufacturer recommends heat-sinking of the carbon-glass resistor leads at a point near where temperature is being measured for most accurate results. This technique, commonly known as "lagging," involves thermal connection of the CGR electrical sense leads to the same surface where the CGR is attached. This may be done by mechanical clamping (taking care not to disrupt the lead insulation), accompanied by use of a thermally conductive grease to provide good thermal bonding. Such steps are necessary because the internal CGR element can be heated via the leads almost as effectively as through its thermal connection to the metal enclosure. The entire point is to avoid heat flow through the leads, which may create temperature errors at the measurement location.

CGRs are the thermometers of choice when measurement is required in high magnetic fields. The silicon diode and germanium resistance thermometers to

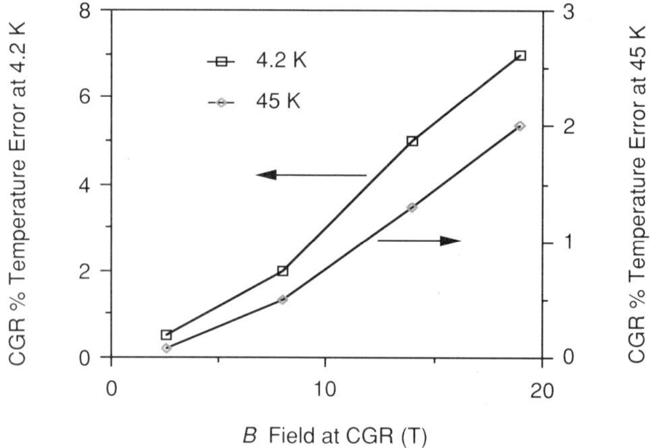

FIGURE 6.6 Well-designed carbon-glass resistor (CGR) thermometers are quite stable and have small errors due to the presence of magnetic fields below ~ 10 T. (Compiled from data provided by Lake Shore Cryotronics, Inc., Westerville, Ohio)

be discussed below should not be used in fields above approximately 2.5 T (Sample, et al., 1982).

Diode Temperature Sensors. Ordinary silicon and gallium arsenide diodes have sufficient sensitivity to temperature to allow their voltage drop (at constant current of ~100 μA) to be a reliable thermometer for many applications. The typical sensitivity ($\Delta V/\Delta T$) of voltage drop versus temperature for silicon diodes is on the order of 2–3 mV/K, increasing as temperature decreases. In fact, there is typically a sudden increase in $\Delta V/\Delta T$ at temperatures in the range of 30 K, resulting in a rather high sensitivity at the lowest temperatures (Talpe, et al., 1987) (see Figure 6.7).

Diode sensors may be used to achieve accuracies of less than 0.1 K at cryogenic temperatures if proper attention is paid to shielding and grounding techniques (Krause, et al., 1986). Since the diode characteristic is highly nonlinear, a sinusoidal current (noise) source will produce an offset in the dc voltage across the diode. In most instances, a single-point ground at the voltmeter reference and a floating current source will suffice for noise control. Additional reduction of noise can be produced by placing a low-leakage bypass capacitor across the diode to shunt the ac currents around the sensor. For significant reduction of 60 Hz noise, Krause et al. recommend a capacitance in the range of 10 to 20 μF. Such a large bypass capacitor will, of course, result in a slower response. If additional high-frequency current components are present, the self-inductance of a large bypass capacitor will not provide a sufficiently low impedance. This situation may be dealt with by shunting the large capacitor with an additional low-capacitance, low-inductance capacitor. The engineer experienced in *rf* techniques will be familiar with these practices (see Figure 6.8).

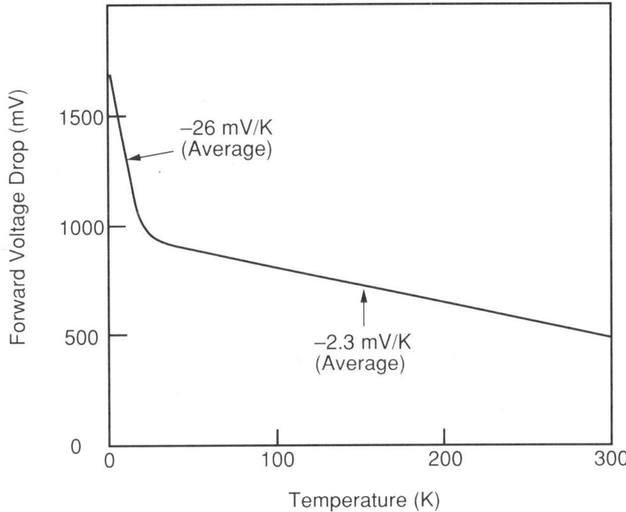

FIGURE 6.7 The voltage-temperature characteristics (with constant current flow) for silicon diodes is fairly linear down to about 20–30 K, where a sharp "knee" is seen prior to an increase in slope. Commercial versions are available, which are interchangeable with relatively small temperature errors. (Compiled from data provided by Lake Shore Cryotronics, Inc., Westerville, Ohio)

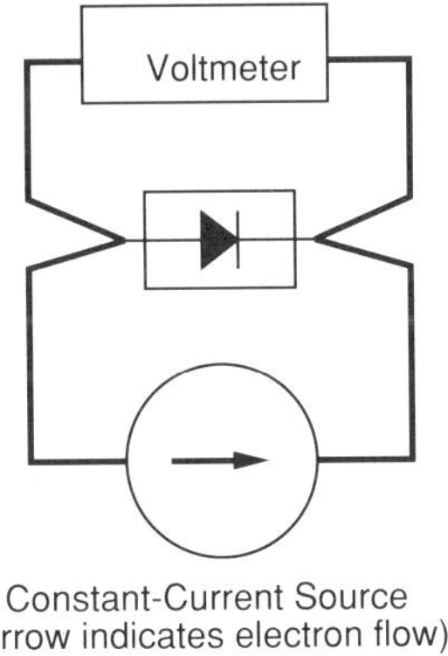

FIGURE 6.8 Silicon diode thermometer; standard four-terminal connection. Diodes designed especially for cryogenic temperature monitoring combine very low forward current (for low dissipation) with small variations between units to allow interchangability.

The Germanium Resistance Thermometer (GRT). The germanium resistance thermometer (GRT) is particularly useful at temperatures below 0.05 K and is commonly used up to 100 K (Krause, et al., 1988). The standard GRT is a germanium chip mounted in a cylindrical copper enclosure. It is important that the germanium chip be mounted in a strain-free manner; four wires are attached for conventional four-terminal measurement of resistance. While a specific device has about two orders of magnitude of useful range, devices can be constructed with useful characteristics below 0.05 K and up to about 100 K. Sensitivity increases at lower temperatures; resistance is typically 10–100 kΩ at low temperatures and as little as a few Ω at higher temperatures. Krause recommends the device for applications below 30 K, where high accuracy is required, since reproducibility of 0.001 K is possible at the lower temperatures (Krause, et al., 1988). On the other hand, this is not the best sensor for use in high magnetic fields.

Rhodium-Iron (Rh-Fe) Resistance Thermometers. The rhodium-iron resistance sensor has an anomalous temperature response near 40 K that leads to good sensitivity at low temperatures. Since the devices are useful from less than 1 K to levels of 800 K with ±0.01 K stability, it is clear that Rh-Fe resistance thermometers are contenders for cryogenic applications. The Rh-Fe sensor is, however, expensive to construct.

A Summary of Cryogenic Thermometers. A summary of the characteristics of several commonly used cryogenic temperature sensors appears in Table 6.1.

Mounting of Cryogenic Thermometers. Regardless of the thermometer type selected, care must be taken to get excellent contact between the temper-

TABLE 6.1 Characteristics of representative cryogenic thermometers.

Type	Temperature Range	Typical Reproducibility	Magnetic Field Limitations
Germanium Resistance Thermometer (GRT)	<0.05 to 100 K	< 0.001 K	B < 2.5 T
Carbon-Glass Resistor (CGR)	1.4 to 330 K	0.001 K	B up to 19 T (or more?)
Platinum Resistance Thermometer (PRT)	14 K to 850° C	0.01 K	Poor Reproducibility
Rhodium-Iron (Rh-Fe) Thermometer	1 to 800 K	0.01 K	Limited Use
Silicon Diode	1 to 473 K	0.03 K	B < 2.5 T
Capacitance Sensor	1 to 330 K	0.3 K	Superior at Highest Fields

(The data is this table is compiled partially from Krause, et al., 1988 and Sample, et al., 1982.)

ature sensor and the material to be measured. In some cases, a good thermal contact can be realized by ordinary mechanical mounting; a tight fit will be sufficient. In many instances, however, it will be necessary to use one of several conductive greases that are commonly used by cryogenics investigators. A minimum amount of grease should be used; just enough to fill the unavoidable gaps. There are also a variety of epoxies that can be used for anchoring, and one can use low-temperature solders such has Woods metal in some instances. Obviously, one should take care not to exceed the manufacturer's recommended maximum temperature for a specific sensor.

As mentioned earlier in the case of CGR mounting, it is often necessary to thermally anchor (i.e., "lag") electrical leads that exit the thermometer sensor. Some sensors, such as the germanium resistance thermometers, depend largely on the leads for thermal contact. Anchoring is generally required for all sensors, since the leads will reach room temperature on the end opposite the cryogenic sensor, that is, at the meter.

Typically, the sensor leads are mounted to a surface that will have a similar temperature to the measured surface, and this is done relatively near the sensor. A variety of means may be used for the anchoring, including GE-7031 varnish (Krause, et al., 1988). When it is possible to select the connecting leads

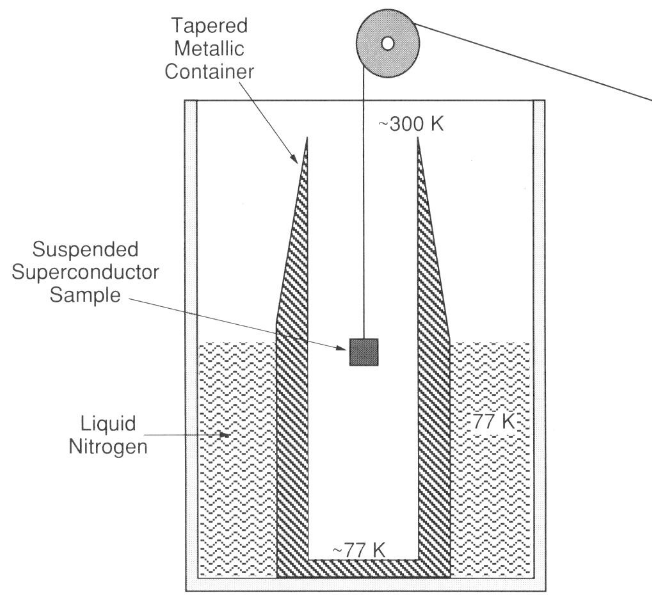

FIGURE 6.9 While a variety of more elaborate (and expensive) methods are available for cooling superconductor samples, this tapered-container cryostat with a pulley provides a smooth temperature gradient (~0.5 K/mm) as the sample is moved above the cryogen surface. (After A. C. Rose-Innes, "A simple system for obtaining temperatures between 77 K and room temperature," *J. Phys. E*, Vol. 21, No. 7, July 1988, pp. 729–730.)

between the cryogenic sensor and the room-temperature monitoring location, one should use a small-gauge wire with low thermal conductivity. Phosphor-bronze, at least as small as AWG 32, is a reasonable choice. If the wire is too small, one can expect problems with breakage as the sensor is moved.

It is useful to have a few criteria in mind when selecting an appropriate sensor for cryogenic measurements. Sample, Brandt, and Rubin propose the following guidelines (provided in abbreviated form here), which are given in greater detail in their paper (Sample, et al., 1982):

a. Use neither germanium resistance nor silicon diode thermometers in magnetic fields above approximately 2.5 T.
b. At temperatures below 20 K, carbon resistance thermometers may be used in fields up to at least 19 T.
c. While chromel P/constantan thermocouples may be used above 10 K in high magnetic fields, avoid the use of Au/(0.07%)Fe wire in fields above ~ 6 T.
d. Low-temperature thermistors may be used for narrow temperature ranges or as multiple sensors with overlapping ranges, even in high magnetic fields.
e. The $SrTiO_3$ capacitance thermometer, while limited by its time-dependent drift to use as a noncalibratible transfer standard, it is the only commercially available sensor that is virtually unaffected by magnetic fields.
f. Platinum resistance thermometers are useful in low-temperature, high-field applications, but suffer somewhat from a lack of information on the reproducibility of magnetoresistance effects.
g. Carbon-glass resistance thermometers, while results are not as reproducible (in zero field) as for germanium resistance thermometers, are available commercially with calibrations from approximately 1.5 to 300+ K and exhibit errors due to magnetoresistance that is correctable to a few tens of mK for fields up to 19 T.

USEFUL MATERIALS FOR CRYOGENIC EXPERIMENTS

While some of the following information is based on the author's experience, the balance was compiled from a very useful paper by A. C. Anderson, who also cites a number of useful references regarding thermal properties of these materials (Anderson, 1980):

Apiezon-N Grease. This material, available in small tubes from several suppliers, is useful for the temporary attachment of superconductor specimens to substrates. At room temperature (~ 300 K), the grease is soft but may be used to mount small specimens onto substrates. Above about 305 K, the grease becomes too soft for use as a paste; below about 270 K, the grease hardens to provide an excellent mechanical bond. At low temperatures, the specific heat varies as $1/T$.

GE 7031 Varnish. This varnish, mentioned previously, is commonly used for thermal contact and mechanical bonding.

Scotchcast #5, #8, and #10 epoxies. These epoxies are used for good thermal contact in bonding applications, as well as for vacuum seals. Resin #8 is reported to exhibit flexibility at room temperature.

Stycast 2850 GT and 2850 FT epoxies. These epoxies are used for vacuum sealing, particularly around electrical feedthroughs. Thermal conductivity at cryogenic temperatures varies approximately as T^2. Containers should be carefully sealed; shelf life is limited.

Stycast 1266 epoxy. This epoxy is used for seals and for bonding. Since magnetic susceptibility is evidently very small, this Stycast 1266 is recommended for use in high magnetic field environments.

G10 and G11 (NEMA). These rigid construction materials are glass-reinforced epoxy-laminates.

Macor ceramic. This ceramic is machinable, has good electrical insulating characteristics, and is used for construction of rigid shapes for cryogenic applications.

Reticulated Glassy Carbon. This rigid material is used for thermal isolation.

Pt-W wire. A resistance wire with a relatively low specific heat.

MEASUREMENT OF RESISTIVITY IN SUPERCONDUCTORS

The measurement of resistivity on superconductor samples is, at least in principle, not different from measurement of the same parameter on metals. The most important difference is that, as temperature falls below T_c, the resistance to be measured is *very* low. Very small resistances are difficult to measure. Zero resistances are, in principal, impossible to measure. At best, one can determine that the resistance is below a certain upper limit. Nevertheless, four-terminal measurements of resistivity are quite useful for the determination of T_c in superconducting samples. Such measurements, like those of susceptibility, may be generally considered to be in two categories, that of dc and ac measurement. In either instance, one is likely (at $T < T_c$) to be measuring nV levels across the sample, with currents at the mA level. Either approach involves a fairly simple four-terminal circuit arrangement, except that a lock-in amplifier is typically used to sense the very small ac signal for alternating current measurements. (See Figure 6.10 for a dc resistance measurement of a Y-Ba-Cu-O film.)

ELECTRICAL CONTACTS

Measurement of such parameters as dc resistance and Hall effect require attachment of electrical leads to the superconductor. A normal-to-superconductor contact involves a proximity effect at the interface; if a thin insulating junction is present at the superconductor surface, tunneling may occur. In short, it is possible to have a measurement dominated by the contact interface.

The superconductor's characteristic lengths, that is, magnetic penetration depth (λ) and coherence length (ξ) are of particular interest in making contacts (Talvacchio, 1989). The coherence distance is the decay length for the quantum-mechanical wave function associated with the condensation of two electrons into a Cooper pair. For normal-superconductor contacts, ξ is effectively the minimum distance between the normal and superconducting regions. For operation at high frequencies, λ essentially replaces the normal skin depth

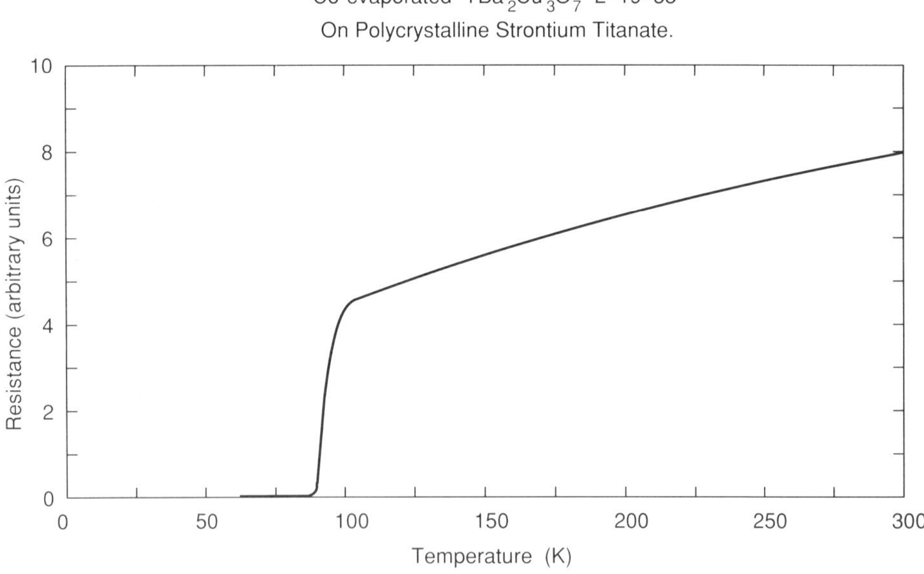

FIGURE 6.10 Example of a typical four-terminal dc resistance measurement of a Y-Ba-Cu-O thin film, this one coevaporated on polycrystalline $SrTiO_3$. A single percolating path that develops near T_c will shunt the electrodes, masking information about the specimen at lower temperatures. (Provided by Group MEE-11, Electronics Research and Exploratory Development, Los Alamos National Laboratory)

δ as the characteristic depth of current flow and magnetic field penetration at the superconductor surface.

In the high-temperature copper-oxides, typified by Y-Ba-Cu-O, the very short coherence length may be exceeded by the dimensions of degraded regions resulting from dislocation strain fields or disorder in a few unit-cells.

Attachment of Contacts

There are a number of techniques for attaching leads to superconductors, and the most popular fall into the categories of soldered or painted-on connections (usually silver) or press-on contacts that may be spring-loaded. The latter are preferable for the measurement of thin films, since paint-on or solder-on contacts tend to alter the surface or even penetrate the film entirely.

Some investigators have reported success using Dupont© no. 4929 silver paste (Munakata, et al., 1987), while others had good results using a technique where platinum was pressed against a $YBa_2Cu_3O_x$ sample (Grader, et al., 1987). Other investigators use a silver epoxy prior to annealing; the primary epoxy constituents disappear during high-temperature processing and only a pure silver contact is left. A contact resistance less than 10 $\mu\Omega/mm^2$ at 77 K is reported for this technique, compared to about 1 Ω/mm^2 for more conventional methods (van der Maas, et al., 1987).

The lowest impedance connections have been achieved by sputter-deposition (in argon) of either gold or silver onto a clean Y-Ba-Cu-O surface at temperatures below 100°C. Typical sputter rate was approximately 1 nm/s. The initial deposition of two to six μm of Au or Ag was then annealed in oxygen for at least one hour at 600°C. Contact surface resistivity (the product of contact resistance and contact area) was on the order of 10^{-10} Ω-cm^2, with indications that 10^{-12} Ω-cm^2 could be achieved (Ekin, et al., 1988).

In contrast to indium contacts on Y-Ba-Cu-O, which exhibit a high concentration of oxygen at the interface which diffuses into the 1-2-3 ceramic, Au or Ag show virtually no oxygen and make an excellent contact to the ceramic surface. The authors of this study point out that silver also behaves as a temperature-mediated passivation buffer, which protects the Y-Ba-Cu-O surface at room temperature but allows oxygen penetration into the 1-2-3 surface at elevated temperatures.

Another interesting approach that essentially welds ("spark-bonds") gold wires to Y-Ba-Cu-O single-crystal samples is reported to have been fairly successful (Iye, et al., 1988). The method (see Figure 6.11) involves the discharge of a capacitor through the wire to the proposed contact location. While the connection exhibits high mechanical strength and a well-defined contact location, the initial contact has rather high resistance, ranging upward from about 10 kΩ, and shows a semiconductive-type behavior with temperature. The resistance can be reduced by at least two approaches. The most direct solution is to place a conductive paste around the junction, but this diminishes some of the advantages of the spark-bonding technique. However, since the high-resistance

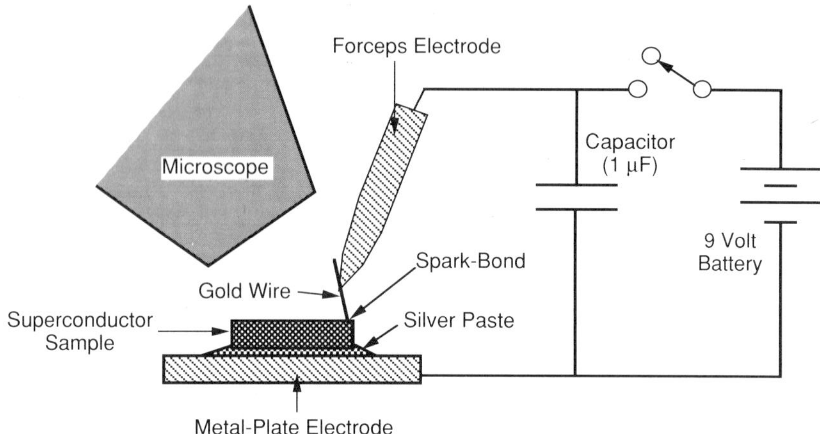

FIGURE 6.11 Spark-bonding (a time-honored technique for applying contacts to semiconductors) is accomplished by application of modest energy to the very small contact area between the 50 μm diameter gold wire and the superconductor surface. Oxygen depletion occurs at the contact site, resulting in high contact resistance that must be lowered by post-annealing in oxygen. (After Y. Iye, T. Tamegai, H. Takeya, and H. Takei, "A simple method for attaching electrical leads to small samples of high-T_c oxides," *Jap. J. Appl. Phys.*, Vol. 27, No. 4, April 1988, pp. L658–660.)

problem is largely a result of oxygen depletion at the junction, reannealing the sample in oxygen can be used to reduce the contact resistance to ~ 1 Ω.

A practical method for forming a variety of contact shapes is the plasma-arc spray of silver onto Y-Ba-Cu-O. This approach has yielded excellent contact resistance, on the order of 10^{-8} Ω-cm at 77 K. The plasma-arc spray method is easy, does not require any postprocessing and, perhaps best of all, is applicable to arbitrary shapes and large areas (Katz, et al., 1988; Maley, et al., 1988).

Other methods for attaching leads to high-T_c superconductor samples include pressed indium contacts and ultrasonic bonding (Iye, et al., 1988). Investigators at AT&T have constructed an ultrasonic probe for preparation of indium-solder contacts onto 1-2-3 ceramics (Peterson, et al., 1988). The device was based on a model BP-4 ultrasonic probe and SG-250 generator purchased from the Blackstone Corporation. The commercial generator produced up to 200 W at 20 kHz. The final probe consisted of an ultrasonic piezoelectric crystal and coupling body; this was attached to a titanium horn that was used to couple energy to the indium solder load. The AT&T investigators point out that indium-solder electrical contacts are rugged, have a low melting point (155°C; as low as 58°C for some alloys, i.e., 49% Bi, 21% In, 18% Pb, and 12% Sn), good oxidation resistance, bond to many materials, and are inexpensive and readily available. (See Figure 6.12 for an ultrasound soldering iron for indium-solder contacts.)

One of the chief difficulties in reproducing surface measurements (even on the same sample) is due to a variation in the surface characteristics; contact

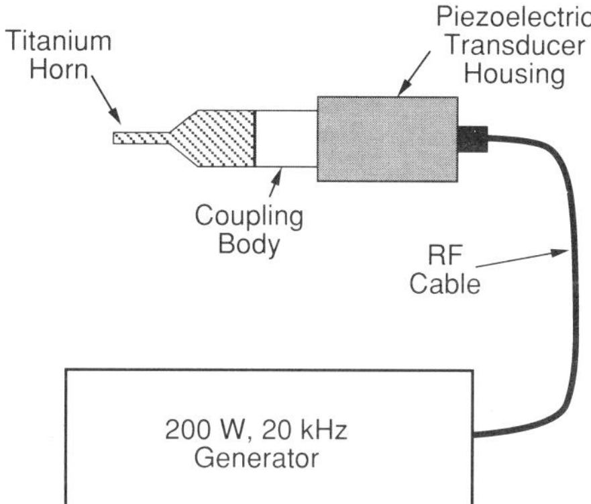

FIGURE 6.12 An ultrasound "soldering-iron" may be used to make indium-solder contacts on ceramic superconductors. A unit similar to the one illustrated here was assembled using a Model BP-4 ultrasonic probe and a Model SG-250 generator, both manufactured by the Blackstone Corporation. See the text for details. (After G. E. Peterson, R. P. Stawicki, and U. C. Paek, "Radio frequency properties of high-T_c superconductors," Proc. 38th Electronic Components Conference, Los Angeles, May 9–11, 1988, pp. 159–167.)

resistance tends to vary with contact pressure and even with the portion of the sample selected for attachment of contacts. There is some evidence that this may be a result of a very thin layer of lossy or even insulating material that forms on the surface of typical ceramic superconductor samples. In an investigation of mechanical polishing it was determined that four-terminal resistance measurements, using painted-on silver contacts, were more reproducible if approximately 10 μm of the superconductor surface was removed prior to application of the electrodes (Kishida, et al., 1988). This was done with emery paper that had a roughness of 10 μm, and followed by cleaning in acetone. A dispersion in the measurement of the zero dc resistance temperature (T_z) was reduced from about 7 K, prior to surface removal, to about 2 K afterwards.

NORMAL CONDUCTORS

While a variety of normal conductors (copper, silver, gold, etc.) may be used at cryogenic temperatures, it is clearly useful to understand how the resistivity of a particular material will change as a function of temperature. In some instances, alloys with a low temperature coefficient of resistivity may be desirable. Recall the simple relationship given earlier for resistivity ρ:

$$\rho = \rho_o\{1 + \alpha(T - T_o)\}$$

where ρ_o is resistivity at temperature T_o, T is the temperature of the material and α is the temperature coefficient of resistivity for the particular material, having units of 1/K.

A particularly useful alloy for use as conductor wire in cryogenic applications is manganin, an alloy of copper, manganese, and nickel. Typical proportions are 84% copper, 12% manganese, and 4% nickel. Manganin has a resistivity approximately 25 times that of copper (Harris, 1952) and a relatively low thermoelectric power against copper (2–3 μV/°C) (Kishida, et al., 1988). While the material is relatively stable when properly annealed, manganin is usually stored in a sealed container; small wire sizes may require a protective coating.

Other useful alloys include copper-nickel (constantan), nickel-chrome (Nichrome), copper-manganese-aluminum (Therlo), and gold-chromium (Harris, 1952).

SINGLE CRYSTALS; A SPECIAL MEASUREMENT PROBLEM

Single crystals of the 1-2-3 copper oxide and other high-T_c superconductors are of enormous interest, since they are expected to represent the pure form of the material. In contrast to other forms of the ceramics, the single, pure crystal of the material does not present the problems associated with grain boundaries, impurities, and multiple crystal-axis orientations. Therefore, the single crystal is desirable for accurate measurements of anisotropy of such parameters as ac and dc resistivity, critical field (H_c) levels, and critical current (J_c).

In addition to the scientific interest in such determinations, engineers involved in applications are interested in the single crystal since it represents the "ultimate" form of the superconductor. The basic point is this: if the single-crystal does not exhibit a sufficiently high critical current for a specific application, one would certainly not expect to achieve the required J_c in a polycrystalline sample of the same material. Thus, with information on the behavior of (pure) single crystals, one can determine the inherent value for a parameter as opposed to a value that may be improved by clever processing techniques yet to be developed. It is worth noting that many of the "single-crystals" of Y-Ba-Cu-O produced to date appear to be contaminated by a variety of spurious materials. Furthermore, virtually all such samples exhibit twinning boundaries, which are believed to be implicated in limiting J_c.

Unfortunately, single crystals to date are difficult to measure, since the samples produced are typically not much larger than about 1 by 1 by 0.1 mm. Clearly, placing contacts on a sample this small, while possible, requires some delicacy. Even so, development of four-probe resistivity measurement holders for very small samples have been reported (Papaioannou, et al., 1988) and spark-bonding of leads to single crystals of Y-Ba-Cu-O has also been successful (Iye, et al., 1988). An additional problem is associated with the fact that the single crystals are highly anisotropic in conductivity, critical current, and critical field. Recall that electrical conductivity is high along the *a-b* planes of the Y-Ba-Cu-O crystal, low along the longer *c*-axis. Fortunately, measurement techniques that are useful for highly anisotropic samples have been developed (Montgomery, 1971) (see Figure 6.13).

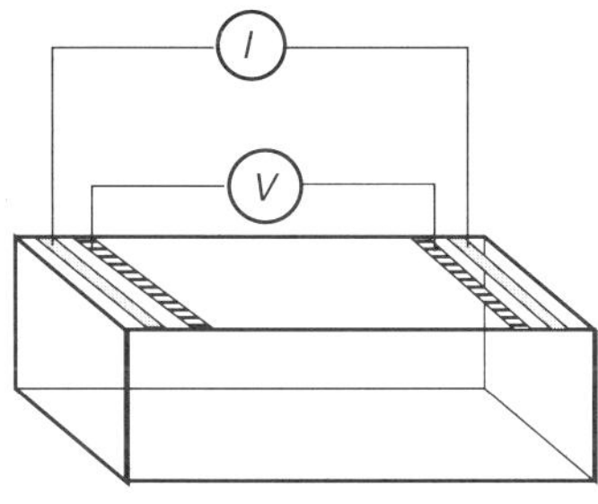

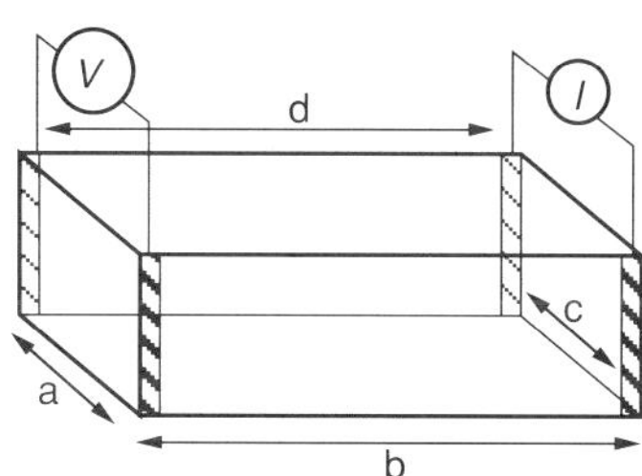

FIGURE 6.13 The standard four-probe resistance measurement approach (left) is contrasted with the Montgomery technique where contacts are placed on the corners of the specimen. The Montgomery method is used to determine the resistivity of anisotropic specimens, and is therefore of interest in characterizing the high-T_c ceramic superconductor single crystals. For details, see the following reference. (After H. C. Montgomery, "Method for Measuring Electrical Resistivity of Anisotropic Materials", *J. Appl. Phys.*, Vol. 42, No. 7, June 1971, pp. 2971–2975.)

THERMAL ANALYSIS

While thermal analysis may not be commonly encountered by the engineer, it is of sufficient importance in the determination of processing parameters for ceramic superconductors that the subject merits some mention here.

Thermogravimetric Analysis

Perhaps the most important of this category of analysis techniques is thermogravimetric analysis (TGA). In TGA, the weight of a ceramic superconductor sample is measured using a sensitive microbalance *during* a heating process. TGA is typically performed with the ceramic sample in a vacuum, but either inert, reducing, or oxidizing atmospheres may be used as the experiment requires. For example, when Y-Ba-Cu-O ceramics are heated in a vacuum, TGA may be used to determine the proportion of oxygen that is lost from the sample. A chief difficulty in performing oxygen-loss experiments on, for example, a 100 mg sample, is that sensitivity in the μg range must be maintained over a wide temperature range. The microbalance must be located in a chamber that has excellent isolation from the variety of thermal effects associated with the heating procedure (Cassel, 1988).

Thermomechanical Analysis

In thermomechanical analysis (TMA), the dimensional changes of the specimen are measured under either zero applied force or a fixed force level. Using TMA, one may obtain data on thermal shrinkage, coefficients of thermal expansion, volume changes, or softening points.

Differential Scanning Calorimetry

Differential scanning calorimetry (DSC) is probably the most commonly used of the thermal analysis techniques in general use. DSC is used to measure heat flow (into or out of) the specimen during specified changes in temperature. Heat flow is, of course, affected by thermal conductivity, phase changes, and specific heat capacity of the specimen. In ceramic superconductors, the measurement of phase changes is of particular importance as a means of detection and identification of impurities.

Differential Thermal Analysis

Differential thermal analysis (DTA) is a procedure where one measures the difference in temperature between the superconducting sample and a specified reference material, as each are heated in the same atmosphere. The information provided is similar to DSC, but DTA is useful over a wider temperature range, albeit at somewhat reduced accuracy over DSC. DTA is particularly useful in the detection of high-temperature phase changes in Y-Ba-Cu-O; data may be used to determine whether the Y-Ba-Cu-O has only one (desired) phase or multiple phases.

NONCONTACT CHARACTERIZATION OF SUPERCONDUCTORS

It is preferable, and in some instances virtually required, to avoid placing contacts on superconductors that must be characterized. As indicated before, thin films are particularly sensitive to the application of contacts. Two basic electromagnetic techniques that perform electrical characterization of superconductors without making direct electrical contact will be discussed, as will two optical techniques.

Eddy-Current Characterization

Sophisticated eddy current techniques have been used to determine the resistivity, T_c, and penetration depth (λ) of metallic and superconducting samples (Schawlow & Devlin, 1959; Dalrymple & Prober, 1984). In addition, very straightforward dual-coil, mutual inductance eddy-current techniques may be used to characterize samples in a qualitative manner, that is, to determine T_c and the "broadness" of the phase transition (Doss, et al., 1988).

Referring to Figure 6.14, the resonant frequency (f_r) of the oscillator is a function of the series capacitance (C_S) and the total inductance, that is,

$$f_r = \frac{1}{2\pi\sqrt{C_S(L_A + L_B \pm 2M)}} \tag{6.1}$$

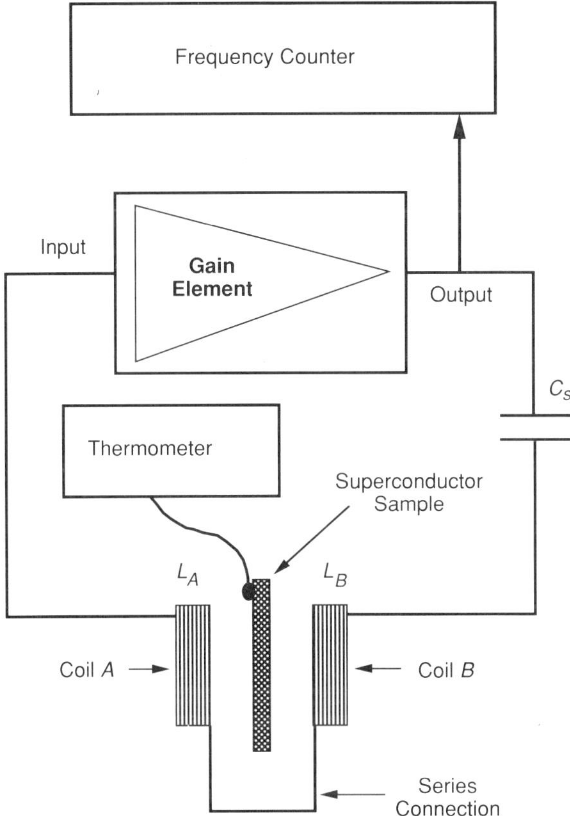

FIGURE 6.14 In this two-coil version of a noncontact characterization technique, eddy currents induced in the superconductor sample modify mutual and self inductance associated with the sensor coils A and B. The variation of net inductance changes with sample conductance, shifting oscillator frequency. (After J. D. Doss, D. W. Cooke, C. W. McCabe, and M. A. Maez, "Noncontact methods used for characterization of high-T_c superconductors," *Rev. Sci. Instrum.*, Vol. 59, No. 4, April 1988, pp. 659–661.)

where L_A and L_B are self-inductances of coils A and B respectively, and M is the mutual inductance. If the sample is placed precisely between the coils but not close enough to either coil to significantly affect effective self-inductance, changes in sample conductivity will, for practical purposes, affect only the mutual inductance. In practice (especially with thin-film samples) it is common to have a configuration where the sample is quite close to one of the sensing coils. In this instance, both mutual and self-inductance (of one coil) are modified by currents induced in the sample. It is, of course, possible to use a single coil for characterization of a sample. The primary advantage of using dual coils is that sample mounting at the center tends to reduce noise induced by microphonics.

The ± sign on $2M$ may be selected by winding the coils either as aiding or opposing, but the $+2M$ "aiding" configuration is preferred. If the $-2M$ "opposing" configuration is used, an increase in sample conductivity will tend to *increase* total inductance by modification of the $2M$ quantity, but self-inductance of the coil nearest to the sample will be *decreased*. The resultant data may be difficult to interpret.

In the case of a series-aiding $(+2M)$ configuration, as the sample conductance increases, the level of induced eddy currents also increases, and this tends to decrease the effective value of M. While the effect on M cannot be directly calculated in closed form for irregularly shaped samples, the change in M for a small spherical sample centered between coils is (Abel, et al., 1964)

$$\Delta M = 16\pi^2 n_s n_p V \chi f_s f_p \tag{6.2}$$

where n_p and n_s are the number of turns per unit length on the primary and secondary sensors respectively, V is the volume of the spherical sample, and χ is the sample susceptance. The parameters f_s and f_p are rather complex geometric factors that reduce to unity when coil diameter is much greater than distance between coils along the center axis.

In practice, this type of system has been used quite effectively between 1 and 30 MHz, and appears to be useful in excess of 1 GHz for single-turn sensing coils. Samples may be characterized ranging from bulk volumes of several cm^3 to films with thickness on the order of 0.5 μm and areas of 10 mm^2 (Doss, 1989). At operating frequencies near 20 MHz, measured frequency shifts of several hundred kHz at T_c are common. Since even ordinary (non-temperature compensated) oscillator circuits at this frequency have noise and drift on the order of a few Hz, resolution is excellent. Over the range of resistivity near and above T_c, the shift in frequency is approximately proportional to sample resistivity unless the system is saturated. For very high or very low values of sample resistivity, a change in resistivity will not be sensed by this eddy-current method. It is necessary to be familiar with the dynamic range for a particular instrument; a very flat response (in frequency shift) to a large change in temperature is often an indication of saturation.

This eddy-current technique, which is affected by changes in sample susceptance, is not particularly applicable for absolute measurements of resistivity, but is quite useful for rapid, non-contact measurement of T_c, the width of the transition region, and for characterization of losses for $T < T_c$.

Similar approaches have, however, been used to determine the London penetration depth λ in thin films (Fiory, et al., 1988) and bulk samples (Schawlow & Devlin, 1959). (See Figure 6.15 for eddy-current data on a pair of $Tl_2Ca_2Ba_2Cu_3O_x$ bulk samples.)

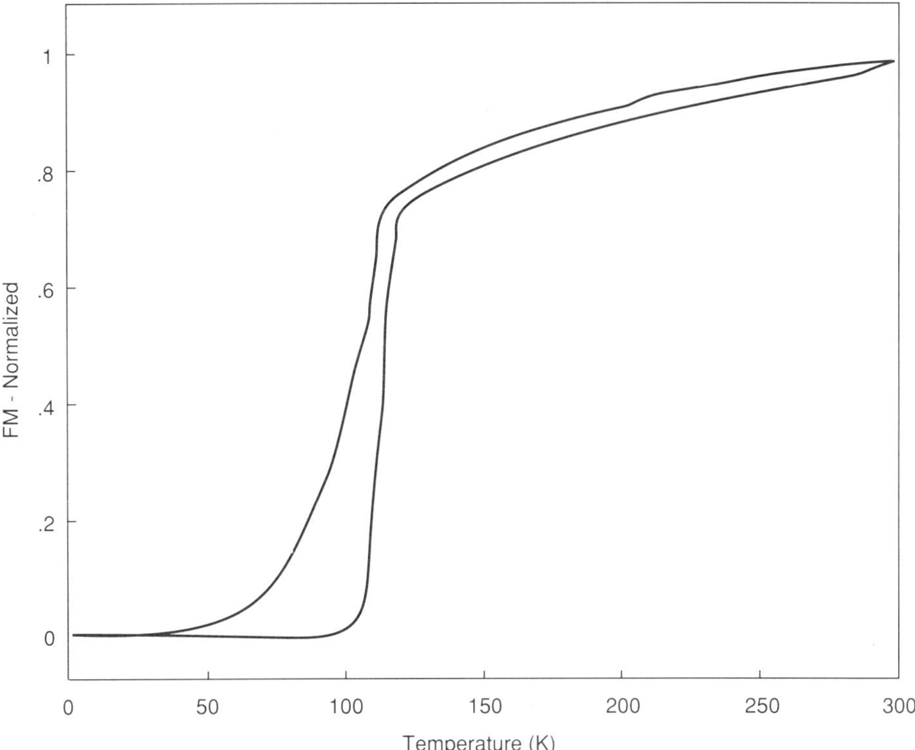

FIGURE 6.15 Eddy-current data on a pair of $Tl_2Ca_2Ba_2Cu_3O_x$ bulk samples. The difference in loss below T_c (due to different processing schedules) is obvious with this 20 MHz *rf* technique, but wouid not be revealed in a dc resistance measurement since a single percolative path would produce zero resistance at ~ 100 K. (Data provided courtesy of Los Alamos National Laboratory)

Resonant-Cavity Characterization of Surface Resistance

There is, of course, a need to determine the surface resistance of superconductors at high frequency, both for theoretical and for practical reasons. This may be accomplished by a variety of microwave techniques, but one of the most popular involves installation of the superconductor sample in some type of resonant structure (see Figure 6.16).

A resonant superconducting 3 GHz niobium cavity (originally designed for particle accelerator applications) has been used by several investigators to determine the characteristics of $YBa_2Cu_3O_x$ (Hein, et al., 1987; Hagen, et al., 1987). Investigators at the David Sarnoff Research Center have introduced a "disk resonator method" for measurement of R_s that consists of a pair of disks separated by a dielectric spacer. One of the disks is the high-T_c superconductor sample, which dominates the loss of the system (Fathy, et al., 1988).

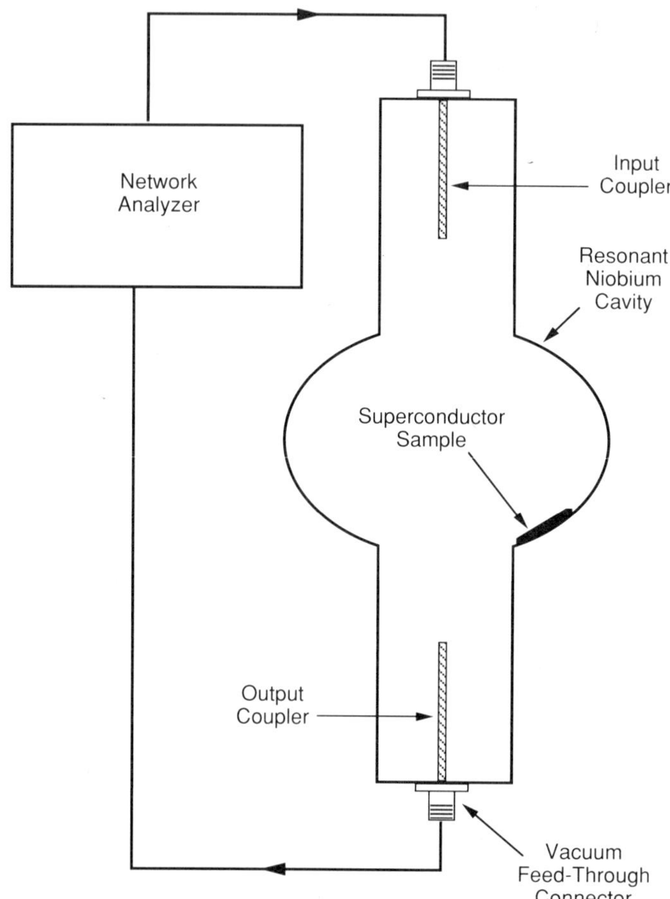

FIGURE 6.16 The surface resistance (R_s) of a superconducting sample can be determined by analysis of its effects on the Q of a superconducting resonant cavity. In this illustration, Q is determined by use of a network analyzer (a sweep generator would serve as well). For determination of the very high Q's seen in superconducting cavities ($> 10^9$!), it is usually necessary to use the "decrement" method illustrated in Figure 6.19.

While the precise configuration of the resonator will determine how cavity losses are related to superconductor-sample R_s, the following discussion will serve to outline the basic approach.

The fundamental approach involves measurement of cavity Q over a wide range of temperatures, usually 4–300 K, with and without the sample in place. The level of perturbation caused by the sample can be used to determine sample R_s.

The general relationship between the cavity (or sample) surface resistance is simply

$$R_s = G\left(\frac{Z_o}{Q_o}\right) \quad 6.3$$

where Z_o is the characteristic impedance of free space (377 Ω), Q_o is the unloaded cavity Q, and G is a cavity geometric factor (Hahn, et al., 1968). When a sample is inserted, G must be modified to take the field perturbation caused by the sample into account. Superconductor samples are generally compared to samples of normal conductor that have the same size and shape. (See Figure 6.17 for a comparison of cavity $1/Q$ versus temperature and Figure 6.18 for resonant cavity characterization of thin films.)

Coupling to the cavity may vary from "critical" to very "weak" and is typically determined by voltage standing-wave-ratio (VSWR) measurements (Ginzton, 1967). The basic relationship for determining unloaded cavity quality factor (Q_o) from bandwidth measurements is familiar to rf engineers:

$$Q_o = \frac{f_r}{f_H - f_L} = \frac{f_r}{\text{bandwidth}} \quad 6.4$$

The bandwidth is measured in the usual manner by determining the upper and lower frequencies (f_H and f_L) at the half-power locations around resonance (f_r), where field is peak field/$(2)^{1/2}$. Caution: when diode detectors are used, nonlinearities may make it difficult to determine the half-power locations.

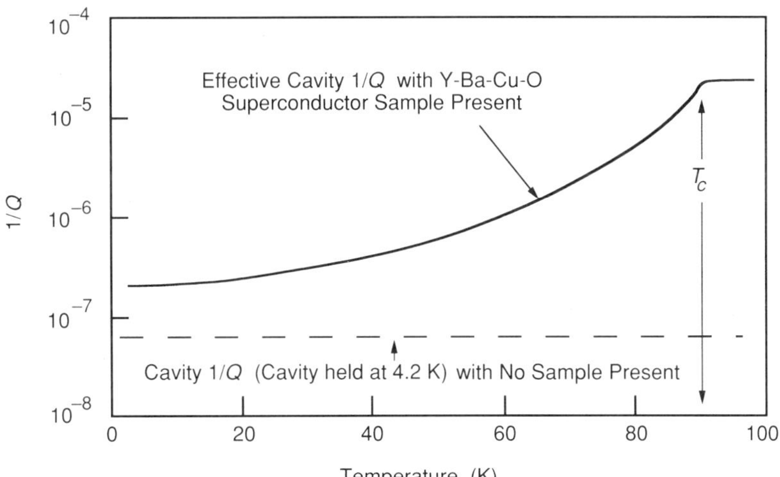

FIGURE 6.17 The presence of a superconducting sample in a superconducting resonant cavity results in losses that are evaluated by comparison to the response for a sample of similar size and geometry whose surface-resistance characteristics are known. In this type of experiment, only the sample temperature is altered; cavity temperature remains well below T_c. Nonsuperconducting cavities may also be used in similar measurements, but the lower values of cavity Q result in less sensitivity to the presence of the relatively small superconductor sample losses.

In instances where the loaded cavity Q (Q_L) is smaller than the unloaded value (Q_o) due to coupling effects from an external circuit that has effective Q_{ext}, the relationship between the Q's is

$$\frac{1}{Q_L} = \frac{1}{Q_0} + \frac{1}{Q_{ext}} \qquad 6.5$$

System Q may be measured most conveniently by use of a network analyzer to determine bandwidth of resonance until Q's exceed about 10^7, where bandwidth measurements may not provide sufficiently high resolution to determine Q_L with the accuracy required. (The unloaded Q (i.e., Q_o) of very clean niobium cavities, cooled in low magnetic fields, commonly exceeds 10^9 and can be as high as 10^{11}!)

At such extremely high Q values, it is advisable to consider techniques where the characteristic transient "filling" or "decay" time of the cavity is determined by pulsed-*rf* methods. This approach is also referred to as the "decrement" method. The "decay time" (Δt) is the period required for the cavity

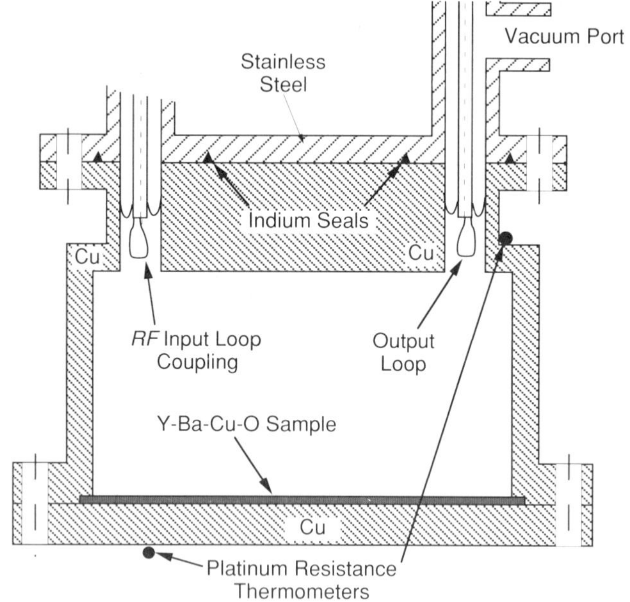

FIGURE 6.18 This sketch illustrates a cylindrical TE_{011} mode copper cavity used to evaluate R_s on a superconducting sample that forms the bottom plate. Investigators at the University of Wuppertal used this design to characterize samples of Y-Ba-Cu-O 50 μm thick films that had been electrophoretically plated on a 12.5 cm diameter silver disc. (After H. Piel, M. Hein, N. Klein, U. Klein, A. Michalke, G. Müeller, and L. Ponto, "Superconducting perovskites in microwave fields," Proc. of Int. Conf. on High Temperature Superconductors, Interlaken, Switzerland, *Physica C*, Vol. 153–155, Superconductivity, June 1988, pp. 1604–1609.)

field to decrease to $1/\epsilon$ of its original value. The loaded cavity quality factor (Q_L) is related to the frequency and Δt by

$$Q_L = \Delta t(\pi f) \qquad 6.6$$

The E and H fields change by a factor of $1/\epsilon$ in (Q_L/π) rf cycles, where ϵ is the base of natural logarithms, that is, 2.71828. Another way of viewing this expression is that $(\Delta t)f$ is the number of rf cycles during the period Δt.

As in the previous instance, it is always necessary (particularly for the high Q's measured in superconducting cavities) to unfold the value of Q_o from the measured value of Q_L and the value for Q_{ext}, with the latter usually determined by VSWR measurements. Q_{ext} in this instance is a single term that includes the losses associated with couplers, external transmission lines, and connectors. The decrement method (see Figure 6.19) is a useful approach at high Q levels, since the characteristic fill and decay times are proportional to Q_L and are generally on the order of ms if the cavity is very weakly coupled to the measurement system. In addition, the decrement method is much less expensive to perform than techniques that require a sophisticated network analyzer or similar instrumentation.

Measurement of Magnetic Susceptibility (χ)

Measurements of magnetic susceptibility are crucial for determining whether a material is actually superconducting, as opposed to merely having a very low

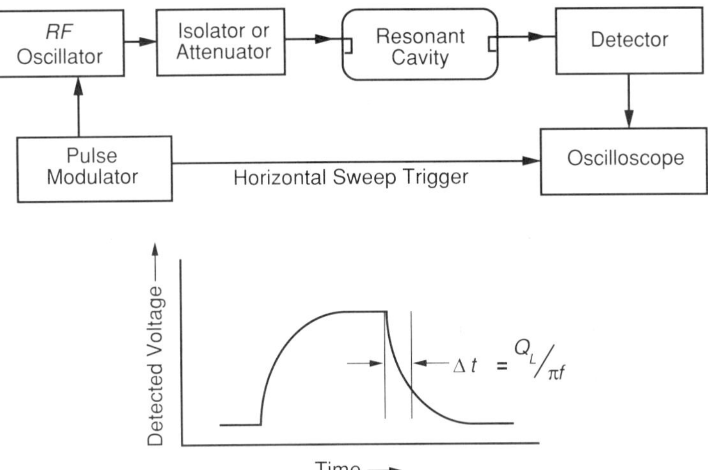

FIGURE 6.19 When cavity Q is too high for reliable measurement of bandwidth by swept-frequency techniques, measurement of the characteristic "fill" or "decay" times will yield Q values. Higher Q yields longer fill times, which makes the measurement more accurate as Q increases. This technique is commonly known as the "decrement method."

resistivity or a sudden drop in resistivity. In addition, a determination of χ is often indicative of the percent volume of a sample that is superconducting. Measurement of dc resistivity, for example, does not provide such an indication, since a single percolative path will indicate zero resistance in a sample that is primarily composed of a nonsuperconducting phase.

There are a variety of methods for determination of susceptibility, which may be broken into categories of dc and ac measurements. Figure 6.20 illustrates an ac susceptibility measurement, where an oscillator drives a pair of field coils (one of which is coupled to the superconducting sample) as well as the reference port on a lock-in amplifier. (Lock-in amplifiers are commonly used in superconductor characterization wherever very low level ac signals must be detected.) The coils are wound so that the induced fields are opposing, resulting in a net voltage of zero across the coils. Also see Figure 6.22.

When the sample susceptibility is close to zero, that is, magnetization M is approximately zero, the sample will have little effect on the signal sensed at the lock-in amplifier. As temperature is decreased to the critical level, a superconducting sample will begin to expel flux, that is, exhibit the Meissner effect. This expulsion of flux will unbalance the coils so that a voltage appears at the lock-in amplifier. The phase will be 90° shifted from the drive voltage, with the direction being dependent on whether the sample is paramagnetic ($\chi \rightarrow$ positive) or, in the case of a superconductor, diamagnetic ($\chi \rightarrow$ negative). In addition, the magnitude of the voltage will be directly proportional to the sample magnetization. The signal is also proportional to the volume of the sample. In practice, susceptometers are calibrated with a specific volume of a material with known susceptibility, such as (paramagnetic) copper sulphate.

A conductive sample will also produce eddy currents in the presence of the ac fields. This will tend to produce misleading results when attempting to measure ac susceptibility. These currents are out of phase with the currents

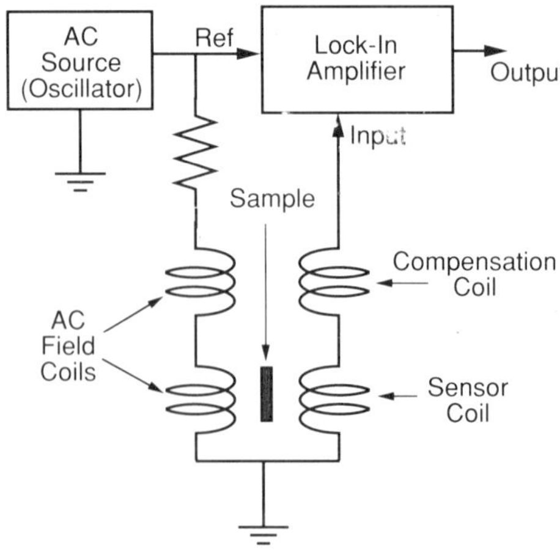

FIGURE 6.20 A basic ac susceptometer technique is illustrated above.

induced in the sensing coil by susceptance effects and can, in principle, be rejected by the lock-in amplifier on that basis. Even so, the requirements for phase locking on the lock-in amplifier may be severe.

A dc susceptibility technique is illustrated in Figure 6.21 (Abrahams & Keffer, 1971). This is the "Gouy balance method." In this approach, the superconductor specimen is suspended in a nonhomogeneous magnetic field. The general relationship between force (F), field (H), and magnetization (M) is

$$F = \frac{1}{2}\nabla \int (M \cdot H) dV = \frac{1}{2}\chi \nabla \int H^2 dV \qquad 6.7$$

over volume V where $\chi = M/H$. If the fields at each end of the sample are H_a and H_b respectively, varying only along the y-axis of the sample,

$$F = \frac{1}{2}\chi A \int \frac{d}{dy}\left(H_y^2\right) dy = \frac{1}{2}\chi A\left(H_a^2 - H_b^2\right) \qquad 6.8$$

so that susceptibility may be determined from the measurement of force produced in a field of known gradient:

$$\chi = \frac{2F}{A\left(H_a^2 - H_b^2\right)} \qquad 6.9$$

where A is the (constant) cross-section of the sample.

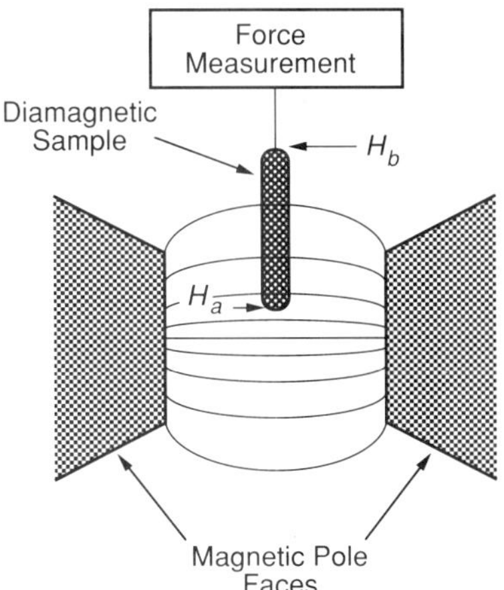

FIGURE 6.21 The susceptibility of diamagnetic samples may be measured by determination of the force developed when the sample is suspended in a nonuniform magnetic field. This approach is also known as the "Gouy balance method." (After E. Abrahams and F. Keffer, "Magnetic Susceptibility," *McGraw-Hill Encyclopedia of Science and Technology*, Vol. 13, McGraw-Hill, New York, 1971, pp. 348–349.) (Also see: C. Kittel, *Introduction to Solid State Physics*, 3rd Ed., John Wiley & Sons, New York, 1966.)

An improved vibrating-sample magnetometer (VSM) has recently been described that is three orders of magnitude more sensitive ($\sim 10^{-8}$ emu) than conventional VSMs with "comparable working space" (Flanders, 1988). This alternating-gradient magnetometer is also reported to be capable of much more rapid measurements ($\sim$ 100 s for a complete hysteresis loop over ± 10 kOe) than SQUID magnetometers, since the latter generate point-by-point data. In the improved VSM, the superconducting specimen is mounted on the tip of a cantilevered rod; the rod includes a piezoelectric element to monitor sample deflection. More conventional VSMs respond to a small voltage that is induced into a sensor coil by an alternating flux from the sample under measurement. A piezoelectric sensor has the advantage (over the inductive sensor) that it is not affected by magnetic fields.

The sample is magnetized by a variable dc H field, and is also exposed to a smaller alternating H field gradient. The alternating force on the sample (due to the alternating gradient) is proportional both to the magnetic moment of the superconducting specimen and to the magnitude of the alternating H field. The alternating H field is operated very near the resonant frequency of the cantilever rod (100–1000 Hz); this increases the sensitivity considerably. Mechanical Q of the system is on the order of 25–250. Investigators report that the signal to noise ratio is on the order of 500 for measurements made on a 25 μm diameter sphere that has a moment of 3.7×10^{-6} emu.

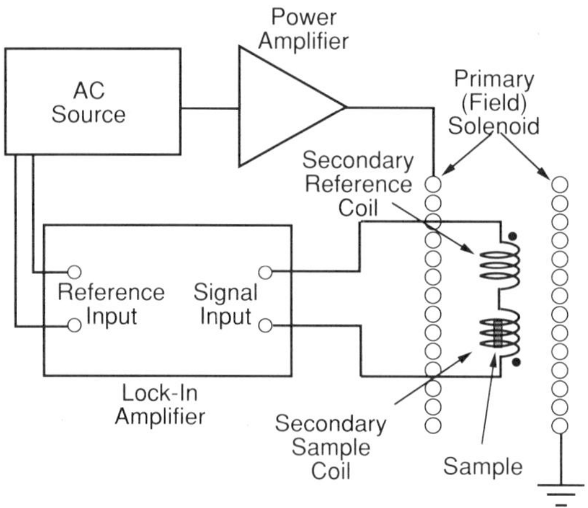

FIGURE 6.22 In this ac susceptometer (cooling system not shown) the secondary coils are wound in opposition so that the output voltage is zero in the absence of a sample. The presence of a sample with a finite susceptibility creates a voltage imbalance drive to the lock-in amplifier which is 90° out of phase with the primary drive signal. The sign of this 90° shift is dependent on whether the sample is paramagnetic or diamagnetic. The amplitude of the signal is proportional to the volume of the sample. Since it is essential to have a highly uniform field at the sample location, extra turns may be added near the ends to flatten the central field. (After W. A. Steyert, "Superconductor Characterization Cryostat: Instrumentation and Methodology Handbook," APD Cryogenics, Allentown, PA, December 1987, 25 pp.)

Optical Characterization of Superconductors

It has already been mentioned that infrared reflection is a time-honored method for measurement of the superconducting energy gap Δ. A variety of other optical techniques are useful in the characterization of superconductors, and two of these are worthy of special mention.

Raman Spectroscopy. Raman spectroscopy is based on the Raman effect, that is, the phenomenon of light scattering from a material where a wavelength change occurs in the process of scattering. The intensity of Raman scattering is typically about three orders of magnitude lower than Rayleigh scattering, which exhibits no change in wavelength. The intensity per molecule of the Raman scattering is independent of the state of the medium, aside from a relatively small effect due to the refractive index of the material (Welsh, 1974). In practical Raman spectroscopy, the specimen under investigation is illuminated by a quartz mercury lamp; the resultant spectrum is a function of the scattering medium. A laser may also be used to illuminate the specimen.

Raman spectroscopy (RS) may be used to identify the impurity phases on the surface of superconductors. Using an Ar^+ laser for illumination, this method is effective to a depth of somewhat less than 100 nm for the incident photon wavelength of 514.5 nm. This effective depth is comparable to the London penetration depth. RS is sensitive to either insulating or metallic impurities on ceramic superconductors at about a 2% level. However, to obtain quantitative information on the impurity levels, it is first necessary to have information on the Raman cross section for each impurity (Cooke, et al., 1988b).

Thermally Stimulated Luminescence. Thermally stimulated luminescence (TSL) is seen in ceramic superconductor samples heated to ~350–400°C if the samples have been previously irradiated by highly ionizing photons (generally gamma rays). Fortunately, the ionizing radiation dose (typically ~10^5 rad of ^{60}Co gammas) necessary for effective TSL is well below the fluence required for damage to the superconductor. The γ irradiation induces metastable electron-hole pairs in insulating impurities that are commonly present on the surface of the superconducting ceramics. Activation energy of the traps is typically in the range of 0.1 to 0.3 eV (Fujiwara, et al., 1988). These metastable pairs recombine on subsequent heating to produce luminescence. In addition to detection of surface insulating impurities, TSL may also be used to detect the non-superconducting tetragonal 1-2-3 phase which results from loss of oxygen.

In a typical TSL apparatus, the luminescence output is recorded as a function of temperature in a light-tight enclosure containing a heater, thermometer, and photomultiplier tube. The resultant plot of luminescence versus temperature, or "glow curve," is a function of the type and amount of insulating impurity present. It is a common practice to integrate to obtain the area under the glow curve, which is proportional to the total light emitted by the specimen during the heat cycle. Fortunately, the insulating impurity phases (i.e., Ba_3CuO_4,

Y_2O_3, $BaCO_3$, $BaCuO_2$, and the green Y_2BaCuO_5 phase) of $YBa_2Cu_3O_x$ each exhibit very characteristic glow curves; the maxima are well defined (Cooke, et al., 1988b).

A significant advantage of this method is that only the insulating impurities exhibit TSL; the superconductor (like all metals) does not have a sufficient band gap to trap the gamma ray-induced electron-hole pairs.

The fortunate result is that a very pure ceramic superconductor exhibits little or no luminescence, while a specimen with surface impurities presents a relatively intense glow curve. This circumstance makes TSL a simple, inexpensive method for rapidly determining the surface quality of bulk superconductors. While TSL is also applicable to superconducting films, it may be necessary to have a film that is thick compared with the effective TSL probe depth, since most insulating substrates will show a strong luminescence that will tend to saturate the measurement (Cooke, et al., 1987 & 1988a).

REFERENCES

Abel, W. R., A. C. Anderson, and J. C. Wheatley, (1964). "Temperature Measurements Using Small Quantities of Cerium Magnesium Nitrate," *Rev. Sci. Instrum.*, Vol. 35, No. 4, April, pp. 444–449.

Abrahams, E., and F. Keffer, (1971). "Magnetic Susceptibility," *McGraw-Hill Encyclopedia of Science and Technology*, Vol. 13, McGraw-Hill, New York, pp. 348–349.

Anderson, A. C., (1980). "Instrumentation at temperatures below 1 K," *Rev. Sci. Instrum.*, Vol. 51, No. 12, December, pp. 1603–1613.

Cassel, B., (1988). "Thermal analysis in SC R&D," *Superconductor Industry*, Fall, pp. 24–29.

Cooke, D. W., H. Remmp, Z. Fisk, J. L. Smith, and M. S. Jahan, (1987). "Thermally stimulated luminescence from rare earth-doped barium copper oxides," *Phys. Rev. B*, Vol. 36, No. 4, August 1, pp. 2287–2289.

Cooke, D. W., M. S. Jahan, J. L. Smith, M. A. Maez, W. L. Hults, I. D. Raistrick, D. E. Peterson, J. A. O'Rourke, S. A. Richardson, J. D. Doss, E. R. Gray, B. Rusnak, G. P. Lawrence, and C. Fortgang, (1988a). "Detection of surface ($\sim$ 1 μm) impurity phases in high-T_c superconductors," *Appl. Phys. Lett.*, Vol. 54, No. 10, March 6, pp. 960–962.

Cooke, D. W., M. T. Trkula, R. E. Meunchausen, N. S. Nogar, J. L. Smith, and M. S. Jahan, (1988b). "Characterization of high-temperature superconductors by optical techniques," LA-UR-88-3760, submitted to *J. Opt. Soc. Am. B*, November.

Couach, M., F. Monnier, A. De Combarieu, and E. Bedin, (1982). "On the stability of carbon glass thermometers cycled at low temperatures under high magnetic fields," *Cryogenics*, Vol. 22, September, pp. 483–484.

Dalrymple, B. J., and D. E. Prober, (1984). "Upper Critical Fields of the Superconducting Layered Compounds $Nb_{1-x}Ta_xSe_2$," *J. Low Temp. Phys.*, Vol. 56, Nos. 5/6, pp. 545–574.

de Bruyn Ouboter, R., (1987). "Superconductivity: Discoveries During the Early Years of Low Temperature Research at Leiden, 1908–1914," *IEEE Trans. Mag.*, Vol. MAG-23, No. 2, March, pp. 355–370.

Doss, J. D., D. W. Cooke, C. W. McCabe, and M. A. Maez, (1988). "Noncontact methods used for characterization of high-T_c superconductors," *Rev. Sci. Instrum.*, Vol. 59, No. 4, April, pp. 659–661.

Doss, J. D., (1989). "Measuring the new superconductors," *Nature*, Vol. 338, No. 6214, March 30, pp. 441–442.

Ekin, J. W., T. M. Larson, and N. F. Bergren, (1988). "High-T_c superconductor/noble-metal contacts with surface resistivities in the 10^{-10} Ω cm^2 range," *Appl. Phys. Lett.*, Vol. 52, No. 21, May 23, pp. 1819–1821.

Fathy, A., D. Kalokitis, and E. Belohoubek, (1988). "Microwave characteristics and characterization of high-T_c superconductors," *Microwave Journal*, Vol. 31, No. 10, October, pp. 75–94.

Fiory, A. T., A. F. Hebard, P. M. Mankiewich, and R. E. Howard, (1988). "Penetration depths of high-T_c films measured by two-coil mutual inductances," *Appl. Phys. Lett.*, Vol. 52, No. 25, June 20, pp. 2165–2167.

Flanders, P. J., (1988). "An alternating gradient magnetometer," *J. Appl. Phys.*, Vol. 63, No. 8, April 15, pp. 3940–3945.

Fujiwara, Y., M. Tonouchi, and T. Tobayashi, (1988). "Thermally stimulated luminescence from high-T_c superconducting Tl-Ba-Ca-Cu-O system," *Jap. J. Appl. Phys.*, Vol. 27, No. 9, September.

Ginzton, E. L., (1967). *Microwave Measurements*, McGraw-Hill, New York.

Grader, G. S., P. K. Gallagher, and E. M. Gyorgy, (1987). "High-temperature resistivity of the $Ba_2YCu_3O_x$ superconductor," *Appl. Phys. Lett.*, Vol. 51, No. 14, October 5, pp. 1115–1117.

Hagen, M., M. Hein, N. Klein, A. Michalke, G. Müller, H. Piel, R. W. Röth, F. M. Mueller, H. Sheinberg and J.-L. Smith, (1987). "Observation of *RF* Superconductivity in $Y_1Ba_2Cu_3O_{9-\delta}$ at 3 GHz," Un. Wuppertal pub. 87-12, April, pp. 01–06.

Hahn, H., H. J. Halama, and E. H. Foster, (1968). "Measurement of Surface Resistance of Superconducting Lead at 2.868 GHz," *J. Appl. Phys.*, Vol. 39, No. 6, May, pp. 2606–2609.

Harris, F. K., (1952). *Electrical Measurements*, John Wiley & Sons, New York.

Hein, M., N. Klein, F. M. Mueller, H. Piel, R. W. Röth, W. Weingarten, and J. O. Willis, (1987). "On the surface resistance of the perovskite superconductors at 3 GHz," Un. Wuppertal pub. WUB 87-15h, June, pp. 01–02.

Hoge, H. J., (1955). "Temperature Measurement in Engineering," in *Temperature; Its Measurement and Control in Science and Industry*, Vol. II, H. C. Wolfe, ed., American Institute of Physics, Reinhold, New York, 1955, p. 305.

Iye, Y., T. Tamegai, H. Takeya, and H. Takei, (1988). "A simple method for attaching electrical leads to small samples of high-T_c oxides," *Jap. J. Appl. Phys.*, Vol. 27, No. 4, April, pp. L658–660.

Katz, J. D., J. O. Willis, M. P. Maley, and R. G. Castro, (1988). "Low resistivity, $YBa_2Cu_3O_7$-to-silver electrical contacts by plasma spraying," submitted to *J. Appl. Phys.*

Kishida, S., H. Tokutaka, K. Nishmori, N. Ishihara, Y. Wanatabe, and Y. Noshiki, (1988). "Effects of preparation conditions and mechanical polishing on the superconducting behavior of high-T_c $Y_1Ba_2Cu_3O_{7-y}$," *Jap. J. Appl. Phys.*, Vol. 27, No. 3, March, pp. L325–L328.

Krause, J. K., and B. C. Dodrill, (1986). "Measurement system induced errors in diode thermometry," *Rev. Sci. Instrum.*, Vol. 57, No. 4, April, pp. 661–665.

Krause, J. K., P. R. Swinehart, and J. R. Bergen, (1988). "Temperature Sensors for Cryogenic Applications," *Sensors*, February.

Lake Shore Cryotronics (Technical Data Sheets), (1988). 64 E. Walnut St., Westerville, Ohio 43081, (614)-891-2243.

Maley, M. P., J. O. Willis, J. D. Katz, R. G. Castro, and R. M. Aiken, (1988). "Low contact-resistivity junctions to ceramic superconductors," *IEEE Trans. Magn.*, Vol 25, No. 2, March, pp. 2053–2055.

McGee, Thomas D., *Principles and Methods of Temperature Measurement*, John Wiley & Sons, New York, 1988.

Montgomery, H. C., (1971). "Method for Measuring Electrical Resistivity of Anisotropic Materials", *J. Appl. Phys.*, Vol. 42, No. 7, June, pp. 2971–2975.

Munakata, F., K. Shinohara, H. Kanesaka, N. Hirosaki, A. Okada, and M. Yamanaka, (1987). "Electrical properties of $YBa_2Cu_3O_y$ at high temperatures," *Jap. J. Appl. Phys.*, Vol. 26, No. 8, August, pp. L1292–L1293.

Papaioannou, J., and J. L. Dye, (1988). "Four-probe single-crystal holder for conductivity measurements," *Rev. Sci. Instrum.*, Vol. 59, No. 3, March, pp. 496–497.

Peterson, G. E., R. P. Stawicki, and U. C. Paek, (1988). "Radio frequency properties of high-T_c superconductors," Proc. 38th Electronic Components Conference, Los Angeles, May 9–11, pp. 159–167.

Rosenbaum, R. L., (1968). "Some Properties of Gold-Iron Thermocouple Wire," *Rev. Sci. Instrum*, Vol. 39, No. 6, June, pp. 890–899.

Rubin, L. G., B. L. Brandt, and H. H. Sample, (1982). "Cryogenic thermometry: a review of recent progress, II" *Cryogenics*, Vol. 22, October, pp. 491–503.

Sample, H. H., B. L. Brandt, and L. G. Rubin, (1982). "Low-temperature thermometry in high magnetic fields. V. Carbon-glass resistors," *Rev. Sci. Instrum.*, Vol. 53, No. 8, August, pp. 1129–1135.

Sample, H. H., L. J. Neuringer, and L. G. Rubin, (1974). "Low temperature thermometry in high magnetic fields. III. Carbon resistors (0.5–4.2 K); thermocouples," *Rev. Sci. Instrum.*, Vol. 45, No. 1, January, pp. 64–73.

Schawlow, A. L., and G. E. Devlin, (1959). "Effect of the Energy Gap on the Penetration Depth of Superconductors," *Phys. Rev.*, Vol. 113, No. 1, January 1, pp. 120–126.

Talpe, J., G. Stolovitzky, and V. Bekeris, (1987). "Cryogenic thermometry and level detection with common diodes," *Cryogenics*, Vol. 27, No. 12, December, pp. 693–695.

Talvacchio, J., (1989). "Electrical contact to superconductors," to appear in *IEEE Trans. CHMT.*

van der Maas, J., V. A. Gasparov, and D. Pavuna, (1987). "Improved low contact resistance in high-T_c Y-Ba-Cu-O ceramic superconductors," *Nature*, Vol. 328, August 13, pp. 603–604.

Welsh, H. L., (1974). "Raman effect and Raman spectroscopy," in *The Encyclopedia of Physics*, R. M. Besancon, ed., Van Nostrand Reinhold, New York, pp. 785–788.

SUGGESTIONS FOR FURTHER READING

Cheong, S-W., Z. Fisk, R. S. Kwok, J. P. Remeika, J. D. Thompson, and G. Gruner, "Electronic anisotropy in single-crystal La_2CuO_4," *Phys. Rev. B*, Vol. 37, No. 10, April 1, 1988, pp. 5916–5919.

Cowey, L., H. Jones, and D. Dew-Hughes, "Technique for short sample testing of high-T_c superconducting wire," *Cryogenics*, Vol. 28, No. 3, March 1988, pp. 181–182.

Goldfarb, R. B., and J. V. Minervini, "Calibration of ac susceptometer for cylindrical specimens," *Rev. Sci. Instrum.*, Vol. 55, No. 5, May 1984, pp. 761–764.

Iye, Y., "Small low-cost platinum resistance thermometers for thermometry in magnetic fields," *Cryogenics*, Vol. 28, No. 3, March 1988, pp. 164–168.

Krause, J. K., P. R. Swinehart, "Temperature Trends; Demystifying Cryogenic Temperature Sensors," *Photonics Spectra*, Optical Publishing, August 1985. (For reprints, contact Lake Shore Cryotronics, Inc., 64 E. Walnut St., Westerville, OH 43081.)

Lounasmaa, O. V., *Experimental Principles and Methods Below 1 K*, Academic Press, New York, 1974.

Martens, J. S., J. B. Beyer, and D. S. Ginley, "Microwave surface resistance of $YBa_2Cu_3O_{6.9}$ superconducting films," *Appl. Phys. Lett.*, Vol. 52, No. 21, May 23, 1988, pp. 1822–1824.

Omega 1987 Complete Temperature Measurement Handbook and Encyclopedia, Omega Engineering, Inc., Stamford, CT, 1987.

Pazoi, B. G., and D. M. Ginsberg, "Low-temperature sample-rotating cryostat with two degrees of freedom for measuring the anisotropy of the upper critical field of superconductors," *Rev. Sci. Instrum.*, Vol. 59, No. 5, May 1988, pp. 776–777.

Richardson, R. C., and E. N. Smith, eds., *Experimental Techniques in Condensed Matter at Low Temperatures*, Addison-Wesley, New York, 1988.

Rose-Innes, A. C., "A simple system for obtaining temperatures between 77 K and room temperature," *J. Phys. E*, Vol. 21, No. 7, July 1988, pp. 729–730.

Rubin, L. G., "The new(?) wave: Digital lock-in amplifiers," *Rev. Sci. Instrum.*, Vol. 59, No. 3, March 1988, pp. 514–515.

Sankawa, I., M. Sato, T. Konaka, M. Kobayashi, and K. Ishihara, "Microwave surface resistance studies of $YBa_2Cu_3O_{7-\delta}$ single crystal," *Jap. J. Appl. Phys.*, Vol. 27, No. 9, September 1988, pp. L1637–L1638.

Sridhar, S., and W. L. Kennedy, "Novel technique to measure the microwave response of high T_c superconductors between 4.2 and 200 K," *Rev. Sci. Instrum.*, Vol. 59, No. 4, April 1988, pp. 531–536.

Staskiewicz, J., A. Czyziewski, C. Szypowski, and W. Gulbinski, "Improvement of the sensitivity of a bridge for the inductive measurement of superconducting critical temperatures," *J. Phys. E*, Vol. 21, 1988, pp. 62–63.

Steyert, W. A., and R. C. Longworth, "A High T_c Superconductor Characterization Cryostat," Poster presentation, Int. Conference on Superconductivity, Drexel University, Philadelphia PA, July 29–30, 1987.

Steyert, W. A., "Superconductor Characterization Cryostat: Instrumentation and Methodology Handbook," APD Cryogenics, Allentown, PA, December 1987, 25 pp.

Wang, N., B. Sadoulet, T. Shutt, J. Beeman, E. E. Haller, A. Lange, I. Park, R. Ross, C. Stanton, and H. Steiner, "A 10 mK Temperature Sensor," *IEEE Trans. on Nucl. Sci.*, Vol. 35, No. 1, Feb. 1988, pp. 55–58.

Wheatley, J. C., "A perspective on the history and future of low-temperature refrigeration," *Physica 109 & 110B*, 1982, pp. 1764–1774.

White, G. K., *Experimental Techniques in Low-Temperature Physics*, 3rd Ed., Clarendon Press, Oxford, 1979.

CHAPTER 7

SAFETY CONSIDERATIONS

He that's secure is not safe.

(Benjamin Franklin, 1748)

People unfamiliar with low temperature work believe the whole field of low temperature is pretty esoteric, but all it really requires is access to liquid helium . . .

(Ivar Giaever, 1974)

INTRODUCTION

The latter quotation given above is certainly not intended as an implication that its distinguished author was casual about safety, but rather to point out the fact that most of us are surprised to discover that working with cryogens is easier than we had imagined. With only a little practice, experiments with liquid helium and liquid nitrogen are generally so trouble-free that it is easy to forget the potential for accident and injury.

Engineers and technicians who deal with superconductors must, inevitably, deal with the challenges of making their systems or circuits operate at cryogenic temperatures. (Until, at least, the advent of room-temperature superconductors!) Cryogens are not, of course, the only hazards that one may encounter in working with superconductors.

This chapter will outline some of the techniques and data required as an introduction. For those who require more detailed information, appropriate references are provided.

CRYOGENS

To a first approximation, one might assert that the conventional superconductors (lead, niobium) will generally be cooled with liquid helium and that $YBa_2Cu_3O_x$ will usually be cooled with liquid nitrogen. Other cryogens, such as neon, may prove to be useful for cooling the copper oxides when lower temperatures are required (for operation in higher magnetic fields, for higher critical current density, etc.). Some properties of these cryogens are given in the following table:

Cryogen	Boiling Point (K at 1 atm)	Spec. Heat J/g-K	Heat of Vap. J/g
Helium	4.2	5.2	20.2
Neon	27.1	1.8	85.7
Nitrogen	77.3	2.0	199.0

One of the more striking facts is that the ratio of (mass) heats of vaporization of neon (4.2) or nitrogen (9.85) to helium. The ratio is even larger if one considers the *volumetric* comparison of nitrogen to helium; the former being 6.45 times as dense, has a volumetric heat of vaporization that is 64 times that of helium. This has obvious implications for the dissipation of a cryogen supply.

The handling of cryogens should be left to those who are thoroughly familiar with such materials, but it is useful for the those of us who must work around these materials to be introduced to a few factors related to the safety of dealing with these substances, either as liquids or gases.

Hazards

The most obvious hazard is that cryogens can cause injury due to the extremely low temperatures involved. This can range from minor frostbite to major injuries such as loss of sight if cryogens come into contact with the eye, particularly the cornea. Long-term contact (i.e., several seconds) can result in severe injury to exposed tissues. Workers should exercise reasonable care and wear protective clothing (gloves, safety glasses, etc.) when there is a potential for exposure to cryogens, and eye protection is particularly recommended.

Some cryogens, not recommended for use in superconductor applications, are either toxic (fluorine, ozone), flammable (hydrogen, methane), or both (carbon monoxide). While the superconductor experiment or application will not be likely to utilize such dangerous cryogens, others are "toxic" in a more subtle fashion. It is a good rule of thumb that typical cryogens expand a factor of about 700 times when converted from the liquid to the gas phase. (For helium, the expansion factor is ~700, for nitrogen, ~650.)

This expansion has obvious implications for rupture of a pressurized dewar, but another hazard may be less apparent. Even an apparently innocuous cryogen such as liquid nitrogen, when converted to gas, will tend to replace a portion of the oxygen in the workplace. Obviously, the larger the nitrogen supply, the greater proportion of oxygen that will be replaced. Asphyxiation can result if the cryogen gas volume is sufficient to replace a significant proportion of the oxygen.

At sea level, the partial pressure of oxygen is about 0.2 atmosphere (21% O_2). Subjects in excellent condition can tolerate a temporary decrease to about 13% O_2, (depending on the gas that is replacing the oxygen!) but individuals in poor health are much less tolerant (Edeskuty, et al., 1971). Actions to be considered include:

1. Limit the amount of liquid cryogen storage in a given room, taking into account the volume of the room and the ventilation system.
2. Design (or modify) the ventilation system with use or storage of cryogens taken into account.
3. Consider use of oxygen-depletion alarms to alert personnel when oxygen content has dropped below a specified level.

Another danger caused by the conversion of liquid cryogen to the gaseous phase is an increase in pressure when the gas is not allowed to expand. The most obvious hazard is the rupture of a dewar, cryostat, or transfer line. For this reason, a well-designed system *always* utilizes vents and/or overpressure relief valves at appropriate locations, and redundancy is recommended in case a single valve (or vent) fails. For relief valves of the type where a pressure-relief plug may become a missile, it is obviously wise to point such devices in a safe direction!

Valves may fail if a system if contaminated by another gas (for example, water vapor) that will freeze at the system temperature, so it is imperative to be familiar with the operation of such systems and to keep all components clean and free of contaminants.

If a particular cryogen is flammable, one should certainly avoid means of ignition (such as sparks). If the cryogen is a fuel, one must take great care to avoid contamination by an oxidant. The knowledge that one is using a non-flammable cryogen should not lead to complacency on the flammability issue. Spencer points out that even liquid nitrogen or helium are capable of condensing oxygen from the atmosphere (oxygen boiling point is 90.1 K at 1 atm.) which can lead to "enrichment or entrapment in unsuspected areas (Spencer, 1983)."

The reader who will be involved in the design of cryogenic systems is referred to detailed treatments on this subject (Cryogenic Safety, 1960; Edeskuty, et al., 1971; Spencer, 1983).

TOXICITY OF MATERIALS; THALLIUM

A noted investigator was reported to have made the following observation, in jest of course:

> . . . nature has mechanisms to limit populations and we do have a highly overpopulated field [working] in superconductivity. Perhaps thallium has come along to solve that problem.
>
> (Superconductor Week, 1988)

Those who are not interested in population control can always come up with some flimsy excuse to avoid making a personal contribution to a worthy cause. For example another prominent scientist responded like this:

> I don't want to get poisoned
>
> (Superconductor Week, 1988)

As new superconductors are developed, it is impossible to predict what elements or compounds will be used. Some of these will be toxic. An example is thallium, the subject of much recent interest. The element and compounds of thallium (i.e. , salts, sulfides) are particularly toxic. An unprepared encounter with such materials can result in severe injury or even death. At best, you may spend many hours carefully cleaning exposed instrumentation so it can be used for important experiments or production.

Although thallium, like lead or arsenic, is quite toxic, there is no reason to avoid use of this element entirely. Perhaps the most appropriate comment concerning the use of thallium in the production of superconducting compounds was made by an industrial hygenist at Argonne National Laboratory:

> The controls and practices that have been described here are not exotic, but instead are textbook industrial hygiene practices for toxic materials. They are effective for control of arsenic, lead, and mercury hazards, and they will be effective for control of thallium hazards as well. Thallium and its compounds are hazardous, but they are far from unmanageable.
>
> (Myers, 1988)

The controls and practices described by Myers include the following:

1. Good ventilation or containment of the thallium
2. Use of impervious gloves when handling thallium
3. Avoid grinding, drilling, cutting, abrading, and so forth, of thallium
4. Employee education in hazards and safe handling practices

Since one cannot be positive of the composition of samples that are produced during research programs, it is also best to be conservative even where solid samples of the thallium compounds are to be measured. A case in point: Unexpectedly, thallium oxide forms on an experimental sample. Thallium oxide is toxic and is also *water soluble* (Khurana, 1988). There are several ways that water-soluble thallium oxide on the surface of a bulk sample of Tl-Ca-Ba-Cu-O could get into an experimenter's body. These include absorption through *unbroken* skin, ingestion, or inhalation.

One should wear a face mask and protective gloves when handling experimental thallium preparations, and follow this by a thorough washing before eating or drinking. If there is any prospect for of airborne contaminants, good ventilation is essential. Myers recommends that a laboratory ventilation hood should have a face velocity in the range of 120–150 ft/min for applications where highly toxic materials are to be exhausted. Note that this is an appropriate velocity range rather than simply a minimum velocity. While lower velocities may be innefective in removing the contaminant effectively, greater velocities may lead to turbulence that will also degrade the effectiveness of the exhaust hood (Myers, 1988).

When uncertain of the hazards of a specific material, ask the supplier for a Material Safety Data (MSD) Sheet. The following selected information on thallium was compiled from a MSD sheet (MSD, 1981):

Material Name: THALLIUM
Other Designations: Tl, CAS #007 440 280
. . . absorption through the skin is a significant hazard. *Skin contact with thallium is dangerous*, especially when repeated or prolonged.

Boiling point,°C		1457
Vapor pressure @	825°C, mm Hg	1*
Melting Point,°C		303*

*Volatilization begins to be significant above 174°C.

Thallium is odorless, giving no warning of exposure to vapors.

Prevent skin contact with thallium or its oxidization products.

Finely divided thallium metal dust potentially explosive if contacted by fire or flame.

Firefighters should wear self-contained breathing apparatus.

Forms toxic compounds on contact with moisture. Thallium oxidizes slowly in 20 C air and more rapidly as temperature increases.

Readily soluble in HNO_3, less so in HNO_4, sparingly in HCl owing to formation of insoluble TlCl.

Violent reaction with F_2, thallium ignites on contact with F_2.

Thallium is cumulative in the body . . . is absorbed through the skin . . . rapidly absorbed by the gastrointestinal tract . . . a mitotic agent and a general cellular poison. Retained . . . in kidney, bone, and soft tissue.

Acute systems include gastrointestinal disorders, a metallic taste, autonomic nervous system impairment, disorientation and convulsions, pain and tingling of extremities, restlessness, vasomotor disturbance, dyspnea, and respiratory failure.

> Chronic symptoms: polyneuritis . . . loss of hair; 10–30 days after intoxication . . . cardiac, renal as well as endocrine disorders.
>
> Provide general and local exhaust ventilation to meet TLV requirements in the workplace. Protective clothing, chemical goggles, rubber gloves and apron may be needed . . . Eyewash stations should be readily accessible to work area. Wear NIOSH approved respirators where required.
>
> Provide preplacement and periodic medical examinations for those regularly exposed to thallium . . .
>
> Thallium metal must be stored under vaseline oil.
>
> Practice rigid personal hygiene. Wash hands and face before eating, drinking, or smoking after handling (thallium). Don't eat or smoke in work areas where thallium is present. Prevent skin contact with thallium and avoid breathing of thallium dust.

It is not the author's intent, in this brief chapter, to consider every possible hazard, toxic or otherwise, that those working with superconductors may encounter. There is no substitute for using common sense; inform yourself of the hazards associated with any unfamiliar material or procedure and follow the recommended methods to reduce any risk to an acceptable level.

On the other hand, once reasonable precautions have been made, do not become obsessed with achieving absolute safety. It cannot be done, of course, contrary to the impression left by some public figures who innocently ask whether space travel can be made "safe."

> If you believe the doctors, nothing is wholesome;
> If you believe the theologians, nothing is innocent;
> If you believe the soldiers, nothing is safe.
>
> (R. A. Talbot, Marquess of Salisbury, 1877)

REFERENCES

Cryogenic Safety, summary report of the Cryogenic Safety Conference, Allentown, PA, Air Products, 1960.

Edeskuty, F. J., R. Reider, and K. D. Williamson, Jr., (1971). "Safety," in *Cryogenic Fundamentals*, G. G. Haselden, ed., Academic Press, New York, pp. 634–672.

Franklin, Benjamin, (1748). Cited in *Familiar Quotations*, J. Bartlett, ed., 15th Ed., 1980, p. 347.

Giaever, I., (1974). "Electron tunneling and superconductivity," *Rev. Mod. Phys.*, Vol 46, No. 2, April, pp. 245–250.

Khurana, A., (1988). "The T_c to Beat is 125 K," *Physics Today*, Vol. 41, No. 4, April, pp. 21–25. (See page 25, col. 3 for comments about thallium hazards.)

MSD (Material Safety Data): thallium, (1981). General Electric Company, Schnectady, NY.

Myers, G. E., (1988). "Thallium Safety," *Superconductor Industry*, Vol. 1, No. 2, Winter, pp. 13–19.

Spencer, E. W., (1983). "Cryogenic Safety," in *CRC Handbook of Laboratory Safety*, 2nd Ed. N. V. Steere, ed., CRC Press, Boca Raton, FL. , pp. 574–578.

Superconductor Week, (1988). Vol. 2, No. 3, March 28, p. 1.

Talbot, R. A. Gascoyne-Cecil, Marquess of Salisbury, (1877). Cited in *Familiar Quotatons*, J. Bartlett, ed., 15th Ed., 1980, p. 609.

RECOMMENDED FOR FURTHER READING

Prick, J. J. G., W. G. Sillevis Smitt, and L. Muller, *Thallium Poisoning*, Elsevier, New York, 1955.

Sax, N. I., and R. J. Lewis, Sr., *Dangerous Properties of Industrial Materials*, 7th Ed. Van Nostrand Reinhold, New York, 1988. (Vol. III, TEI-1000 thru TEM-250.)

Stokinger, H. E., "The Metals," in *Patty's Industrial Hygiene and Toxicology, Vol. 2A—Toxicology*, John Wiley & Sons, New York, 1981.

Wade, K., and A. J. Banister, "Aluminum, Gallium, Indium and Thallium," in *Comprehensive Inorganic Chemistry*, Vol. 1, Pergamon, Oxford, 1973.

APPENDIX A

REVIEW OF BASIC ELECTRICAL AND MAGNETIC THEORY

INTRODUCTION

Since this book is primarily concerned with the electric and magnetic properties of superconductors, it is perhaps worthwhile to review some of the basic concepts of electromagnetic fields that are relevant. In addition, a few comments on dielectric loss are provided since dielectric substrates for superconductors are of considerable practical importance.

UNITS

One of the more tedious tasks for the engineer (who is typically trained in rationalized mks or Standard International units) is interpretation of the Gaussian units commonly used in the literature on superconductivity. One who intends to read this literature should invest some effort to become familiar with the variations introduced by these systems of units.

CURRENT FLOW AND TEMPERATURE-MEDIATED RESISTIVITY

Ohm's law is a reasonable place to begin for a reference that is concerned with electrical conductivity of materials. Consider a homogeneous cube of conductive material with highly conductive electrodes placed on opposite faces to generate a current flow when a potential V is placed across its faces (see Figure A.1). The magnitude of current flow is simply

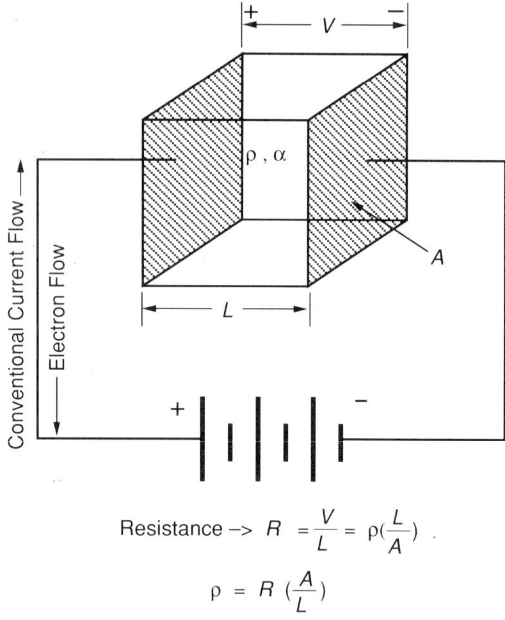

FIGURE A.1 The resistivity (ρ) of regular polyhedron can be determined by measurement of voltage (V) and current (I). Determination of ρ at several temperatures will yield the thermal coefficient of resistivity, α.

$$I = \frac{V}{R} \tag{A.1}$$

where R is the resistance of the cube and V is the potential across the cube. The resistance R is determined by the cube dimensions and resistivity ρ of the material, that is,

$$R = \frac{\rho L}{A} \tag{A.2}$$

where L is the length of the cube and A is the area of a side. Since we are considering a cube, $A = L^2$,

$$R_{\text{cube}} = \frac{\rho}{L} \tag{A.3}$$

The question is somewhat complicated by the fact that ρ is typically a function of temperature, largely due to phonons (quantized vibrations in the crystal lattice) that scatter electrons more effectively at higher temperatures. Over a fairly wide range of temperature, this relationship may be approximated by

$$\rho = \rho_o\{1 + \alpha(T - T_o)\} \quad \text{A.4}$$

where ρ_o is resistivity at temperature T_o, T is the temperature of the material, and α is the temperature coefficient of resistivity for the particular material, having units of 1/K.

For example, the value of ρ_o for copper is approximately 1.77×10^{-6} Ω-cm at 293 K, and α is 0.0028 K^{-1}. While ordinary metals have a positive temperature coefficient and are adequately described by this approximation, the expression clearly does not apply to superconducting phenomena at temperatures near or below T_c.

A more general definition of conductivity can be derived from the equation of motion of the electrons. If the electron mass is m, the velocity v, E_o the electric field at the average electron position, for a radial frequency ω, the relationship is

$$m\frac{dv}{dt} + mgv = eE_o\epsilon^{-j\omega t} \quad \text{A.5}$$

where g is a damping constant related to the average rate of collisions which involve a significant transfer of momentum. While the calculation of g is not straightforward, an empirical value for copper is on the order of 3×10^{13} collisions/s (Jackson, 1963). The solution for steady state electron velocity is

$$v = \frac{e}{m(g - j\omega)} E_o\epsilon^{-j\omega t} \quad \text{A.6}$$

which yields an expression for electrical conductivity that is complex at frequencies where ω is comparable to or larger than g

$$\sigma = \frac{e^2 n_o}{m(g - j\omega)} \quad \text{A.7}$$

Conductivity σ is related to the current density vector $\mathbf{J}$ and the electric field vector $\mathbf{E}$ by:

$$\mathbf{J} = \sigma\mathbf{E} \quad \text{A.8}$$

where resistivity ρ is simply the inverse of σ:

$$\rho = \frac{1}{\sigma} \quad \text{A.9}$$

In expression A.7, n_o is the density of conduction electrons (approximately 10^{23} per cm^3 for copper). For the Cu damping value quoted previously ($g =$

3×10^{13} collisions/s), the frequency crossover point would be approximately 5×10^{12} Hz. For typical metallic conductors, the imaginary term can be ignored for frequencies below about 10^{10}–10^{11} Hz, that is, 10–100 GHz. The fact that conductivity is complex does require consideration at infrared and, of course, at shorter wavelengths than *IR*.

At any nonzero temperature, the random motion of free electrons in a conductor will create noise. The "Johnson" (thermal) noise within a specified bandwidth ($f_H - f_L$) across the terminals of a resistance (R) operating at an absolute temperature (T) is given by

$$v_n(\text{rms}) = \sqrt{4k_B TR(f_H - f_L)} \quad \text{A.10}$$

where k_B is Boltzmann's constant. Note that the noise produced is independent of the center frequency, depending instead on the square root of the bandwidth. For example, the rms noise voltage developed in a bandwidth of 10 kHz across a 1 kΩ resistor at room temperature (300 K) is 406 nV.

THE MAGNETIC FIELDS (B, H, M, A)

The magnetic field H is defined in isotropic media as

$$\mathbf{H} = \frac{\mathbf{B}}{\mu} \quad \text{A.11}$$

where μ is the permeability of the medium. In media with a nonzero magnetization, M,

$$(\mathbf{H} + \mathbf{M}) = \frac{\mathbf{B}}{\mu_o} \quad \text{A.12}$$

where $\mu_o = 4\pi \times 10^{-7}$ H/m. H and M have the same SI units, that is, A/m. In cgs, H has units of oersteds (Oe). For reference:

$$1 \text{ A/m} = 4\pi \times 10^{-3} \text{ Oe}$$

Since, in SI units, the permeability is the product of the relative permeability and the permeability in vacuum, that is, $\mu = \mu_r \mu_o$, it follows that

$$\mathbf{B} = \mu\mathbf{H} = \mu_o(\mathbf{H} + \mathbf{M}) \quad \text{A.13}$$

then

$$\frac{\mu}{\mu_o} = \left(1 + \frac{M}{H}\right) \quad \text{A.14}$$

and since $\mu_r = \mu/\mu_o$

$$\mu_r - 1 = \frac{M}{H} \qquad \text{A.15}$$

The magnetic flux density **B** at a point can be produced by a current distribution. If the volume element of the current density **J** is a distance r from the point in question,

$$\mathbf{B} = \frac{\mu}{4\pi} \int_v \frac{\mathbf{J} \times \mathbf{r}}{r^2} dv \qquad \text{A.16}$$

where **r** is a unit vector directed from the current volume element to the point where **B** is measured, m is the (uniform) permeability of the medium, and dv is the volume element. If the integration is carried out over the entire current-carrying volume, one finds the flux density at the point of interest due to the current flow. Since the divergence of **B** is always zero, it is possible to express **B** as the curl of a second vector, that is,

$$\mathbf{B} = \nabla \times \mathbf{A} \qquad \text{A.17}$$

The vector **A** is commonly known as the vector potential; it is a vector that is also a potential function.

MAGNETIC SUSCEPTIBILITY, χ

Since the ratio M/H is the measure of how susceptible a particular material is to polarization by an applied magnetic field, the quantity is commonly referred to as magnetic susceptibility χ, that is,

$$\chi = \mu_r - 1 = \frac{M}{H} \qquad \text{A.18}$$

Susceptibility (χ) and magnetization (M) are positive for paramagnetic and ferromagnetic materials but are negative for diamagnetic materials (such as superconductors at temperatures below T_c). The relative permeability (μ_r) is equal to one for a vacuum, greater than one for paramagnetic and ferromagnetic materials, and less than one for diamagnetic materials. If the material is perfectly diamagnetic, $\mu_r = 0$. See Table A.1.

In time-varying magnetic fields, one may measure a complex susceptibility, that is,

$$\chi = \chi' - j\chi'' \qquad \text{A.19}$$

where $j = (-1)^{1/2}$.

TABLE A.1 Representative values of μ_r for typical diamagnetic, paramagnetic, and ferromagnetic materials. In contrast to the relatively weak diamagnetism in materials such as copper, a superconductor operated well below the critical transition temperature exhibits nearly complete diamagnetism, so that μ_r approaches zero.

Substance	μ_r	Group Type
Silver	0.99998	Diamagnetic
Lead	0.999983	Diamagnetic
Copper	0.999991	Diamagnetic
Water	0.999991	Diamagnetic
Aluminum	1.00002	Paramagnetic
Palladium	1.0008	Paramagnetic
Cobalt	250	Ferromagnetic
Nickel	600	Ferromagnetic
Iron	5000	Ferromagnetic
Supermalloy	1,000,000	Ferromagnetic

Susceptibility (χ) in SI units is simply the ratio M/H, but in Gaussian or cgs emu unit systems, investigators typically plot $4\pi\chi$, that is, $-4\pi M/H$. (Details of other conventions are discussed further in the chapter on superconducting phenomena.) The background for these units will be outlined briefly as follows.

Recall that in SI units, we use the expression:

$$\mathbf{B} = \mu\mathbf{H} = \mu_o(\mathbf{H} + \mathbf{M}) \tag{A.20}$$

and

$$\mu = \mu_o\left(1 + \frac{H}{M}\right) \tag{A.21}$$

Since $\mu/\mu_o = \mu_r$ (relative permeability),

$$\mu_r = 1 + \frac{H}{M} \tag{A.22}$$

In Gaussian or cgs emu unit systems, the equivalent expression is

$$\mathbf{B} = \mu\mathbf{H} = \mathbf{H} + 4\pi\mathbf{M} \tag{A.23}$$

In this latter case,

$$\mu = 1 + 4\pi\frac{M}{H} = 1 + 4\pi\chi \tag{A.24}$$

(Although some authors use the convention that $\chi = 4\pi M/H$ instead of M/H.) The SI relative permeability μ_r is not defined in cgs or Gaussian units. It is also worth noting that, in Gaussian or cgs units, the magnetic permeability (μ) is dimensionless, while in the SI and rationalized mks units, μ has dimensions of Wb/A-m. A consequence is that **B** and **H** have different dimensions (Wb/m^2 or T (**B**) and A-meter (**H**)) in SI units, but **B** and **H** may have the same dimensions (gauss or Oe) in cgs or Gaussian units.

A consequence of the use of cgs or Gaussian systems is that the diamagnetic effect seen in superconductors (below the critical temperature) may be plotted as **M** (emu/mol), **M** (emu/g), or $4\pi\mathbf{M}$ (gauss). Susceptibility may be plotted as $4\pi M/H$ or $\chi/4\pi$. In SI units, the maximum range for $\mathbf{M}/\mathbf{H}(\chi)$ is ± 1.

ENERGY IN STATIC ELECTRIC AND MAGNETIC FIELDS

In a volume ν with a dielectric constant ϵ, the stored electric energy (W_e) is

$$W_e = \frac{\epsilon}{2}\int_{\nu} E^2 d\nu \qquad \text{A.25}$$

For a capacitor C (F), charged to a constant voltage V (volts), the energy stored is

$$W_C = \frac{CV^2}{2} \qquad \text{A.26}$$

In a volume ν with a permeability μ, the stored magnetic energy (W_M) is

$$W_M = \frac{\mu}{2}\int_{\nu} H^2 d\nu \qquad \text{A.27}$$

The energy stored per unit volume is

$$\frac{\mu H^2}{2} = \frac{B^2}{2\mu} \qquad \text{A.28}$$

For an inductor L (H) with a current I(A) flowing, the energy stored is

$$W_L = \frac{LI^2}{2} \qquad \text{A.29}$$

THE HALL EFFECT

The force **F** on a charge q moving with velocity **v**, under the influence of an electric field **E** and a magnetic field **B** is defined in the Lorentz relation:

$$\mathbf{F} = q\mathbf{E} + q(\mathbf{v} \times \mathbf{B}) \tag{A.30}$$

If a constant current I is flowing through a conducting strip, the difference in potential between opposite sides of the strip (see locations M and N in Figure A.2) will be zero unless a magnetic field is applied normal to the surface shown in the drawing. The force due to the B field will be $q(\mathbf{v} \times \mathbf{B})$ as indicated above. Depending on the polarity of current flow (and the direction of $\mathbf{B}$), charge will be displaced toward either the upper or lower side of the conducting strip. The magnitude of the resultant (Hall) voltage is

$$V_H = vBd \tag{A.31}$$

where v is the average drift velocity of the charge carriers, B is the magnitude of the magnetic field vector, and d is the width of the conducting strip. The drift velocity is related to the current flow (I), density of charge carriers (n), the carrier charge (q), and the cross-sectional area (A) of the conductor by

$$v = \frac{I}{nqA} \tag{A.32}$$

This yields a simple expression for current density J:

$$J = \frac{I}{A} = nqv \tag{A.33}$$

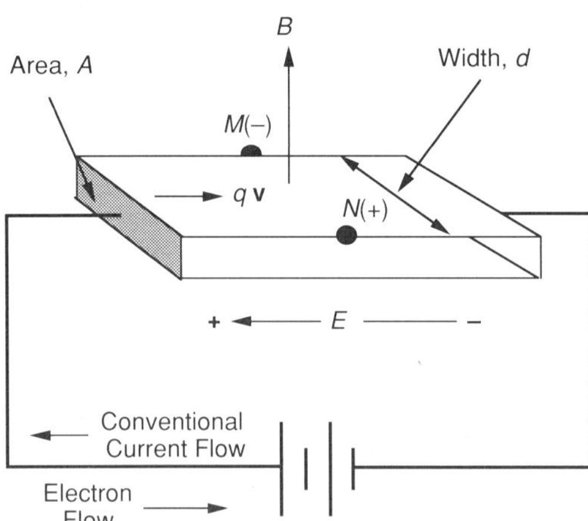

FIGURE A.2 The Hall potential, measured between points M and N, is zero in the absence of a B field. The Hall voltage reverses polarity if the polarity of $\mathbf{E}$ is changed or if the direction of $\mathbf{B}$ is reversed. Hall devices are used for measurement of $\mathbf{B}$ fields, and are used to determine the density and polarity of charge carriers. Superconductors exhibit a vanishing Hall effect as temperature decreases below the critical transition temperature (T_c).

The former expression for drift velocity may be used in the expression for V_H:

$$V_H = \frac{BId}{nqA} \tag{A.34}$$

The quantity $1/nq$ is generally referred to as the Hall coefficient. Since it is possible to measure all of these quantities except for n, the expression may be used to determine n, that is, the density of charge carriers in a metal.

Aside from determination of the polarity and density of the charge carrier, Hall detectors are commonly used to measure magnetic fields. Hall effect measurements are of some importance in the characterization of superconductors where the dc Hall effect disappears (London, 1950; Blatt, 1964).

TIME-VARYING FIELDS

Maxwell's Equations

The Maxwell equations will be given below in integral form and in vector notation.

From Ampere's law, the general case, the line integral of **H** around a closed path equals the current encircled through the surface S:

$$\oint_c \mathbf{H} \cdot d\mathbf{l} = \int_S \left(\mathbf{J} + \frac{\partial \mathbf{D}}{\partial t} \right) \cdot d\mathbf{S} = \mathbf{I}_{\text{total}} \tag{A.35}$$

or

$$\nabla \times \mathbf{H} = \mathbf{J} + \frac{\partial \mathbf{D}}{\partial t} \tag{A.36}$$

If the current density ($\mathbf{J} = \sigma\mathbf{E}$) is dropped, the expression applies to lossless space and is equal to the displacement current:

$$\oint \mathbf{H} \cdot d\mathbf{l} = \int_S \left(\frac{\partial \mathbf{D}}{\partial t} \right) \cdot d\mathbf{S} = I_{\text{displ}} \tag{A.37}$$

From Faraday's induction law, the emf that is the line integral around the circuit equals the negative rate of change of flux on the surface enclosed:

$$\oint_c \mathbf{E} \cdot d\mathbf{l} = -\int_S \frac{\partial \mathbf{B}}{\partial t} \cdot d\mathbf{S} \tag{A.38}$$

or

$$\nabla \times \mathbf{E} = -\frac{\partial \mathbf{B}}{\partial t} \qquad \text{A.39}$$

From Gauss, for electric flux,

$$\oint_S \mathbf{D} \cdot d\mathbf{s} = \int_v \rho \, dv \qquad \text{A.40}$$

or

$$\nabla \cdot \mathbf{D} = \rho \qquad \text{A.41}$$

which implies that electric flux lines always end on charges.

For magnetic flux,

$$\oint_s \mathbf{B} \cdot d\mathbf{s} = 0 \qquad \text{A.42}$$

or

$$\nabla \cdot \mathbf{B} = 0 \qquad \text{A.43}$$

implies that the net magnetic flux through any closed surface S is always zero. (Recall that divergence is defined as the net outward flux of a vector field per unit volume.)

The Faraday induction law will be given further consideration, since it is quite relevant to several subjects to be dealt with in the subsequent discussion of superconductivity. The familiar expression

$$E_{\text{emf}} = -\frac{\partial \phi}{\partial t} \qquad \text{A.44}$$

expresses the fact that the induced emf is equal in the rate of change in flux enclosed by the path. An electrostatic field is conservative, that is,

$$\oint_c E \cdot dl = 0 \qquad \text{A.45}$$

In the case of a time-varying field, with a contour/surface that does not change in time,

$$\oint_c \mathbf{E} \cdot d\mathbf{l} = -\frac{\partial \phi}{\partial t} = -\frac{\partial}{\partial t} \int_S \mathbf{B} \cdot d\mathbf{S} = -\int_S \frac{\partial \mathbf{B}}{\partial t} \cdot d\mathbf{S} \qquad \text{A.46}$$

This may also be expressed in vector notation as

$$\nabla \times \mathbf{E} \equiv \mathrm{curl}(\mathbf{E}) = -\frac{\partial \mathbf{B}}{\partial t} = -\mu_o \frac{\partial \mathbf{H}}{\partial t} \qquad \text{A.47}$$

Conducting Media

In conducting media, where **J** is replaced by $\sigma\mathbf{E}$, **H** by $\mathbf{B}/\mu$, and **D** by $\epsilon\mathbf{E}$

$$\nabla^2\mathbf{B} = \mu\sigma\frac{\partial \mathbf{B}}{\partial t} + \mu\epsilon\frac{\partial^2 \mathbf{B}}{\partial t^2} \qquad \text{A.48}$$

and

$$\nabla^2\mathbf{E} = \mu\sigma\frac{\partial \mathbf{E}}{\partial t} + \mu\epsilon\frac{\partial^2 \mathbf{E}}{\partial t^2} \qquad \text{A.49}$$

which are equations for damped waves. If conductivity is very small or zero, the first term on the right hand side is dropped and the result is

$$\nabla^2\mathbf{E} = \mu\epsilon\frac{\partial^2 \mathbf{E}}{\partial t^2} \qquad \text{A.50}$$

which is representative of undamped waves, with velocity $(\mu\epsilon)^{-1/2}$. For $\mu = \mu_o$ and $\epsilon = \epsilon_o$, this quantity is equal to c, the velocity of light in a vacuum. On the other hand, if conductivity is sufficiently large, the second term on the right may be omitted, leaving

$$\nabla^2\mathbf{E} = \mu\sigma\frac{\partial \mathbf{E}}{\partial t} \qquad \text{A.51}$$

This expression is used to calculate eddy currents and predict skin depth in conductors.

LOSS IN DIELECTRICS; COMPLEX PERMITTIVITY

While the necessity for a review of electromagnetic properties of conductors is an obvious one for an introduction to superconductivity, the reader may wonder why a review of dielectric behavior is necessary. There are several reasons. First, films of the new copper-oxide superconductors are typically mounted on a ceramic substrate such as MgO, and the dielectric qualities of the substrate may be important in high-frequency characterization of the superconductor. Secondly, techniques for determining the microwave characteristics of superconductors often require use of a dielectric support structure. Thirdly, many high-frequency applications of superconductors will find the superconductor in intimate contact with a dielectric. While the dc leakage currents will not be a factor in most of these cases, the ac dielectric losses and the dielectric constant may play a critical role in measurements and applications of superconductors above 100 MHz.

Recall that the definition of capacitance (C) of a parallel plate capacitor with plates of equal area A separated by a distance d is (ignoring fringe effects)

$$C = \kappa \epsilon_o \frac{A}{d} \tag{A.52}$$

where κ is the relative dielectric constant, with typical values ranging from 2 to 10 for many dielectrics, and ϵ_o is the permittivity of free space (8.85×10^{-12} F/m). Another symbol that is commonly used to represent the relative dielectric constant is ϵ_r. While fairly constant at low frequencies, the dielectric "constant" is a function of frequency, and many materials exhibit several maxima and minima at levels above a few hundred MHz.

The loss in a dielectric is typically due to a nonzero conductivity σ and a polarization damping effect. The expression for relative complex dielectric permittivity ϵ is:

$$\epsilon_r = \epsilon' - j\,\epsilon'' = \epsilon'\left(1 - j\frac{\epsilon''}{\epsilon'}\right) = \kappa\left(1 - j\frac{\epsilon''}{\epsilon'}\right) \tag{A.53}$$

where ϵ'' represents the polarization damping losses. It is a common practice to add a conductive loss factor $(-j\,\sigma/\omega)$ to this expression and then combine all losses into a single term:

$$\epsilon_r = \epsilon' - j\,\epsilon'' - j\left(\frac{\sigma}{\omega}\right) = \kappa(1 - j\,\tan\delta_t) \tag{A.54}$$

where κ (also ϵ') is the relative dielectric constant and σ is the electrical conductivity. This expression could also be written in terms of the dielectric constant ϵ:

$$\epsilon = \epsilon_r\,\epsilon_o = \kappa\epsilon_o(1 - j\,\tan\delta_t) \tag{A.55}$$

In an ideal (lossless) dielectric, displacement current will lead applied voltage by precisely 90°. The presence of either conduction or dielectric loss alters this phase to (90°$-\ \delta_t$). If there are no dc conduction losses, the angle is defined as

$$\delta_t = \tan^{-1}\left(\frac{\epsilon''}{\epsilon'}\right) \tag{A.56}$$

The tan (δ_t) term, usually referred to as the "loss tangent" or "dissipation factor," defines a combination of dielectric and resistive loss where a macroscopic representation is sufficient. If losses were zero (as in an ideal vacuum dielectric), the phase angle δ_t would also be zero, and ϵ'' would be zero.

The power factor (*PF*) is defined as sin (δ_t). For small values of δ_t, tan (δ_t) is approximately equal to sin (δ_t), and sin (δ_t) is nearly equal to δ_t. This approximation is useful for most practical dielectrics.

For small power factors, the quality factor (Q) of the capacitor is $1/PF$. The equivalent circuits for determining loss in a dielectric involve connecting either a series (R_s) or parallel (R_p) resistor to the equivalent capacitor. When the power factor is low, the value of either resistance is determined by capacitive reactance X_c and power factor *PF*.

$$F \text{ or } R_s \gg X_c,\ R_s = PF(X_c) = \frac{PF}{\omega C} \propto \frac{PF}{\kappa} \qquad \text{A.57}$$

$$F \text{ or } R_p \gg X_c,\ R_p = \frac{X_c}{PF} = (PF\,\omega C)^{-1} \propto \frac{1}{\kappa PF} \qquad \text{A.58}$$

A small value for *PF* will produce either a small R_s or a large R_p, both of which are desirable for reducing losses. Some typical values of *PF* and relative dielectric constant are tabulated in Table A.2 for several materials.

In many applications, it is important to know the loss per unit volume of a dielectric in an applied field. One must first determine the actual field in the dielectric, which will generally be different from the external field. The

TABLE A.2 Dielectric constant and power factor for several insulating materials.

Material	Dielectric Constant (κ)	Power Factor (*PF*)
Corning 707 glass (R&W)	4.0	12×10^{-4} (at 10^8 Hz)
Glass (electrical) (T)	4.5–7.0	0.002–0.016
Steatite materials (T)	6.1	0.002–0.004
Fuzed quartz (R&W)	3.78	1×10^{-4} (at 10^8 Hz)
Fuzed quartz (Fink)	3.8	1×10^{-5} (at 1 MHz)
Titania (R&W)	100	2.5×10^{-4} (at 10^8 Hz)
Rutile (titanium dioxide) (T)	90–1170	6×10^{-4}
Ceramic Alsimag 393 (R&W)	4.95	10×10^{-4} (at 10^8 Hz)
Polystyrene (R&W)	2.55	1×10^{-4} (at 10^8 Hz)
Polystyrene (T)	2.4–2.9	2×10^{-4}
Mica (electrical) (T)	5–9	0.0001–0.0007
GE Clear silica glass	3.81	*LF/3.8 $= 1 \times 10^{-4}$ (at 10^6 Hz)

References for data:
(R&W): (Rano, et al., 1965)
(T): (Terman, et al., 1955)
(Fink): (Fink & McKenzie, 1975)

general case of unit volume dissipation (P_v) for a field strength E_d in the dielectric is

$$P_v = E_d^2 \omega \kappa \epsilon_0 PF \qquad \text{A.59}$$

This expression shows that for a given field in the dielectric, dissipation increases with frequency and is proportional to κPF, commonly referred to as the *loss factor*. Clearly, a minimum value of dielectric constant (κ), as well as a low value of PF is desirable to minimize dielectric losses. Some references list power factors, a few list the loss factor.

COMPLEX MAGNETIC SUSCEPTIBILITY

It has already been mentioned that magnetic susceptibility, like impedance, has a complex nature when time-varying fields are considered, that is,

$$\chi = \chi' - j\chi'' \qquad \text{A.60}$$

Other considerations should be reviewed in light of the application of superconductors in nonstatic conditions. A few of these will be reviewed in following sections.

SURFACE RESISTANCE (R_S) AND SKIN DEPTH (δ)

The concept of surface resistance is crucial to understanding the behavior of conductors (including superconductors) at high frequencies (see Figure A.3).

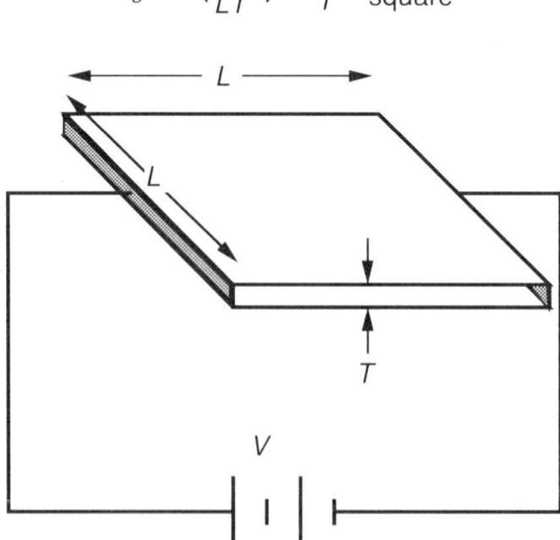

FIGURE A.3 Surface resistance for films with thickness T is simply a function of thickness and resistivity ρ. Surface resistance is the same for any square size.

On a plane with conductivity σ, one can consider the surface to have an impedance that is complex;

$$Z_s = R_s + jX_s \qquad \text{A.61}$$

The surface-resistance concept is used to calculate the power density in the thin surface layer:

$$P_s = \frac{R_s H_s^2}{2} \qquad \text{A.62}$$

where H_s is the surface magnetic field intensity. The magnitudes of the real and imaginary portions of the impedance are identical:

$$R_S = X_S = \frac{1}{\sigma\delta} \qquad \text{A.63}$$

A basic consideration is skin depth (δ), a measure of the exponential penetration of plane electromagnetic waves into normal conductors. (At depth δ, the electromagnetic wave is attenuated to $1/\epsilon$ times the surface value, where ϵ is the base of natural logarithms, 2.71828.)

Under "normal" conditions, the skin depth is inversely proportional to the square root of frequency ($\omega = 2\pi f$), electrical conductivity (σ), or resistivity ($\rho = 1/\sigma$) and magnetic permeability (μ) of the medium penetrated by the electromagnetic wave, that is,

$$\delta = \sqrt{\frac{2}{\omega\mu\sigma}} = \sqrt{\frac{1}{\pi f\mu\sigma}} = \sqrt{\frac{\rho}{\pi f\mu}} \qquad \text{A.64}$$

Recall that, for most materials, $\mu \sim \mu_o$; ferromagnetic metals or superconductors are notable exceptions. At sufficiently high frequencies and low temperatures, skin depth (anomalous) becomes proportional to $\omega^{-1/3}$ instead of $\omega^{-1/2}$ in the more conventional case. In particular, when the electron mean free path ι becomes comparable to skin depth (Gregory, et al., 1973),

$$\delta_{\text{anom}} \propto \left(\frac{\iota}{\sigma\mu\omega}\right)^{1/3} \qquad \text{A.65}$$

Since the surface resistivity (Ω/square) of a conductor is related to skin depth and conductivity by the expression $R_s = (\sigma\delta)^{-1}$, R_s has the following relationship to frequency depending on the electron mean free path ι:

for ι small compared to skin depth:

$$R_s \propto \omega^{1/2} \qquad \text{A.66}$$

for ι comparable to skin depth:

$$R_s \propto \omega^{1/3} \qquad \text{A.67}$$

for ι large compared to skin depth:

$$R_s \propto \omega^{2/3} \qquad \text{A.68}$$

To be more specific, for $\iota \gg \delta$ (Piel, 1986):

$$R_s = \left(\frac{8}{9}\right)\left(\frac{\sqrt{3}\mu_o^2\omega^2\iota}{16\pi\sigma}\right)^{1/3} \qquad \text{A.69}$$

An additional parameter, λ, the London penetration depth, was introduced in the chapter on superconducting phenomena. This is a phenomenon associated with superconducting behavior, which must be taken into account along with δ when calculating either surface resistivity or actual penetration depth. For sufficiently low temperatures, the "superconducting" surface resistance is proportional to ω^2.

The relationship between skin depth (δ; normal or anomalous), London penetration depth (λ), and the mean free path (ι) of conduction electrons is rather complex. The reader is referred the chapter on superconducting phenomena and the original studies for details (Reuter & Sondheimer, 1948; Pippard, 1947).

RESONANT STRUCTURES

Quality Factor (Q)

For resonant structures, such as cavities, the losses indicated by the quality factor (Q) are an important consideration whether the cavity is intended for use in a microwave amplifier circuit or as a particle accelerator component. The most basic definition of Q is in terms of the stored energy U, radial frequency ω, and the losses from joule heating P_w:

$$Q = \frac{U\omega}{P_w} \qquad \text{A.70}$$

but Q can also be expressed in terms of a geometry factor or geometry constant (G) derived from the cavity and the surface resistance R_s:

$$Q = \frac{G}{R_s} \qquad \text{A.71}$$

The geometry factor is independent of frequency and has values of a few hundred Ω depending upon cavity configuration. In practice, the Q of a cavity is measured by dividing the resonant frequency by the half-power (3 db) bandwidth:

$$Q = \frac{f_r}{\Delta f_{-3db}} \qquad \text{A.72}$$

In the case of superconducting cavities, where Q may exceed 10^9, it may prove quite difficult to measure bandwidth. In this instance, one may use the "decrement" method, that is, determine Q by measuring the fill or decay time of the cavity. This technique is described in the chapter on measurements.

If the cavity under consideration is for a particle accelerator application, which must maintain an accelerating electric field of at least E_a in an unloaded state, (i.e., no loading from the particle beam) it is useful to define the shunt impedance per unit length (R_{sh}) in the following manner:

$$R_{sh} = \frac{E_a^2}{P} \qquad \text{A.73}$$

where P is the power per unit length. R_{sh} is proportional to Q.

When operating cavities with high Q, it is critical to avoid introduction other materials (such as lossy dielectrics) into regions of high field since the overall Q may be degraded. In general, coupling (for measurement) to such cavities should be "loose," that is, undercoupled if the Q of such systems is to be determined accurately without an accurate determination of the coupling coefficient. If the coefficient of coupling is known with precision, probes may be critically coupled or overcoupled.

Modes of Resonance

A coaxial transmission line typically supports TEM waves, where both the electric and magnetic components are orthogonal to the direction of propagation. TE or TM waves can propagate in a hollow waveguide but TEM waves cannot; either the electric or magnetic field must have a component in the direction of propagation.

Within a closed cavity, a variety of transverse-electric (TE) and transverse-magnetic (TM) modes can exist. The term "transverse" refers to the direction of wave propagation, as the direction of transmission of energy in waveguides or coaxial transmission lines. The different modes of operation generally (but not always) occur at widely separated frequencies These modes simply represent different field configurations that satisfy the boundary conditions for the particular cavity configuration, and may be "launched" by exciting the cavity in a specific manner. Occasionally, two different modes will occur quite close to the same frequency. If this is troublesome, "mode separators" may be introduced into the cavity. Separators are modifications in the configuration of the cavity that shift the frequency of one mode (usually the unwanted mode) more

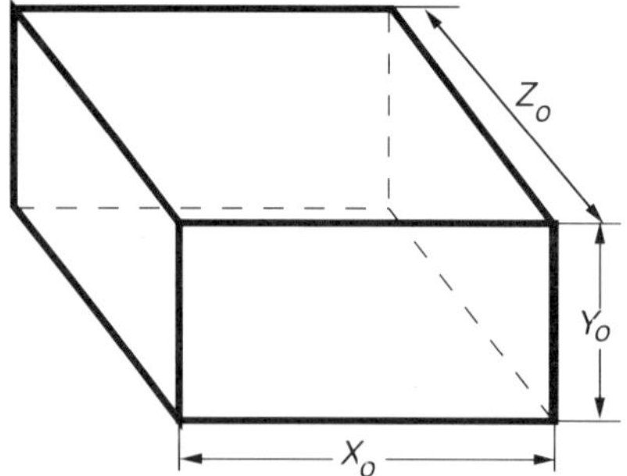

FIGURE A.4 A closed, conducting cavity has a dominant resonant mode and, at least in theory, an infinite number of higher-frequency modes. The dimensions indicated on this closed rectangular cavity correspond to those used in the text to illustrate the relationship between the resonant mode frequencies and the mode numbers m, n, and p. These mode numbers represent the number of half-wave field variations along the x, y, and z directions, respectively.

than another (usually the fundamental). In any case, the object is to increase the frequency difference between the fundamental and the unwanted modes.

The different modes are referred to by identification of the configuration of the electric and magnetic fields (TE, TM) and a set of integers which represent mode numbers (see Figure A.4). These integers, m, n, and p give the number of half-wave variations of the field in specified directions. The directions corresponding to the m, n, and p integers are usually given as x, y, and z. The general form for expressing a mode in a closed cavity is then either TE_{mnp} or TM_{mnp}. If the cavity is rectangular, filled with a material with dielectric constant ϵ and magnetic permittivity μ, and has dimensions along the x, y, and z directions of x_o, y_o, and z_o, respectively, the resonant frequencies may be calculated by

$$f_{\text{res}} = \frac{1}{\sqrt{4\mu\epsilon}} \sqrt{\left(\frac{m}{x_o}\right)^2 + \left(\frac{n}{y_o}\right)^2 + \left(\frac{p}{z_o}\right)^2} \qquad \text{A.74}$$

The expression for simple cylindrical cavity resonant modes is also quite simple, and requires reference to a table of Bessel functions (Seeger, 1986). More complex cavity shapes require computer solutions to find the various resonant modes.

REFERENCES

Blatt, J. M., (1964). *Theory of Superconductivity*, Academic, New York.

Fink, D. G., and A. A. McKenzie, (1975). *Electronic Engineers' Handbook*, McGraw-Hill, pp. 6–36 to 6–37.

Gregory, W. D., W. N. Mathews, Jr., and E. A. Edelsack, (1973). *The Science and Technology of Superconductivity*, 2 Vols., Plenum, New York.

Jackson, John David, (1963). *Classical Electrodynamics*, 1st Ed. John Wiley & Sons, New York.

London, F., (1950). *Superfluids, Macroscopic Theory of Superconductivity*, Vol. 1, John Wiley & Sons, New York.

Piel, H., (1986). "Fundamental features of superconducting cavities for high energy accelerators," Proc. CERN Accelerator School, Advanced Accelerator Physics, Oxford, England, September 16–27, Wuppertal publication #WUB-86-14, November.

Pippard, A. B., (1947). "The surface impedance of superconductors and normal metals at high frequencies," *Proc. R. Soc. A*, 191, pp. 370–415.

Ramo, S., Whinnery, J. R., and T. Van Duzer, (1965). *Fields and Waves in Communications Electronics*, John Wiley & Sons, pp. 330–334.

Reuter, G. E. H., and E. H. Sondheimer, (1948). "The theory of the anomalous skin effect in metals," *Proc. R. Soc. A*, 195, pp. 336–364.

Seeger, J. A., (1986). *Microwave Theory, Components and Devices*, Prentice-Hall, Englewood Cliffs, New Jersey.

Terman, F. E., R. A. Helliwell, J. M. Petit, D. A. Watkins, and W. R. Rambo, (1955). *Electronic and Radio Engineering*, 4th Ed., McGraw-Hill, pp. 25–26.

SUGGESTED FOR FURTHER READING

Bolz, R. E., and G. L. Tuve, *CRC Handbook of Tables for Applied Engineering Science*, 2nd Ed., CRC, Boca Raton, FL, 1987, pp. 244 and 262.

Corson, D. R., and P. Lorrain, *Introduction of Electromagnetic Fields and Waves*, W. H. Freeman, San Francisco, 1962.

Krauss, J. D., and K. R. Carver, *Electromagnetics*, 2nd Ed., McGraw-Hill, New York, 1973.

Plonsey, R., and R. E. Collin, *Principles and Applications of Electromagnetic Fields*, McGraw-Hill, New York, 1961.

Solymar, L., and D. Walsh, *Lectures on the Electrical Properties of Materials*, 4th Ed., Oxford Science Publications, Oxford, New York, 1988.

APPENDIX B

UNITS AND CONVERSIONS

MAGNETIC FIELDS AND UNITS: A BRIEF REVIEW

Most U.S. engineers are trained in SI units, and some will be unfamiliar with the cgs or Gaussian units used by the scientists who generate most of the literature on superconductivity. In SI or rationalized mks units,

$$\mathbf{B} = \mu\mathbf{H} = \mu_o(\mathbf{H} + \mathbf{M}) \qquad \text{B.1}$$

where **B** is the magnetic induction, μ is the permeability of the material, μ_o is the permeability of free space, **H** is the applied magnetic field strength, and **M** is the magnetization of the material.

Further, in SI units

$$\mu = \mu_r \mu_o \qquad \text{B.2}$$

where μ_r is the (dimensionless) relative permeability, so that

$$\frac{\mu}{\mu_o} = \mu_r = \left(1 + \frac{\mathbf{M}}{\mathbf{H}}\right) \text{ or } \frac{\mathbf{M}}{\mathbf{H}} = \mu_r - 1 \qquad \text{B.3}$$

This dimensionless ratio, **M**/**H**, is often referred to as the magnetic susceptibility, and replaced by the symbol χ. In time-varying magnetic fields, χ is a complex quantity, that is,

$$\chi = \chi' + j\chi'' \tag{B.4}$$

where $j = (-1)^{1/2}$.

Virtually all materials show some level of magnetic properties, but these effects are generally relatively weak except for ferromagnetic materials and superconductors. Materials of interest may be categorized as ferromagnetic (iron, cobalt, nickel), paramagnetic (aluminum, palladium), or diamagnetic (silver, lead, copper). A ferromagnetic material exhibits relatively strong magnetization (**M**) in the same direction as the applied **H** field, and thus has a relative permeability (μ_r) that is much greater than 1.

Paramagnetic materials also develop a magnetization in the same direction as the applied field, but the effect is weak; μ_r is slightly greater than 1.

Ordinary diamagnetic effects are also weak, with μ_r slightly less than 1. Clearly, for diamagnetic materials, where μ_r is less than 1, **M** is negative. If a material is perfectly diamagnetic, that is, $\mu_r = 0$, then $\mathbf{M} = -\mathbf{H}$, and $\chi = -1$. Representative values of μ_r for a few materials are given in Table A.1.

GAUSSIAN AND CGS CONVENTIONS

Using Gaussian or cgs emu units,

$$\mathbf{B} = \mu\mathbf{H} = \mathbf{H} + 4\pi\mathbf{M} \tag{B.5}$$

and

$$\mu = 1 + \frac{4\pi\mathbf{M}}{\mathbf{H}} \tag{B.6}$$

In these units, μ is dimensionless; no μ_o or μ_r is defined. One common feature with SI units is that $\mu = 0$ for perfect diamagnetism. This is equivalent to stating that $\mathbf{M} = -\mathbf{H}/4\pi$ for total expulsion of flux. Using these units, one plots $-4\pi\mathbf{M}$ versus **H** to illustrate the diamagnetic properties of superconductors.

If susceptibility is defined as $\chi = \mathbf{M}/\mathbf{H}$, then $\chi = -1/4\pi$ for a total Meissner effect. For this convention, the susceptibility axis may be given in terms of $4\pi\chi$ which has a value of -1 for perfect diamagnetism. Other "units" used commonly are emu/cm^3, emu/g, and emu/mol.

CONVERSION OF EXPRESSIONS FROM GAUSSIAN TO SI UNITS

The list of operations that follow are useful for "translating" a Gaussian expression to the equivalent SI expression (Jackson, 1963). (Obviously, the equivalent inverse operation may be carried out to translate an SI expression to Gaussian.)

Summary of Conversion Factors

To convert Gaussian expressions to SI:

Replace velocity of light c:

$$c \rightarrow \frac{1}{\sqrt{\mu_o \epsilon_o}} \qquad \text{B.7}$$

Replace electric field $\mathbf{E}$ (make equivalent substitution for voltage or potential):

$$\mathbf{E} \rightarrow \mathbf{E}\sqrt{4\pi\epsilon_o} \qquad \text{B.8}$$

Replace displacement vector $\mathbf{D}$:

$$\mathbf{D} \rightarrow \mathbf{D}\sqrt{\frac{4\pi}{\epsilon_o}} \qquad \text{B.9}$$

Replace charge q (make equivalent substitution for charge density, current, current density):

$$q \rightarrow \frac{q}{\sqrt{4\pi\epsilon_o}} \qquad \text{B.10}$$

Replace magnetic induction $\mathbf{B}$:

$$\mathbf{B} \rightarrow \mathbf{B}\sqrt{\frac{4\pi}{\mu_o}} \qquad \text{B.11}$$

Replace magnetic field $\mathbf{H}$:

$$\mathbf{H} \rightarrow \mathbf{H}\sqrt{4\pi\mu_o} \qquad \text{B.12}$$

Replace magnetization $\mathbf{M}$:

$$\mathbf{M} \rightarrow \mathbf{M}\sqrt{\frac{\mu_o}{4\pi}} \qquad \text{B.13}$$

Replace conductivity σ:

$$\sigma \rightarrow \frac{\sigma}{4\pi\epsilon_o} \qquad \text{B.14}$$

Replace dielectric constant ϵ:

$$\epsilon \rightarrow \frac{\epsilon}{\epsilon_o} \qquad \text{B.15}$$

Replace permeability μ:

$$\mu \rightarrow \frac{\mu}{\mu_o} \qquad \text{B.16}$$

Replace resistance R (make equivalent substitute for impedance in general):

$$R \rightarrow R4\pi\epsilon_o \qquad \text{B.17}$$

Replace inductance L:

$$L \rightarrow L4\pi\epsilon_o \qquad \text{B.18}$$

Replace capacitance C:

$$C \rightarrow \frac{C}{4\pi\epsilon_o} \qquad \text{B.19}$$

Example; Skin Depth

For example, the Gaussian expression (some authors omit μ since, in Gaussian units, it is close to 1 for most metals) for skin depth δ_G is

$$\delta_G = \frac{c}{\sqrt{2\pi\mu\omega\sigma}} \qquad \text{B.20}$$

The replacements to be made to convert to the equivalent SI expression (δ_{SI}) are

$$c \rightarrow \frac{1}{\sqrt{\mu_o\epsilon_o}} \qquad \text{B.21}$$

$$\sigma \rightarrow \frac{\sigma}{4\pi\epsilon_o} \qquad \text{B.22}$$

$$\mu \rightarrow \frac{\mu}{\mu_o} \qquad \text{B.23}$$

TABLE B.1

Quantity	Gaussian or cgs emu	Conversion Factor*	Gaussian
B (induction)	gauss	10^{-4}	Wb/m^2, T
H (field strength)	oersted (Oe)	$10^3/4\pi$	A/m
M (magnetization)	emu**/cm^3	10^3	A/m
χ (susceptibility)	dimensionless	4π	dimensionless
μ (permeability)	dimensionless	$4\pi \times 10^{-7}$	Wb/A-m
W (energy density)	erg/cm^3	10^{-1}	J/m^3

*(multiply cgs units by the conversion factor to obtain SI units)
**(While emu is a designation, not a unit, it is common practice in physics papers on superconductivity to substitute emu/cm^3 for gauss)

which gives the result:

$$\delta_{SI} = \frac{1/\sqrt{\mu_o \epsilon_o}}{\sqrt{2\pi(\mu/\mu_o)\omega(\sigma/4\pi\epsilon_o)}} = \sqrt{\frac{2}{\omega\mu\sigma}} \qquad \text{B.24}$$

MAGNETIC PARAMETERS AND UNITS

Standard International (SI) units are generally used in this reference; other unit systems (usually cgs emu) will only be used where it is necessary to avoid confusion with standard practices in other commonly used sources. Another exception, commonly made by those who use SI units, will be occasional reference to gauss (instead of tesla or weber/m^2) for magnetic flux density.

Of particular interest are the units for measuring magnetic field (**H**), induction (**B**), and magnetization (**M**).

SI dimensions for induction **B** are webers/square meter (Wb/m^2) or the equivalent, tesla (T). **B**, in cgs units, may also be measured in gauss, where

$$1\ \text{T} = 1\ \text{Wb/m}^2 = 10^4\ \text{gauss}$$

A conversion table for some of the more useful magnetic units appears in Table B.1.

REFERENCES

Jackson, John David, (1963). *Classical Electrodynamics*, John Wiley & Sons, New York.

SUGGESTED FURTHER READING

Corson, D. R., and P. Lorrain, *Introduction of Electromagnetic Fields and Waves*, W. H. Freeman, San Francisco, 1962.

Ginzton, E. L., *Microwave Measurements*, McGraw-Hill, New York, 1967.

Kip, Arthur F., *Fundamentals of Electricity and Magnetism*, McGraw-Hill, New York, 1969.

Krauss, J. D., and K. R. Carver, *Electromagnetics*, 2nd Ed., McGraw-Hill, New York, 1973.

Plonsey, R., and R. E. Collin, *Principles and Applications of Electromagnetic Fields*, McGraw-Hill, New York, 1961.

Seeger, J. A., *Microwave Theory, Components and Devices*, Prentice-Hall, Englewood Cliffs, New Jersey, 1986.

APPENDIX C

GLOSSARY OF TERMS AND SYMBOLS

Certain symbols will have different meanings depending on the context in which they are used. For example, "δ" will specify skin depth for discussions of high-frequency behavior of conductors, but it is a dimensionless number for discussions of oxygen stoichiometry related to $YBa_2Cu_3O_{7-\delta}$ superconductors.

In another instance, the symbol "Δ" is used either for the Laplacian operator or to represent the magnitude of the superconductor energy gap. While this situation is less than ideal, it appears to be a better choice than selecting symbols different than those in common use. Such terms will be clearly defined in the text, with special care taken where the context of use does not make the identification of the symbol obvious.

GENERAL CONSTANTS

c, velocity of light in vacuum, 2.998×10^8 m/s.

h, Planck's constant, the ratio of a photon's energy to its frequency, 6.63×10^{-34} J-s or 4.14×10^{-15} eV-s.

k_B, Boltzmann's constant, 1.38×10^{-23} J/K or 8.625×10^{-5} eV/K.

N_A, Avogadro's number, 6.02×10^{23} atoms/g-atom.

e, electron charge, 1.602×10^{-19} C.

m_e, electron rest mass, 9.11×10^{-31} kg (5.49×10^{-4} amu).

m_p, proton rest mass, 1.6725×10^{-27} kg (1.00728 amu).

ELECTROMAGNETISM (STANDARD INTERNATIONAL UNITS)

$\mathbf{H}$, Magnetic field strength, A/m.

ϕ, Magnetic flux, Wb.

$\mathbf{B}$, ($\mathbf{B} = \mu\mathbf{H}$) Magnetic flux density (induction), T or Wb/m^2. (In Gaussian or cgs emu, it is common to substitute emu/cm^3 for gauss, even though "emu" is a designation of a unit system rather than a unit itself.)

$\mathbf{A}$, Vector potential, Wb/m, defined by $\mathbf{B} = \nabla \times \mathbf{A}$.

$\mathbf{M}$, Magnetization of material (A/m), $\mathbf{B} = \mu\mathbf{H} = \mu_o(\mathbf{H}+\mathbf{M})$ (Note: in Gaussian units, $\mathbf{B}=\mathbf{H} + 4\pi\mathbf{M}$).

χ, Magnetic susceptibility, $\chi = \mathbf{M}/\mathbf{H}$, since $\mu_r = 1 + \mathbf{M}/\mathbf{H}$.

μ, Magnetic permeability of material, H/m.

μ_o, Magnetic permeability of vacuum $= 4\pi 10^{-7}$ H/m.

μ_r, Relative magnetic permeability, $\mu_r = \mu/\mu_o$, dimensionless.

$\mathbf{E}$, electric field strength, V/m.

$\mathbf{D}$, Electric displacement ($\mathbf{D} = \epsilon\mathbf{E}$ in dielectric media).

$\mathbf{P}$, Poynting vector, electromagnetic power flow per unit area, W/cm^2, defined for time-varying $\mathbf{E}$ and $\mathbf{H}$ as $\mathbf{P} = \mathbf{E} \times \mathbf{H}$.

ϵ, Permittivity of dielectric media.

ϵ_o, Permittivity of vacuum $= 8.854 \times 10^{-12}$ F/m.

I, Electric current, A.

J, Electric current density, A/m^2 (or A/cm^2).

R, Resistance, Ω.

R_s, Surface resistance, Ω/square.

ρ, Resistivity, Ω-cm.

α, Temperature coefficient of resistivity, defined by $\rho = \rho_o\{1 + \alpha(T - T_o)\}$, where ρ is resistivity at temperature T_o.

σ, Electrical conductivity, mhos (Ω^{-1}).

C, Capacitance, F.

L, Inductance, H.

W, Energy, J.

P, Power, W.

f, Frequency, Hz (engineering notation, through microwave range).

ν, Frequency, Hz (quantum notation, applied to photon frequency).

ω, Radial frequency, ($\omega = 2\pi f$) radians/s.

q, Charge, C.

δ, Skin depth, proportional to inverse square root of frequency $\delta = (2/\omega\mu_o\sigma)^{1/2}$ or $(2\rho/\omega\mu_o)^{1/2}$.

δ_{anom}, Anomalous skin depth, proportional to inverse cube root of frequency.

ι, Mean free path of conduction electron.

SUPERCONDUCTIVITY-RELATED TERMS DEFINED

λ or λ_L, (London) penetration depth of H fields at surface of superconductor.

λ_o or $\lambda(0)$, Penetration depth at a temperature of 0 K.

ξ_o, Brian Pippard's coherence length for superconducting electron pairs, applies primarily to pure metals.

$\xi(T)$ or ξ_{GL}, Ginzberg-Landau coherence or "pair-correlation" length, very similar to ξ_o.

κ, Ginzberg-Landau kappa parameter, ratio of penetration depth λ to correlation length $\xi(T)$ that is, $\kappa = \lambda/\xi(T)$.

T_c, Critical temperature of superconductor (transition temperature for zero magnetic field).

T_{co}, T_c-"onset," Temperature at first detectable point of departure from normal to superconducting behavior.

T_z, T_{cf}, T_c-"final," Highest temperature where dc resistivity is zero.

Λ, Defined as $\Lambda = m/n_s e^2 = 4\pi\lambda^2$, where m is the effective mass of the superconducting charge carrier (electron), n_s is the supercurrent electron density, e is the electron charge, and λ is a characteristic penetration depth previously defined.

ϕ_o, Magnetic flux quanta (also referred to as the fluxon or fluxoid, although some authors may use the term fluxoid to refer to multiple integral values of ϕ_o):

$$\phi_o = h/2e = 2.07 \times 10^{-15} \text{ Wb (SI units)}$$

$$\phi_o = hc/2e = 2.07 \times 10^{-7} \text{ gauss-cm}^2 \text{ (cgs units)}$$

Θ_D, Debye temperature.

ω_D, Debye vibrational frequency ($\Theta_D = h\omega_D/2\pi\kappa$).

γ, Density of electron states near the Fermi surface (the Sommerfeld gamma).

H_c, Critical thermodynamic magnetic field, that is, the level that destroys superconducting behavior at a specific temperature. (Type I materials exhibit one critical field, Type II materials exhibit three, H_{c1}, H_{c2}, and H_{c3}. See text for details.)

$H_c(0)$, Critical magnetic field for a temperature of absolute zero (0 K).

J_c, Critical current that, when exceeded, destroys superconducting behavior. Silsbee's hypothesis (for pure elemental superconductors) states that the critical current J_c is that current density just sufficient to produce the critical field at the surface of a superconductor.

j, The imaginary number $(-1)^{1/2}$; (nonengineering texts usually use i).

Type I superconductors are usually very pure metallic elements; they exhibit a very abrupt shift from $\mu \sim 1$ to complete diamagnetism, and generally have relatively low values of H_c and J_c. Type I materials exhibit an "intermediate" state where superconducting and normal behavior exist in different regions, even though the external field is less than H_c. This behavior is a function of sample shape and strength of the external magnetic field. (Also called "First Kind," "Pippard," or "soft.")

Type II superconductors are generally alloys or compounds (except for niobium and vanadium) and are characterized by a gradual change from perfect diamagnetism to normal behavior as the external H field is increased toward the critical level, the interval known as the "mixed" state. The materials generally have a much higher value of critical magnetic field (H_{c2}) and J_c than Type I materials. (Also called "Second Kind," "London," or "hard.")

A15 is a crystal structure associated with a family of Type II superconductors including Nb_3Sn, Nb_3Os, Nb_3OIr, Nb_3Pt, Ta_3Sn, V_3Si, V_3Ge, Mo_3Si, Mo_3Ge, and several other intermetallic compounds. Also known as the β-W (beta-Tungsten) family.

High-T_c superconductors generally refers to materials with critical transition temperatures (T_c) above $\sim$ 23 K. Examples include the copper oxides typified by La-Ba-Cu-O ($T_c \sim 30$ K) and by $YBa_2Cu_3O_{7-\delta}$ ($T_c \sim 90$ K), where δ ranges from approximately 0 to 0.6. These compounds are Type II materials, and are also commonly referred to as "oxygen-deficient perovskites," "superconducting ceramics," and "cuprates," and the latter as "1-2-3."

"1-2-3," Refers to $YBa_2Cu_3O_{7-\delta}$ superconductor, where selected rare earths (Gd, Eu, etc.) may be substituted for yttrium (Y).

Ceramic, A material (usually an insulator) formed by sintering materials in powder form.

Sintering, Heating without melting.

MATHEMATICAL OPERATIONS; VECTOR NOTATION

Gradient of a Scalar Field. For a scalar function $f(x,y,z)$, the "directional derivative" or gradient of the scalar field is defined as

$$\text{grad}(f) = \nabla f = \frac{\partial f}{\partial x}\mathbf{i} + \frac{\partial f}{\partial y}\mathbf{j} + \frac{\partial f}{\partial z}\mathbf{k} \qquad \text{C.1}$$

where **i**, **j**, and **k** are unit vectors along the cartesian coordinates x, y, and z respectively.

In this regard, the "del" operator is defined as

$$\nabla = \frac{\partial}{\partial x}\mathbf{i} + \frac{\partial}{\partial y}\mathbf{j} + \frac{\partial}{\partial z}\mathbf{k} \qquad \text{C.2}$$

Divergence of a Vector Field. For a vector function $\mathbf{v}(x,y,z)$ with components v_1, v_2, and v_3,

$$\text{div}(\mathbf{v}) = \nabla \cdot \mathbf{v} = \frac{\partial v_1}{\partial x} + \frac{\partial v_2}{\partial y} + \frac{\partial v_3}{\partial z} \qquad \text{C.3}$$

Curl of a Vector Field. For a vector $\mathbf{v}(x,y,z) = v_1\mathbf{i} + v_2\mathbf{j} + v_3\mathbf{k}$, the curl is defined as

$$\text{curl}(\mathbf{v}) = \nabla \times \mathbf{v} = \begin{bmatrix} \mathbf{i} & \mathbf{j} & \mathbf{k} \\ \dfrac{\partial}{\partial x} & \dfrac{\partial}{\partial y} & \dfrac{\partial}{\partial z} \\ v_1 & v_2 & v_3 \end{bmatrix} \qquad \text{C.4}$$

$$= \left(\frac{\partial v_3}{\partial y} - \frac{\partial v_2}{\partial z}\right)\mathbf{i} + \left(\frac{\partial v_1}{\partial z} - \frac{\partial v_3}{\partial x}\right)\mathbf{j} + \left(\frac{\partial v_2}{\partial x} - \frac{\partial v_1}{\partial y}\right)\mathbf{k}$$

where **i**, **j**, and **k** are unit vectors along the x, y, and z axes respectively.

The Laplacian Operator. Δ is defined as

$$\nabla^2 = \Delta = \frac{\partial^2}{\partial x^2} + \frac{\partial^2}{\partial y^2} + \frac{\partial^2}{\partial z^2} \qquad \text{C.5}$$

PREFIXES FOR UNITS IN THE SI SYSTEM

Prefix	*Symbol*	*Power*	*Example*
tera	T	10^{12}	THz
giga	G	10^{9}	GHz
mega	M	10^{6}	MHz
kilo	k	10^{3}	kg
hecto	h	10^{2}	hm
deca	da	10^{1}	dam
deci	d	10^{-1}	dm
centi	c	10^{-2}	cm
milli	m	10^{-3}	mA
micro	μ	10^{-6}	μH
nano	n	10^{-9}	ns
pico	p	10^{-12}	pF
femto	f	10^{-15}	fA
atto	a	10^{-18}	aA

APPENDIX D

SUPERCONDUCTOR-RELATED PRODUCTS, SERVICES, PUBLICATIONS, AND ASSOCIATIONS

While considerable care has been taken to provide accurate information, a few organizations will (alas!) cease to exist and others will acquire new addresses and telephone numbers. For example, some of the (617) area codes in Massachusetts will have been changed to (508) by the time this is published. Product lines also change on a regular basis. Readers are invited to supply new (and corrected) information regarding these and additional sources for use in future editions of this reference. This information is provided as a service to the reader; no endorsement is intended either by the author or the publisher. No references have been purposely omitted; it is hoped that this listing will be more complete in future editions.

The information is separated into four basic groups and is presented in the following order:

Providers of goods and services
General publications
Technical, scientific, and trade journals
Societies and associations

PROVIDERS OF GOODS AND SERVICES

Advanced Composite Materials Corporation
1525 S. Buncombe Road
Greer, SC 29651-9208
Ph: (803)-877-0123
(Substrates for superconductors, MgO single crystal, etc.)

AESAR (See Johnson Matthey/AESAR)

Air Products
Industrial Gas Division
Box 538
Allentown, PA 18195
Ph: (215)-481-6747

Aldrich Chemical Company, Inc.
P. O. Box 355
Milwaukee, WI 53201
Ph: for orders . . . (800)-558-9160
(Specialty materials, Y-Ba-Cu-O, etc.)

Alfa Products—(Morton Thiokol, Inc.)
152 Andover Street
Danvers, MA 01923
Ph: (508)-777-1970 or (800)-343-0660
(Research chemicals & materials: barium, bismuth, calcium, yttrium, thallium, etc.)

American Magnetics, Inc.
P. O. Box 2509
112 Flint Road
Oak Ridge, TN 37831-2509
Ph: (615)-482-1056
(Magnets, superconducting magnets, cryogenic systems, superconducting dipstick, etc.)

Anatech Ltd.
5510 Vine Street
Alexandria, VA 22310
Ph: (703)-971-9200
(Monoenergetic filamentless oxygen ion source . . .)

Anderson Physics Laboratories, Inc.
406-T N. Busey Avenue
Urbana, Illinois 61801
Ph: (217)-367-1340 or (800)-635-2578
(Materials for superconductivity research; ultra-high purity inorganic chemicals . . .)

Andonian Cryogenics, Inc.
26 Farwell Street
Newtonville, MA 02160
Ph: (617)-969-8010
(Research cryostats, modular cryogenic instrumentation . . .)

APD Cryogenics, Inc.
1919 Vultee Street
Allentown, PA 18103
Ph: (215)-791-6700, (215)-791-6750, California:(408)-988-5525, in Berkshire, England: (07356) 79373
(Superconductor Characterization Cryostat, Heli-Tran® liquid transfer system, closed-cycle refrigerators, vacuum shrouds)

Applied Superconductivity Test Laboratory
General Dynamics Space Systems Division
P. O. Box 85377
San Diego, CA 92138
Ph: (619)-542-4747 or 542-4876
(Superconducting testing services, cryomagnetic systems, etc.)

ASC: Applied SuperConetics, Inc.
11045 Sorrento Valley Court
San Diego, CA 92121
Ph: (619)-452-3400 or (800)-338-2724
(Superconducting magnets)

Atomergic Chemetals Corp.
91 Caryolyn Boulevard
Farmingdale, NY 11735
Ph: (516)-694-9000
(Substrates, crystal growth system, deposition materials, etc.)

Babcock & Wilcox
P. O. Box 785
Lynchburg, VA 24505
Ph: (804)-522-5250 or 522-5345
(Nb particle accelerator cavities . . .)

F. W. Bell
6120 Hanging Moss Road
Orlando, FL 32807
Ph: (407)-678-6900
(Hall generators for cryogenic applications, Gaussmeters)

BTi (Biomagnetic Technologies)
4174 Sorrento Valley Boulevard
P. O. Box 210079
San Diego, CA 92121
Ph: (619)-453-6300
BTi in Europe: S. H. E. GmbH, Aachen W. Germany, (0241)-15-50-37
BTi in Japan: Niki Glass Co., Ltd., Takanawa, (03)-456-4700
BTi in Australia: Sydney, (02)-85-7761
(SQUID measurement systems, cryogenic probes . . .)

Cabot
Electronic Materials & Refractory Metals Division
300 Holly Road
Boyertown, PA 19512
Ph: (800)-531-3677 or (800)-531-3676

CERAC, Inc.
Box 1178
Milwaukee, WI 53201
Ph: (414)-289-9800
(Superconductor materials . . .)

Ceramaseal (a division of CERAMX)
P. O. Box 260
New Lebanon, NY 12125
Ph: (518)-794-7800
In Europe:
Interceram Manufacturing Ltd.,
Unit 17, Whitegate Industrial Estate
Wrexham, Clwyd UK LL13 8UG
(Crogenic products; ultra-high vacuum electrical feedthrough connectors)

Cerametac, Inc.
2425 South 900 West
Salt Lake City, UT 84119
Ph: (801)-972-2455
(Superconductor materials, ceramics, powders, shapes . . .)

Ceramics Process Systems Corp.
840 Memorial Drive
Cambridge, MA 02139
Ph: (617)-354-2020
(Superconductor materials, powders, ceramic substrates, etc.)

Colorado Superconductor Inc.
P. O. Box 8223
Fort Collins, CO 80526
Ph: (303)-491-9106

Commercial Crystal Laboratories, Inc.
4406 Arnold Avenue
Naples, FL 33942
Ph: (813)-643-6959
(Superconductor substrates: strontium titanate ($SrTiO_3$), magnesium oxide (MgO), YSZ (ZrO_2-Y_2O_3), spinel ($MgO \cdot Al_2O_3$), sapphire (Al_2O_3), YAG ($Y_3Al_5O_{12}$), rutile (TlO_2))

Commonwealth Scientific Corporation (CS)
500 Pendleton St.
Alexandria, VA 22314
Ph: (703)-548-0800
(Ion beam codepositon, sputter deposition . . .)

Conductus, Inc.
2200 Geng Road, Suite 205
Palo Alto, CA 94303
Ph: (415)-354-4314
(High-T_c films and devices, superconducting electronics . . .)

Cooke Vacuum Products, Inc.
13 Merritt Street
Norwalk, CT 06854
Ph: (203)-853-9500
(Multiple-target RF-DC cosputtering systems, electron beam depositon, customized deposition sources, etc.)

Cortest Labs, Inc.
11115 Mills Road, Suite 102
Cypress, TX 77429
Ph: (713)-890-7575

Coulter Corporation
P. O. Box 2145
Hialeagh, FL 33012-0145
Ph: (800)-526-6932 or (305)-885-0131
(Instrumentation; colloid particle behavior, aggregation, surface binding reactions . . . automated submicron particle size analysis . . .)

CRICERAM
Pechiney Corp.
475 Steamboat Road
Greenwich, CT 06830
Ph: (203)-625-8815
or
CRICERAM
B. P. 16
38560 Jarrie
France
Ph: (33)-76.68.82.57.
(Materials, powders)

Cryofab, Inc.
540 N. Michigan Avenue
P. O. Box 485
Kenilworth, NJ 07033
Ph: (201)-686-3636
(Cryogenic instruments, custom fabrication, repairs . . .)

Cryogenic Consultants, Limited
Metrostore Building, 231 The Vale
London W3 7QS
UK
Ph: UK: (0174)36049, USA: (914)-986-4090,
Germany: 07022-51767, Netherlands: 02990-28908,
Japan: 0482-55-2012, Switzerland: 056-45 41 45
(Superconducting magnets, SQUID susceptometers, cryogenic systems)

Cryo Industries of America, Inc.
24 Keewaydin Drive
Salem, NH 03079
Ph: (603)-893-2060
(Cryogenic systems, temperature sensors, vacuum systems . . .)

Cryomagnetics, Inc.
P. O. Box 548
739 Emory Valley Road
Oak Ridge, TN 37831-0548
Ph: (615)-482-9551
(Custom superconducting magnets and systems)

CVC Products, Inc.
525 Lee Road
P. O. Box 1886
Rochester, NY 14603
Ph: (716)-458-2550
(High-T_c materials, high-vacuum deposition for superconductor films . . .)

EG&G (See Princeton Applied Research)
Electronics Space Products International (ESPI)
5310 Derry Avenue
Agoura Hills, CA 91301
Ph: local: (818)-991-6724, USA: (800)-638-2581, CA: (800)-848-7873 (MgO single crystals, high-purity materials, Forms: rod, wire, sheet, shot, foil, bar, pellets, powder, evap. sources, sputtering targets, compounds)

EMCORE
35 Elizabeth Avenue
Somerset, NJ 08873
Ph: (201)-271-2090
(MOCVD reactors, gas source components, material characterization)

ENERJET
Advanced Technology Products Division
(Kurt J. Lesker Co.)
1515 Worthington Avenue
Clairton, PA 15025
Ph: (412)-233-4200, USA: (800)-245-1656, PA: (800)-242-0599,
CA: (800)-848-5386, Canada: (800)-544-3940
(Magnetron sputter guns, etc.)

Fluoramics, Inc.
103 Pleasant Avenue
Upper Saddle River, NJ 07458
Ph: (201)-825-8110
(Y-Ba-Cu-O superconductor samples)

General Atomics, Inc.
(Formerly GA Technologies Inc.)
P. O. Box 85608
San Diego, CA 92138-5608
Ph: (619)-455-2816 or (800)-224-1269
(Custom superconducting magnets to 12 T & 4m, NMR magnets, high-T_c ceramics)

General Dynamics
Space Systems Division
MZ 92-8260, P. O. Box 85990
San Diego, CA 92138
Ph: (619)-542-9000 or 542-4713
(Superconducting magnet systems; See also Applied Superconductivity Test Laboratory)

George Associates
P. O. Box 960
Berkeley, CA 94701
Ph: (415)-843-3587
(Magnetic susceptibility instrumentation)

GMW Associates
P. O. Box 2578
Redwood City, CA 94064
Ph: (415)-368-4484
(Magnet and ion beam products)

Goodfellow Metals Limited
Cambridge Science Park
Milton Road
Cambridge CB4 4DJ
UK
Ph: Cambridge (0223) 69671
(Films, coatings, sputtering targets, etc.)

Gorham Advanced Materials Institute (GAMI)
P. O. Box 250
Gorham, ME 04038
Ph: (207)-892-5445
or
209 West Central Street, Suite 226
Natick, MA 01760
Ph: (617)-651-9907
(Materials processing, ceramics, market studies . . .)

W. R. Grace & Co.
Davison Chemical Division
P. O. Box 2117
Baltimore, MD 21203
Ph: (800)-638-0670
(Wide range of high-T_c materials, rare earths, specialty chemicals)

R. G. Hansen & Associates
631 Chapala Street
Santa Barbara, CA 93101
Ph: (805)-564-3388 or (805)-966-1063
(Cryogenics systems . . .)

Hi T_c Superconco, Inc.
(Subsidiary of Lambertville Ceramic Mfg. Co.)
P. O. Box 128
245 N. Main Street
Lambertville, NJ 08530
Ph: (609)-397-2010
(Superconductor materials, sputtering targets, *RF* cavities, superconductor bearings, etc.)

HITEC—Materials
Ing. Kesachtkar GmbH & Co KG
Gustav-Schönleberstr. 4,
P. O. B. 3666, D-7500 Karlsruhe 21
Federal Republic of Germany
Ph: (07 21) 59 06 64
(Superconductor powders, bulk ceramics, etc.)

Howard Associates
156 Mountain Avenue
Summit, NJ 07901
Ph: (201)-277-4312
(*Howard Superconductivity Index*)

HYPRES, Inc.
500 Executive Boulevard
Elmsford, NY 10523
Ph: (914)-592-1190
(Picosecond signal processors, 70 GHz oscilloscopes . . .)

ICE, Dept. of Chemistry
University of Wisconsin
1101 University Avenue
Madison, WI 53706
Ph: (608)-262-3033
(High-T_c demonstration kits)

Intermagnetics General Corp.
1875 Thomaston Avenue
Waterbury, CT 06704
Ph: (203)-753-5215
(Superconducting wire, multifilamentary, tapes, magnet wire . . .)

International Superconductor Corp.
271 Madison Avenue, Suite 908
New York, NY 10016
Ph: (212)-796-9416

IWECO, Inc.
Medical and Cryogenic Division
8350 Mosley
Houston, TX 77075
Ph: (713)-943-2000
(Cryogenic freezers, phase separators, specialty gases, . . .)

Janis Research Company, Inc.
2 Jewel Drive
P. O. Box 696
Wilmington, MA 01887
Ph: (508)-657-8750
(Vibrating sample magnetometers, superconducting magnets, . . .)

Johnson Matthey/AESAR
P. O. Box 1087
Seabrook, NH 03874
Ph: (800)-343-1990, in NH: 474-5511
(Superconductor research materials, bismuth, thallium, barium, etc.)

Kali-Chemie Corporation
41 W. Putnam Avenue
Greenwich, CT 06830
Ph: (203)-629-7900 or (212)-949-1651
and
Kali-Chemie AG
Postfach 220
D-3000 Hannover 1
West Germany
Ph: (511)-857-2698
(Superconducting materials, 1-2-3, bismuth & thallium compounds)

Keithly Instruments, Inc.
28775 Aurora Road
Cleveland, OH 44139
Ph: (216)-248-0400
(Instruments, measurements)

Koch Process Systems, Inc.
20 Walkup Drive
Westborough, MA 01581
Ph: (617)-366-9111
(Manufacturing services; cryogenics . . .)

Kristallhandel Kelpin (KHK)
6906 Leimen/HD
Federal Republic of Germany
Ph: 06224/725 58
(High-T_c sputter targets, single-crystal substrates)

Laboratoire de Cristallographie
University of Geneva
24, quai E. Ansermet
CH-1211 Genève
Switzerland
Ph: 0041-22-21 92 55
(PC software for simulation of X-ray & neutron-powder diffraction patterns: LAZY PULVERIX-PC)

Lake Shore Cryotronics, Inc.
64 East Walnut Street
Westerville, OH 43081
Ph: (614)-891-2243
(Cryogenic thermometry, superconductor measurements, eddy-current noncontact superconductor characterization, calibrations)

Lambda Physik GmbH
Hans-Böckler-Str. 12
D-3400 Göttingen
Federal Republic of Germany
Ph: (0551)-69 38-0
and
Lambda Physik Inc.
289 Great Road
Acton, MA 01720
Ph: (617)-263-1100
(Excimer lasers . . .)

Kurt J. Lesker Co.
1515 Worthington Avenue
Clarion, PA 15025
Ph: (412)-233-4200
(Sputter deposition . . .)

Leybold Inficon Inc.
6500 Fly Road
East Syracuse, NY 13057
Ph: (315)-437-0377 or (315)-434-1100
(Vacuum technology, vacuum process engineering, measurements, thin-film deposition control, etc.)

Linear Research Inc.
5231 Cushman Place, Suite 21
San Diego, CA 92110
Ph: (619)-299-0719
(Instruments, ac resistance measurements . . .)

MacroQuan Software
272 N. El Camino Real
Suite B-110
Encinitas, CA 92024
Ph: (619)-632-1994
(JJ Simulator: Josephson junction computer simulation software)

Materials Research Corporation (MRC)
Route 303
Orangeburg, NY 10962
Ph: (914)-359-4200
(Superconduting materials; powders, formulations . . .)

Microscience Inc.
41 Accord Park Drive
Norwell, MA 02061
Ph: (617)-871-0308
(Thin-film research sysem; IBEX-2000)

MMR Technologies, Inc.
1400 Shoreline Boulevard, #A5
Mountain View, CA 94043-1312
Ph: (415)-962-9620
(Temperature controllers, dewars, resistivity & Hall measurements, etc.)

Morton Thiokol, Inc. (See Alfa Products)

MTM (Meyer Tool & Manufacturing) Cryolab
9221 South Kilpatrick
Oak Lawn, IL 60453
Ph: (312)-425-9080
(Cryogenic, vacuum engineering & fabrication)

National Superconductor, Inc.
13968 Van Ness Avenue
Gardena, CA 90249
Ph: (213)-323-3923
(Meissner levitation kit, resistivity measurement kit, magnets . . .)

Outokumpu Copper
Outokumpu Oy, Pori Works
P. O. Box 60, SF-28101 Pori
Finland
Ph: (358)-39-826111
(Multifilament superconductors)

Oxford Instruments Limited
Eynsham, Oxford OX8 1TL
UK
Ph: (0865) 882855
and
Oxford Instruments North America Inc.
3A Alfred Circle
Bedford, MA 01730
Ph: (617)-275-4350
(Superconducting magnets, "Cryospares," electronics, ion beam systems . . .)

F. E. Penna, Co.
Box 3253
Casper, WY 82602-3253
Ph: (307)-577-5032
(Materials—high-purity scandium oxide . . .)

Perkin-Elmer Corporation
Superconductor Operations Center
761 Main Avenue
Norwalk, CT 06859-0280
Ph: (203)-834-6491
(Materials, deposition systems, instrumentation, surface physics . . .)

Physical Dynamics Incorporated (PDI)
335 Paint Branch Drive
College Park, MD 20742
Ph: (301)-454-8039
(Superconductivity testing services; microwave absorption, dc/ac magnetic susceptibility, etc.)

Princeton Applied Research (EG&G)
P. O. Box 2565
Princeton, NJ 08543-2565
Ph: (609)-452-2111
(Model 4500 vibrating sample magnetometer . . .)

Princeton Measurements Corp.
371 Wall Street
Princeton, NJ 08540
Ph: (609)-924-7885
In Europe:
Instrumat, s.a.r.l.
Avenue des Andes
Zone d'Activités de Courtaboeuf
Bolte Postale No. 86
91943 Les Ulis
France
Ph: (1) 69.28.27.34
(Alternating gradient force magnetometer . . .)

Process Products Corporation
37 Flagship Drive
N. Andover, MA 01845
Ph: (508)-689-3828
(Rapid thermal processors, chemical vapor deposition . . .)

Quadratech Advanced Materials
P. O. Box 257
Alfred Station, NY 14803
Ph: (607)-587-8013
(Superconductor materials, powders, crucibles, substrates, etc.)

Quantum Design
11578 Sorrento Valley Road, Suite 30
San Diego, CA 92121
Ph: (619)-481-4400
(Superconductor instrumentation, SQUID magnetometers, contract research, etc.)

RIBER Division of Instruments SA, Inc.
6 Olsen Avenue
Edison, NJ 08820
Ph: (201)-494-8660, in France: (1)-47.08.92.50
(UHV Deposition system for superconductor thin films)

RMC—Cryosystems
1802 West Grant Road, Suite 122
Tucson, AZ 85745
Ph: (602)-882-4228
(Closed-cycle refrigerators, other cryo systems . . .)

SAES Getters USA Inc.
1122 E. Cheyenne Mountain Boulevard
Colorado Springs, CO 80906
Ph: (303)-576-3200
and
SAES Getters S.p.A. Milan
Via Gallarte, 215
20151 Milano, Italy
Ph: (2)306.541
(Targets for superconductor film deposition)

Saphikon, Inc.
51 Power Street
Milford, NH 03055
Ph: (603)-673-5831
("Grown to shape" sapphire tube, rod, ribbon, filaments, specials . . .)

Sargent-Welch Scientific Co.
7300 North Linden Avenue
Skokie, IL 67007
Ph: (800)-SARGENT
(OXITEC superconductivity demonstration kits, superconductor video . . .)

Sievers Research, Inc.
1930 Central Avenue
Suite C
Boulder, CO 80301
Ph: (303)-444-2009
(Nitric oxide analyzer . . .)

SPEX Industries, Inc.
3880 Park Avenue
Edison, NJ 08220
Ph: (201)-549-7144
(Analyzed high-purity oxides, compounds, metals . . .)

Spire Corp.
Patriots Park
Bedford, MA 01730
Ph: (617)-275-6000
(Metalorganic chemical vapor deposition MOCVD systems . . .)

STREM Chemicals, Inc.
7 Mulliken Way, Dexter Industrial Park
P. O. Box 108
Newburyport, MA 01950
Ph: (617)-462-3191
(Organometallics; barium, bismuth, calcium, copper, strontium, thallium, yttrium . . .)

Supercon, Inc.
830 Boston Turnpike
Shrewsbury, MA 01545
Ph: (508)-842-0174
(Superconductor conductors, wires, tapes, conventional and high-T_c, etc.)

Superconductive Components, Inc.
1145 Chesapeake Avenue
Columbus, OH 43212
Ph: (614)-486-0261
(Hi-T_c ceramic powders, shapes, machined parts, wires, films, etc.)

Superconductive Materials Corporation
(American CHEMET Corporation)
Deerfield, IL 60015
Ph: (312)-948-0800 or (406)-227-5302
(Processing technology, 1-2-3 materials . . .)

Superconductive Technologies, Inc.
4630 Indiana Street
Goldon, CO 80403
Ph: (312)-677-0560
(Oxitec™ Superconductor Lecture/Demonstration Kits)

Teledyne Wah Chang Albany
P. O. Box 460
Albany, OR 97321-0136
Ph: (503)-926-4211
(Metals and fabrication)

US Inc.
The Pruneyard Tower Two, Suite 405
Campbell, CA 95008
Ph: (408)-371-6900
(Planar magnetron sputter sources)

Walker Scientific, Inc.
Walker Magnetics Group
Rockdale Street
Worcester, MA 01606
Ph: (508)-852-3674 or (508)-853-3232
(Laboratory electromagnets, regulated magnet power supplies)

Wang NMR, Inc.
7074A Commerce Circle
Pleasanton, CA 94566
Ph: (415)-463-0302
(Superconductor magnet system technology, NMR magnets, engineering services)

Zeamer Systems Group, Inc.
45 South Street, Bldg. F.
Hopkinton, MA 01748
Ph: (508)-435-2383
(Custom superconductor testing systems, magnetic susceptibility, J_c, Hall, etc.)

GENERAL PUBLICATIONS

Advanced Coatings & Surface Technology
(See heading below under Technical Insights, Inc.)

The American Ceramic Society, Inc. (See also CCI below)
Book Service Department
757 Broodsedge Plaza Drive
Westerville, OH 43081-6136
Ph: (614)-890-4700
(High-T_c publications)

Arbitech Communications, Inc.
4010 Moorpark Avenue, Suite 105
San Jose, CA 95117
Ph: (408)-433-5803
(Superconductivity World Database . . .)

ASM International
Metals Park, OH 44073-9989
Ph: (216)-338-5151
(Japan Materials Reports)

Battelle, Columbus Division
505 King Avenue
Columbus, OH 43201-2693
Ph: (614)-424-4090
or
Battelle-Europe Geneva
7, Route de Drize
1227 Carouge-Geneva
Switzerland
Ph: 41-22-270.270
(Business assessment programs . . .)

The Cambridge Report on Superconductivity
One Kendall Square, Suite 2200
Cambridge, MA 02139
Ph: (617)-494-6506 or (800)-527-0230
(Superconductor Newsletter)

CCI (Ceramics Correspondence Institute)
Manager of Continuing Education
American Ceramic Society, Inc.
757 Brooksedge Plaza Drive
Westerville, OH 43081-6136
Ph: (614)-890-4700

Falmouth Associates, Inc.
170 U. S. Route 1
Falmouth, ME 04105
Ph: (207)-781-3632
(Superconductor market research services)

Future Technology Surveys, Inc.
123 W. Washington Street
Madison, GA 30650
Ph: (404)-342-9638
(Survey on High-T_c Superconductor Applications, etc.)

High-T_c Update
(Ellen O. Feinberg)
12 Physics
Ames Laboratory
Iowa State University
Ames, Iowa 50011-3020
Ph: (515)-294-3877
(USDOE supported high-T_c newsletter)

High-Tech Ceramics
Business Communications Company, Inc.
25 Van Zant Street
Norwalk, CT 06855
Ph: (203)-853-4266

Howard Superconductivity Index
Howard Associates
156 Mountain Avenue
Summit, NJ 07901
Ph: (201)-277-4312

INSPEC Key Abstract: *High-Temperature Superconductors . . .*
INSPEC Dept.
IEEE Service Center
445 Hoes Lane
P. O. Box 1331
Piscataway, NJ 08855-1331
Ph: (201)-562-5554

Institute for Scientific Information
ASCATOPICS Coordinator
3501 Market Street
Philadelphia, PA 19104
Ph: (800)-523-1850, ext. 1585 or (215)-386-0100
(Computer search services)

New Technology Week
King Communications Group, Inc.
627 National Press Building
Washington, D. C. 20045
Ph: (202)-638-4260

OTA Study: *Commercializing High-T_c Superconductivity*
(GPO Stock Number 053-003-01112-3)
Superintendent of Documents, Government Printing Office
Washington, D. C. 20402-9325
Ph: (202)-783-3238

Pasha Publications
1401 Wilson Boulevard, Suite 900
Arlington, VA 22209-9970
Ph: (800)-424-2908 or (703)-528-1244
(*The Superconductivity Directory*: "comprehensive guide to who's who in superconductivity")

SCAA (Superconductor Applications Association)
24781 Camino Villa Avenue
El Toro, CA 92630
Ph: (800)-854-8263 or (714)-586-8727

Solid State & Superconductivity Abstracts
Cambridge Scientific Abstracts
7200 Wisconsin Avenue
Bethesda, MD 20814
Ph: (301)-961-6750

Superconductivity Vol. 1, January–June 1987
Superconductivity Vol. II, July–December 1987
(Papers from *Physical Review Letters* and *Physical Review B*)
American Physical Society
Publications Liaison Office
500 Sunnyside Boulevard
Woodbury, NY 11797

Superconductivity (a study . . .)
Business Communications Company, Inc.
25 Van Zant Street
Norwalk, CT 06855
Ph: (203)-853-4266

Superconductivity Flash Report
Doral Plaza, Suite 626
155 N. Michigan Avenue
Chicago, IL 60601
Ph: (312)-565-0979
(Superconductor newsletter)

Superconductivity News•
Superconductivity Publications, Inc.
65 Jackson Drive, Suite 2000
Cranford, NJ 07016
Ph: (201)-709-0504
(Superconductor newsletter)

The Superconductor Advisory Newsletter
P. O. Box C
Tenafly, NJ 07670
Ph: (212)-431-8293

Superconductors: Exploiting Electronics and Computer Applications
SEAI Technical Publications
P. O. Box 590
Madison, GA 30650
Ph: (404)-342-9638

Superconductor Week
Atlantic Information Services, Inc.
1050 17th Street N.W., Suite 480
Washington, D. C. 20036
Ph: (202)-775-9008 or (800)-521-4323
(Newsletter)

Technical Insights, Inc.
P. O. Box 1304
Fort Lee, NJ 07024-9967
Ph: (201)-568-4744
(Monthly report: *Advanced Coatings & Surface Technology*)

World Business Publications Ltd.
15 Selvage Lane, Mill Hill
London NW7 3SS
UK
Ph: (0144)65141
(*High Performance Materials* report, *New Materials World*)

World Scientific High T_c Newsletter
World Scientific Publishing Co. Inc.
687 Hartwell Street
Teaneck, NJ 07666
Ph: (201)-837-8859
or
World Scientific Publishing Co. Pte. Ltd.
Farrer Road
P. O. Box 128
Singapore 9128
Ph: 2786188

World Scientific Publishing Co. Inc.
687 Hartwell Street
Teaneck, NJ 07666
Ph: (800)-227-7562 or (201)-837-8858
(Various high-T_c conference proceedings . . .)

TECHNICAL, SCIENTIFIC, AND TRADE JOURNALS

Angewandte Chemie
VCH Publishers Inc.
220 E. 23rd Street, Suite 909
New York, NY 10010-4606
Ph: (800)-422-8824 or (305)-428-5566
or
VCH Verlagsgesellschaft (Reader Service)
Postfach 1260/1280
D-6940 Weinheim
Federal Republic of Germany
Ph: (06201)-602-0

Cryogenics
Georgetown Cryogenics Information Center
3530 W. Place N. W.
Washington, D.C. 20007
Ph: (202)-337-5076

Cryogenics & Low-Temperature Physics
Plenum Publishing Corp.
233 Spring Street
New York, NY 10013-1578
or
Plenum Publishing Corp.
89/90 Middlesex Street
London E1 7EZ
UK

Journal of Crystal Growth
Elsevier Science Publishing Co., Inc.
52 Vanderbilt Avenue
New York, NY 10017
or
Elsevier Science Publishers B. V.
North-Holland Physics Publishing Division
P. O. Box 103
1000 AC Amsterdam
The Netherlands

Journal of Electron Microscope Technology
Alan R. Liss, Inc.
41 E. 11th Street
New York, NY 10003
Ph: (212)-475-7700

Journal of Materials Research & MRS Bulletin
Materials Research Society
9800 McKnight Road, Suite 327
Pittsburg, PA 15237
Ph: (412)-367-3003

Journal of Superconductivity
Plenum Publishing Corp.
233 Spring Street
New York, NY 10013-1578
or
Plenum Publishing Corp.
89/90 Middlesex Street
London E1 7EZ
UK

Physica C
Elsevier Science Publishing Co. Inc.
52 Vanderbilt Avenue
New York, NY 10017
Ph: (212)-370-5520
or
Elsevier Science Publishers B. V.
North-Holland Physics Publishing Division
P. O. Box 103
1000 AC Amsterdam
The Netherlands

Physics Today
American Institute of Physics
Marketing Services
335 E. 45th Street
New York, NY 10017

Solid State and Superconductivity Abstracts
(formerly *Solid State Abstracts Journal*)
Cambridge Scientific Abstracts
7200 Wisconsin Avenue
Bethesda, MD 20814
Ph: (301)-961-6756

Superconductor Industry
Rodman Publishing
Box 555
Ramsey, NJ 07446
Ph: (201)-825-2552

Superconductor Science & Technology
USA, Canada, and Mexico:
American Institute of Physics, Dept N/M
335 E. 45th Street
New York, NY 10017
USA
Ph: (516)-349-7800
Japan:
Maruzen Co. Ltd
3-10 Nihonbashi 2-Chome
Chou-Ku, Tokyo 103
Japan
Ph: 272 7211
United Kingdom:
Journals Marketing Dept.
IOP Publishing Ltd., Techno House
Redcliffe Way
Bristol, BS1 6NX
UK
Ph: (0272) 297481

Superconductivity Theory and Applications
Elsevier Science Publishing Co. Inc.
52 Vanderbilt Avenue
New York, NY 10017
or
Elsevier Science Publishing Co. Inc.
P. O. Box 1663, Grand Central Station
New York, NY 10163-1663
Ph: (212)-370-5520

Supercurrents
P. O. Box 889
Belmont, CA 94002
Ph: (415)-595-3808

SOCIETIES AND ASSOCIATIONS

The American Ceramic Society, Inc.
757 Broodsedge Plaza Drive
Westerville, OH 43081-6136
Ph: (614)-890-4700

American Chemical Society
1155 16th Street N.W.
Washington, D.C. 20036
Ph: (202)-872-4700

American Physical Society
335 E. 45th Street
New York, NY 10017-3483

ASM International
Metals Park, OH 44073

Council on Superconductivity for American Competitiveness (CSAC)
1050 Thomas Jefferson Street N.W., 6th Floor
Washington, D.C. 20007
Ph: (202)-965-4070

The Metallurgical Society
420 Commonwealth Drive
Warrendale, PA 15086
Ph: (412)-776-9000

Superconductor Applications Association
24781 Camino Villa Avenue
El Toro, CA 92630
Ph: (800)-854-8263 or in CA: (714)-586-8727

World Congress on Superconductivity
ET Room 1105
Houston Lighting and Power Co.
P. O. Box 1700
Houston, TX 77001
Ph: (713)-229-7385

APPENDIX E

TEMPERATURE SCALES; CONVERSIONS

Virtually all work on superconductivity is reported with temperature on the Kelvin (K) scale, except for processing where one normally uses Celsius (C). Relationships to other scales are given below:

Kelvin (K) to Celsius (C):

$$C = K - 273.15 \quad \text{E.1}$$

Celsius (C) to Kelvin:

$$K = C + 273.15 \quad \text{E.2}$$

Fahrenheit (F) to Celsius:

$$F = \left(\frac{9}{5}\right)C + 32 \quad \text{E.3}$$

Celsius to Fahrenheit:

$$C = \frac{5}{9}(F - 32) \quad \text{E.4}$$

Rankine (R) to Kelvin:

$$K = \left(\frac{5}{9}\right)R \quad \text{E.5}$$

Kelvin to Rankine:

$$R = \left(\frac{9}{5}\right)K \tag{E.6}$$

Rankine to Fahrenheit:

$$F = R - 459.69 \tag{E.7}$$

APPENDIX F

PEROVSKITES

ORIGINS

The original mineral bearing the name perovskite, (calcium titanate—$CaTiO_3$), was described in the early nineteenth century by Gustav Rose, who named the ceramic after a Russian mineralogist, one Count L. A. von Perovski (Hazen, 1988). Although several of the naturally occurring minerals are rare, the production of synthetic perovskites is of considerable importance.

EXISTING APPLICATIONS

The term "perovskite," particularly prior to the recent developments in high-temperature superconductivity, is not familiar to most electrical engineers. This is in spite of the fact that production of synthetic perovskites for electronic applications is a $20 billion per year industry. Perhaps the most useful of more than 150 synthetic perovskites is barium titanate ($BaTiO_3$), which is manufactured in large quantities for its dielectric qualities and excellent piezoelectric characteristics (Moore, 1971). Common applications include capacitors, loudspeakers, microphones, voltage-surge arrestors, and buzzers.

PEROVSKITE STRUCTURE

Consider the abbreviated list of "ideal" perovskites below . . . notice the patterns:

calcium titanate	$CaTiO_3$
silver zinc fluoride	$AgAnF_3$
lithium barium fluoride	$LiBaF_3$
cesium cadmium bromide	$CsCdBr_3$
strontium titanate	$SrTiO_3$

The list could go on for several pages, but the few compounds given illustrate some of the general features of the basic perovskite structure:

1. The chemical formula for an ideal perovskite is ABX_3, where:
2. A and B are metals, and are positively charged (cations)
3. X is a nonmetal and is negatively charged (anion)
4. X can be either oxygen or a member of the halogen family, that is, fluorine, bromine, or chlorine.

A perovskite, in its "ideal" form, is a cubic structure with a relatively large metallic atom (A) at its center, smaller metallic atoms (B) at its corners, and nonmetallic atoms (X) on midpoints between the corners. The bulk crystal is typically either the cubic, octahedral, or dodecahedral form. See Figure F.1 for an illustration of the basic unit cell. If you count the relative number of atoms listed above, it may appear that the cube should have the formula: AB_8X_{12}. However, since the basic "unit cell" of the perovskite is not generally considered in isolation, one should consider how the B atoms at the corners and the X atoms along the sides are shared with adjacent (usually identical) unit cells in the bulk crystalline material. When this sharing of B and X atoms is taken into account, the correct formula for the bulk perovskite is ABX_3.

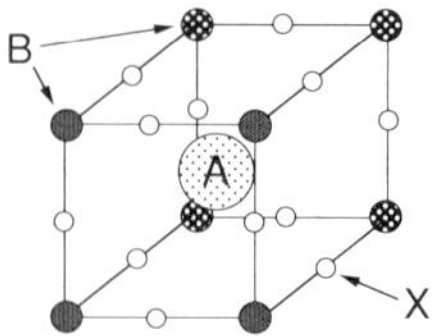

FIGURE F.1 The basic unit of the "ideal" perovskite is a cube with the largest metallic cation (−) at the center, the smallest located at the corners. The nonmetallic anions (+), typically oxygen or fluorine, lie along the center points between the corners. At first glance, the numbers of atoms in the basic structural unit may seem to be at odds with the statement that there is a single A and B atom for every three nonmetallic X atoms. Note, however, that the basic perovskite cube shares B and X atoms with adjacent cubes.

VARIATIONS FROM THE IDEAL FORM

The ideal perovskite structure, such as calcium titanate ($CaTiO_3$), is an insulator. Since all the atomic sites are filled, strong ionic bonds hold the electrons firmly in their positions. This strong bonding also results in a hard mineral (or synthetic) that has a relatively high melting temperature. Since the ideal structure is symmetric, it naturally has three-dimensional isotropy. There are, however, variations that lead to marked difference in electrical, mechanical, and optical properties, including anisotropy. These properties may either be useful (as in piezoelectrics) or rather a nuisance (as in the new ceramic superconductors.)

A-B Ion Size Variations

When the central "A" ion is slightly too small compared to the "B" ion at the corners, this can lead to a shifting in the position of the B and X ions. In this instance, the resulting unit cell may appear more like a cluster of polyhedrons than a cube; the structure may "tilt" or "twist" resulting in an electric polarization of the entire crystal or in piezoelectric behavior.

Mixed A or B Sites

Rather than having a single element associated with the A and/or B sites, it is possible to have a variety of characteristics when more than one element can occupy these sites. An example of mixed B sites is calcium uranium oxide (Ca_2CaUO_6) where the A sites are always calcium, but the B sites are occupied by either calcium or uranium (Hazen, 1988). At very high pressures and temperatures, such as are found in the earth's interior, either magnesium or iron may fill the "A" site in magnesium-iron silicate $(Mg,Fe)SiO_3$.

Stoichiometry

The substitution of one ion for another does not exhaust the list of ways that perovskites can be altered. It is not uncommon to find materials with missing X atoms. This is, in fact, a very important factor in some of the new high-temperature superconductors.

Twinning

In any perovskite specimen that is not of the pure cubic form, a phenomenon called twinning, is commonly seen. This involves, for example, the alignment of the shorter side of one cell with the slightly larger side of an adjacent cell. Twinning is almost always seen in the orthorhombic $YBa_2Cu_3O_{7-\delta}$ superconductor, and is believed to have significant effects on electrical, magnetic, optical, and mechanical properties.

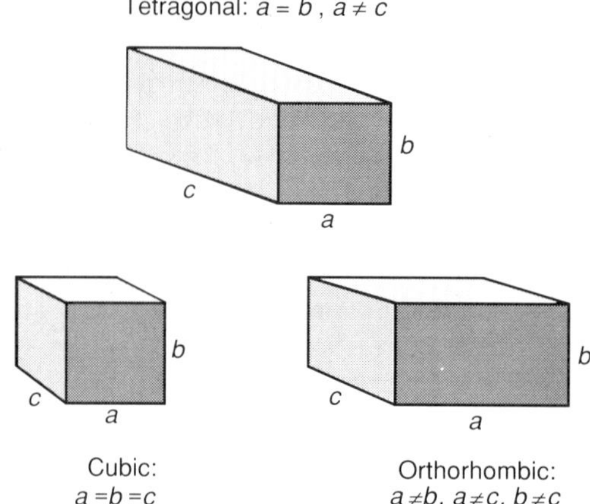

FIGURE F.2 While the meaning of the term "cube" is obvious enough in reference to crystalline structure, the terms "orthorhombic" and "tetragonal" will be less familiar to some readers. The relation between edge dimensions for these latter configurations is illustrated above. For reference, the $YBa_2Cu_3O_x$ superconductor unit-cell is orthorhombic in its superconducting state ($x \sim 7$) but shifts to a tetragonal structure when the oxygen content drops low enough ($x \sim 6.5$) for the compound to lose its superconducting properties. In contrast, the Ba-K-Bi-O superconductor has a simple cubic structure.

Summary of Types of Variations

The following variations may be commonly found in perovskites:

- Mixed A or B cations
- Off-stoichiometry (missing X anions)
- Cation off-centering
- Octahedral tilting
- Twinning
- Impurities
- Combinations of the above

$YBa_2Cu_3O_{7-\delta}$: AN OXYGEN-DEFICIENT PEROVSKITE

In the $YBa_2Cu_3O_{7-\delta}$ ceramic superconductor, the A sites in a unit cell are filled by barium on the end subunits and by yttrium in the center, while copper occupies all of the B (corner) sites. In this 1-2-3 ceramic superconductor, the already imperfect perovskite is further modified by the fact that several of the oxygen (X) sites are unoccupied. The precise number of oxygen sites which are unoccupied is, in fact, crucial to the superconducting behavior of the compound (see Figure F.3).

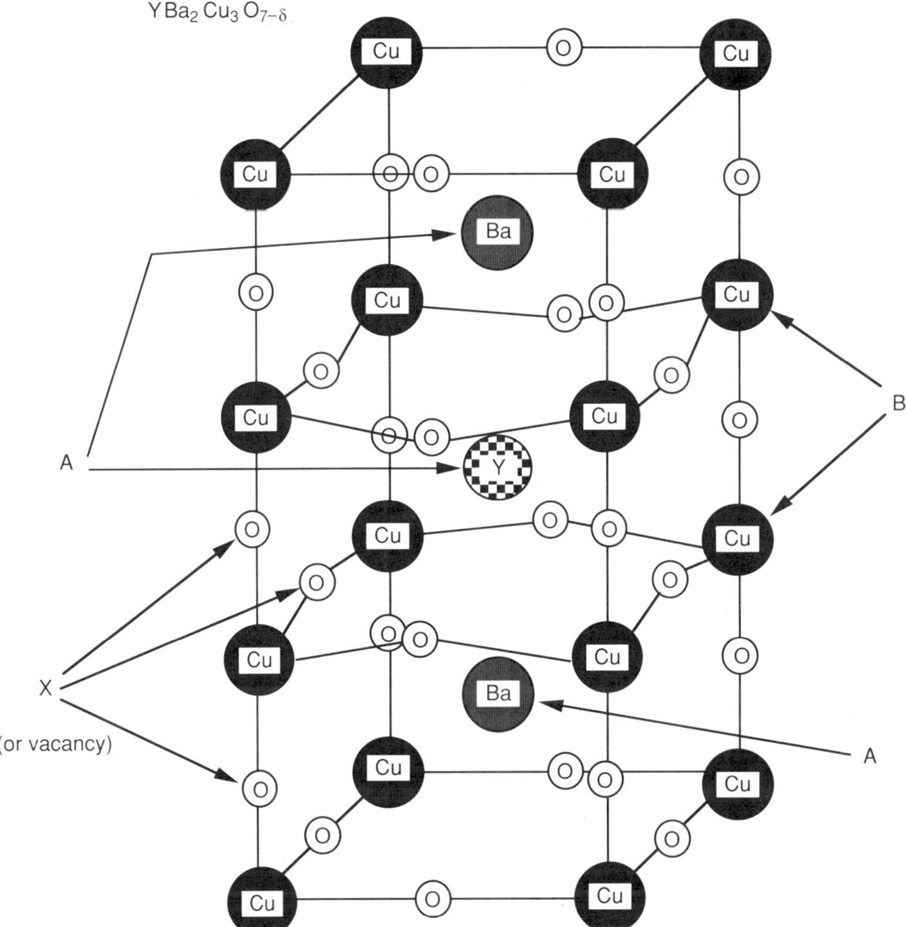

FIGURE F.3 The "1-2-3" $YBa_2Cu_3O_{7-\delta}$ superconducting ceramic has an orthorhombic unit cell, comprised of three perovskite-like subcells that are not quite cubic. While the "B" atoms are copper, the "A" central atoms alternate between barium on the end subcells and yttrium (or one of several suitable rare earths) on the central subcell. The 1-2-3 superconductor is also notable for being oxygen-deficient; not all of the oxygen locations are filled. When enough oxygen is removed for $(7-\delta)$ to become approximately 6.5, the structure becomes tetragonal and loses its superconducting properties.

It is clear from the figure that the superconducting perovskite-like structure differs from the ideal perovskite in a variety of ways. One of the most significant results is that the material is not isotropic. This anisotropy has unfortunate results for the superconducting characteristics of this ceramic. The supercurrent flows primarily (if not entirely) in the copper-oxide planes that are perpendicular to the long axis of the crystal structure. The corollary to this statement is that the superconductor does not superconduct as effectively along the long axis, that is, perpendicular to the Cu-O planes.

(The long axis is the c axis, the short axes are a and b—this a and b notation is not to be confused with the A, B, and C elements in the ABX_3 chemical formula for perovskites!)

Such anisotropy is, of course, hardly a property one is seeking for common superconductor applications. It is possible, of course, that an unusual application might benefit from anisotropic superconductivity.

There has even been speculation that layers of perovskites under great pressure in the earth's mantle may be sufficiently conducting to provide electric current flow that would generate the earth's magnetic field (Lawren, 1988).

For a more detailed description of perovskites, the reader is referred to the excellent book by Robert M. Hazen listed in the Further Reading section.

REFERENCES

Hazen, R. M., (1988). "Perovskites," *Scientific American*, Vol. 258, No. 6, June, pp. 74–81.

Lawren, Bill, (1988). "The Biggest Superconductor," *Omni*, Vol. 10, No. 4, January, p. 33.

Moore, P. B., (1971). "Perovskite," *McGraw-Hill Encyclopedia of Science & Technology*, Vol. 10, McGraw-Hill, pp. 33–34.

SUGGESTED FOR FURTHER READING

Hazen, R. M., *The Breakthrough—The Race for the Superconductor*, Summit Books, New York, 1988.

APPENDIX G

NOVEL SUPERCONDUCTORS

"GAPLESS" SUPERCONDUCTORS

While they may not currently be of paramount interest to engineers, it is worthwhile to point out (in the interest of completeness) that there is a class of superconductors referred to as "gapless"; these materials actually have a very low density of states, rather than a complete absence of states, in the "gap" region. Examples of materials that are gapless but exhibit zero resistance below a critical temperature include lead and indium film with magnetic impurities added. The energy region at H_{c2} for Type II superconductors may also be considered as a gapless region (Lynton, 1974).

f-ELECTRON SYSTEMS

Unusual superconducting phenomena are exhibited by alloys of actinide elements with partially filled f-electron shells and rare earth elements. Some of these novel phenomena are:

a. The coexistence of antiferromagnetism and superconductivity.
b. Return to a normal (resistive) state at a second critical temperature which is lower than the transition temperature where resistance disappeared.
c. Superconductivity caused by the presence of magnetic fields.
d. Superconductivity caused by electrons with large "effective" mass.

There are two basic types of these f-electron superconductors (Maple, 1986):

1. Those caused by magnetic interaction of localized f-electrons magnetic moments and the spins and momenta of conduction electrons.
2. Those related to a single band of electrons that is a hybrid of localized and "itinerant" electrons.

For further reading, see the excellent review by Brian Maple.

REFERENCES

Lynton, E. A., (1974). *Superconductivity*, Methuen, London.

Maple, M. B., (1986). "Novel types of superconductivity in f-electron systems," *Physics Today*, March, pp. 72–80.

APPENDIX H

MISCELLANEOUS FEATURES OF SUPERCONDUCTORS

SPECIFIC HEAT

The discovery of an exponential term in the specific heat of superconductors suggested the existence of the energy gap, and this was soon confirmed by measurements of electromagnetic absorption (Tinkham, 1986).

The specific heats will be considered in general terms. In both the normal state and the superconducting state, the specific heat is a sum of electronic (C_{ne} or C_{se}) and lattice (C_{nl} or C_{sl}) terms.

In the normal state (Atkins, 1958),

$$C_n = C_{ne} + C_{nl} = aT + bT^3 \qquad \text{H.1}$$

For metals, the form is, more specifically:

$$C_v = \gamma T + \left(\frac{12\pi^4 R}{5}\right)\left(\frac{T}{\Theta_D}\right)^3 \qquad \text{H.2}$$

where R is the gas constant, γ is the density of electron states near the Fermi surface (the Sommerfeld gamma), and Θ_D is the Debye temperature. If a material follows this relationship, a plot of C_v/T versus T^2 will provide a straight line with a slope of $12\pi^4 R/5\Theta_D^3$ and an intercept equal to γ on the C_v/T axis:

$$\frac{C_v}{T} = \gamma + T^2\left(\frac{12\pi^4 R}{5\Theta_D^3}\right) \qquad \text{H.3}$$

Since the linear electron term predominates in the normal state, this expression is reduced to

$$C_n = \gamma T \tag{H.4}$$

In the superconducting state

$$C_s = C_{se} + C_{sl} = cT^3 + bT^3 \qquad \text{for } 0.7T_c < T < T_c \tag{H.5}$$

which may be reduced to

$$C_s = \frac{3\gamma T^3}{T_c^2} \qquad \text{for } 0.7T_c < T < T_c \tag{H.6}$$

and

$$C_s = C_{se} + C_{sl} = A \exp\left(\frac{-\beta T_c}{T}\right) + bT^3 \qquad \text{for } 0 < T < 0.7T_c \tag{H.7}$$

where the constant b is the same in each case.

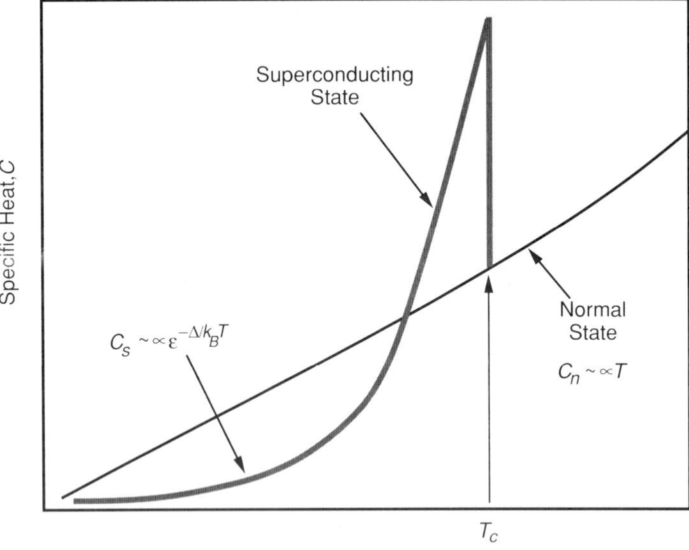

FIGURE H.1 In its normal state, the typical metallic superconductor has a specific heat that varies in almost a linear fashion with temperature if held in its normal state by, for example, a strong magnetic field. If the sample is allowed to make the superconducting transition, the specific heat has a discontinuous increase at T_c and then falls exponentially as temperature decreases.

For temperatures very far below T_c the electronic term predominates, so that

$$c_s \propto \epsilon^{-\Delta/kT} \tag{H.8}$$

where Δ represents the gap in the spectrum of allowable energy states. The indication of the energy gap (Δ) by specific heat measurements is a hallmark of BCS-type behavior in a superconductor.

As mentioned earlier, BCS theory predicts that the energy gap increases from a value of zero at $T = T_c$, to its maximum at a temperature of 0 K:

$$2\Delta = 3.53kT_c \tag{H.9}$$

Entropy is less in the superconducting state than in the normal, that is, the superconducting state is more ordered than the normal. Further,

$$\gamma T_c^2 = 2\mu_o H(0)^2 \tag{H.10}$$

Recall that $H(0)$ is the critical magnetic field at $T = 0$ K. The critical field H_c is related to T_c, T, μ_o, and γ by

$$H_c = T_c\left\{1 - \left(\frac{T}{T_c}\right)^2\right\}\sqrt{\frac{\gamma}{2\mu_o}} \tag{H.11}$$

THERMAL CONDUCTIVITY, VARIATIONS BELOW T_c

In metals (in their normal phase), thermal conductivity (k_n) is the sum of an electron (k_e) and a phonon (k_p) contribution, that is,

$$k_n = k_e + k_p \tag{H.12}$$

The electron component may be related to residual electrical resistivity (ρ_r) by (Lynton, 1969)

$$k_e^{-1} = \eta T^2 + \left(\frac{\rho_r}{LT}\right) \tag{H.13}$$

where L is the Lorentz number (2.44×10^{-8} W-Ω/deg^2) and η is a constant for the specific material, which is proportional to $\Theta_D^{-1/2}$. The first term (ηT^2), which predominates at high temperatures, represents electron scattering by phonons. The second term represents effects of scattering from imperfections in the lattice, and predominates below the temperature where k_e has a maximum:

$$T_{k_e\max} = \left(\frac{\rho_r}{2\eta L}\right)^{1/3} \tag{H.14}$$

In a pure metal in its normal phase, thermal conductivity tends to increase as temperature decreases and reaches the peak previously defined which is typically a few K. From this peak, thermal conductivity falls for further decreases in temperature. For a pure metallic superconductor, thermal conductivity falls significantly as the temperature is decreased below T_c, and is often much smaller than the corresponding (normal) value for the same material at temperatures much lower than T_c. (The "normal" measurements below T_c may be made with the aid of a magnetic field greater than H_c.) (See Figure H.2 for the normal versus superconducting phase of thermal conductivity for a pure metal.)

For temperatures very far below T_c, the relation of the normal to the superconducting thermal conductivity in very pure metallic superconductors is approximately:

$$k_n = k_s\left(\frac{\eta}{T^2}\right) \quad \text{H.15}$$

where the constant η, which may have a value > 100, is a function of the material under consideration. Thus, the ratio of normal to superconducting thermal conductivity may exceed two orders of magnitude at very low temperatures. This has led to the design of heat switches, where a magnetic field is used to switch a conductor between the high and low states of thermal conductivity by moving the metal between its normal and superconducting states respectively.

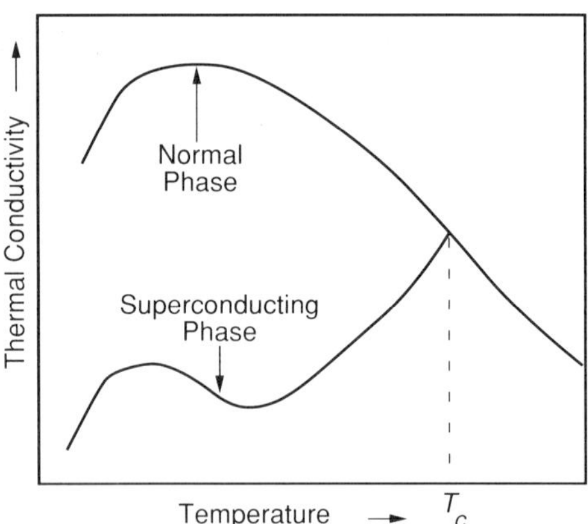

FIGURE H.2 In the case of a Type I superconductor, thermal conductivity is usually much less in the superconducting state than it would be if the same sample was forced to remain normal below T_c by application of a magnetic field. The low thermal conductivity in the superconducting state is due to the inability of a highly ordered state to transmit disorder. (This situation may be reversed in certain alloys and compounds, i.e., Type II superconductors.)

For an alloy or compound, the behavior just described may be reversed, that is, the superconducting thermal conductivity may be higher than the normal-state values.

The notion that infinite electrical conductivity and extremely low thermal conductivity may occur in the same material may seem surprising at first, since metals in their normal phase exhibit excellent thermal and electrical conductivity at the same time, largely due to the presence of a high density of free electrons. The difference is related to the high order of the superconducting state; the transfer (conduction) of heat is essentially a transfer of disorder, or, to be more technical, of entropy. A highly ordered state cannot transmit disorder, thus the thermal conductivity of a superconductor must decrease as temperature approaches 0 K.

THERMOELECTRIC EFFECTS

Since all thermoelectric effects are intimately related to electron transport properties in materials, it is not surprising that such measurements are of considerable interest to those involved in characterizing superconductors. In particular, a superconducting metal shows no evidence of thermoelectric effects, including the Thomson effect, the Peltier effect, or the Seebeck effect. The Hall constant is also zero in the perfectly superconducting state.

The Thomson effect occurs when a temperature gradient is forced across an electrical conductor in the presence of current flow. A reversible generation of heat is produced, called the Thomson heat.

Ordinary Joule heating is irreversible, that is, the conductor is always a generator of heat, regardless of the direction of current flow. With the Thomson effect the conductor can be either a generator or absorber of heat, as a function of the direction of current flow and the direction of the imposed thermal gradient.

The Seebeck effect is the phenomena exploited in thermocouple temperature measurements, that is, a difference in potential developed between pairs of metallic junctions (e.g., constantan and copper) held at different temperatures. The reverse effect, a cooling or heating of the junctions when a current is made to flow, is the Peltier effect.

A fundamentally important parameter is the open-circuit voltage produced between two junctions per unit temperature. Generally, the voltage produced depends on the properties of each metal used. If, however, one of the materials is a superconductor (operated well below T_c), it will make no contribution to the voltage produced. This is a method for determining the properties of a nonsuperconducting sample, or, conversely, if the properties of the non-superconducting material are known, to determine whether the other material is behaving like a superconductor over a range of temperatures.

Generally, $E_A(T)$ refers to the absolute thermal emf of material A at temperature T and is typically measured in μV. The "thermoelectric power" is the

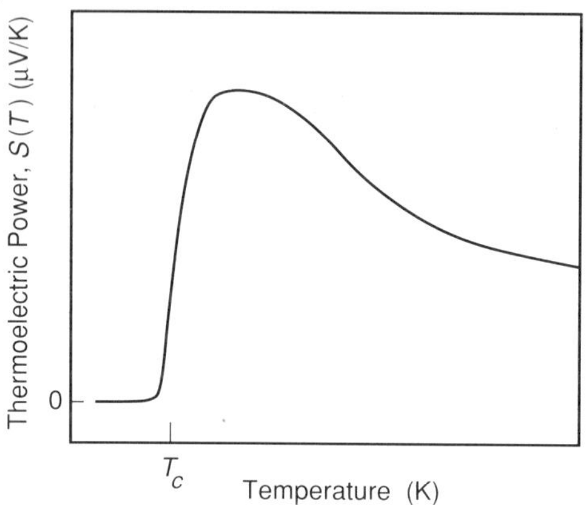

FIGURE H.3 As a material enters the superconducting state, thermoelectric power $S(T)$ drops sharply at T_c and approaches a zero value well above 0 K.

derivative with respect to temperature of this function, that is,

$$S_A(T) = \frac{dE_A(T)}{dT} \qquad \text{H.16}$$

and is generally presented in units of μV/K. A zero value for thermopower below T_c is diagnostic of superconducting behavior, so $S_A(T)$ is routinely measured on the new copper-oxide compounds. (See Figure H.3 for the thermoelectric power of a superconductor versus temperature.)

REFERENCES

Atkins, K. R., (1958). "Superfluids," in *Handbook of Physics*, E. U. Condon and H. Odishaw, eds., McGraw-Hill, New York.

Lynton, E. A., (1969). *Superconductivity*, Methuen, London.

Tinkham, M., (1986). "Superconductivity, 75th Anniversary," *Physics Today*, March, pp. 22–23.

LIST OF FIGURES

A BRIEF HISTORY OF SUPERCONDUCTIVITY — Page

THE PHENOMENA OF SUPERCONDUCTIVITY — Page

HIGH-TEMPERATURE SUPERCONDUCTORS Page

APPLICATIONS OF SUPERCONDUCTORS Page

PROCESSING OF HIGH-TEMPERATURE SUPERCONDUCTORS Page

MEASUREMENTS Page

APPENDIX A REVIEW OF BASIC ELECTRICAL AND MAGNETIC THEORY **Page**

APPENDIX F PEROVSKITES **Page**

APPENDIX H MISCELLANEOUS FEATURES OF SUPERCONDUCTORS **Page**

SELECTED BIBLIOGRAPHY

BOOKS, MONOGRAPHS, CONFERENCE PROCEEDINGS

Al'tov, V. A., V. B. Zenkevich, M. G. Kremlev, and V. V. Sychev, *Stabilization of Superconducting Magnetic Systems*, Trans. by G. D. Archard, Plenum, New York, 1977.

Barron, R., *Cryogenic Systems*, McGraw-Hill, New York, 1966.

Blatt, J. M., *Theory of Superconductivity*, Academic, New York, 1964.

Brechna, H., *Superconducting Magnet Systems*, Springer-Verlag, Berlin, 1973.

Bremer, John W., *Superconductive Devices*, McGraw-Hill, New York, 1962.

Brodsky, M. B., R. C. Dynes, K. Kitazawa, and H. L. Tuller, eds., *High-Temperature Superconductors*, Proc. Mat. Res. Soc., Vol. 99, 1988.

Bumby, J. R., *Superconducting Rotating Electrical Machines*, Clarendon, Oxford, 1983.

Chilton, Frank, ed., "Superconductivity," *Proc. International Conference on the Science of Superconductivity*, Stanford, USA, North-Holland, Amsterdam, 1971.

Cohen, Morrel H., ed., *Superconductivity in Science and Technology*, University of Chicago, Chicago, 1968.

Coleman, R. V., "Solid State Physics," Vol. 11 in *Methods of Experimental Physics* series, L. Marton, ed., Academic, New York, 1974.

Collings, E. W., *Design and Fabrication of Conventional and Unconventional Superconductors*, Noyes Publications, Park Ridge, NJ, 1984.

Corson, D. R., and P. Lorrain, *Introduction of Electromagnetic Fields and Waves*, W. H. Freeman, San Francisco, 1962.

Cramer, K. R., and Shih-I Pai, *Magnetofluid Dynamics for Engineers and Applied Physicists*, McGraw-Hill (Scripta), New York, 1973.

Cryogenic Safety, summary report of the Cryogenic Safety Conference, Allentown, PA, Air Products, 1960.

Dalven, R., *Introduction to Applied Solid State Physics*, Plenum, New York, 1980.

De Gennes, P. G., *Superconductivity of Metals and Alloys*, Trans. by P. A. Pincus, W. A. Benjamin, New York, 1966.

Deaver, B., and J. Ruvalds, *Advances in Superconductivity*, NATO Advanced Science Series, Plenum, New York, 1983.

Feynman, R. P., R. B. Leighton, M. Sands, *The Feynman Lectures on Physics, Quantum Mechanics* Vol. 3, Addison-Wesley, Reading, MA, 1965.

Fishlock, D., *A Guide to Superconductivity*, Elsevier, New York, 1969.

Foner, S., and B. B. Swartz, *Superconductor Materials Science, Metallurgy, Fabrication and Applications*, Plenum, New York, 1981.

Ginzton, E. L., *Microwave Measurements*, McGraw-Hill, New York, 1967.

Gregory, W. D., W. N. Mathews, Jr., and E. A. Edelsack, *The Science and Technology of Superconductivity*, 2 Vols., Plenum, New York, 1973.

Halley, J. W., ed., *Theories of High Temperature Superconductivity*, Addison-Wesley, Redwood City, CA, 1988.

Harper, J. M. E., R. C. Colton, L. C. Feldman, *Thin Film Processing and Characterization of High-Temperature Superconductors*, American Vacuum Society Series No. 3, American Institute of Physics Conference Proceedings No. 165, AIP Publication, New York, 1988.

Harris, F. K., *Electrical Measurements*, John Wiley & Sons, New York, 1952.

Hatfield, W. E., and J. H. Miller, Jr., ed., *High-Temperature Superconducting Materials, Preparation, Properties and Processing*, Marcel Dekker, New York, 1988.

Hazen, R. M., *The Breakthrough—The Race for the Superconductor*, Summit Books, New York, 1988.

Huebener, R. P., *Magnetic Flux Structures in Superconductors*, Springer-Verlag, Berlin, 1979.

Hunt, V. D., *Superconductivity Sourcebook*, John Wiley & Sons, New York, 1988.

Jackson, John David, *Classical Electrodynamics*, John Wiley & Sons, New York, 1963.

Kip, Arthur F., *Fundamentals of Electricity and Magnetism*, McGraw-Hill, New York, 1969.

Koonce, Calvin Scott, *Superconducting Properties of Semiconductors*, Ph. D. Dissertation, University of California, Berkeley, 1967. Available through University Microfilms, Ann Arbor, Michigan.

Krauss, J. D., and K. R. Carver, *Electromagnetics*, 2nd Ed., McGraw-Hill, New York, 1973.

Kuper, Charles G., *An Introduction to the Theory of Superconductivity*, Clarendon, Oxford, 1968.

London, F., *Superfluids, Macroscopic Theory of Superconductivity*, Vol. 1, John Wiley & Sons, New York, 1950.

Lounasmaa, O. V., *Experimental Principles and Methods Below 1 K*, Academic, New York, 1974.

Lynton, E. A., *Superconductivity*, Methuen, London, 1962.

Mayo, J. L., *Superconductivity—The Threshold of a New Technology*, TAB Books, Blue Ridge Summit, PA, 1988.

Mayo, W. E., *Processing & Applications of High-T_c Superconductors*, Metallurgical Society, Warrendale, PA, 1988.

Metzger, Robert M., ed., *High-Temperature Superconductivity: The First Two Years*, Gordon & Breach, New York, 1988.

Newhouse, Vernon L., *Applied Superconductivity*, 2 Vols., John Wiley & Sons, New York, 1975.

Petley, B. W., *An Introduction to the Josephson Effects*, Mills & Boon, London, 1971.

Plonsey, R., and R. E. Collin, *Principles and Applications of Electromagnetic Fields*, McGraw Hill, New York, 1961.

Pollock, D. D., *The Theory and Properties of Thermocouple Elements*, ASTM Special Technical Publication 492, American Society for Testing and Materials, Philadelphia, 1971.

Rechowicz, M., *Electric power at low temperatures*, Clarendon, Oxford, 1975.

Reed, R. P., and A. F. Clark, *Materials at Low Temperatures*, American Society for Metals, Metals Park, OH, 1983.

Richardson, R. C., and E. N. Smith, ed., *Experimental Techniques in Condensed Matter at Low Temperatures*, Addison-Wesley, New York, 1988.

Rickayzen, G., *Theory of Superconductivity*, Wiley-Interscience, New York, 1965.

Roberts, B. W., *Superconductive Materials and Some of Their Properties*, NBS Technical Note 482, May 1969.

Rose-Innes, A. C., and E. H. Rhoderick, *Introduction to Superconductivity*, 2nd Ed., Pergamon, Oxford, 1978.

Rosenberg, H. M., *Low Temperature Solid State Physics*, Clarendon, Oxford, 1963.

Roy, D. K., *Quantum Mechanical Tunnelling and its Applications*, World Scientific, Philadelphia, 1986.

Sax, N. I., and R. J. Lewis, Sr., *Dangerous Properties of Industrial Materials*, 7th Ed., Van Nostrand Reinhold, New York, 1988.

Schriber, S. O., ed., *Workshop on Superconducting Accelerators for Medium Energy Proton LINACs*, Los Alamos National Laboratory publication LA-UR-87-1160, March 1987.

Schrieffer, J. R., *Theory of Superconductivity*, W. A. Benjamin, New York, 1964.

Schwartz, Brian B., and Simon Foner, eds., *Superconductor Applications: SQUIDs and Machines*, Plenum, New York, 1977.

Seeger, J. A., *Microwave Theory, Components and Devices*, Prentice-Hall, Englewood Cliffs, NJ, 1986.

Sherman, A., *Chemical Vapor Deposition for Microelectronics*, Noyes Publications, Park Ridge, NJ, 1987.

Shoenberg, D., *Superconductivity*, Cambridge University, London, 1938.

Solymar, L., *Superconductive Tunneling and Applications*, Wiley-Interscience, New York, 1972.

Solymar, L., and D. Walsh, *Lectures on the Electrical Properties of Materials*, 4th Ed., Oxford Science Publications, Oxford, NY, 1988.

Superconductivity (Vol. 1, January–June 1987, Vol. II, July–December 1987), Papers from *Physical Review Letters* and *Physical Review B*, American Physical Society, Woodbury, NY.

Tinkham, M., *Introduction to Superconductivity*, Robert E. Krieger, Malabar, FL, 1985.

Tinkham, M., *Superconductivity—Documents on Modern Physics* series, Gordon & Breach, New York, 1965.

Van Duzer, T., and C. W. Turner, *Principles of Superconductive Devices and Circuits*, Elsevier, New York, 1981.

von Laue, M., *Theory of Superconductivity*, Trans. by L. Meyer and W. Band, Academic, New York, 1952.

Wallace, P. R., ed., *Superconductivity*, Proceedings of the Advanced Summer Study Institute, McGill University, Montréal, Vols. 1 & 2, Gordon & Breach, New York, 1969.

Weinstock, H., and W. C. Overton, Jr., *Squid Applications to Geophysics*, Society of Exploratory Geophysicists, Tulsa, 1981.

White, G. K., *Experimental Techniques in Low-Temperature Physics*, 3rd Ed., Clarendon, Oxford, 1979.

Williams, J., *A Designer's Guide to: Innovative Linear Circuits*, Cahners, Newton, MA, 1985.

Wilson, M. S., *Superconducting Magnets*, Monographs on Cryogenics series, Clarendon, Oxford, 1986.

Wolfe, Hugh C., ed., *Temperature; Its Measurement and Control in Science and Industry*, Vol. 2, American Institute of Physics, Reinhold, New York, 1955.

SCIENTIFIC AND TRADE JOURNAL ARTICLES

Abel, W. R., A. C. Anderson, and J. C. Wheatley, "Temperature Measurements Using Small Quantities of Cerium Magnesium Nitrate," *Rev. Sci. Instrum.*, Vol. 35, No. 4, April 1964, pp. 444–449.

Adachi, H., K. Setsune, and K. Wasa, "Superconductivity in $(La_{0.9}Sr_{0.1})_2CuO_4$ single-crystal films," *Phys. Rev. B*, Vol. 35, No. 16, June 1, 1987, pp. 8824–8825

Adachi, H., Y. Ichikawa, K. Setsune, S-I Hatta, K. Hirochi, and K. Wasa, "Preparation and properties of superconducting Bi-Sr-Ca-Cu-O thin films," *Jap. J. Appl. Phys.*, Vol. 27, No. 4, April 1988, pp. L643–L645.

Adachi, S., O. Inoue, and S. Kawashima, "Superconducting properties in a Bi-Sr-Ca-Cu-O system," *Jap. J. Appl. Phys.*, Vol. 27, No. 3, March 1988, pp. L344–L346.

Adkins, C. J., "Two-particle Tunneling between Superconductors," *Phil. Mag.*, Vol. 8, No. 90, June 1963, pp. 1051–1061.

Aizaki, N., K. Terashima, J-I Fujita, and S. Matsui, " $YBa_2Cu_3O_y$ Superconducting Thin Film Obtained by Laser Annealing," *Jap. J. Appl. Phys.*, Vol. 27, No. 2, February 1988, pp. L231–L233.

Akoh, H., F. Shinoki, M. Takahashi, and S. Takada, "S-N-S Josephson junction consisting of Y-Ba-Cu-O/Au/Nb thin films," *Jap. J. Appl. Phys.*, Vol. 27, No. 4, April 1988, pp. L519–L521.

Allen, P. B., "Isotope shift controversies," *Nature*, Vol. 335, No. 6189, September 29, pp. 396–397.

Allgeier, C., and J. S. Schilling, "Meissner effect in Y-Ba-Cu-O and La-Sr-Cu-O high-temperature superconductors," *Phys. Rev. B*, Vol. 35, No. 16, June 1, 1987, pp. 8791–8793.

Anacker, W., "Josephson Computer Technology: An IBM Research Project," *IBM J. Res. Develop.*, Vol. 24, No. 2, March 1980, pp. 107–112.

Anderson, A. C., "Instrumentation at temperatures below 1 K," *Rev. Sci. Instrum.*, Vol. 51, No. 12, December 1980, pp. 1603–1613.

Anderson, P. W., "Theory of flux creep in hard superconductors," *Phys. Rev. Lett.*, Vol. 9, No. 7, October 1, 1962, pp. 309–311.

Anderson, P. W., and Y. B. Kim, "Hard superconductivity: Theory of the motion of Abrikosov flux lines," *Rev. Mod. Phys.*, Vol. 36, January 1964, pp. 39–43.

Anderson, P. W., and E. Abrahams, "Superconductivity theories narrow down," *Nature*, Vol. 327, No. 6121, June 4, 1987, p. 363.

Andrienko, A. V., V. I. Ozhogin, L. V. Podd'Yakov, and A. Yu. Yakubovskii, "Surface impedance measurements of energy gap in high-T_c superconductors," Proc. Int. Conf. on High Temperature Superconductors, Interlaken, Switzerland, *Physica C*, Vol. 153–155, Superconductivity, June 1988, pp. 665–666.

Arnett, P. A., and D. J. Herrel, "Power Design for Gigabit Josephson Logic Systems," *IEEE Trans. Microwave Theory & Tech.*, Vol. MTT-28, No. 5, May 1980, pp. 500–508.

Bagley, B. G., L. H. Greene, J.-M. Tarascon, and G. W. Hull, "Plasma oxidation of the high-T_c superconducting perovskites," *Appl. Phys. Lett.*, Vol. 51, No. 8, August 24, 1987, pp. 622–624.

Baller, T. S., G. N. A. van Veen, and H. A. M. van Hal, "The influence of substrate material and annealing procedure on the properties of superconducting thin films," *Appl. Phys.*, Vol. A46, No. 3, July 1988, pp. 215–220.

Bando, Y., T. Kijima, Y. Kitami, J. Tanaka, F. Izumi, and M. Yokoyama, "Structure and composition analysis of high-T_c superconducting Bi-Ca-Sr-Cu-O oxide by high-resolution analytical electron microscopy," *Jap. J. Appl. Phys.*, Vol. 27, No. 3, March 1988, pp. L358–L360.

Bao, Z. L., F. R. Wang, Q. D. Jiang, S. Z. Wang, Z. Y. Ye, K. Wu, C. Y. Li, and D. L. Yin, "YBaCuO superconducting thin films with zero resistance at 84 K by multilayer deposition," *Appl. Phys. Lett.*, Vol. 51, No. 12, September 21, 1987, pp. 946–947.

Bardeen, J., L. N. Cooper, and J. R. Schreiffer, "Theory of Superconductivity," *Phys. Rev.*, Vol. 108, No. 5, December 1, 1957, pp. 1175–1204.

Batlogg, B., A. P. Ramirez, R. J. Cava, R. B. van Dover, and E. A. Rietman, "Electronic properties of $La_{2-x}Sr_xCuO_4$ high-T_c superconductors," *Phys. Rev. B*, Vol. 35, No. 10, April 1, 1987, pp. 5340–5342.

Batlog, B., R. J. Cava, A. Jayaraman, R. B. van Dover, G. A. Kourouklis, S. Sunshine, D. W. Murphy, L. W. Rupp, H. S. Chen, A. White, K. T. Short, A. M. Mujsce, and E. A. Rietman, "Isotope effect in the high-T_c superconductors $Ba_2YCu_3O_7$ and $Ba_2EuCu_3O_7$," *Phys. Rev. Lett.*, Vol. 58, No. 22, June 1, 1987, pp. 2333–2336.

Batlogg, B., G. Kourouklis, W. Weber, R. J. Cava, A. Jayaraman, A. E. White, K. T. Short, L. W. Rupp, and E. A. Reitman, "Nonzero isotope effect in $La_{1.85}Sr_{0.15}CuO_4$," *Phys. Rev. Lett.*, Vol. 59, No. 8, August 24, 1987, pp. 912–914.

Bean, C. P., "Magnetization of hard superconductors," *Phys. Rev. Lett.*, Vol. 8, No. 250, March 15, 1962, pp. 250–253.

Bean, C. P., "Magnetization of high-field superconductors," *Rev. Mod. Phys.*, Vol. 36, January 1964, pp. 31–39.

Bean, C. P., and J. D. Livingston, "Surface barrier in Type II superconductors," *Phys. Rev. Lett.*, Vol. 12, No. 14, January 6, 1964, pp. 14–16.

Beasly, M. R., "Advanced Superconducting Materials for Electronic Applications," *IEEE Trans. Electron Devices*, Vol. ED-27, No. 10, October 1980, pp. 2009–2015.

Bednorz, J. G., and K. A. Müller, "Possible High-T_c Superconductivity in the Ba-La-Cu-O System," *Z. Phys. B—Condensed Matter 64*, 1986, pp. 189–193.

Bednorz, J. G., and K. A. Müller, "Perovskite-type oxides—The new approach to high-T_c superconductivity," *Rev. Mod. Phys.*, Vol. 60, No. 3, July 1988, pp. 585–600.

Bégin, G., P. R. Chritchlow, and R. Roberge, "Technological impact study of superconductivity at liquid nitrogen temperature for an electric utililty," pres. at 1988 Conf. on Electrical Applications of Superconductivity, Orlando, FL, September 21–23, 1988.

Berlincourt, T. G., "Type II superconductivity," *Rev. Mod. Phys.*, Vol. 36, January 1964, pp. 19–26.

Bertin, C. L., and K. Rose, "Radiant-Energy Detection by Superconducting Films," *J. Appl. Phys.*, Vol. 39, No. 6, May 1968, pp. 2561–2568.

Bertin, C. L., and K. Rose, "Comparison of Superconducting and Semoconducting Bolometers," *J. Appl. Phys.*, Vol. 42, No. 1, January 1971, pp. 163–166.

Bertin, C. L., and K. Rose, "Enhanced-Mode Radiation Detection by Superconducting Films," *J. Appl. Phys.*, Vol. 42, No. 2, February 1971, pp. 631–642.

Beyermann, W. P., B. Alavi, and G. Grüner, "Surface impedance measurements in $La_{1.8}Ba_{0.2}CuO_{4-y}$," *Phys. Rev. B*, Vol. 35, No. 16, June 1, 1987, pp. 8826–8828.

Beyers, R., G. Lim, E. M. Engler, V. Y. Lee, M. L. Ramirez, R. J. Savoy, and R. D. Jacowitz, "Annealing treatment effects on structure and superconductivity in $Y_1Ba_2Cu_3O_{9-x}$," *Appl. Phys. Lett.*, Vol. 51, No. 8, August 24, 1987, pp. 614–616.

Bhargava, R. N., S. P. Herko, and W. N. Osborne, "Improved High-T_c Superconductors," *Phys. Rev. Lett.*, Vol. 59, No. 13, September 28, 1987, pp. 1468–1471.

Biegel, B. A., and R. Singh, "Use of high-temperature superconductors in high speed electronic switches with current gain," *High-Temperature Superconductors*, Vol. 99, Proc. Mat. Res. Soc. Symp., Boston, November 30–December 4, 1987, pp. 873–876.

Bitter, F., "Ultrastrong magnetic fields," *Scientific American*, Vol. 213, No. 1, July 1965, pp. 65–73.

Bömmel, H. E., "Ultrasonic Attenuation in Superconducting Lead," *Phys. Rev.*, Vol. 96, No. 1, October 1, 1954, pp. 220–221.

Bourne, L. C., M. L. Cohen, W. N. Creager, M. F. Crommie, A. M. Stacy, and A. Zettl, "Onset of Superconductivity," in Y-Ba-Cu-O at 100 K," *Phys. Lett. A*, Vol. 120, No. 9, March 23, 1987, pp. 494–496.

Bourne, L. C., M. F. Crommie, A. Zettl, H.-C. zur Loye, S. W. Keller, K. L. Leary, A. M. Stacy, K. J. Chang, M. L. Cohen, and D. E. Morris, "Search for isotope effect in superconducting Y-Ba-Cu-O," *Phys. Rev. Lett.*, Vol. 58, No. 22, June 1, 1987, pp. 2337–2339.

Bourne, L. C., A. Zettl, K. J. Chang, M. L. Cohen, A. M. Stacy, and W. K. Ham, "Elasticity studies of $La_{2-x}Sr_xCuO_4$," *Phys. Rev. B*, Vol. 35, No. 16, June 1, 1987, pp. 8785–8787.

Boyce, J. B., F. Bridges, and T. Claeson, "X-ray-absorbtion studies of $YBa_2Cu_3O_{7-\delta}$ and $GdBa_2Cu_3O_{7-\delta}$ superconductors," *Phys. Rev. B*, Vol. 36, No. 10, October 1, 1987, pp. 5251–5257.

Bozovic, I., D. Mitzi, M. Beasly, A. Kapitulnik, T. Geballe, S. Perkowitz, G. L. Carr, B. Lou, R. Sudharsanan, and S. S. Yom, "Vibrational spectra and lattice instabilities in the high-T_c superconductors $YBa_2Cu_3O_7$ and $GdBa_2Cu_3O_7$," *Phys. Rev. B*, Vol. 36, No. 7, September 1, 1987, pp. 4000–4002.

Braginski, A. I., "Material constraints on electronic applications of oxide superconductors," Proc. Int. Conf. on High Temperature Superconductors, Interlaken, Switzerland, *Physica C*, Vol. 153–155, Superconductivity, June 1988, pp. 1598–1603.

Braginski, A. I., M. G. Forrester, J. Talvacchio, and G. R. Wagner, "Prospects for thin-film electronic devices of high-T_c superconductors," Proc. 5th Int. Workshop on Future Electron Devices—High Temperature Superconducting Devices, June 2–4, 1988, pp. 171–179.

Brown, S. E., J. D. Thompson, J. O. Willis, R. M. Aikin, E. Zirngiebl, J. L. Smith, Z. Fisk, and R. B. Schwarz, "Magnetic and superconducting properties of $RBa_2Cu_3O_x$ compounds," *Phys. Rev. B*, Vol. 36, No. 4, August 1, 1987, pp. 2298–2300.

Buchhold, T. A., "Superconductive Machinery," in *Applied Superconductivity*, V. L. Newhouse, ed., 2 Vols., John Wiley & Sons, New York, 1975.

Bunshah, R. F., "Physical vapor deposition of metals, alloys and ceramics," in *New Trends in Materials Processing*, American Society for Metals, 1967, pp. 200–267.

Campbell, P., "A superconductivity primer," *Nature*, Vol. 330, November 5, 1987, pp. 21–24.

Camps, R. A., J. E. Evetts, B. A. Glowacki, S. B. Newcomb, R. E. Somekh, and W. M. Stobbs, "Microstructure and critical current of superconducting $YBa_2Cu_3O_{7-x}$," *Nature*, Vol. 329, September 17, 1987, pp. 229–232.

Caplin, D., "The electrons are paired," *Nature*, Vol. 326, No. 6116 April 6, 1987, pp. 827–828.

Caplin, D., "Contrasts in critical current," *Nature*, Vol. 335, No. 6187, September 15, 1988, p. 204.

Carr, P. H., "Potential Microwave Applications of High Temperature Superconductors," *Microwave Journal*, Vol. 30, No. 12, December 1987, pp. 91–94.

Cassel, B., "Thermal analysis in SC R&D," *Superconductor Industry*, Fall 1988, pp. 24–29.

Cava, R. J., R. B. van Dover, B. Batlogg, and E. A. Rietman, "Bulk Superconductivity at 36 K in $La_{1.8}Sr_{0.2}CuO_4$," *Phys. Rev. Lett.*, Vol. 58, No. 4, January 26, 1987, pp. 408–410.

Cava, R. J., B. Batlogg, R. B. van Dover, D. W. Murphy, S. Sunshine, T. Siegrist, J. P. Remeka, E. A. Rietman, S. Zahurak, and G. P. Espinosa, "Bulk Superconductivity at 91 K in Single-Phase Oxygen-Deficient Perovskite $Ba_2YCu_3O_{9-\delta}$," *Phys. Rev. Lett.*, Vol. 58, No. 16, April 20, 1987, pp. 1676–1679.

Cava, R. J., B. Batlogg, C. H. Chen, E. A. Rietman, S. M. Zahurak, and D. Werder, "Oxygen stoichiometry, superconductivity and normal-state properties of $YBa_2Cu_3O_{7-\delta}$," *Nature*, Vol. 329, October 1, 1987, pp. 423–425.

Cava, R. J., B. Batlogg, C. H. Chen, E. A. Rietman, S. M. Zahurak, and D. Werder, "Single-phase 60-K bulk superconductor in annealed $Ba_2YCu_3O_{7-\delta}(0.3 < \delta < 0.4)$ with correlated oxygen vacancies in the Cu-O chains," *Phys. Rev. B*, Vol. 36, No. 10, October 1, 1987, pp. 5719–5722.

Cava, R. J., B. Batlogg, J. J. Krajewski, R. Farrow, L. W. Rupp, Jr., A. E. White, K. Short and T. Kometani, "Superconductivity near 30 K without copper: $Ba_{0.6}K_{0.4}BiO_3$ perovskite," *Nature*, Vol. 332, No. 6167, April 28, 1988, pp. 814–816.

Cava, R. J., B. Batlogg, J. J. Krajewski, L. W. Rupp, L. F. Schneemeyer, T. Siegrist, R. B. vanDover, P. Marsh, W. F. Peck, Jr., P. K. Gallagher, S. H. Glarum, J. H. Marshall, R. C. Farrow, J. V. Waszczak, R. Hull, and P. Trevor, "Superconductivity near 70 K in a new family of layered copper oxides," *Nature*, Vol. 336, No. 6196, November 1988, pp. 211–214.

Chaddah, P., G. Ravi Kumar, A. K. Grover, C. Radhakrishnamurty, and G. V. Subba Rao, "Critical state model and the magnetic behavior of high-T_c superconductors," pres. Int. Conf. on Critical Currents in High-T_c Superconductors, Snowmass Village, CO, August 16–19, 1988.

Chan, H. W. K., and T. Van Duzer, "Josephson Nonlatching Logic Circuits," *IEEE J. Solid-State Circuits*, Vol. SC-12, No. 1, February 1977, pp. 73–79.

Cheetham, A. K., A. M. Chippendale, and S. J. Hibble, "Control of copper valence in $Bi_2Sr_{2-x}CaCu_2O_8$," *Nature*, Vol. 333, No. 6168, May 5, 1988, p. 21.

Chen, G., and W. A. Goddard III, "The magnon pairing mechanism of superconductivity in cuprate ceramics," *Science*, Vol. 239, No. 4842, February 19, 1988, pp. 899–902.

Chen, J. T., L. E. Wenger, C. J. McEwan, and E. M. Logothetis, "Observation of the Reverse ac Josephson Effect in Y-Ba-Cu-O at 240 K," *Phys. Rev. Lett.*, Vol. 58, No. 19, May 11, 1987, pp. 1972–1975.

Chen, J.-W., S. J. Keating, C. Y. Keating, X. Wu, J. Xu, P. E. Reyes-Morel, and T. Y. Tien, "Structural behaviour and superconductivity of $YBa_2Cu_3O_x$," *Solid State Commun.*, Vol. 63, No. 11, 1987, pp. 997–1001.

Chen, Xiao-Dong, Sang Young Leeh, J. P. Golbeng, Sung-Ik Lee, R. D. McMichael, Yi Song, Tae W. Noh, and J. R. Gaines, "Practical preparation of copper oxide superconductors," *Rev. Sci. Instrum.* 58(9), September 1987, pp. 1565–1571.

Cheong, S-W., S. E. Brown, Z. Fisk, R. S. Kwok, J. D. Thompson, E. Zirngiebl, and G. Gruner, "Normal-state properties of $ABa_2Cu_3O_{7-y}$ compounds ($A=Y$ and Gd): Electron-electron correlations," *Phys. Rev. B*, Vol 36, No. 7, September 1, 1987, pp. 3913–3916.

Cheong, S-W., Z. Fisk, R. S. Kwok, J. P. Remeika, J. D. Thompson, and G. Gruner, "Electronic anisotropy in single-crystal La_2CuO_4," *Phys. Rev. B*, Vol. 37, No. 10, April 1, 1988, pp. 5916–5919.

Chrisey, D. B., G. P. Summers, W. G. Maisch, E. A. Burke, W. T. Elam, H. Herman, J. P. Kirkland, and R. A. Neiser, "Catastrophic loss of superconductivity in ion-irradiated films of $YBa_2Cu_3O_{y-\delta}$," *Appl. Phys. Lett.*, Vol. 53, No. 11, September 12, 1988, pp. 1001–1003.

Chrisey, D. B., W. G. Maisch, G. P. Summers, A. R. Knudson, and E. A. Burke, "The influence of radiation damage on the superconducting properties of thin film $YBa_2Cu_3O^{+}_{y-\delta}$," Accept. for pub. in *IEEE Trans. Nucl. Sci.*, December 1988.

Chu, C. W., P. H. Hor, R. L. Meng, L. Gao, Z. J. Huang, and Y. Q. Wang, "Evidence for Superconductivity above 40 K in the La-Ba-Cu-O System," *Phys. Rev. Lett.*, Vol. 58, No. 4, January 26, 1987, pp. 405–407.

Chu, C. W., P. H. Hor, R. L. Meng, Z. J. Huang, L. Gao, Y. Y. Xue, Y. Y. Sun, Y. Q. Wang, and J. Bechtold, "New materials and high temperature superconductivity," Proc.

Int. Conf. on High Temperature Superconductors, Interlaken, Switzerland, *Physica C*, Vol. 153–155, Superconductivity, June 1988, pp. 1138–1143.

Clark, G. J., A. D. Marwick, R. H. Koch, and R. B. Laibowitz, "Effects of radiation damage on ion-implanted thin films of metal-oxide superconductors," *Appl. Phys. Lett.*, Vol. 51, No. 2, July 13, 1987, pp. 139–141.

Clark, G. J., F. K. Legoues, A. D. Marwick, R. B. Laibowitz, and R. H. Koch, "Radiation damage in thin films of high-T_c superconductors," *High-Temperature Superconductors*, Vol. 99, Proc. Mat. Res. Soc. Symp., Boston, November 30–December 4, 1987, pp. 127–132.

Clarke, J., "Advances in SQUID Magnetometers," *IEEE Trans. Electron Devices*, Vol. ED-27, No. 10, October 1980, pp. 1896–1908.

Clarke, J., "SQUIDs, brains and gravity waves," *Physics Today*, March 1986, pp. 36–44.

Clarke, J., "Small-scale analog applications of high-transition-temperature superconductors," *Nature*, Vol. 333, No. 6168, May 5, 1988, pp. 29–35.

Clougherty, D. P., and K. H. Johnson, "Thermodynamic critical fields in high-T_c superconductivity," Proc. Int. Conf. on High Temperature Superconductors, Interlaken, Switzerland, *Physica C*, Vol. 153–155, Superconductivity, June 1988, pp. 699–700.

Cohn, D. R., L. Bromberg, W. Halverson, B. Lax, and P. P. Woskov, "Possible high-frequency cavity and waveguide applications of high temperature superconductors," Massachussets Institute of Technology, PFC/JA/87-49, October 1, 1987.

Collocott, S. J., G. K. White, S. X. Dou, and R. K. Williams, "Thermal properties of the high-T_c superconductors $La_{1.85}Sr_{0.15}CuO_4$ and $YBa_2Cu_3O_7$," *Phys. Rev. B*, Vol. 36, No. 10, October 1, 1987, pp. 5684–5685.

Cook, R. F., T. R. Dinger, and D. R. Clarke, "Fracture toughness measurements of $YBa_2Cu_3O_x$ single crystals," *Appl. Phys. Lett.*, Vol. 51, No. 6, August 10, 1987, pp. 454–456.

Cooke, D. W., H. Remmp, Z. Fisk, J. L. Smith, and M. S. Jahan, "Thermally stimulated luminescence from rare-earth-doped barium copper oxides," *Phys. Rev. B*, Vol. 36, No. 4, August 1, 1987, pp. 2287–2289.

Cooper, J. R., B. Alavi, L-W Zhou, W. P. Beyermann, and G. Grüner, "Thermoelectric power of some high-T_c oxides," *Phys. Rev. B*, Vol. 35, No. 16, June 1, 1987, pp. 8794–8796.

Cooper, L. N., "Theory of Superconductivity," *Am. J. Phys.*, February 1960, pp. 91–101.

Cooper, L. N., "Origins of the Theory of Superconductivity," *IEEE Trans. Magn.*, Vol. MAG-23, No. 2, March 1987, pp. 376–379.

Corak, W. S., B. B. Goodman, C. B. Satterthwaite, and A. Wexler, "Atomic Heats of Normal and Superconducting Vanadium," *Phys. Rev.*, Vol. 102, No. 3, May 1, 1956, pp. 656–661.

Couach, M., F. Monnier, A. De Combarieu, and E. Bedin, "On the stability of carbon glass thermometers cycled at low temperatures under high magnetic fields," *Cryogenics*, Vol. 22, September 1982, pp. 483–484.

Cowey, L., H. Jones, and D. Dew-Hughes, "Technique for short sample testing of high-T_c superconducting wire," *Cryogenics*, Vol. 28, No. 3, March 1988, pp. 181–182.

Cox, T., "Superconductivity discovers the chemical bond," *New Scientist*, Vol. 118, No. 1618, June 1988, pp. 64–66.

Crabtree, G. W., J. Z. Liu, A. Umezawa, W. K. Kwok, C. H. Sowers, S. K. Malik, B. W. Veal, D. J. Lam, B. Brodsky, and J. W. Downey," *Phys. Rev. B*, Vol. 36, No. 17, September 1, 1987, pp. 4021–4024.

Crowe, J. W., "Trapped Flux Superconducting Memory," *IBM J. Res. Develop.*, Vol. 1, 1957, pp. 295–303.

Dahl, P. F., "Kammerlingh Onnes and the discovery of superconductivity: The Leyden years, 1911–1914," *Historical Studies in the Physical Sciences*, Vol. 15, Part 1, 1984, pp. 1–37.

Dahl, P. F., "Superconductivity after world war I and circumstances surrounding the discovery of a state $B = 0$," *Historical Studies in the Physical and Biological Sciences*, Vol. 16, Part 1, 1986, pp. 1–58.

Dai, U., G. Deutscher, and R. Rosenbaum, "Critical current of $Y_1Ba_2Cu_3O_4$ in strong applied fields," *Appl. Phys. Lett.*, Vol. 51, No. 6, August 10, 1987, pp. 460–462.

Dalrymple, B. J., and D. E. Prober, "Upper Critical Fields of the Superconducting Layered Compounds $Nb_{1-x}Ta_xSe_2$," *J. Low Temp. Phys.*, Vol. 56, Nos. 5/6, 1984, pp. 545–574.

Damento, M. A., K. A. Gschneidner, Jr., and R. W. McCallum, "Preparation of single crystals of superconducting $YBa_2Cu_3O_{y-x}$ from CuO melts," *Appl. Phys. Lett.*, Vol. 51, No. 9, August 31, 1987, pp. 690–691.

Datta, T., C. P. Poole, Jr., H. A. Farach, C. Almasan, J. Estrada, D. U. Gubser, and S. A. Wolf, "New approach to characterizing the high-temperature superconducting transition," *Phys. Rev. B*, Vol. 37, No. 13, May 1, 1988, pp. 7843–7845.

D'Addario, L., "Saturation of the SIS mixer by out-of-band signals," *IEEE Trans. Microwave Theory & Tech.*, Vol. 36, No. 6, June 1988, pp. 1103–1105.

de Bellefon, A., D. Broskiewicz, R. Bruére-Dawson, P. Espigat, B. Mettout, N. Perrin, D. Limange, L. C. L. Yuan, and G. Waysand, "Applications of superheated superconducting detectors," *IEEE Trans. Nucl. Sci.*, Vol. 35, No. 1, February 1988, pp. 73–77.

de Bruyn Ouboter, R., "Superconductivity: Discoveries During the Early Years of Low Temperature Research at Leiden, 1908–1914," *IEEE Trans. Magn.*, Vol. MAG-23, No. 2, March 1987, pp. 355–370.

Delaney, M. A., R. S. Withers, A. C. Anderson, J. B. Green, and Robert W. Mountain, "Superconductive Delay Line with Integral MOSFET Taps," *IEEE Trans. Magn.*, Vol. MAG-23, No. 2, March 1987, pp. 791–795.

de Lima, O. F., J. Mattson, C. H. Sowers, and M. B. Brodsky, "Superconducting thin films based on $La_{2-x}Sr_xCuO_4$," *Appl. Phys. Lett.*, Vol. 51, No. 5, August 3, 1987, pp. 369–370.

Deutscher, G., "Superconducting glass and related properties," Proc. Int. Conf. on High Temperature Superconductors, Interlaken, Switzerland, *Physica C*, Vol. 153–155, Superconductivity, June 1988, pp. 15–20.

Dhong, S. H., and T. Van Duzer, "Minimum-Width Control-Current Pulse for Josephson Logic Gates," *IEEE Trans. Electron Devices*, Vol. ED-27, No. 10, October 1980, pp. 1965–1973.

Dijkkamp, D., and T. Venkatesan, "Preparation of Y-Ba-Cu oxide superconductor thin films using pulsed laser evaporation from high T_c bulk material," *Appl. Phys. Lett.*, Vol. 51, No. 8, August 24, 1987, pp. 619–621.

Dimos, D., P. Chaudhari, J. Mannhart, and F. K. LeGoues, "Orientation dependence of grain-boundary critical currents in $YBa_2Cu_3O_{7-\delta}$ bicrystals," *Phys. Rev. Lett.*, Vol. 61, No. 2, July 11, 1988, pp. 219–222.

Dinger, T. R., T. K. Worthington, W. J. Gallagher, and R. L. Sandstrom, "Direct observation of electronic anisotropy in single-crystal $Y_1Ba_2Cu_3O_{7-x}$," *Phys. Rev. Lett.*, Vol. 58, No. 25, June 22, 1987, pp. 2687–2690.

Dinger, T. R., and S. W. Tozer, "Old behaviour in new materials," *Nature*, Vol. 332, No. 6161, March 17, 1988, p. 204.

Dominec, J., L. Smrcka, P. Vasek, and S. Geurten, "Stability of $YBa_2Cu_3O_{7-y}$ Superconductor in Water," *Solid State Commun.*, Vol. 65, No. 5, 1988, pp. 373–374.

Donaldson, G. B., and H. Faghihi-Nejad, "An Ellipsometric Study of RF Sputter Oxidation of Lead-Indium Alloys," *IEEE Trans. Electron Devices*, Vol. ED-27, No. 10, October 1980, pp. 1988–1997.

Doss, J. D., D. W. Cooke, C. W. McCabe, and M. A. Maez, "Noncontact methods used for characterization of high-T_c superconductors," *Rev. Sci. Instrum.*, Vol. 59, No. 4, April 1988, pp. 659–661.

Doss, J. D., "Measuring the new superconductors," *Nature*, Vol. 338, No. 6214, March 30, 1989, pp. 441–442.

Dou, S. X., A. J. Bourdillon, C. C. Sorrell, S. P. Ringer, K. E. Easterling, N. Savvides, J. B. Dunlop, and R. B. Roberts, "Electron microscopy and microanalysis of a $YBa_2Cu_3O_x$ superconducting oxide," *Appl. Phys. Lett.*, Vol. 51, No. 7, August 17, 1987, pp. 535–537.

Driessen, A., R. Griessen, N. Koeman, E. Salomons, R. Brouwer, D. G. de Groot, K. Heeck, H. Hemmes, and J. Rector, "Pressure dependence of the T_c of $YBa_2Cu_3O_7$ up to 170 kbar," *Phys. Rev. B*, Vol. 36, No. 10, October 1, 1987, pp. 5602–5605.

Dunlap, B. D., M. V. Nevitt, M. Slaski, T. E. Klippert, Z. Sungaila, A. G. McKale, D. W. Capone, R. B. Poeppel, and B. K. Flandermeyer, "Heat capacity of superconducting $La_{1.85}Sr_{0.15}CuO_4$," *Phys. Rev. B*, Vol. 35, No. 13, May 1, 1987, pp. 7210–7212.

Durny', R., J. Hautala, S. Ducharme, B. Lee, O. G. Symko, P. C. Taylor, and D. J. Zheng, "Microwave absorption in the superconducting and normal phases of Y-Ba-Cu-O," *Phys. Rev. B*, Vol. 36, No. 4, August 1, 1987, pp. 2361–2363.

Eaglesham, D. J., and C. J. Humphreys, "New phases in the superconducting Y:Ba:Cu:O system," *Appl. Phys. Lett.*, Vol. 51, No. 6, August 10, 1987, pp. 457–459.

Eatough, M. O., D. S. Ginley, B. Morosin, and E. L. Venturini, "Orthorhombic-tetragonal transition in high-temperature superconductor $YBa_2Cu_3O_7$," *Appl. Phys. Lett.*, Vol. 51, No. 5, August 3, 1987, pp. 367–368.

Ekin, J. W., "Superconductors," in *Materials at Low Temperatures*, R. P. Reed and A. F. Clark, eds., American Society for Metals, Metals Park, OH, 1983.

Ekin, J. W., A. J. Panson, A. I. Braginski, M. A. Janoko, M. Hong, J. Kwo, S. H. Liou, D. W. Capone II, and B. Flandermeyer, "Transport Critical-Current Characteristics of $Y_1Ba_2Cu_3O_x$," Proc. Mat. Res. Soc. Symp. on High Temperature Superconductors, April 23–24, 1987. *Mat. Res. Soc.*, Vol. EA-11, June 1987, pp. 223–226.

Ekin, J. W., T. M. Larson, and N. F. Bergren, "High T_c superconductor/noble-metal contacts with surface resistivities in the 10^{-10} Ω-cm^2 range," *Appl. Phys. Lett.*, Vol. 52, No. 21, May 23, 1988, pp. 1819–1821.

Emery, V. J., "Magnetic effects explored," *Nature*, Vol. 333, No. 6168, May 5, 1988, pp. 14–15.

Endo, U., S. Koyama, and T. Kawai, "Preparation of the high-T_c phase of the Bi-Sr-Ca-Cu-O superconductor," *Jap. J. Appl. Phys.*, Vol. 27, No. 8, August 1988, pp. L1476–L1479.

Enokihara, A., H. Higashino, K. Setsune, T. Mitsuyu, and K. Wasa, "Superconductivity in 2-μm wide strip line of Gd-Ba-Cu-O thin film fabricated by low-temperature process," *Jap. J. Appl. Phys.*, Vol. 27, No. 8, August 1988, pp. L1521–1523.

Enomoto, Y., T. Murakami, and M. Suzuki, "Infrared optical detector using superconducting oxide thin film," Proc. Int. Conf. on High Temperature Superconductors, Interlaken, Switzerland, *Physica C*, Vol. 153–155, Superconductivity, June 1988, pp. 1592–1597.

Eryu, O., K. Murakami, K. Takita, K. Masuda, H. Uwe, H. Kudo, and T. Sakudo, "Y-Ba-Cu oxide films formed with pulsed-laser induced fragments," *Jap. J. Appl. Phys.*, Vol. 27, No. 4, April 1988, pp. L628–L631.

Esaki, L., "Long journey into tunneling," *Rev. Modern Phys.*, Vol. 46, No. 2, April 1974, pp. 237–244.

Faltens, T. A., W. K. Ham, S. W. Keller, K. J. Leary, J. N. Michaels, A. M. Stacy, H.-C. zur Loye, D. E. Morris, T. W. Barbee III, L. C. Bourne, M. L. Cohen, S. Hoen, and A. Zettl, "Observation of an oxygen isotope shift in the superconducting transition temperature of $La_{1.85}Sr_{0.15}CuO_4$," *Phys. Rev. Lett.*, Vol. 59, No. 8, August 24, 1987, pp. 915–918.

Farrell, D. E., B. S. Chandrasekhar, M. R. DeGuire, M. M. Fang, V. G. Kogan, J. R. Clem, and D. K. Finnemore, "Superconducting properties of aligned crystalline grains of $Y_1Ba_2Cu_3O_{7-\delta}$," *Phys. Rev. B*, Vol. 36, No. 7, September 1, 1987, pp. 4025–4035.

Fathy, A., D. Kalokitis, and E. Belohoubek, "Microwave characteristics and characterization of high-T_c superconductors," *Microwave Journal*, Vol. 31, No. 10, October 1988, pp. 75–94.

Feigelson, R. S., D. Gazit, D. K. Fork, and T. H. Geballe, "Superconducting Bi-Ca-Sr-Cu-O fibers grown by the laser-heated pedestal growth method," *Science*, Vol. 240, No. 4859, June 17, 1988, pp. 1642–1645.

Felici, R., J. Penfold, R. C. Ward, E. Olsi, and C. Matacotta, "Determination of the magnetic penetration depth of the high-T_c superconductor YBa_2Cu_3-O_{7-x} by polarized neutron reflection," *Nature*, Vol. 329, October 8, 1987, pp. 523–525.

Fietz, W. A., M. R. Beasley, J. Silcox, and W. W. Webb, "Magnetization of superconducting Nb-25% Zr wire," *Phys. Rev.*, Vol. 136, No. 2A, October 19, 1964, pp. A335–A345.

Fink, H. J., "Inherent low-frequency losses of the superconducting surface sheath," *Phys. Rev. Lett.*, Vol. 16, No. 11, March 14, 1966, pp. 447–450.

Fiory, A. T., A. F. Hebard, P. M. Mankiewich, and R. E. Howard, "Penetration depths of high T_c films measured by two-coil mutual inductances," *Appl. Phys. Lett.*, Vol. 52, No. 25, June 20, 1988, pp. 2165–2167.

Fisher, R. A., J. E. Gordon, and N. E. Phillips, "Specific heat clues for theory," *Nature*, Vol. 330, December 17, 1987, pp. 601–602.

Fisher, R. A., J. E. Gordon, S. Kim, N. E. Phillips, and A. M. Stacy, "Specific heat of $YBa_2Cu_3O_7$," Proc. Int. Conf. on High Temperature Superconductors, Interlaken, Switzerland, *Physica C*, Vol. 153–155, Superconductivity, June 1988, pp. 1092–1095.

Fisk, Z., J. D. Thompson, E. Zirngiebl, J. L. Smith, and S.-W. Cheong, "Superconductivity of Rare Earth-Barium-Copper Oxides," *Solid State Commun.*, Vol. 62, No. 11, 1987, pp. 743–744.

Fitzgerald, K., "Superconductivity: Fact vs. Fancy," *IEEE Spectrum*, Vol. 25, No. 5, May 1988, pp. 33–41.

Flanders, P. J., "An alternating gradient magnetometer," *J. Appl. Phys.*, Vol. 63, No. 8, April 15, 1988, pp. 3940–3945.

Flükiger, R., T. Müller, W. Goldacker, T. Wolf, E. Seibt, I. Apfelstedt, H. Küpfer, and W. Schauer, "Metallurgy and critical currents in $YBa_2Cu_3O_7$ wires," Proc. Int. Conf. on High Temperature Superconductors, Interlaken, Switzerland, *Physica C*, Vol. 153–155, Superconductivity, June 1988, pp. 1574–1579.

Foner, S., and T. P. Orlando, "Superconductors: The Long Road Ahead," *Technology Review*, Vol. 91, No. 2, February–March 1988, pp. 36–47.

Forgan, T., "Finding characteristic wavelengths," *Nature*, Vol. 329, October 8, 1987, pp. 483–485.

Fork, D. K., J. B. Boyce, F. A. Ponce, R. I. Johnson, G. B. Anderson, G. A. N. Connel, C. B. Eom, and T. H. Geballe, "Preparation of oriented Bi-Ca-Sr-Cu-O thin films using pulsed laser deposition," *Appl. Phys. Lett.*, Vol. 53, No. 4, July 25, 1988, pp. 337–339.

Francois, M., E. Walker, J.-L. Jorda, K. Yvon, and P. Fischer, "Structure of the high-temperature superconductor $Ba_2YCu_3O_7$ by X-ray and neutron powder diffraction," *Solid State Commun.*, Vol. 63, No. 12, 1987, pp. 1149–1153.

Frost, H. M., J. H. Gieske, D. E. Peterson, and E. F. Brown, "*In situ* probing of superconducting ceramics during sintering," Workshop on Processing of Ceramic Superconductors, Los Alamos National Laboratory, Pub. #LA-UR-88-2633, August 15–17, 1988.

Fujii, H., H. Kawanaka, W. Ye, S. Orimo, and H. Fukuba, "Effect of hydrogen absorption on superconductivity in $YBa_2Cu_3O_{6.91}$ and $GdBa_2Cu_3O_{6.89}$," *Jap. J. Appl. Phys.*, Vol. 27, No. 4, April 1988, pp. L525–L528.

Fujiwara, Y., M. Tonouchi, and T. Tobayashi, "Thermally stimulated luminescence from high-T_c superconducting Tl-Ba-Ca-Cu-O system," Accepted for publication in *Jap. J. Appl. Phys.*, Vol. 27, No. 9, September 1988.

Fukutomi, M., J. Machida, Y. Tanaka, T. Asano, H. Maeda, and K. Hoshino, "Sputter deposition of BiSrCaCuO thin films," *Jap. J. Appl. Phys.*, Vol. 27, No. 4, April 1988, pp. L632–633.

Fukutomi, M., J. Machida, Y. Tanaka, T. Asano, T. Yamamoto, and H. Maeda, "New technique for preparation of BiSrCaCuO thin films with T_c of 100 K and above," *Jap. J. Appl. Phys.*, Vol. 27, No. 8, August 1988, pp. L1484–L1486.

Fulde, P., "Current theories of the high-T_c superconducting materials," Proc. Int. Conf. on High Temperature Superconductors, Interlaken, Switzerland, *Physica C*, Vol. 153–155, Superconductivity, June 1988, pp. 1769–1774.

Gallagher, W. J., R. L. Sandstrom, T. R. Dinger, T. M. Shaw, and D. A. Chance, "Identification and preparation of single phase 90 K oxide superconductor and structural determination by lattice imaging," *Solid State Commun.*, Vol. 63, No. 2, 1987, pp. 147–150.

Gao, L., Z. J. Huang, R. L. Meng, P. H. Hor, J. Bechtold, Y. Y. Sun, C. W. Chu, Z. Z. Sheng, and A. M. Hermann, "Bulk superconductivity in $Tl_2CaBa_2Cu_2O_{8+\delta}$ up to 120 K," *Nature*, Vol. 332, No. 6165, April 14, 1988, pp. 623–624.

Gao, Y., T. J. Wagener, J. H. Weaver, B. Flandermeyer, and D. W. Capone II, "Reaction and intermixing at metal-superconductor interfaces: Fe/$YBa_2Cu_3O_{6.9}$," *Appl. Phys. Lett.*, Vol. 51, No. 13, September 28, 1987, pp. 1032–1034.

Garland, M. M., "Fabrication of Y-Ba-Cu-O mixed phase superconductors," *Appl. Phys. Lett.*, Vol. 51, No. 13, September 28, 1987, pp. 1030–1031.

Garwin, L., and P. Campbell, "New superconductors in perspective," *Nature*, Vol. 330, December 17, 1987, pp. 611–614.

Geballe, T. H., and J. K. Hulm, "Superconductivity—The state that came in from the cold," *Science*, Vol. 239, No. 4838, January 22, 1988, pp. 367–375.

Gheewala, T. R., "Design of 2.5-Micrometer Josephson Current Injection Logic (CIL)," *IBM J. Res. Dev.*, Vol. 24, No. 2, March 1980, pp. 130–154.

Gheewala, T. R., "Josephson-Logic Devices and Circuits," *IEEE Trans. Electron Devices*, Vol. ED-27, No. 10, October 1980, pp. 1857–1869.

Giaever, I., "Energy Gap in Superconductors Measured by Electron Tunneling," *Phys. Rev. Lett.*, Vol. 5, No. 4, August 15, 1960, pp. 147–148.

Giaever, I., "Electron tunneling and superconductivity," *Rev. Mod. Phys.*, Vol. 46, No. 2, April 1974, pp. 245–250.

Gieske, J. H., and Frost, H. M., "Sound velocity measurements in green-body ceramics as a function of sintering temperature," Proc. Review of Progress in QNDE, La Joya, CA, August 1–5, 1988.

Ginley, D. S., E. L. Venturini, J. F. Kwak, R. J. Baughman, M. J. Carr, P. F. Hlava, J. E. Schirber, and B. Morosin, "A 120 K Bulk Superconductor $(Tl_1Ba_1Ca_1)Cu_2O_x$," to be published in *Physica C*; cited in *High-T_c Update*, Vol. 2, No. 7, April 1, 1988, p. 5.

Ginley, D. S., J. F. Kwak, R. P. Hellmer, R. J. Baughman, E. L. Venturini, and B. Morosin, "Sequential E-beam Evaporated Films of $Tl_2CaBa_2Cu_2O_y$," cited in *High-T_c Update*, Vol. 2, No. 10, May 15, 1988, p. 4.

Ginsberg, D. M., "Resource Letter Scy-1 on Superconductivity," *Am. J. Phys.*, Vol. 32, No. 2, February 1964, pp. 1–5.

Ginsberg, D. M., S. E. Inderhees, M. B. Salamon, N. Goldenfeld, J. P. Rice, and B. G. Pazol, "Specific heat of a single crystal of $YBa_2Cu_3O_{7-\delta}$: Fluctuation effects in a bulk superconductor in zero magnetic field and in a magnetic field," Proc. Int. Conf. on High Temperature Superconductors, Interlaken, Switzerland, *Physica C*, Vol. 153–155, Superconductivity, June 1988, pp. 1082–1085.

Glover, R. E., III, and M. Tinkham, "Conductivity of superconducting films for photon energies between 0.3 and $40kT_c$," *Phys. Rev.*, Vol. 108, No. 2, October 15, 1957, pp. 243–256.

Golben, J. P., Sung-Ik Lee, Sang Young Lee, Yi Song, Tae W. Noh, Xiao-Dong Chen, and J. R. Gaines, "Superconductivity of the single-phase compound $Er_1Ba_2Cu_3O_{9-\delta}$: Transport and structural properties," *Phys. Rev. B*, Vol. 35, No. 16, June 1, 1987, pp. 8705–8708.

Goldfarb, R. B., and J. V. Minervini, "Calibration of ac susceptometer for cylindrical specimens," *Rev. Sci. Instrum.*, Vol. 55, No. 5, May, 1984, pp. 761–764.

Goldfarb, R. B., A. F. Clark, A. I. Braginski, and A. J. Panson, "Evidence for two superconducting components in oxygen-annealed single-phase Y-Ba-Cu-O," *Cryogenics*, Vol. 27, September 1987, pp. 475–480.

Goncharov, I. N., and L. Miu, "Improved soldered joints of technical current-carrying superconductors," *Cryogenics*, Vol. 28, No. 3, March 1988, pp. 183–184.

Goodman, B. B., "Type II or London superconductors," *Rev. Mod. Phys.*, Vol. 36, January 1964, pp. 12–19.

Gordon, J. E., "Small oven for obtaining temperatures to 1350°C," *Rev. Sci. Instrum.*, Vol. 59, No. 3, March 1988, pp. 497–498.

Gorshunov, B. P., G. V. Koslov, S. I. Krasnosvobodtsev, E. V. Pechen, A. M. Prokhorov, A. S. Prokhorov, O. I. Syrotynsky, and A. A. Volkov, "Submillimetre properties of high-T_c superconductors," Proc. Int. Conf. on High Temperature Superconductors, Interlaken, Switzerland, *Physica C*, Vol. 153–155, Superconductivity, June 1988, pp. 667–668.

Gorter, C. J., "Some remarks on superconductivity of the second kind," *Rev. Mod. Phys.*, Vol. 36, January 1964, pp. 27–31.

Gorter, C. J., "Superconductivity until 1940 in Leiden and as seen from there," *Rev. Mod. Phys.*, Vol. 36, January 1964, pp. 3–7.

Goto, E., "The parametron, a digitial computing element which utilizes parametric oscillation," Proc. IRE, Vol. 47, August 1959, pp. 1309–1316.

Gough, C. E., M. S. Colclough, E. M. Forgan, R. G. Jordan, M. Keene, C. M. Muirhead, A. I. M. Rae, N. Thomas, J. S. Abell, and S. Sutton, "Flux quantization in a high-T_c superconductor," *Nature*, Vol. 326, No. 6116, April 6, 1987, p. 855.

Gough, C. E., S. K. Khamas, T. S. M. Maclean, M. J. Mehler, N. McNalford, and M. M. Harmer, "Critical currents in a high-T_c superconducting short dipole antenna," *IEEE Trans. Magn.*, Vol. 25, No. 2, March 1989, pp. 1313–1314.

Grader, G. S., P. K. Gallagher, and E. M. Gyorgy, "High-temperature resistivity of the $Ba_2YCu_3O_x$ superconductor," *Appl. Phys. Lett.*, Vol. 51, No. 14, October 5, 1987, pp. 1115–1117.

Grader, G. S., H. M. O'Bryan, and W. W. Rhodes, "Improved press forging of $Ba_2YCu_3O_x$ superconductor," *Appl. Phys. Lett.*, Vol. 52, No. 21, May 23, 1988, pp. 1831–1833.

Grader, G. S., P. K. Gallagher, and A. T. Fiory, "Hall coefficient and oxygen stoichiometry in $YBa_2Cu_3O_{7-\delta}$ ceramics at elevated temperatures," *Phys. Rev. B.*, Vol. 38, No. 1, July 1, 1988, pp. 844–847.

Grader, G. S., E. M. Gyorgy, L. G. Van Uitert, W. H. Gordkiewicz, T. R. Kyle, and M. Eibschutz, "Persistient currents in Tl-Ba-Ca-Cu-O superconductors," *Appl. Phys. Lett.*, Vol. 53, No. 4, July 25, 1988, pp. 319–323.

Grant, P. M., R. B. Beyers, E. M. Engler, G. Lim, S. S. Parkin, M. L. Ramirez, V. Y. Lee, A. Nazzal, J. E. Vazquez, and R. J. Savoy, "Superconductivity above 90 K in the compound $YBa_2Cu_3O_x$: Structural, transport, and magnetic properties," *Phys. Rev. B*, Vol. 35, No. 13, May 1, 1987, pp. 7242–7244.

Grant, P. M., S. S. P. Parkin, V. Y. Lee, E. M. Engler, M. L. Ramirez, J. E. Vazquez, G. Lim, and R. L. Greene, "Effect for Superconductivity in La_2CuO_4," *Phys. Rev. Lett.*, Vol. 58, No. 23, June 8, 1987, pp. 2482–2485.

Gray, K. E., "Experimental Aspects of High-T_c Superconductors," 3rd Workshop on RF Superconductivity, Argonne National Laboratory, September 14–19, 1987.

Greaves, C., "Infinite stacks of copper oxide," *Nature*, Vol. 334, No. 6179, July 21, 1988, pp. 193–194.

Green, J. B., L. N. Smith, A. C. Anderson, S. A. Reible, and R. S. Withers, "Analog Signal Correlator using Superconductive Integrated Components," *IEEE Trans. Magn.*, Vol. MAG-23, No. 2, March 1987, pp. 895–898.

Greiner, J. H., C. J. Kircher, S. P. Klepner, S. K. Lahri, A. J. Warnecke, S. Basavaiah, E. T. Yen, J. M. Baker, P. R. Brosious, H.-C. W. Huang, M. Murakami, and I. Ames, "Fabrication Process for Josephson Integrated Circuits," *IBM J. Res. Dev.*, Vol. 24, No. 2, March 1980, pp. 195–205.

Griessen, R., "Pressure dependence of high-T_c superconductors," *Phys. Rev. B*, Vol. 36, No. 10, October 1, 1987, pp. 5284–5290.

Gross, R., J. Bosch, R. P. Huebener, J. Mannhart, C. C. Tsuei, M. Scheuermann, M. M. Oprysko, and C. C. Chi, "Spatially resolved observation of the critical current in high-T_c superconducting films," *Nature*, Vol. 332, No. 6167, April 28, 1988, pp. 818–819.

Guéret, P., A. Moser, P. Wolf, "Investigations for a Josephson Computer Main Memory with Single-Flux-Quantum Cells," *IBM J. Res. Dev.*, Vol. 24, No. 2, March 1980, pp. 155–166.

Gunn, M., and J. Porter, "Out of the freezer, into the fire," *New Scientist*, Vol. 118, No. 1618, June 1988, pp. 58–63.

Guo, Y., J.-M. Langlois, and W. A. Goddard III, "Electronic structure and valence-bond band structure of cuprate superconducting materials," *Science*, Vol. 239, No. 4842, February 19, 1988, pp. 896–899.

Gurvitch, M., and A. T. Fiory, "Preparation and substrate reactions of superconducting Y-Ba-Cu-O films," *Appl. Phys. Lett.*, Vol. 51, No. 13, September 28, 1987, pp. 1027–1029.

Gutsmiedl, P., G. Wolff, and K. Andres, "Evidence for antiferromagnetic interactions in the high-temperature superconductor $La_{1.85}Sr_{0.15}CuO_{4-y}$," *Phys. Rev. B*, Vol. 36, No. 1, September 1, 1987, pp. 4043–4046.

Hagen, M., M. Hein, N. Klein, A. Michalke, G. Müller, H. Piel, and R. W. Röth, "Observation of RF Superconductivity in $Y_1Ba_2Cu_3O_{9-\delta}$ at 3 GHz, Wuppertal pub. #87-12, April 1987 (To be published in the *Journal of Magnetism and Magnetic Materials*).

Hagen, M., M. Hein, N. Klein, A. Michalke, G. Müller, H. Piel, R. W. Röth, F. M. Mueller, H. Sheinberg, and J.-L. Smith, "Observation of RF Superconductivity in $Y_1Ba_2Cu_3O_{9-\delta}$ at 3 GHz," Un. Wuppertal pub. 87-12, April 1987, pp. 01–06.

Hahn, H., H. J. Halama, and E. H. Foster, "Measurement of Surface Resistance of Superconducting Lead at 2. 868 GHz," *J. Appl. Phys.*, Vol. 39, No. 6, May 1968, pp. 2606–2609.

Hakuraku, Y., F. Sumiyoshi, and T. Ogushi, "Role of added fluorine to enhance the electromagnetic properties of superconducting Y-Ba-Cu-O compounds," *Appl. Phys. Lett.*, Vol. 52, No. 18, May 2, 1988, pp. 1528–1530.

Haldar, R., Y. Z. Lu, and B. C. Giessen, "$EuBa_2Cu_3O_x$ produced by oxidation of a rapidly solidified precursor alloy: An alternative preparation method for high-T_c ceramic superconductors," *Appl. Phys. Lett.*, Vol. 51, No. 7, August 17, 1987, pp. 538–539.

Harada, Y., H. Nakane, N. Miyamoto, U. Kawabe, E. Goto, and T. Soma, "Basic operations of the quantum flux parametron," *IEEE Trans. Magn.*, Vol. MAG-23, No. 5, September 1987, pp. 3801–3807.

Harada, K., N. Fujimori, and S. Yazu, "Y-Ba-Cu-O thin film on Si substrate," *Jap. J. Appl. Phys.*,' Vol. 27, No. 8, August 1988, pp. L1524–L1526.

Harter, W. G., A. M. Hermann, and Z. Z. Sheng, "Levitation effects involving high T_c thallium based superconductors," *Appl. Phys. Lett.*, Vol. 53, No. 12, September 19, 1988, pp. 1119–1121.

Hashimoto, T., K. Fueki, A. Kishi, T. Azumi, and H. Koinuma, "Thermal expansion coefficients of high-T_c superconductors," *Jap. J. Appl. Phys.*, Vol. 27, No. 2, February 1988, pp. L214–L216.

Hashimoto, T., T. Kosaka, Y. Yoshida, K. Fueki, and H. Koinuma, "Superconductivity and substrate interaction of screen-printed Bi-Sr-Ca-Cu-O films," *Jap. J. Appl. Phys.*, Vol. 27, No. 3, March 1988, pp. L384–L386.

Hattori, T., N. Higemoto, S. Kanazawa, and M. Kobayashi, "Magnetic shielding using high-T_c superconductor," *Jap. J. Appl. Phys.*, Vol. 27, No. 6, June 1988, pp. L1120–L1122.

Hayakawa, H., "Josephson computer technology," *Physics Today*, March 1986, pp. 46–52.

Hazen, R. M., "Perovskites," *Scientific American*, Vol. 258, No. 6, June 1988, pp. 74–81.

Hazen, R. M., L. W. Finger, R. J. Angel, C. T. Prewitt, N. L. Ross, C. G. Hadidiacos, P. J. Heaney, D. R. Veblen, Z. Z. Sheng, and A. M. Hermann, "100-K superconducting phases in the Tl-Ca-Ba-Cu-O system," *Phys. Rev. Lett.*, Vol. 60, No. 16, April 19, 1988.

Hein, M., N. Klein, F. M. Mueller, H. Piel, R. W. Röth, W. Weingarten, and J. O. Willis, "On the surface resistance of the perovskite superconductors at 3 GHz," Un. Wuppertal pub. WUB 87-15h, June 1987, pp. 01–02.

Henkels, W. H., and H. H. Zappe, "An Experimental 64-Bit Decoded Josephson NDRO Random Access Memory," *IEEE J. Solid-State Circuits*," Vol. SC-13, No. 5, October 1978, pp. 591–600.

Hermann, A. M., Z. Z. Sheng, D. C. Vier, S. Schultz, and S. B. Oseroff, "Magnetization of the 120 K Tl-Ca/Ba-Cu-O Superconductor," Dept. of Physics, University of Arkansas, Fayetteville, AR, cited in *High-T_c Update*, Vol. 2, No. 6, March 15, 1988, p. 4.

Hermann, A. M., Z. Z. Sheng, D. C. Vier, S. Schultz, and S. B. Oseroff, "Magnetization of the 120 K Tl-Ca-Ba-Cu-O superconductor," *Phys. Rev. B*, Vol. 37, No. 16, June 1, 1988, pp. 9742–9744.

Herr, S. L., K. Kamarás, C. D. Porter, M. G. Doss, and D. B. Tanner, "Optical properties of $La_{1.85}Sr_{0.15}CuO_4$: Evidence for strong electron-phonon and electron-electron interactions," *Phys. Rev. B*, Vol. 36, No. 1, July 1, 1987, pp. 733–735.

Hewat, E. A., M. Dupuy, P. Bodet, J. J. Capponi, C. Chaillout, J. L. Hodeau, and M. Marezio, "Superstructure of the superconductor $Bi_2Sr_2CaCu_2O_8$ by high-resolution electron microscopy," *Nature*, Vol. 333, No. 6168, May 5, 1988, pp. 53–54.

Hidaka, Y., M. Oda, M. Suzuki, Y. Maeda, Y. Enomoto, and T. Murakami, "Large anisotropy of the upper critical magnetic field in single crystal Bi-(Sr,Ca)-Cu-O," *Jap. J. Appl. Phys.*, Vol. 27, No. 4, April 1988, pp. L538–L541.

Hidaka, T., T. Matsui, and Y. Nakagawa, "Ba isotope effect in a $Ba_2YCu_3O_{7-\delta}$ superconductor," *Jap. J. Appl. Phys.*, Vol. 27, No. 4, April 1988, pp. L553–L555.

Hikita, M., Y. Tajima, A. Katslui, Y. Hidaka, T. Iwata, and S. Tsurumi, "Electrical properties of high-T_c superconducting single-crystal $Eu_1Ba_2Cu_3O_y$," *Phys. Rev. B*, Vol. 36, No. 13, November 1, 1987, pp. 7199–7202.

Hinks, D. G., B. Dabrowski, J. D. Jorgensen, A. W. Mitchell, D. R. Richards, S. Pei, and D. Shi, "Synthesis, structure and superconductivity in the $Ba_{1-x}K_xBiO_{3-y}$ system," *Nature*, Vol. 333, No. 6176, June 30, 1988, pp. 836–838.

Hinks, D. G., D. R. Richards, B. Dabrowski, D. T. Marx, and A. W. Mitchell, "The oxygen isotope effect in $Ba_{0.625}K_{0.375}BiO_3$," *Nature*, Vol. 335, No. 6189, September 29, 1988, pp. 419–421.

Hoddeson, L., G. Baym, S. Heims, and H. Schubert, "Collective Phenomena," Chapter 8 in *Out of the Crystal Maze: A History of Solid State Physics, 1900–1960*, L. Hoddeson, E. Braun, J. Teichmann, and S. Weart, eds., Oxford University, Oxford, 1989.

Honda, T., T. Wada, M. Sakai, M. Miyajima, N. Nishikawa, S.-I. Uchida, K. Uchinokura, and S. Tanaka, "Preparation of high-T_c (105 K) superconducting phase in the Bi-Sr-Ca-K-Cu-O system," *Jap. J. Appl. Phys.*, Vol. 27, No. 4, April 1988, pp. L545–L547.

Hong, M., S. H. Liou, J. Kwo, and B. A. Davidson, "Superconducting Y-Ba-Cu oxide films by sputtering," *Appl. Phys. Lett.*, Vol. 51, No. 9, August 31, 1987, pp. 694–696.

Hor, P. H., L. Gao, R. L. Meng, Z. J. Huang, Y. Q. Wang, K. Forster, J. Vassilious, C. W. Chu, M. K. Wu, J. R. Ashburn, and C. J. Torng, "High-Pressure Study of the New Y-Ba-Cu-O Superconducting Compound System," *Phys. Rev. Lett.*, Vol. 58, No. 9, March 2, 1987, pp. 911 912.

Hor, P. H., R. L. Meng, Y. Q. Wang, L. Gao, Z. J. Juang, J. Bechtold, K. Forster, and C. W. Chu, "Superconductivity above 90 K in the Square-Planar Compound System $ABa_2Cu_3O_{6+x}$ with A = Y, La, Nd, Sm, Eu, Gd, Ho, Er, and Lu," *Phys. Rev. Lett.*, Vol. 58, No. 18, May 4, 1987, pp. 1891–1894.

Huang, C. Y., "Some aspects of high-temperature superconductors; A review," Lockheed Missiles & Space Company, Inc. report 7454w, 1988.

Huang, C. Y., L. J. Dries, P. H. Hor, R. L. Meng, C. W. Chu, and R. B. Frankel, "Observation of possible superconductivity at 230 K," *Nature*, Vol. 328, No. 6129, July 30, 1987, pp. 403–404.

Huang, C. Y., Y. Shapira, E. J. McNiff, Jr., P. N. Peters, B. B. Schwartz, M. K. Wu, R. D. Shull, and C. K. Chiang, "Magnetic hysteresis of high-temperature $YBa_2Cu_3O_x$-AgO Superconductors: Explanation of magnetic suspension," *Mod. Phys. Lett. B*, Vol. 2, No. 7, 1988, pp. 869–874.

Hundley, M. F., A. Zettl, A. Stacy, and M. L. Cohen, "Transport properties of the superconducting oxide $La_{1.85}Sr_{0.15}CuO_4$," *Phys. Rev. B*, Vol. 35, No. 16, June 1, 1987, pp. 8800–8803.

Hyland, G. J., "On the fluorination dependence of T_c in $YBa_2Cu_3O_{7-\delta}$," *Jap. J. Appl. Phys.*, Vol. 27, No. 4, April 1988, pp. L598–L599.

Ichikawa, Y., H. Adachi, T. Mitsuyu, and K. Wasa, "Effect of overcoating with dielectric films on the superconductive properties of the high-T_c Y-Ba-Cu-O films," *Jap. J. Appl. Phys.*, Vol. 27, No. 3, March 1988, pp. L381–L383.

Ihara, H., N. Terada, M. Jo, M. Hirabayashi, M. Tokumoto, Y. Kimura, T. Matsubara, and R. Sugise, "Possibility of superconductivity at 65°C in Sr-Ba-Y-Cu-O system," *Jap. J. Appl. Phys.*, Vol. 26, No. 8, August 1987, pp. L1413–L1415.

Ihara, H., R. Sugise, M. Hirabayashi, N. Terada, M. Jo, K. Hayashi, A. Negishi, M. Tokumoto, Y. Kimura, and T. Shimomura, "A new high-T_c $TlBa_2Ca_3Cu_4O_{11}$ superconductor with T_c >120 K," *Nature*, Vol. 334, No. 6182, August 11, 1988, pp. 510–511.

Imai, S., T. Tozawa, G-I Oya, T. Sugai, Y. Takeuti, and N. Mikoshiba, "Submillimeter-wave response of Nb-YBaCuO point-contact Josephson junctions," *Jap. J. Appl. Phys.*, Vol. 27, No. 4, April 1988, pp. L522–L524.

Inam, A., X. D. Wu, T. Venkatesan, S. B. Ogale, C. C. Chang, and D. Dijkkamp, "Pulsed laser etching of high T_c superconducting films," *Appl. Phys. Lett.*, Vol. 51, No. 14, October 5, 1987, pp. 1112–1114.

Inderhees, S. E., M. B. Salamon, T. A. Friedmann, and D. M. Ginsberg, "Measurement of the specific-heat anomaly at the superconducting transition of $YBa_2Cu_3O_{7-\delta}$," *Phys. Rev. B*, Vol. 36, No. 4, August 1, 1987, pp. 2401–2403.

Inoue, O., S. Adachi, and S. Kawsashima, "Formation and deposition of superconducting $BiSrCaCu_2O_x$ ceramics," *Jap. J. Appl. Phys.*, Vol. 27, No. 3, March 1988, pp. L347–L349.

Ishida, T., and H. Mazaki, "Complex susceptibility of high-T_c superconductor $ErBa_2Cu_3$-O_{6+x}," *Jap. J. Appl. Phys.*, Vol. 26, No. 8, August 1987, pp. L1296–1298.

Ishida, T., and H. Mazaki, "Complex susceptibility of the oxide superconductor BiSrCa-Cu_2O_x," *Jap. J. Appl. Phys.*, Vol. 27, No. 4, April 1988, pp. L531–532.

Ishikawa, M., Y. Nakazawa, T. Takabatake, A. Kishi, R. Kato, and A. Maesono, "Specific heat study on $Ba_2YCu_3O_7$ samples with a double superconducting transition," Proc. Int. Conf. on High Temperature Superconductors, Interlaken, Switzerland, *Physica C*, Vol. 153–155, Superconductivity, June 1988, pp. 1089–1091.

Itoh, T., H. Uchikawa, and H. Sakata, "Synthesis of Tl-Ba-Ca-Cu-O superconductor and its properties," *Jap. J. Appl. Phys.*, Vol. 27, No. 4, April 1988, pp. L559–560.

Iwasa, Y., and D. B. Montgomery, "High-field superconducting magnets," in *Applied Superconductivity*, Vol. 1, V. L. Newhouse, ed., Academic, 1975, pp. 387–487.

Iye, Y., "Small low-cost platinum resistance thermometers for thermometry in magnetic fields," *Cryogenics*, Vol. 28, No. 3, March 1988, pp. 164–168.

Iye, Y., T. Tamegai, H. Takeya, and H. Takei, "A simple method for attaching electrical leads to small samples of high-T_c oxides," *Jap. J. Appl. Phys.*, Vol. 27, No. 4, April 1988, pp. L658–660.

Iye, Y. , T. Tamegai, T. Sakakibara, T. Goto, N. Miura, H. Takeya, and H. Takei, "The anisotropic superconductivity of $RBa_2Cu_3O_{7-x}$ (R : Y, Gd and Ho) single crystals," Proc. Int. Conf. on High Temperature Superconductors, Interlaken, Switzerland, *Physica C*, Vol. 153–155, Superconductivity, June 1988, pp. 26–31.

Jiang, X., H. Yu, Z. Zhang, N. Zhu, H. Qi, G. Shu, Y. Tian, D. Pang, X. Zeng, and Z. Yang, "Effect of crystal structure on superconductivity on Y-Ba-Cu-O system compounds," *Appl. Phys. Lett.*, Vol. 51, No. 8, August 24, 1987, pp. 625–627.

Jin, S., R. C. Sherwood, T. H. Tiefel, R. B. van Dover, D. W. Johnson, Jr., and G. S. Grader, "Stress and field dependence of critical current in $Ba_2YCu_3O_{y-\delta}$ superconductors," *Appl. Phys. Lett.*, Vol. 51, No. 11, September 14, 1987, pp. 855–857.

Jin, S., T. H. Tiefel, R. C. Sherwood, G. W. Kammlott, and S. M. Zahurak, "Fabrication of dense $Ba_2YCu_3O_{y-\delta}$ superconductor wire by molten oxide processing," *Appl. Phys. Lett.*, Vol. 51, No. 12, September 21, 1987, pp. 943–945.

Jin, S., A. Fastnacht, T. H. Tiefel, and R. C. Sherwood, "Transport critical current in rare-earth-substituted superconductors $RBa_2Cu_3O_{7-\delta}$ (R = Gd, Dy, Sm, Ho, Y)," *Phys. Rev. B*, Vol. 37, No. 10, April 1, 1988, pp. 5828–5830.

Jin, S., T. H. Tiefel, R. C. Sherwood, R. B. van Dover, M. E. Davis, G. W. Kammlott, and R. A. Fastnacht, "Melt-textured growth of polycrystalline $YBa_2Cu_3O_{7-\delta}$ with high transport J_c at 77 K," *Phys. Rev. B*, Vol. 37, No. 13, May 1, 1988, pp. 7850–7853.

Johnston, D. C., H. Prakash, W. H. Zachariasen, and R. Viswanathan, "High Temperature Superconductivity in the Li-Ti-O Ternary System," *Mat. Res. Bull.*, Vol. 8, 1973, pp. 777–784.

Jordan, H. E., "Feasibility study of electric motors constructed with high temperature superconducting materials," pres. at 1988 Conf. on Electrical Applications of Superconductivity, Orlando, FL, September 21–23, 1988.

Josephson, B. D., "Possible new effects in superconductive tunneling," *Phys. Lett.*, Vol. 1, No. 7, July 1, 1962, pp. 251–253.

Josephson, B. D., "The discovery of tunneling supercurrents," *Rev. Mod. Phys.*, Vol. 46, No. 2, April 1974, pp. 251–254.

Joyce, C., "Superconductors: America stalls," *New Scientist*, Vol. 118, No. 1617, June 16, 1988, pp. 36–38.

Kaiser, D. L., F. Holtzberg, B. A. Scott, and T. R. McGuire, "Growth of $YBa_2Cu_3O_x$ single crystals," *Appl. Phys. Lett.*, Vol. 51, No. 13, September 28, 1987, pp. 1040–1042.

Kaneko, T., H. Yoshida, S. Abe, H. Morita, K. Noto, and H. Fujimori, "Pressure Dependence of High-T_c Superconducting Oxides," *Jap. J. Appl. Phys.*, Vol. 26, No. 8, August 1987, pp. L1374–L1376.

Kapitulnik, A., "Properties of films of high-T_c perovskite superconductors," Proc. Int. Conf. on High Temperature Superconductors, Interlaken, Switzerland, *Physica C*, Vol. 153–155, Superconductivity, June 1988, pp. 520–526.

Karpinski, J., E. Kaldis, and S. Rusiecki, "High pressure phase diagrams (1–3000 bar O_2) of the (Y-Ba-Cu-O)-O_2 systems," Proc. E-MRS, (& *J. Less Common Metals*), Strasbourg, November 8–10, 1988.

Karpinski, J., E. Kaldis, S. Rusiecki, E. Jilek, P. Fischer, P. Bordet, C. Chaillout, J. Chenavas, J. L. Hodeau, and M. Marezio, "Two new bulk superconducting phases in the Y-Ba-Cu-O system: $YBa_2Cu_{3.5}O_{7+x}$:(T_c = 40 K) and $YBa_2Cu_4O_{8+x}$:(T_c = 80 K)," Proc. E-MRS, (& *J. Less Common Metals*), Strasbourg, November 8–10, 1988.

Karpinski, J., E. Kaldis, E. Jilek, S. Rusiecki, and B. Bucher, "Bulk synthesis of the 81-K superconductor $YBa_2Cu_4O_8$ at high oxygen pressure," *Nature*, Vol. 336, December 15, 1988, pp. 660–662.

Karthikeyan, J., K. P. Sreekumar, M. B. Kurup, D. S. Patil, P. V. Anantapadmanabhan, N. Venkatramani, and V. K. Rohatgi, "Plasma sprayed superconducting $Y_1Ba_2Cu_3O_{7-x}$ coatings," *J. Phys. D*, Vol. 21, No. 7, July 1988, pp. 1246–1249.

Kato, T., K. Aihara, J. Kuniya, T. Kamo, and S.-P. Matsuda, "Phase transition in $YBa_2Cu_3O_{7-x}$ through hydrogen ion implantation," *Jap. J. Appl. Phys.*, Vol. 27, No. 4, April 1988, pp. L564–L566.

Katz, J. D., J. O. Willis, M. P. Maley, and R. G. Castro, "Low resistivity, $YBa_2Cu_3O_7$-to-silver electrical contacts by plasma spraying," submitted to *J. Appl. Phys.*, 1988.

Kawabe, U., "Application of high-temperature superconductors to SQUID and other devices," Proc. Int. Conf. on High Temperature Superconductors, Interlaken, Switzerland, *Physica C*, Vol. 153–155, Superconductivity, June 1988, pp. 1586–1591.

Kedves, F. J., S. Meszaros, K. Vad, G. Halasz, D. Keszei, and L. Mihaly, "Estimation of maximum electrical resistivity of high T_c superconducting ceramics by the Meissner effect," *Solid State Commun.*, Vol. 63, No. 11, 1987, pp. 991–992.

Khachaturyan, K., E. R. Weber, P. Tejedor, A. Stacy, and A. Portis, "Changes in microwave absorption of new high T_c superconductors with small magnetic fields," *High-Temperature Superconductors*, Vol. 99, Proc. Mat. Res. Soc. Symp., Boston, November 30–December 4, 1987, pp. 383–386.

Khurana, A., "The T_c to Beat is 125 K," *Physics Today*, Vol. 41, No. 4, April 1988, pp. 21–25.

Kijima, T., J. Tanaka, Y. Bando, M. Onoda, and F. Izumi, "Identification of a high-T_c superconducting phase in the Bi-Ca-Sr-Cu-O System," *Jap. J. Appl. Phys.*, Vol. 27, No. 3, March 1988, pp. L369–L371.

Kikuchi, M., N. Kobayashi, H. Iwasaki, D. Shindo, T. Oku, A. Tokiwa, T. Kajitani, K. Hiraga, Y. Syono, and Y. Muto, "Synthesis and superconductivity of a new high-T_c Tl-Ba-Ca-Cu-O phase," *Jap. J. Appl. Phys.*, Vol. 27, No. 6, June 1988, pp. L1050–L1053.

Kim, Y. B., C. F. Hempstead, and A. R. Strnad, "Magnetization and critical supercurrents," *Phys. Rev.*, Vol. 129, No. 2, January 15, 1963, pp. 528–535.

Kim, Y. B., C. F. Hempstead, and A. R. Strnad, "Resistive states of hard superconductors," *Rev. Mod. Phys.*, Vol. 36, January 1964, pp. 43–45.

Kim, B. F., J. Bohandy, T. E. Phillips, W. J. Green, E. Agostinelli, F. J. Adrian, K. Moorjani, L. J. Swartzendruber, R. D. Shull, L. H. Bennett, and J. S. Wallace, "Superconducting thin films of Bi-Sr-Ca-Cu-O obtained by laser ablation processing," *Appl. Phys. Lett.*, Vol. 53, No. 4, July 25, 1988, pp. 319–320.

Kirschner, I., and J. Bankuti, "Superconductivity in La-Ba-Cu-O metallic oxide compounds above 50 K," *Phys. Rev. B*, Vol. 36, No. 4, August 1, 1987, pp. 2313–2315.

Kirschner, I., M. Lamm, T. Porjesz, J. Matrai, GY. Kovacs, T. Trager, T. Karman, J. György, and G. Zsolt, "First experiments with a high-T_c superconducting magnet," Proc. Int. Conf. on High Temperature Superconductors, Interlaken, Switzerland, *Physica C*, Vol. 153–155, Superconductivity, June 1988, pp. 1417–1418.

Kishida, S., H. Tokutaka, K. Nishmori, N. Ishihara, Y. Wanatabe, and Y. Noshiki, "Effects of preparation conditions and mechanical polishing on the superconducting behavior of high-T_c $Y_1Ba_2Cu_3O_{7-y}$," *Jap. J. Appl. Phys.*, Vol. 27, No. 3, March 1988, pp. L325–L328.

Kisu, T., K. Enpuku, K. Yoshida, M. Takeo, and K. Yamafuji, "Spatial distribution of trapped magnetic flux in the high T_c superconductor $Y_1Ba_2Cu_3O_{7-\delta}$," *Jap. J. Appl. Phys.*, Vol. 26, No. 8, August 1987, pp. L1348–L1350.

Kitagawa, T., S. Shibata, H. Okazaki, T. Kimura, and Y. Enomoto, "High-T_c superconducting film on silicon substrate with ZrO_2 buffer layer," *Jap. J. Appl. Phys.*, Vol. 27, No. 6, June 1988, pp. L1113–L1115.

Kitazawa, K., H. Takagi, K. Kishio, T. Hasegawa, S. Uchida, S. Tajima, S. Tanaka, and K. Fueki, "Electronic properties of cuprate superconductors," Proc. Int. Conf. on High Temperature Superconductors, Interlaken, Switzerland, *Physica C*, Vol. 153–155, Superconductivity, June 1988, pp. 9–14.

Klein, N., G. Müller, H. Piel, and J. Schurr, "Superconducting microwave resonators for physics experiments," presented at 1988 ASC *IEEE Trans. Magn.*, Vol. 25, No. 2, March 1989.

Kobayashi, T., K. Nomura, F. Uchikawa, T. Masumi, and Y. Uehara, "Superconducting Bi-Sr-Ca-Cu-O thick films by the Sol-Gel method," *Jap. J. Appl. Phys.*, Vol. 27, No. 10, October 1988, pp. L1880–L1882.

Koch, H., R. Cantor, J. F. March, H. Eickenbusch, and R. Schöllhorn, "Point-contact characteristics obtained with single-phase Y-Ba-Cu-O superconductors," *Phys. Rev. B*, Vol. 36, No. 1, July 1, 1987, pp. 722–725.

Koch, R. H., and P. Chaudhari, "Improving thin-film techniques," *Nature*, Vol. 332, No. 6166, April 21, 1988, pp. 682–683.

Koinuma, H., M. Kawasaki, S. Nagata, K. Takeuchi, and K. Fueki, "Preparation of high-T_c Bi-Sr-Ca-Cu-O superconducting thin films by AC sputtering," *Jap. J. Appl. Phys.*, Vol. 27, No. 3, March 1988, pp. L376–L377.

Koinuma, H., Y. Takemura, T. Hashimoto, K. Takeuchi, and K. Fueki, "Reversible resistivity control of $Ba_2YCu_3O_{7-\delta}$ thin films by laser annealing," *Jap. J. Appl. Phys.*, Vol. 27, No. 4, April 1988, pp. L652–L654.

Komatsu, T., K. Imai, R. Sato, K. Matusita, and T. Yamashita, "Preparation of high-T_c superconducting Bi-Ca-Sr-Cu-O ceramics by the melt quenching method," *Jap. J. Appl. Phys.*, Vol. 27, No. 4, April 1988, pp. L533–L535.

Komatsu, T., R. Sato, K. Imai, K. Matusita, and T. Yamashita, "High-T_c superconducting glass ceramics based on the Bi-Ca-Sr-Cu-O system," *Jap. J. Appl. Phys.*, Vol. 27, No. 4, April 1988, pp. L550–L552.

Komatsu, T., O. Tanaka, K. Matusita, and T. Yamashita, "On the new substrate materials for high-T_c superconducting Ba-Y-Cu-O thin films," *Jap. J. Appl. Phys.*, Vol. 27, No. 9, September 1988, pp. L1686–L1689.

Kotani, S., N. Fujimaki, S. Morohashi, S. Ohara, T. Imamura, and S. Hasuo, "High-Speed Unit-Cell for Josephson LSI Circuits," *IEEE Trans. Magn.*, Vol. MAG-23, No. 2, March 1987, pp. 869–874.

Krause, J. K., P. R. Swinehart, and J. R. Bergen, "Temperature Trends; Demystifying Cryogenic Temperature Sensors," Photonics Spectra, Optical Publishing, August 1985. (For reprints, contact Lake Shore Cryotronics, Inc., 64 E. Walnut St., Westerville, OH 43081)

Krause, J. K., and B. C. Dodrill, "Measurement system induced errors in diode thermometry," *Rev. Sci. Instrum.*, Vol. 57, No. 4, April 1986, pp. 661–665.

Krause, J. K., P. R. Swinehart, and J. R. Bergen, "Temperature Sensors for Cryogenic Applications," *Sensors*, February 1988.

Kreilick, T. S., and E. Gregory, "Further improvements in current density by reduction of filament spacing in multifilamentary Nb-Ti superconductors," *Cryogenics*, Vol. 27, July 1987, pp. 401–403.

Kresin, V. Z., "On the relation between the energy gap and the critical temperature," *Solid State Commun.*, Vol. 63, No. 8, 1987, pp. 725–727.

Kroger, H., "Josephson Devices Coupled by Semiconductor Links," *IEEE Trans. Electron Devices*, Vol. ED-27, No. 10, October 1980, pp. 2016–2026.

Kubo, Y., Y. Shimakawa, T. Manako, T. Satoh, and H. Igarashi, "Superconducting phases with T_c above 100 K in the Tl-Ba-Ca-Cu-O system," *Jap. J. Appl. Phys.*, Vol. 27, No. 4, April 1988, pp. L591–593.

Kumakura, H., H. Shimizu, K. Takahashi, K. Togano, and H. Maeda, "Upper critical field of new oxide superconductor Bi-Sr-Ca-Cu-O," *Jap. J. Appl. Phys.*, Vol. 27, No. 4, April 1988, pp. L668–L669.

Kunzler, J. E., E. Buehler, F. S. L. Hsu, and J. H. Wernick, "Superconductivity in Nb_3Sn at high current density in a magnetic field of 88 kgauss," *Phys. Rev. Lett.*, Vol. 6, No. 3, February 1, 1961, pp. 89–91.

Kunzler, J. E., "Recollection of Events Associated with the Discovery of High Field-High Current Superconductivity," *IEEE Trans. Magn.*, Vol. MAG-23, No. 2, March 1987, pp. 396–402.

Küpfer, H., S. M. Green, C. Jiang, Yu Mei, H. L. Luo, R. Meier-Hirmer, and C. Politis, "Weak link problem and intragrain current density in polycrystalline $Bi_1Ca_1Sr_1Cu_2O_x$ and $Tl_2Ca_2Ba_2Cu_3O_{10}$," *Z. Phys. B—Condensed Matter*, Vol. 71, 1988, pp. 63–67.

Küpfer, H., W. Wiech, I. Apfelstedt, R. Flükiger, R. Meier-Hirmer, T. Wolf, and H. Scheurer, "Influence of fast neutron irradiation on inter- and intragrain properties of ceramic $YBa_2Cu_3O_7$," presented at 1988 ASC (To be published in *IEEE Trans. Magn.*, MAG-25, 1989), San Francisco, August 21–25, 1988, (MM-11), Abstracts, p. 54.

Kwak, J. F., E. L. Venturini, D. S. Ginley, and W. Fu, "Grain decoupling at low magnetic fields in ceramic $YBa_2Cu_3O_{7-\delta}$," in *Novel Superconductivity*, S. A. Wolf and V. Z. Kresin, eds., Plenum, 1987, pp. 983–991.

Kwasnitza, K., V. Plotzner, M. Waldmann, and E. Widmer, "Currents, magnetization and ac-losses of $YBa_2Cu_3O_7$ superconductors in rapidly changing magnetic fields," Proc. Int. Conf. on High Temperature Superconductors, Interlaken, Switzerland, *Physica C*, Vol. 153–155, Superconductivity, June 1988, pp. 1565–1566.

Kwok, W. K., G. W. Crabtree, D. G. Hinks, D. W. Capone, J. D. Jorgensen, and K. Zhang, "Normal and superconducting-state properties of $La_{1.85}Sr_{0.15}CuO_4$," *Phys. Rev. B*, Vol. 35, No. 10, April 1, 1987, pp. 5343–5346.

Kwok, W. K., G. W. Crabtree, A. Umezawa, B. W. Veal, J. D. Jorgensen, S. K. Malik, L. J. Nowiki, A. P. Paulikas, and L. Nunez, "Electronic behavior of oxygen deficient $YBa_2Cu_3O_{7-\delta}$" (Mat. Sci. Div., Argonne Nat. Lab—preprint).

Laegreid, T., K. Fossheim, E. Sandvold, and S. Lulsrud, "Specific heat anomaly at 220 K connected with superconductivity at 90 K in ceramic $YBa_2Cu_3O_{7-\delta}$," *Nature*, Vol. 330, December 17, 1987, pp. 637–638.

Laibowitz, R. B., R. H. Koch, P. Chaudhari, and R. J. Gambino, "Thin superconducting oxide films," *Phys. Rev. B*, Vol. 35, No. 16, June 1, 1987, pp. 8821–8823.

Larbalestier, D., G. Fisk, B. Montgomery, and D. Hawksworth, "High-Field Superconductivity," *Physics Today*, March 1986, pp. 24–33.

Larbalastier, D. C., S. E. Babcock, X. Cai, M. Daeumling, D. P. Hampshire, T. F. Kelly, L. A. Lavanier, P. J. Lee, and J. Seuntjens, "Weak links and the poor tranport critical currents of the 1-2-3 compounds," Proc. Int. Conf. on High Temperature Superconductors, Interlaken, Switzerland, *Physica C*, Vol. 153–155, Superconductivity, June 1988, pp. 1580–1585.

Laundrie, A., "Superconducting ICs generate and detect signals to 100 GHz," *Microwaves & RF*, Vol. 27, No. 9, September 1988, pp. 163–164.

Leary, K. J., H.-C. zur Loye, S. W. Keller, T. A. Faltens, W. K. Ham, J. N. Michaels, and A. M. Stacy, "Observation of an oxygen isotope effect in $YBa_2Cu_3O_7$," *Phys. Rev. Lett.*, Vol. 59, No. 11, September 14, 1987, pp. 1236–1239.

LeBlanc, M. A., B. C. Belanger, and R. M. Fielding, "Paramagnetic helical current flow in Type-II superconductors," *Phys. Rev. Lett.*, Vol. 14, No. 17, April 26, 1965, pp. 704–707.

Lee, W. Y., V. Y. Lee, J. Salem, T. C. Huang, R. Savoy, D. C. Bullock, and S. S. P. Parkin, "Superconducting Tl-Ca-Ba-Cu-O thin films with zero resistance at temperatures up to 120 K," *Appl. Phys. Lett.*, Vol. 53, No. 4, July 25, 1988, pp. 329–331.

Leung, E. M. W., R. E. Baily, and P. H. Michels, "Using a small hybrid pulse power transformer unit as a component of a high-current opening switch for a railgun," *IEEE Trans. Magn.*, Vol. 25, No. 2, March 1989, pp. 1779–1782.

Levi, B. G., and B. Schwarzschild, "Super Collider Magnet Program Pushes Toward Prototype," *Physics Today*, Vol. 41, No. 4, April 1988, pp. 17–21.

Lin, C., G. Lu, Z.-X. Liu, Y.-X. Sun, J. Lan, G.-Z. Li, S.-Q. Feng, C.-D. Wei, Z.-Z. Gan, Z.-Y. Liu, Z.-M. Zheng, and B.-X. Lin, "Magnetic properties of the high-T_c superconductor $YBa_2Cu_3O_{9-x}$," *Solid State Commun.*, Vol. 63, No. 12, 1987, pp. 1129–1133.

Lindley, D., "Superconductor pessimism in the U.S.," *Nature*, Vol. 333, No. 6176, June 30, 1988, p. 789.

Little, W. A., "Superconductivity at room temperature," *Scientific American*, Vol. 212, No. 2, February 1965, pp. 21–27.

Liu, R. S., Y. T. Huang, P. T. Wu, and J. J. Chu, "Epitaxial growth of high-T_c Bi-Ca-Sr-Cu-O superconducting layer by LPE process," *Jap. J. Appl. Phys.*, Vol. 27, No. 8, August 1988, pp. L1470–1472.

Loe, K. F., and E. Goto, "Analysis of flux input and output Josephson pair device," *IEEE Trans. Magn.*, Vol. MAG-21, No. 2, March 1985, pp. 884–887.

London, F., and H. London, "The Electromagnetic Equations of the Supraconductor," *Proc. R. Soc. A*, Vol. A149, 1935, pp. 71–88.

Macfarlane, J. C., R. Driver, and R. B. Roberts, "Novel electromagnetic effects in high-temperature superconductors," *Appl. Phys. Lett.*, Vol. 51, No. 13, September 28, 1987, pp. 1038–1039.

Macfarlane, J. C., R. Driver, R. B. Roberts, E. C. Horrigan, and C. Andrikidis, "Electromagnetic shielding properties of yttrium barium cuprate superconductor," Proc. Int. Conf. on High Temperature Superconductors, Interlaken, Switzerland, *Physica C*, Vol. 153–155, Superconductivity, June 1988, pp. 1423–1424.

Maeda, A., T. Yabe, K. Uchinokura, and S. Tanaka, "High temperature superconductivity at 90 K in orthorhombic $LaBa_2Cu_3O_y$," *Jap. J. Appl. Phys.*, Vol. 26, No. 8, August 1987, pp. L1368–L1370.

Maeda, H., Y. Tanaka, M. Fukutomi, and T. Asano, "A New high-T_c Oxide Superconductor Without a Rare Earth Element," *Jap. J. Appl. Phys.*, Vol. 27, No. 2, February 1988, pp. L209–L210.

Maeda, A., T. Yabe, H. Ikuta, Y. Nakayama, T. Wada, S. Okuda, T. Itoh, M. Izumi, K. Uchinokura, S.-I. Uchida, and S. Tanaka, "Physical properties of an 80 K-superconductor: Bi-Sr-Ca-Cu-O ceramics," *Jap. J. Appl. Phys.*, Vol. 27, No. 4, April 1988, pp. L661–L664.

Mai, Z., L. Chen, X. Chu, D. Dai, Y. Ni, Y. Huang, Z. Xiao, P. Ge, and Z. Zhao, "The Microregion Compositional Variation in $Y_1Ba_2Cu_3O_{7-x}$ Material," *Phys. Lett. A*, Vol. 127, No. 5, February 29, 1988, pp. 297–299.

Makous, J. L., L. Maritato, C. M. Falco, J. P. Cronin, G. P. Rajendran, E. V. Uhlmann, and D. R. Uhlmann, "Superconducting and structural properties of sputtered thin films of $YBa_2Cu_3O_{7-x}$," *Appl. Phys. Lett.*, Vol. 51, No. 25, December 21, 1987, pp. 2164–2166.

Maletta, H., A. P. Malozemoff, D. C. Cronemeyer, C. C. Tsuei, R. L. Greene, J. G. Bednorz, and K. A. Müller, "Diamagnetic shielding and Meissner effect in the high T_c superconductor," *Solid State Commun.*, Vol. 62, No. 5, 1987, pp. 323–326.

Maley, M., "Superconducting Ceramic Properties: Critical Current Densities And AC Losses," Proc. EPRI Workshop on High Temperature Conductivity, Washington, D.C., October 22–23, 1987.

Maley, M. P., J. O. Willis, J. D. Katz, R. G. Castro, and R. M. Aiken, "Low contact-resistivity junctions to ceramic superconductors," presented at 1988 ASC, (To be published in *IEEE Trans. Magn.*, MAG-25, 1989), San Francisco, August 21–25, 1988, (ME-5), Abstracts, p. 18.

Malik, M. K., V. D. Nair, A. R. Biswas, R. V. Raghavan, P. Chaddah, P. K. Mishra, G. Ravi, Kumar, and B. A. Dasannacharya, "Texture formation and enhanced critical currents in $YBa_2Cu_3O_7$," *Appl. Phys. Lett.*, Vol. 52, No. 18, May 2, 1988, pp. 1525–1527.

Malozemoff, A. P., "Computer applications of high temperature superconductivity," Proc. Int. Conf. on High Temperature Superconductors, Interlaken, Switzerland, *Physica C*, Vol. 153–155, Superconductivity, June 1988, pp. 1049–1054.

Mansfield, J. F., S. Chevacharoenkul, and A. I. Kingon, "Space group and chemical analysis of Y_2BaCuO_{5-x} by convergent beam electron diffraction and X-ray energy-dispersive spectroscopy," *Appl. Phys. Lett.*, Vol. 51, No. 13, September 28, 1987, pp. 1035–1037.

Maple, M. B., "Novel types of superconductivity in f-electron systems," *Physics Today*, March 1986, pp. 72–80.

Marshall, D. B., R. E. DeWames, P. E. D. Morgan, and J. J. Ratto, "Flux penetration in high-T_c superconductors: Implications for magnetic suspension and shielding," Preprint: submitted to *Phys. Rev. Lett.*

Martens, J. S., J. B. Beyer, and D. S. Ginley, "Microwave surface resistance of YBa_2-$Cu_3O_{6.9}$ superconducting films," *Appl. Phys. Lett.*, Vol. 52, No. 21, May 23, 1988, pp. 1822–1824.

Martens, J. S., and J. S. Beyer, "Measure RF losses of superconductors in planar circuits," *Microwaves & RF*, Vol. 27, No. 9, September 1988, pp. 150–156.

Matisoo, J., "The Tunneling Cryotron—A Superconductive Logic Element Based on Electron Tunneling," *Proc. IEEE*, Vol. 55, No. 2, February 1967, pp. 172–180.

Matisoo, J., "Overview of Josephson Technology Logic and Memory," *IBM J. Res. Dev.*, Vol. 24, No. 2, March 1980, pp. 113–129.

Matsuda, M., A. Kikuchi, T. Maeda, M. Ishii, Y. Iwai, M. Takata, and T. Yamashita, "Observation of $GdBa_2Cu_3O_{7-\delta}$ ceramic microstructure," *Jap. J. Appl. Phys.*, Vol. 27, No. 4, April 1988, pp. L529–L530.

Matsui, Y., H. Maeda, Y. Tanaka, and S. Horiuchi, "High-resolution electron microscopy of modulated structure in the new high-T_c superconductors of the Bi-Sr-Ca-Cu-O system," *Jap. J. of Appl. Phys.*, Vol. 27, No. 3, March 1988, pp. L361–364.

Matsumoto, T., H. Aoki, A. Matsushita, M. Uehara, N. Mori, H. Takahashi, C. Murayama, and H. Maeda, "Effect of magnetic field and high pressure on the superconductivity of the new high-T_c oxide Bi-Sr-Ca-Cu-O," *Jap. J. Appl. Phys.*, Vol. 27, No. 4, April 1988, pp. L600–L602.

Matsushita, T., "Strong flux pinning by cracks in films of superconducting oxide," *Jap. J. Appl. Phys.*, Vol. 27, No. 9, September 1988, pp. L1712–L1714.

Matsuzaki, K., A. Inoue, H. Kimura, K. Aoki, and T. Masumoto, "High T_c superconductor prepared by oxidization of a liquid-quenched $Yb_1Ba_2Cu_3$ alloy foil in air," *Jap. J. Appl. Phys.*, Vol. 26, No. 8, August 1987, pp. L1310–L1312.

Matthias, B. T., "Superconductivity," *Scientific American*, Vol. 197, No. 5, November 1957, pp. 92–103.

Matthias, B. T., T. H. Geballe, S. Geller, and E. Corenzwit, "Superconductivity of Nb_3Sn," *Phys. Rev.*, Vol. 95, No. 6, September 15, 1954, p. 1435.

Matthias, B. T., T. H. Geballe, R. H. Willens, E. Corenzwit, and G. W. Hull, Jr., "Superconductivity in Nb_3Ge," *Phys. Rev.*, Vol. 139, No. 5A, August 30, 1965, pp. A1501–A1503.

Matthias, B. T., T. H. Geballe, L. D. Longinotti, E. Corenzwit, G. W. Hull, R. H. Willens, and J. P. Maita, "Superconductivity at 20 Degrees Kelvin," *Science*, Vol. 156, May 5, 1967, pp. 645–646.

Mattis, D. C., "Anomalous skin effect in a magnetic field," *Phys. Rev.*, Vol. 111, No. 2, July 15, 1958, pp. 403–411.

Mattis, D. C., and J. Bardeen, "Theory of the anomalous skin effect in normal and superconducting metals," *Phys. Rev.*, Vol. 111, No. 2, July 15, 1958, pp. 412–417.

McDonald, D. G., R. L. Peterson, C. A. Hamilton, R. E. Harris, and R. L. Kautz, "P:icosecond Applications of Josephson Junctions," *IEEE Trans. Electron Devices*, Vol. ED-27, No. 10, October 1980, pp. 1945–1965.

McGee, Thomas D., *Principles and Methods of Temperature Measurement*, John Wiley & Sons, New York, 1988.

McGinnis, D. P., J. B. Beyer, and J. E. Nordman, "A modified superconducting current injection transistor and distributed amplifier design," *IEEE Trans. Magn.*, Vol. 25, No. 2, March 1989, San Francisco, August 21–25, pp. 1262–1265.

Mehran, F., S. E. Barnes, T. R. McGuire, W. J. Gallagher, R. L. Sandstrom, T. R. Dinger, and D. A. Chance, "Paramagnetic resonance of Cu^{2+} ions in the superconductor $Y_{0.1}Ba_{0.8}CuO_x$," *Phys. Rev. B*, Vol. 36, No. 1, July 1, 1987, pp. 740–742.

Meingast, C., M. Daeumling, P. J. Lee, and D. C. Larbalestier, "Proximity effect depression of the critical temperature in two-phase Nb-Ti superconductors," *Appl. Phys. Lett.*, Vol. 51, No. 9, August 31, 1987, pp. 688–689.

Mendelssohn, K., "On different types of superconductivity," *Rev. Mod. Phys.*, Vol. 36, January 1964, pp. 50–51.

Mendelssohn, K., "Prewar work on superconductivity as seen from Oxford," *Rev. of Mod. Phys.*, Vol. 36, January 1964, pp. 7–12.

Meyer, H. M., III, T. J. Wagener, D. M. Hill, Y. Gao, S. G. Anderson, S. D. Krahn, J. H. Weaver, B. Flandermeyer, and D. W. Capone II, "Spectroscopic evidence for passivation of the $La_{1.85}Sr_{0.15}CuO_4$ surface with gold," *Appl. Phys. Lett.*, Vol. 51, No. 14, October 5, 1987, pp. 1118–1120.

Migliori, A., T. Chen, B. Alavi, and G. Grüner, "Ultrasound Anomaly at T_c in YBa_2-Cu_3O_y," *Solid State Commun.*, Vol. 63, No. 9, 1987, pp. 827–829.

Mints, R. G., and A. L. Rakhmanov, "The current-carrying capacity of twisted multifilimentary superconducting composites," *J. Phys. D*, Vol. 20, 1988, pp. 826–830.

Miyahara, K., M. Mukaida, M. Tokumitsu, S. Kubo, and K. Hohkawa, "Abrikosov Vortex Memory with Improved Sensitivity and Reduced Write Levels," *IEEE Trans. Magn.*, Vol. MAG-23, No. 2, March 1987, pp. 875–878.

Mogro-Camparo, A., L. G. Turner, M. F. Garbauskas, and R. W. Green, "Superconducting thin films of Bi-Ca-Sr-Cu-O by sequential evaporation," *Appl. Phys. Lett.*, Vol. 53, No. 4, July 25, 1988, pp. 327–328.

Mohamed, M. A.-K., W. A. Miner, J. Jung, J. P. Franck, and S. B. Woods, "Decay of trapped flux in the high-T_c superconducting compound $Y_1Ba_2Cu_2O_{6.5+\delta}$," *Phys. Rev. B*, Vol. 37, No. 10, April 1, 1988, pp. 5834–5836.

Montgomery, H. C., "Method for Measuring Electrical Resistivity of Anisotropic Materials," *J. Appl. Phys.*, Vol. 42, No. 7, June 1971, pp. 2971–2975.

Moodenbaugh, A. R., M. Suenaga, T. Asano, R. N. Shelton, H. C. Ku, R. W. McCallum, and P. Klavins, "Superconductivity near 90 K in the La-Ba-Cu-O System," *Phys. Rev. Lett.*, Vol. 58, No. 18, May 4, 1987, pp. 1885–1887.

Moorjani, K., J. Bohandi, F. J. Adrian, B. F. Kim, R. D. Schull, C. K. Chiang, L. J. Swartzendruber, and L. H. Bennett, "Superconductivity in bulk and thin films of $La_{1.85}Sr_{0.15}CuO_{4-\delta}$ and $Ba_2YCu_3O_{7-\delta}$," *Phys. Rev. B*, Vol. 36, No. 7, September 1, 1987, pp. 4036–4038.

Moreland, J., L. F. Goodrich, J. W. Ekin, T. E. Capobianco, A. F. Clark, A. I. Braginski, and A. J. Panson, "Josephson effect above 77 K in a YBaCuO break junction," *Appl. Phys. Lett.*, Vol. 51, No. 7, August 17, 1987, pp. 540–541.

Morisue, M., M. Kaneko, and H. Hosoya, "A Content Addressable Memory Circuit using Josephson Junctions," *IEEE Trans. Magn.*, Vol. MAG-23, No. 2, March 1987, pp. 743–746.

Mott, N., "Is there an explanation?," *Nature*, Vol. 327, No. 6119, May 21, 1987, pp. 185–186.

Moulthrop, A., and M. S. Muha, "Superconducting stepper motors," *Rev. Sci. Instrum.*, Vol. 59, No. 4, April 1988, pp. 649–650.

Mueller, F. M., S. P. Chen, M. L. Prueitt, J. F. Smith, J. L. Smith, and D. Wohllenben, "Coherent Twin Boundaries in 1-2-3 Superconducting Oxides," (Submitted to *Phys. Rev. Lett.*, November 12, 1987.

Mueller, O., "RF components at low temperatures," *RF Design*, January 1989, pp. 29–39.

Müller, K. A., and J. Georg Bednorz, "The Discovery of a Class of High-Temperature Superconductors," *Science*, Vol. 237, September 4, 1987, pp. 1133–1139.

Mukaida, M., M. Yamamoto, Y. Tazoh, K. Kuroda, and K. Hohkawa, "Synthesis of Superconducting $YBa_2Cu_3O_{7-\delta}$ Thin Films by Electron Beam Deposition," *Jap. J. Appl. Phys.*, Vol. 27, No. 2, February 1988, pp. L211–L213.

Mukhanov, O. A., V. K. Semenov, and K. K. Likharev, "Ultimate Performance of the RSFQ Logic Circuits," *IEEE Trans. Magn.*, Vol. MAG-23, No. 2, March 1987, pp. 759–762.

Mulder, G. B. J., H. H. J. ten Kate, H. J. G. Krooshoop, and L. J. M. van de Klundert, "Thermally and magnetically controlled superconducting rectifiers," *IEEE Trans. Magn.*, Vol. 25, No. 2, March 1989, pp. 1819–1822.

Munakata, F., K. Shinohara, H. Kanesaka, N. Hirosaki, A. Okada, and M. Yamanaka, "Electrical properties of $YBa_2Cu_3O_y$ at high temperatures," *Jap. J. Appl. Phys.*, Vol. 26, No. 8, August 1987, pp. L1292–L1293.

Murphy, D. W., D. W. Johnson, Jr., S. Jin, and R. E. Howard, "Processing techniques for the 93 K superconductor $Ba_2YCu_3O_7$," *Science*, Vol. 241, No. 4868, August 19, 1988, pp. 922–930.

Nagashima, T., K. Watanabe, H. Saito, and Y. Fukai, "Superconductivity in $Tl_{1.5}SrCaCu_2O_x$," *Jap. J. Appl. Phys.*, Vol. 27, No. 6, June 1988, pp. L1077–L1079.

Naito, M., D. P. E. Smith, M. D. Kirk, B. Oh, M. R. Hahn, K. Char, D. B. Mitzi, J. Z., Sun, D. J. Webb, M. R. Beasly, O. Fischer, T. H. Geballe, R. H. Hammond,

A. Kapitulnik, and C. F. Quate, "Electron-tunneling studies of thin films of high-T_c superconducting La-Sr-Cu-O," *Phys. Rev. B*, Vol. 35, No. 13, May 1, 1987, pp. 7228–7231.

Nakagawa, H., I. Kurosawa, S. Takada, and H. Hayakawa, "Josephson 4-bit Digital Counter Circuit Made by Nb/Al-oxide/Nb Junctions," *IEEE Trans. Magn.*, Vol. MAG-23, No. 2, March 1987, pp. 739–742.

Nakamori, T., H. Abe, Y. Takahashi, T. Kanamori, and S. Shibata, "Preparation of superconducting Bi-Ca-Sr-Cu-O printed thick films using a coprecipitation of oxalates," *Jap. J. Appl. Phys.*, Vol. 27, No. 4, April 1988, pp. L649–L651.

Nakane, H., T. Nishino, M. Hirano, K. Takagi, and U. Kawabe, "DC-SQUID Using High-Critical-Temperature Oxide Superconductors," *Jap. J. Appl. Phys.*, Vol. 26, No. 10, October 1987, pp. L1581–L1582.

Nakanishi, S., M. Kogachi, H. Sasakura, N. Fukuoka, S. Minamigawa, K. Nakahigashi, and A. Yanase, "The orthorhombic to tetragonal transition under controlled oxygen content in high-T_c superconductor $YBa_2Cu_3O_y$," *Jap. J. Appl. Phys.*, Vol. 27, No. 3, March 1988, pp. L329–L332.

Nakao, M., H. Kuwahara, R. Yuasa, H. Mukaida, and A. Mizukami, "Magnetron sputtering of Bi-Ca-Sr-Cu-O thin films with superconductivity above 80 K," *Jap. J. Appl. Phys.*, Vol. 27, No. 3, March 1988, pp. L378–L380.

Nakao, K., N. Miura, K. Tatsuhara, S.-I. Uchida, H. Takagi, T. Wada, and S. Tanaka, "Superconductivity in $YBa_2Cu_3O_{7-x}$ in a 100 tesla magnetic field," *Nature*, Vol. 332, No. 6167, April 28, 1988, pp. 816–818.

Narayan, J., V. N. Shukla, S. J. Lukasiewicz, N. Biunno, R. Singh, A. F. Schreiner, and S. J. Pennycook, "Microstructure and properties of $YBa_2Cu_3O_{9-\delta}$ superconductors with transitions at 90 and near 290 K," *Appl. Phys. Lett.*, Vol. 51, No. 12, September 21, 1987, pp. 940–942.

Nasu, H., H. Myoren, Y. Ibara, S. Makida, Y. Nishiyama, T. Kato, T. Imura, and Y. Osaka, "Formation of high-T_c superconducting $BiSrCaCu_2O_x$ films on ZrO_2/Si(100)," *Jap. J. Appl. Phys.*, Vol. 27, No. 4, April 1988, pp. L634–L635.

Nasu, H., S. Makida, Y. Ibara, T. Kato, T. Imura, and Y. Osaka, "Preparation of $BiSrCaCu_2O_x$ films with T_c >77 K by pyrolysis of organic acid salts," *Jap. J. Appl. Phys.*, Vol. 27, No. 4, April 1988, pp. L536–L537.

Naughton, M. J., P. M. Chaikin, C. W. Chu, P. H. Hor, and R. L. Meng, "Critical fields of $(La_{0.925}Ba_{0.075})_2CuO_{4-y}$," *Solid State Commun.*, Vol. 62, No. 8, 1987, pp. 531–533.

Neffe, J., "German maglev trains way ahead of rivals," *Nature*, Vol. 335, No. 6189, September 29, p. 388.

Neumeier, J. J., Y. Dalichaouch, J. M. Ferreira, R. R. Hake, B. W. Lee, M. B. Maple, M. S. Torikachvili, K. N. Yang, and H. Zhou, "Thulium barium copper oxide—A 90-K superconductor with a potential 1-MG upper critical field," *Appl. Phys. Lett.*, Vol. 51, No. 5, August 3, 1987, pp. 371–373.

Neumeier, J. J., M. B. Maple, and M. S. Torikachvili, "Pressure dependence of the superconducting transition temperature of $(Y_{1-x}Pr_x)Ba_2Cu_3O_{7-\delta}$ Compounds; Evidence for 4f electron hybridization," *Physica C*, Vol. 156, 1988, pp. 574–578.

Ni, B., T. Munakata, T. Matsushita, M. Iwakuma, K. Funaki, M. Takeo, and K. Yamafuji, "AC inductive measurements of intergrain and intragrain currents in high-T_c superconductors," *Jap. J. Appl. Phys.*, Vol. 27, No. 9, September 1988, pp. 177–181.

Niemann, R. C., R. C. Bossert, J. A. Carson, N. H. Engler, J. D. Gonczy, E. T. Larson, T. H. Nico, and T. Ohmori, "Superconducting super collider second generation dipole magnet cryostat design," *IEEE Trans. Magn.*, Vol. 25, No. 2, March 1989, pp. 1615–1619.

Nikonov, A. A., G. V. Sotnikov, A. S. Tokarev, and I. V. Kurchatov, "Influence of radiation defects on superconductivity in $YBa_2Cu_3O_{7-\delta}$ films," *IEEE Trans. Magn.*, Vol. 25, No. 2, March 1989, pp. 2579–2582.

Nishino, T., H. Nakane, Y. Tarutani, M. Hirano, T. Aida, S. Kominami, and U. Kawabe, "Light detection by superconducting weak link fabricated with high-critical-temperature oxide-superconductor film," *Jap. J. Appl. Phys.*, Vol. 26, No. 8, August 1987, pp. L1320–L1322.

Noakes, D. R., "High-T_c Superconductor at TRIUMF," *IEEE Nuclear and Plasma Sciences Society News*, June 1987, pp. 25–26.

Nobumasa, H., K. Shimizuh, Y. Kitano, and T. Kawai, "Formation of a 100 K superconducting Bi(Pb)-Sr-Ca-Cu-O film by spray pyrolysis," *Jap. J. Appl. Phys.*, Vol. 27, No. 9, September 1988, pp. L1669–L1671.

Noel, H., P. Gougeion, J. Padiou, J. C. Levet, M. Potel, O. Laborde, and P. Monceau, "Anisitropy of the superconducting magnetic field H_{c2} of a single crystal of $TmBa_2Cu_3O_{7-\delta}$," *Solid State Commun.*, Vol. 63, No. 10, 1987, pp. 915–917.

Ogale, S. B., D. Dijkkamp, T. Venkatesan, X. D. Wu, and A. Inam, "Current transport in high-T_c polycrystaline films of Y-Ba-Cu-O," *Phys. Rev. B*, Vol. 36, No. 13, November 1, 1987, pp. 7210–7213.

Ogasawara, T., "Feasibility study on large-scale, high current density superconductor by dynamic stabilization," *Cryogenics*, Vol. 27, No. 12, December 1987, pp. 673–677.

Ogushi, T., G. N. Suresha, Y. Honjo, Y. Ozono, Y., I. Kawano, and T. Numata, "Possibility of Superconductivity with High T_c in La-Sr-Nb-O System," *J. Low Temp. Phys.*, Vol. 69, Nos. 5/6, 1987, pp. 451–457.

Oh, B., M. Naito, S. Arnason, P. Rosenthal, R. Barton, M. R. Beasley, T. H. Geballe, R. H. Hammond, and A. Kapitulnik, "Critical current densities and transport in superconducting $YBa_2Cu_3O_{7-\delta}$ films made by electron beam coevaporation," *Appl. Phys. Lett.*, Vol. 51, No. 11, September 14, 1987, pp. 852–854.

Oh, B., K. Char, A. D. Kent, M. Naito, M. R. Beasley, T. H. Geballe, R. H. Hammond, A. Kapitulnik, and J. M. Graybeal, "Upper critical field, fluctation conductivity, and dimensionality of $YBa_2Cu_3O_{7-x}$," *Phys. Rev. B*, Vol. 37, No. 13, May 1, 1988, pp. 7861–7864.

Ohta, M., K. Takahashi, and M. Kosuge, "Superconductor with a layered structure in the Bi-Sr-Ca-Cu oxides," *Jap. J. Appl. Phys.*, Vol. 27, No. 4, April 1988, pp. L567–L568.

Oka, K., K. Nakane, M. Ito, M. Saito, and H. Unoki, "Phase-equilibrium diagram in the ternary system Y_2O_3-BaO-Cu-O," *Jap. J. Appl. Phys.*, Vol. 27, No. 6, June 1988, pp. L1065–L0167.

Okada, M., A. Okayama, T. Morimoto, T. Matsumoto, K. Aihara, and S. Matsuda, "Fabrication of Ag-Sheathed Ba-Y-Cu Oxide Superconductor Tape," *Jap. J. Appl. Phys.*, Vol. 27, No. 2, February 1988, pp. L185–L187.

Okura, K., K. Ohmatsu, H. Takei, H. Hitotsuyanagi, and T. Nakahara, "Superconductivity of $Ba_2Y(Cu,Ti)_3O_y$ Oxide," *Jap. J. Appl. Phys.*, Vol. 27, No. 4, April 1988, pp. L655–L657.

Omori, M., S. Tomiyoshi, H. Yamauchi, T. Kamiyama, and T. Fukuda, "Preparation of oriented orthorhombic $LnBa_2Cu_3O_{7-x}$ Polycrystals," *Jap. J. Appl. Phys.*, Vol. 26, No. 8, August 1987, pp. L1421–L1423.

Onnes, K., "Report on the Researches Made in the Leiden Cryogenic Laboratory between the Second and Third International Congress of Refrigeration," Suppl. No. 34b, 1913, in *Superconductivity—Selected Reprints*, American Institute of Physics, New York, 1964, pp. 8–23.

Ono, A., H. Nozaki, and Y. Ishizawa, "Preparation and properties of $Ba_2YCu_3O_{7-y}$ single crystals grown using an indium oxide flux," *Jap. J. Appl. Phys.*, Vol. 27, No. 3, March 1988, pp. L340–L343.

Oota, A., Y. Kiyoshima, and T. Hioki, "Electrical, magnetic and superconducting properties of (Y,Sc)-(Ba,Sr)-Cu oxide," *Jap. J. Appl. Phys.*, Vol. 26, No. 8, August 1987, pp. L1356–L1358.

Orenstein, J., G. A. Thomas, D. H. Rapkine, C. G. Bethea, B. F. Levine, R. J. Cava, E. A. Rietman, and D. W. Johnson, Jr., "Normal-state gap transition in Cu-O superconductors," *Phys. Rev. B*, Vol. 36, No. 1, July 1, 1987, pp. 729–732.

Orlando, T. P., K. A. Delin, S. Foner, E. J. McNiff, Jr., J. M. Tarascon, L. H. Green, W. R. McKinnon, and G. W. Hull, "Upper critical fields of high-T_c superconducting $La_{2-x}Sr_xCuO_{4-y}$: Possibility of 140 tesla," *Phys. Rev. B*, Vol. 35, No. 10, April 1, 1987, pp. 5347–5349.

Orlando, T. P., K. A. Delin, S. Foner, E. J. McNiff, Jr., J. M. Tarascon, L. H. Greene, W. R. McKinnon, and G. W. Hull, "Upper critical fields of high-T_c superconducting $Y_{2-x}Ba_xCuO_{4-y}$," *Phys. Rev. B*, Vol. 35, No. 13, May 1, 1987, pp. 7249–7251.

Orlando, T. P., K. A. Delin, S. Foner, E. J. McNiff, Jr., J. M. Tarascon, L. H. Greene, W. R. McKinnon, and G. W. Hull, "Upper critical fields and anistropy limits of high-Tc superconductors $R_1Ba_2Cu_3O_{7-y}$, where R = Nd, Eu, Gd, Dy, Ho, Er, and Tm and $YBa_2Cu_3O_{7-y}$," *Phys. Rev. B*, Vol. 36, No. 4, August 1, 1987, pp. 2394–2397.

Ourmazd, A., and J. C. H. Spence, "Detection of oxygen ordering in superconducting cuprates," *Nature*, Vol. 329, October 1, 1987, pp. 425–427.

Oussena, M., S. Senoussi, G. Collin, J. M Broto, H. Rakoto, S. Askenazy, and J. C. Ousset, "Longitudinal and transverse magnetoresistances of $YBa_2Cu_3O_7$ in very high field up to 430 kG," *Phys. Rev. B*, Vol. 36, No. 7, September 1, 1987, pp. 4014–4017.

Ovshinsky, S. R., R. T. Young, D. D. Allred, G. DeMaggio, and G. A. Van der Leeden, "Superconductivity at 155 K," *Phys. Rev. Lett.*, Vol. 58, No. 24, June 15, 1987, pp. 2579–2581.

Padamsee, H., "High T_c superconductors and their impact on niobium based accelerator applications," Cornell University publication CLNS 87/112, pres. at 1987 Electron Beam Melting and Refining Conference, Reno, NV.

Padamsee, H., "Superconducting structures for electron accelerators," Presented at MIT DOE/EPRI Workshop on Novel Applications of High Temperature Superconductors, Salem, MA, June 1988.

Padamsee, H., K. Green, J. Gruschus, J. Kirchgessner, D. Moffat, D. L. Rubin, J. Sears, and Q. S. Shu, "Microwave Superconductivity for Particle Accelerators—How the High T_c Superconductors Measure Up?," Cornell Univ. pub. #CLNS 88/835, pres. at Conf. on Superconductivity and Applications, Institute on Superconductivity, Buffalo, NY, April 18–20, 1988.

Pan, S. K., A. R. Kerr, M. J. Feldman, and A. W. Kleinsasser, "SIS technology boosts sensitivity of mm-wave receivers," *Microwaves & RF*, Vol. 27, No. 9, September 1988, pp. 139–146.

Pande, C. S., A. K. Singh, L. Toth, D. U. Gubser, and S. Wolf, "Domainlike effects observed in the high-temperature superconductor Y-Ba-Cu-O," *Phys. Rev. B*, Vol. 36, No. 10, October 1, 1987, pp. 5669–5671.

Panson, A. J., A. I. Braginski, J. R. Gavaler, J. K. Hulm, M. A. Janocko, H. C. Pohl, A. M. Stewart, J. Talvacchio, and G. R. Wagner, "Effect of compositional variation and annealing in oxygen on superconducting properties of $Y_1Ba_2Cu_3O_{8-y}$," *Phys. Rev. B*, Vol. 35, No. 16, June 1, 1987, pp. 8774–8777.

Papaioannou, J., and J. L. Dye, "Four-probe single-crystal holder for conductivity measurements," *Rev. Sci. Instrum.*, Vol. 59, No. 3, March 1988, pp. 496–497.

Parks, R. D., "Quantum effects in superconductors," *Scientific American*, Vol. 213, No. 4, October 1965, pp. 57–67.

Parmigiani, F., G. Chiarello, N. Ripamonti, H. Goretzki, and U. Roll, "Observation of carboxylic groups in the lattice of sintered $Ba_2YCu_3O_{7-y}$ high-T_c superconductors," *Phys. Rev. B*, Vol. 36, No. 13, November 1, 1987, pp. 7148–7150.

Pazoi, B. G., and D. M. Ginsberg, "Low-temperature sample-rotating cryostat with two degrees of freedom for measuring the anisotropy of the upper critical field of superconductors," *Rev. Sci. Instrum.*, Vol. 59, No. 5, May 1988, pp. 776–777.

Peterreins, T., F. Pröbst, F. V. Feilitzsch, and H. Kraus, "Phonon mediated positron resolved detection of α-particles with superconducting tunnel junctions," *IEEE Trans. Nucl. Sci.*, Vol. 35, No. 1, February 1988, pp. 70–72.

Petersen, D. A., H. Ko, and T. Van Duzer, "Dynamic Behavior of a Josephson Junction Latching Comparator for use in a High-Speed Analog-to-Digital Converter," *IEEE Trans. Magn.*, Vol. MAG-23, No. 2, March 1987, pp. 891–894.

Peterson, G. E., R. P. Stawicki, and U.-C. Paek, "Radio frequency properties of high-T_c superconductors," Proc. 38th Electronic Components Conf., Los Angeles, May 9–11, 1988, pp. 159–167.

Peterson, G. E., R. P. Stawicki, and U.-C. Paek, "Comparison of the low-temperature resistance of high-T_c superconductors and copper at 0. 5 to 20 MHz," *Advanced Ceramic Materials*, Vol. 3, No. 5, 1988, pp. 522–523.

Peterson, I., "A New Recipe for Superconductivity," *Science News*, Vol. 133, No. 8, February, 1988, p. 116.

Peterson, J. F., and H. A. Steinherz, "Vacuum pump technology: A short course on theory and operations, Part I," *Solid State Technol.*, December 1981.

Peterson, J. F., and H. A. Steinherz, "Vacuum pump technology: A short course on theory and operations, Part II," *Solid State Technol.*, January 1982.

Piel, H., "Recent Progress in RF Superconductivity," *IEEE Trans. Nucl. Sci.*, Vol. NS-32, No. 5, October 1985, pp. 3365–3569.

Piel, H., M. Hein, N. Klein, U. Klein, A. Michalke, G. Müeller, and L. Ponto, "Superconducting perovskites in microwave fields," Proc. Int. Conf. on High Temperature Superconductors, Interlaken, Switzerland, *Physica C*, Vol. 153–155, Superconductivity, June 1988, pp. 1604–1609.

Pippard, A. B., "The surface impedance of superconductors and normal metals at high frequencies," *Proc. R. Soc. A*, Vol. 191, 1947, pp. 370–415.

Pippard, A. B., "Early Superconductivity Research (Except Leiden)," *IEEE Trans. Magn.*, Vol. MAG-23, No. 2, March 1987, pp. 371–375.

Poirier, M., G. Quirion, K. R. Poeppelmeier, and J. P. Thiel, "Microwave study of the high-T_c superconductor $La_{1.8}Sr_{0.2}CuO_4$," *Phys. Rev. B*, Vol. 36, No. 7, September 1, 1987, pp. 3906–3909.

Pond, J. M., J. H. Classen, and W. L. Carter, "Kinetic Inductance Microstrip Delay Lines," *IEEE Trans. Magn.*, Vol. MAG-23, No. 2, March 1987, pp. 903–906.

Pool, R., "Copperless Superconductivity," *Science*, Vol. 240, No. 4859, June 17, 1988, p. 1614.

Pool, R., "New superconductors come through," *Science*, Vol. 240, No. 4859, June 17, 1988, pp. 1613–1615.

Pool, R., "Superconductor credits bypass Alabama," *Science*, Vol. 241, No. 4866, August 5, 1988, pp. 655–657.

Pool, R., "A testable theory of superconductivity," *Science*, Vol. 242, No. 4875, October 7, 1988, p. 31.

Przbysz, J., and R. D. Blaugher, "Josephson Data Latch for Frequency Agile Shift Registers," *IEEE Trans. Magn.*, Vol. MAG-23, No. 2, March 1987, pp. 777–780.

Przyslupski, P., J. Igalson, J. Rauluszkiewicz, and T. Skos'kiewicz, "High-T_c superconductivity in $La_{1.8}Sr_{0.2}CuO_4$ and $Y_{1.2}Ba_{0.8}CuO_4$ systems," *Phys. Rev. B*, Vol. 36, No. 1, July 1, 1987, pp. 743–744.

Radcliffe, W. J., J. C. Gallop, C. D. Langham, M. Gee, and M. Stewart, "Measurements of microwave impedance of YBaCuO superconductors," Proc. Int. Conf. on High Temperature Superconductors, Interlaken, Switzerland, *Physica C*, Vol. 153–155, Superconductivity, June 1988, pp. 635–636.

Radhakrishnan, V., and V. L. Newhouse, "Noise Analysis for Amplifiers with Superconducting Input," *J. Appl. Phys.*, Vol. 42, No. 1, January 1971, pp. 129–132.

Ramakrishna, B. L., E. W. Ong, and Z. Iqbal, "Low-field microwave absorption in high-T_c oxides," *High-Temperature Superconductors*, Vol. 99, Proc. Mat. Res. Soc. Symp., Boston, November 30–December 4, 1987, pp. 455–458.

Raveau, B., C. Michel, M. Hervieu, and J. Provost, "Crystal chemistry of perovskite superconductors," Proc. Int. Conf. on High Temperature Superconductors, Interlaken, Switzerland, *Physica C*, Vol. 153–155, Superconductivity, June 1988, pp. 3–8.

Reeves, M. E., T. A. Friedmann, and D. M. Ginsberg, "Specific-heat measurements on two high-transition-temperature superconducting oxides: $La_{1.85}Ba_{0.15}CuO_4$ and $La_{1.8}Sr_{0.2}CuO_4$," *Phys. Rev. B*, Vol. 35, No. 13, May 1, 1987, pp. 7207–7209.

Reeves, M. E., D. S. Citrin, B. G. Pazol, T. A. Friedmann, and D. M. Ginsberg, "Specific heat of $GdBa_2Cu_3O_{7-\delta}$ in the normal and superconducting states," *Phys. Rev. B*, Vol. 36, No. 13, November 1, 1987, pp. 6915–6919.

Reilly, J. J., M. Suenaga, J. R. Johnson, P. Thompson, and A. R. Moodenbaugh, "Superconductivity in $H_xYBa_2Cu_3O_7$," *Phys. Rev. B*, Vol. 36, No. 10, October 1, 1987, pp. 5694–5697.

Rettori, C., D. Davidov, I. Belaish, and I. Felner, "Magnetism and critical fields in the high-T_c superconductors $YBa_2Cu_3O_{7-x}S_x$ ($x = 0,1$): An ESR Study," *Phys. Rev. B*, Vol. 36, No. 7, September 1, 1987, pp. 4028–4031.

Reuter, G. E. H., and E. H. Sondheimer, "The theory of the anomalous skin effect in metals," *Proc. R. Soc. A*, 195, 1948, pp. 336–364.

Rice, T. M., "Oxide Superconductors—Clues from copper-free samples," *Nature*, Vol. 332, No. 6167, April 28, 1988, p. 780.

Richards, P. L., "Analog superconducting electronics," *Physics Today*, March 1986, pp. 54–62.

Richards, P. L., and Tek-Ming Shen, "Superconductive Devices for Millimeter Wave Detection, Mixing, and Amplification," *IEEE Trans. Electron Devices*, Vol. ED-27, No. 10, October 1980, pp. 1909–1920.

Richardson, T. J., and L. C. DeJonghe, "Aluminum cladding of High T_c superconductor by thermocompression bonding," *Appl. Phys. Lett.*, Vol. 53, No. 23, December 5, 1988, pp. 2342–2343.

Robinson, A. L., "An Oxygen Key to the New Superconductors," *Research News*, May 29, 1987, pp. 1063–1065.

Robinson, G. M., "Superconductors: high-tech sweepstakes," *Design News*, Vol. 44, No. 12, June 20, 1988, p. 64.

Robyn, D., W. W. Fletcher, and J. A. Alic, "Bringing superconductivity to market," *Issues in Science & Technology*, Vol. 5, No. 2, Winter 1988–1989, pp. 38–45.

Rose, K., C. L. Bertin, and R. M. Katz, "Radiation Detectors," in *Applied Superconductivity*, Vol. 1, V. L. Newhouse, ed., Academic, 1975, pp. 267–308.

Rose-Innes, A. C., "A simple system for obtaining temperatures between 77 K and room temperature," *J. Phys. E*, Vol. 21, No. 7, July 1988, pp. 729–730.

Rosen, H., E. M. Engler, T. C. Strand, V. Y. Lee, and D. Bethune, "Raman studies of lattice modes in the high-critical temperature superconductor Y-Ba-Cu-O," *Phys. Rev. B*, Vol. 36, No. 1, July 1, 1987, pp. 726–728.

Rosenbaum, R. L., "Some Properties of Gold-Iron Thermocouple Wire," *Rev. Sci. Instrum.*, Vol. 39, No. 6, June 1968, pp. 890–899.

Rowell, J. M., "Superconducting Tunneling Spectroscopy and the Observation of the Josephson Effect," *IEEE Trans. Magn.*, Vol. MAG-23, No. 2, March 1987, pp. 380–389.

Rubin, L. G., B. L. Brandt, and H. H. Sample, "Cryogenic thermometry: a review of recent progress, II," *Cryogenics*, Vol. 22, October 1982, pp. 491–503.

Rubin, L. G., "The new(?) wave: Digital lock-in amplifiers," *Rev. Sci. Instrum.*, Vol. 59, No. 3, March 1988, pp. 514–515.

Sadoulet, B., "Cryogenic detectors of particles: Hopes and challenges," *IEEE Trans. Nucl. Sci.*, Vol. 35, No. 1, February 1988, pp. 47–54.

Sample, H. H., Neuringer, L. J., and L. G. Rubin, "Low temperature thermometry in high magnetic fields. III. Carbon resistors (0.5–4.2 K); thermocouples," *Rev. Sci. Instrum.*, Vol. 45, No. 1, January 1974, pp. 64–73.

Sample, H. H., B. L. Brandt, and L. G. Rubin, "Low-temperature thermometry in high magnetic fields. V. Carbon glass resistors," *Rev. Sci. Instrum.*, Vol. 53, No. 8, August 1982, pp. 1129–1135.

Sandstrom, R. L., W. J. Gallagher, T. R. Dinger, R. H. Koch, R. B. Laibowitz, A. W. Kleinsasser, R. J. Gambino, B. Bumble, and M. F. Chisholm, "Reliable single-target sputtering process for high-temperature superconducting films and devices," *Appl. Phys. Lett.*, Vol. 53, No. 5, August 1, 1988, pp. 444–446.

Sankawa, I., M. Sato, T. Konaka, M. Kobayashi, and K. Ishihara, "Microwave surface resistance studies of $YBa_2Cu_3O_{7-\delta}$ single crystal," *Jap. J. Appl. Phys.*, Vol. 27, No. 9, September 1988, pp. L1637–L1638.

Schawlow, A. L., and G. E. Devlin, "Effect of the Energy Gap on the Penetration Depth of Superconductors," *Phys. Rev.*, Vol. 113, No. 1, January 1, 1959, pp. 120–126.

Schenck, A., "μSR investigations of the magnetic flux penetration in the mixed state of the high temperature superconductors," Proc. Int. Conf. on High Temperature Superconductors, Interlaken, Switzerland, *Physica C*, Vol. 153–155, Superconductivity, June 1988, pp. 1127–1132.

Schlesinger, Z., R. T. Collins, M. W. Shafer, and E. M. Engler, "Normal-state reflectivity and superconducting energy-gap measurement of $La_{2-x}Sr_xCuO_4$," *Phys. Rev. B*, Vol. 36, No. 10, October 1, 1987, pp. 5275–5278.

Schmitt, R. W., "The discovery of electron tunneling into superconductors," *Physics Today*, Vol. 14, No. 12, December 1961, pp. 38–41.

Schneemeyer, L. F., R. B. van Dover, S. H. Glarum, S. A. Sunshine, R. M. Fleming, B. Batlogg, T. Siegrist, J. H. Marshall, J. V. Waszcsak, and L. W. Rupp, "Growth of superconducting single crystals in the Bi-Sr-Ca-Cu-O system from alkali chloride fluxes," *Nature*, Vol. 332, No. 6163, March 31, 1988, pp. 422–424.

Schrieffer, J. R., X.-G. Wenh, and S.-C. Zhang, "The current theoretical situation in high-T_c superconductivity," Proc. Int. Conf. on High Temperature Superconductors, Interlaken, Switzerland, *Physica C*, Vol. 153–155, Superconductivity, June 1988, pp. 21–25.

Segre, C. U., B. Dabrowski, D. G. Hinks, K. Zhang, J. D. Jorgensen, M. A. Beno, and Ivan K. Schuller, "Oxygen ordering and superconductivity in $La(Ba_{2-x}La_x)Cu_3O_{7+\delta}$," *Nature*, Vol. 329, September 17, 1987, pp. 227–229.

Senoussi, S., M. Oussena, and M. Ribault, "Critical fields and the critical current density of $La_{1.8}Sr_{0.15}CuO_4$," *Phys. Rev. B*, Vol. 36, No. 7, September 1, 1987, pp. 4003–4006.

Sera, M., S. Shamoto and M. Sato, "Electron tunneling studies of high-T_c superconductor $YBa_2Cu_3O_{7-\delta}$," *Solid State Comm.*, Vol. 65, No. 9, 1988, pp. 997–999.

Shapiro, S., "Millimeter and Submillimeter Detectors and Devices," in *The Science and Technology of Superconductivity*, W. D. Gregory, W. N. Mathews, Jr., and E. A. Edelsack, eds., Vol. 2, Plenum, New York, 1973, pp. 631–652.

Sheng, Z. Z., and A. M. Hermann, "Bulk superconductivity at 120 K in the Tl-Ca/Ba-Cu-O system," *Nature*, Vol. 332, No. 6160, March 1988, pp. 138–139.

Sheng, Z. Z., and A. M. Hermann, "Superconductivity in the rare-earth-free Tl-Ba-Cu-O system above liquid nitrogen temperature," *Nature*, Vol. 332, No. 6159, March 3, 1988, pp. 55–58.

Sheng, Z. Z., W. Kiehl, J. Bennett, A. El Ali, D. Marsh, G. D. Mooney, F. Arammash, J. Smith, D. Viar, and A. M. Hermann, "New 120 K Tl-Ca-Ba-Cu-O superconductor," *Appl. Phys. Lett.*, Vol. 52, No. 20, May 16, 1988, pp. 1738–1740.

Shibata, S., T. Kitagawa, H. Okazaki, and T. Kimura, "*C*-axis-oriented superconducting oxide film by the sol-gel method," *Jap. J. Appl. Phys.*, Vol. 27, No. 4, April 1988, pp. L646–L648.

Shih, I., and C. X. Qiu, "Chemical etching of Y-Cu-Ba-O thin films," *Appl. Phys. Lett.*, Vol. 52, No. 18, May 2, 1988, pp. 1523–1524.

Shimizu, N., Y. Harada, N. Miyamoto, and E. Goto, "A new A/D converter with quantum flux parametron," *IEEE Trans. Magn.*, Vol. 25, No. 2, March 1989, pp. 865–868.

Shinohara, K., F. Munakata, and M. Yamanaka, "Preparation of Y-Ba-Cu-O superconducting thin films by chemical vapor deposition," *Jap. J. Appl. Phys.*, Vol. 27, No. 9, September 1988, pp. L1683–L1685.

Shu, Q. S., K. Gendreau, W. Hartung, J. Kirchgessner, D. Moffat, R. Noer, H. Padamsee, D. L. Rubin, and J. Sears, "Influence of consensed gases on field emission and the performance of superconducting cavities," presented at 1988 ASC, San Francisco (To be published in *IEEE Trans. Magn.*, MAG-25, 1989).

Siegrist, T., S. M. Zahurak, D. W. Murphy, and R. S. Roth, "The parent stucture of the layered high-temperature superconductors," *Nature*, Vol. 334, No. 6179, July 21, 1988, pp. 231–232.

Singh, P. B., S. K. Agarwal, B. Jayaram, R. K. Nayar, K. C. Nagpal, and A. V. Narlikar, "Effect of compaction on the superconducting transition of $YBa_2Cu_3O_{9-y}$ compound," *Jap. J. Appl. Phys.*, Vol. 27, No. 6, June 1988, pp. L1116–1119.

Sleight, A. W., "Oxide superconductors: A chemist's view," *High-Temperature Superconductors*, Vol. 99, Proc. Mat. Res. Soc. Symp., Boston, November 30–December 4, 1987, pp. 3–8.

Sleight, A. W., J. L. Gillson, and P. E. Bierstedt, "High-Temperature Superconductivity in the $BaPb_{1-x}Bi_xO_3$ System," *Solid State Commun.*, Vol. 117, 1975, pp. 27–28.

Soderman, D. A., and K. Rose, "Microwave Studies of Thin Superconducting Films," *J. Appl. Phys.*, Vol. 39, No. 6, May 1968, pp. 2610–2617.

Spencer, E. W., "Cryogenic Safety," in *CRC Handbook of Laboratory Safety*, 2nd Ed., N. V. Steere, ed., CRC, Boca Raton, FL, 1983, pp. 574–578.

Sridhar, S., "Microwave Technology and Materials Research," *Microwave Journal*, June 1987, pp. 117–123.

Sridhar, S., "Microwave response of thin-film superconductors," *J. Appl. Phys.*, Vol. 63, No. 1, January 1, 1988, pp. 159–166.

Sridhar, S., C. A. Shiffman, and H. Hamdeh, "Electrodynamic response of $Y_1Ba_2Cu_3O_y$ and $La_{1.85}Sr_{0.15}CuO_{4-\delta}$ in the superconducting state," *Phys. Rev. B*, Vol. 36, No. 4, August 1, 1987, pp. 2301–2304.

Sridhar, S., and W. L. Kennedy, "Novel technique to measure the microwave response of high T_c superconductors between 4. 2 and 200 K," *Rev. Sci. Instrum.*, Vol. 59, No. 4, April 1988, pp. 531–536.

Stacy, A. M., J. V. Badding, M. J. Geselbracht, W. K. Ham, G. F. Holland, R. L. Hoskins, S. W. Keller, C. F. Millikan, and H.-C. zur Loye, "High-Temperature Superconductivity in Y-Ba-Cu-O: Identification of a Copper-Rich Superconducting Phase," *J. Am. Chem. Soc.*, Vol. 109, No. 8, 1987, pp. 2528–2530.

Staskiewicz, J., A. Czyziewski, C. Szypowski, and W. Gulbinski, "Improvement of the sensitivity of a bridge for the inductive measurement of superconducting critical temperatures," *J. Phys. E*, Vol. 21, 1988, pp. 62–63.

Stephens, R. B., "Critical current limitations in ceramic oxide superconductors," General Atomics Pub GA-A19385, pres. at Int. Conf. on Critical Currents in High-Temperature Superconductors, Snowmass Village, CO, August 16–19, 1988.

Steyert, W. A., "Superconductor Characterization Cryostat: Instrumentation and Methodology Handbook," APD Cryogenics, Allentown, PA, December 1987. 25 pp.

Steyert, W. A., and R. C. Longworth, "A High T_c Superconductor Characterization Cryostat," Poster presentation, Int. Conf. on Superconductivity, Drexel University, Philadelphia, July 29–30, 1987

Stoffel, N. G., J. M. Tarascon, Y. Chang, M. Onellion, D. W. Niles, and G. Margaritondo, "Effects of oxygen stoichiometry on the electronic structure of $YBa_2Cu_3O_x$," *Phys. Rev. B*, Vol. 36, No. 7, September 1, 1987, pp. 3986–3989.

Stritzker, B., W. Zander, F. Dworchak, U. Poppe, and K. Fischer, "Electron irradiation of $YBa_2Cu_3O_7$ at low temperatures," *High Temperature Superconductors,* Proc. Mat. Res. Soc. Symp., Vol. 99, December 30, 1987, pp. 491–495.

Subramanian, M. A., J. C. Calabrese, C. C. Torardi, J. Gopalakrishman, T. R. Askew, R. B. Flippen, K. J. Morrissey, U. Chowdhry, and A. W. Sleight, "Crystal structure of the high-temperature superconductor $Tl_2Ba_2CaCu_2O_8$," *Nature*, Vol. 332, No. 6163, March 31, 1988, pp. 420–422.

Sumiyama, A., T. Yoshitomi, H. Endo, J. Tsuchiya, N. Kijima, M. Mizuno, and Y. Oguri, "Superconductivity of $Bi_{0.25-y}Sr_{0.25-y}Ca_{2y}Cu_{0.5}O_x$ (y =0. 1, 0. 125, 0. 15)," *Jap. J. Appl. Phys.*, Vol. 27, No. 4, April 1988, pp. L542–L544.

Sun, J. Z., D. J. Webb, M. Naito, K. Char, M. R. Hahn, J. W. P. Hsu, A. D. Kent, D. B. Mitzi, B. Oh, M. R. Beasly, T. H. Geballe, R. H. Hammon, and A. Kapitulnik, "Superconductivity and Magnetism in the High-T_c Superconductor Y-Ba-Cu-O," *Phys. Rev. Lett.*, Vol. 58, No. 15, April 13, 1987, pp. 1574–1576.

"Superconductivity research nets hot results," *Design News*, Vol. 44, No. 10, May 23, 1988.

Swinbanks, D., "Japanese industry dreams of superconductor future," *Nature*, Vol. 329, October 1, 1987, p. 378.

Swinbanks, D., "High-critical-temperature superconductor made of glass," *Nature*, Vol. 332, No. 6165, April 28, 1988, p. 575.

Swinbanks, D., "No rare earths in new film," *Nature*, Vol. 332, No. 6167, April 28, 1988, p. 770.

Syono, Y., K. Hiraga, N. Kobayashi, M. Kikuchi, K. Kusaba, T. Kajitani, D. Shindo, S. Hosoya, A. Tokiwa, S. Terada, and Y. Muto, "An X-ray diffraction and electron microscopic study of a new high-T_c superconductor based on the Bi-Ca-Sr-Cu-O system," *Jap. J. Appl. Phys.*, Vol. 27, No. 4, April 1988, pp. L569–L572.

Tajima, Y., M. Hikita, T. Ishii, H. Fuke, K. Sugiyama, M. Date, A. Yamagishi, A. Katsui, Y. Hidaka, T. Iwata, and S. Tsurumi, "Upper critical field and resistivity of single-crystal $EuBa_2Cu_3O_y$: Direct measurements under high field up to 50 T," *Phys. Rev. B*, Vol. 37, No. 13, May 1, 1988, pp. 7956–7959.

Takahashi, K., M. Nakao, D. R. Dietderich, H. Kumakura, and K. Togano, "Preparation and properties of Tl-Ca-Ba-Cu-O," *Jap. J. Appl. Phys.*, Vol. 27, No. 8, August 1988, pp. L1457–L1459.

Takayama-Muromachi, E., Y. Uchida, A. Ono, F. Izumi, M. Onoda, Y. Matsui, K. Kosuda, S. Takekawa, and K. Kato, "Identification of the superconducting phase in the Bi-Ca-Sr-Cu-O System," *Jap. J. Appl. Phys.*, Vol. 27, No. 3, March 1988, pp. L365–L368.

Takayama-Muromachi, E., Y. Uchida, Y. Matsui, M. Onoda, and K. Kato, "On the 110 K superconductor in the Bi-Ca-Sr-Cu-O system," *Jap. J. Appl. Phys.*, Vol. 27, No. 4, April 1988, pp. L556–L558.

Takita, K., H. Akinaga, H. Katoh, T. Uchino, T. Ishigaki, and H. Asano, "Anisotropic upper critical fields of orthorhombic single phase $YBa_2Cu_3O_{7-\delta}$ studied using preferentially oriented pellets," *Jap. J. Appl. Phys.*, Vol. 26, No. 8, August 1987, pp. L1323–L1325.

Talpe, J. G. Stolovitzky, and V. Bekeris, "Cryogenic thermometry and level detection with common diodes," *Cryogenics*, Vol. 27, No. 12, December 1987, pp. 693–695.

Talvacchio, J., "Electrical contact to superconductors," to appear in *IEEE Trans. CHMT* (1989).

Tamegai, T., A. Watanabe, I. Oguro, and Y. Iye, "Structures and upper critical fields of high T_c superconductors (RE)$Ba_2Cu_3O_x$," *Jap. J. Appl. Phys.*, Vol. 26, No. 8, August 1987, pp. L1304–1306

Tanaka, S., "Research on high-T_c superconductivity in Japan," *Physics Today*, December 1987, pp. 53–57.

Tanaka, Y., M. Fukutomi, T. Asano, and H. Maeda, "Effects of synthesis conditions on the properties of a superconducting Bi-Sr-Ca-Cu-O system," *Jap. J. Appl. Phys.*, Vol. 27, No. 4, April 1988, pp. L548–L549.

Tateno, J., and N. Masaki, "Measurement of RF Superconductivity in $YBa_2Cu_3O_{7-x}$ with the Standing-Wave Method at 9 GHz," *Jap. J. Appl. Phys.*, Vol. 26, No. 10, October 1987, pp. L1654–L1656.

Terada, N., H. Ihara, M. Jo, M. Hirabayashi, Y. Kimura, K. Matsutani, K. Hirata, E. Ohno, R. Sugise, and F. Kawashima, "Sputter synthesis of $Ba_2YCu_3O_y$ as-deposited superconducting thin films from stoichiometric target—A mechanism for compositional deviation and its control," *Jap. J. Appl. Phys.*, Vol. 27, No. 4, April 1988, pp. L639–L642.

Terasaki, I., Y. Nakayama, K. Uchinokura, A. Maeda, T. Hasegawa, and S. Tanaka, "Superconducting films of $YBa_2Cu_3O_x$ and Bi-Sr-Ca-Cu-O fabricated by electron-beam deposition with a single source," *Jap. J. Appl. Phys.*, Vol. 27, No. 8, August 1988, pp. L1480–L1483.

Thomas, G. A., A. J. Millis, R. N. Bhatt, R. J. Cava, and E. A. Rietman, "Far-Infrared reflectivity of $La_{2-x}Sr_xCuO_4$," *Phys. Rev. B*, Vol. 36, No. 1, July 1, 1987, pp. 736–739.

Thompson, J. R., S. T. Sekula, D. K. Christen, B. C. Sales, L. A. Boatner, and Y. C. Kim, "Magnetization and susceptibility studies of the superconductive compound $Gd_1Ba_2Cu_3O_z$," *Phys. Rev. B*, Vol. 36, No. 1, July 1, 1987, pp. 718–721.

Tinkham, M., "Superconductivity, 75th Anniversary," *Physics Today*, March 1986, pp. 22–23.

Tinkham, M., "Critical currents in high T_c superconductors," *Helv. Phys. Acta*, Vol. 61, 1988, pp. 443–446.

Togano, K., H. Kumakura, K. Fukutomi, and K. Tachikawa, "Superconducting properties of the $Ba_{1-x}Y_xCuO_{3-y}$ system," *Appl. Phys. Lett.*, Vol. 51, No. 2, July 13, 1987, pp. 136–138.

Togano, K., H. Kumkura, H. Maeda, K. Takahashi, and M. Kakao, "Preparation of high-T_c Bi-Sr-Ca-Cu-O superconductors," *Jap. J. Appl. Phys.*, Vol. 27, No. 3, March 1988, pp. L323–L324.

Tomita, M., T. Hayashi, H. Takaoka, Y. Ishii, Y. Enomoto, and T. Murakami, "Cross-sectional TEM observation of $Ba_2YCu_3O_{7-x}$ film on $SrTiO_3$," *Jap. J. Appl. Phys.*, Vol. 27, No. 4, April 1988, pp. L636–L638.

Tonouchi, M., Y. Sakaguchi, K. Hashimoto, Y. Yoshizako, and T. Kobayashi, "Germanium thin film growth onto high T_c superconducting films," *IEEE Trans. Magn.*, Vol. 25, No. 2, March 1989), pp. 957–960.

Torardi, C. C., M. A. Subramanian, J. C. Calabrese, J. Gopalakrishnan, K. J. Morrissey, T. R. Askew, R. B. Flippen, U. Chowdhry, and A. W. Sleight, "Crystal Structure of $Tl_2Ba_2Ca_2Cu_3O_{10}$, a 125 K Superconductor," *Science*, Vol. 240, No. 4852, April 29, 1988, pp. 661–663.

Tsai, J. S., Y. Kubo, and J. Tabuchi, "Josephson Effects in the Ba-Y-Cu-O Compounds," *Phys. Rev. Lett.*, Vol. 58, No. 19, May 11, 1987, pp. 1979–1981.

Tsaur, B.-Y., M. S. Dilorio, and A. J. Strauss, "Preparation of superconducting $YBa_2Cu_3O_x$ thin films by oxygen annealing of multilayer metal films," *Appl. Phys. Lett.*, Vol. 51, No. 11, September 14, 1987, pp. 858–860.

Uchiyama, T., and T. Mamiya, "Low-power persistent switch for superconducting magnet," *Rev. Sci. Instrum.*, Vol. 58, No. 11, November 1987, pp. 2192–2193.

Uehara, M., Y. Asada, H. Maeda, and K. Ogawa, "Magnetic properties of $BiSrCaCu_2O_x$ superconductors," *Jap. J. Appl. Phys.*, Vol. 27, No. 4, April 1988, pp. L665–667.

Uher, C., and A. B. Kaiser, "Thermal transport properties of $YBa_2Cu_3O_7$ superconductors," *Phys. Rev. B*, Vol. 36, No. 10, October 1, 1987, pp. 5680–5683.

van Bentum, P. J. M., H. van Kempen, L. E. C. van de Leemput, J. A. A. J. Perenboom, L. W. M. Schreurs, and P. A. Teunissen, "High-field measurements on the high-T_c superconductors $La_{1.85}Sr_{0.15}CuO_{4-\delta}$ and $YBa_2Cu_3O_{7-\delta}$," *Phys. Rev. B*, Vol. 36, No. 10, October 1, 1987, pp. 5279–5283.

van Bentum, P. J. M., H. F. C. Hoevers, H. Van Kempen, L. E. C. Van De Leemput, M. J. M. F. De Nivelle, L. W. M. Schreurs, R. T. M. Smokers, and P. A. A. Teunissen, "Determination of the energy gap in $YBa_2Cu_3O_{7-\delta}$ by tunneling, far infrared reflection and Adreeev Reflection," Proc. Int. Conf. on High Temperature Superconductors, Interlaken, Switzerland, *Physica C*, Vol. 153–155, Superconductivity, June 1988, pp. 1718–1723.

van der Maas, J., V. A. Gasparov, and D. Pavuna, "Improved low contact resistance in high-T_c Y-Ba-Cu-O ceramic superconductors," *Nature*, Vol. 328, August 13, 1987, pp. 603–604.

van Dover, R. B., L. F. Schneemeyer, E. M. Gyorgy, and J. V. Waszczak, "Critical Current Densities in Single Crystal $Bi_{2.2}Sr_2Ca_{0.8}Cu_2O_{8+\delta}$," cited in High-$T_c$ Update, Vol. 2, No. 9, May 1, 1988, submitted to *Appl. Phys. Lett.*, April 1, 1988.

Van Duzer, T., "Josephson Digital Devices and Circuits," *IEEE Trans. on Microwave Theory & Techniques*, Vol. MTT-28, No. 5, May 1980, pp. 490–500.

Vaslow, D. F., G. H. Dieckmann, D. D. Eli, A. B. Ellis, D. S. Holmes, A. Lefkow, M. MacGregor, J. E. Nordman, M. F. Petras, and Y. Yang, "Superconducting Bi-Ca-Sr-Cu-O thin films by spray pyrolysis of metal acetates," *Appl. Phys. Lett.*, Vol. 53, No. 4, July 25, 1988, pp. 324–326.

Venkatesan, T., S. B. Ogale, C. C. Chang, and D. Dijkkamp, "Pulsed laser etching of high T_c superconducting films," *Appl. Phys. Lett.*, Vol. 51, No. 14, October 5, 1987, pp. 1112–1114.

Verhoeven, J. D., A. J. Bevolo, R. W. McCallum, E. D. Gibson, and M. A. Noack, "Auger study of grain boundaries in large-grained $YBa_2Cu_3O_x$," *Appl. Phys. Lett.*, Vol. 52, No. 9, February 29, 1988, pp. 745–747.

Voronel, A. V., D. Linsky, A. Kisliuk, and S. Drislikh, "Heat capacity and equilibration time near T_c of $YBa_2Cu_3O_7$," Proc. Int. Conf. on High Temperature Superconductors, Interlaken, Switzerland, *Physica C*, Vol. 153–155, Superconductivity, June 1988, pp. 1086–1088.

Wadayama, Y., K. Kudo, A. Nagata, K. Ikeda, S. Hanada, and O. Izumi, "Superconductivity and microstructural observation of $Er_{0.5}Y_{0.5}Ba_2Cu_3O_{7-y}$," *Jap. J. Appl. Phys.*, Vol. 27, No. 4, April 1988, pp. L561–L563.

Waldrop, M. M., "The 1987 Nobel Prize in Physics," *Research News*, October 23, 1987, pp. 481–482.

Waldrop, M. M., "New superconductors come through," *Science*, Vol. 240, No. 4859, June 17, 1988, pp. 1613–1615.

Wang, N., B. Sadoulet, T. Shutt, J. Beeman, E. E. Haller, A. Lange, I. Park, R. Ross, C. Stanton, and H. Steiner, "A 10 mK Temperature Sensor," *IEEE Trans. Nucl. Sci.*, Vol. 35, No. 1, February 1988, pp. 55–58.

Wang, X-l, T. Nanba, M. Ikezawa, Y. Isikawa, K. Mori, K. Kobayashi, K. Kasai, K. Sato, and T. Fukase, "Optical Properties of Polycrystalline $YBa_2Cu_3O_{7-\delta}$," *Jap. J. Appl. Phys.*, Vol. 26, No. 8, August 1987, pp. L1391–L1393.

Webb, W. W., "Magnetometers and Interference Devices," in *The Science and Technology of Superconductivity*, W. D. Gregory, and W. N. Mathhews, Jr., eds., Plenum, New York, 1973.

Weber, S., "Found! A practical way to turn out Josephson junction chips," *Electronics*, February 19, 1987, pp. 49–52.

Wheatley, J. C., "A perspective on the history and future of low-temperature refrigeration," *Physica 109 & 110B*, 1982, pp. 1764–1774.

Wiesmann, H., R. R. Corderman, A. R. Moodenbaugh, M. W. Ruckman, and M. Strongin, "Penetration depth measurements in $YBa_2Cu_3O_{7-\delta}$ at 4 MHz," *High-Temperature Superconductors*, Vol. 99, Proc. Mat. Res. Soc. Symp., Boston, November 30–December 4, 1987, pp. 387–390.

Willis, J. O., Z. Fisk, J. D. Thompson, S.-W. Cheong, R. M. Aikin, J. L. Smith, and E. Zirngiebl, "Superconductivity above 90 K in Magnetic Rare Earth-Barium-Copper Oxides," *J. Magnetism and Magnetic Materials*, Vol. 67, 1987, pp. L139–L142.

Willis, J. O., J. R. Cost, R. D. Brown, J. D. Thompson, and D. E. Peterson, "Radiation damage in $YBa_2Cu_3O_7$ by fast neutrons," High-Temperature Superconductors, Vol. 99, Proc. Materials Research Society Symposium, Boston, November 30–December 4, 1987, pp. 391–394.

Willis, J. O., J. R. Cost, R. D. Brown, J. D. Thompson, and D. E. Peterson, "Radiation damage in $YBa_2Cu_3O_7$ by fast neutrons," Proc. Mat. Res. Soc. Symp., Vol. 99, 1988, pp. 391–394.

Willis, J. O., M. E. McHenry, M. P. Maley, and H. Sheinberg, "Magnetic shielding by superconducting Y-Ba-Cu-O hollow cylinders," presentation at 1988 ASC, (To be published in *IEEE Trans. Magn.*, MAG-25, 1989), San Francisco, August 21–25, 1988, (MT-3), Abstracts, p. 76.

Wolsky, A. M., R. F. Giese, and E. J. Daniels, "The new superconductors: Prospects for applications," *Scientific American*, Vol. 260, No. 2, February 1989, pp. 60–69.

Worthington, T. K., W. H. Gallagher, D. L. Kaiser, F. H. Holtzberg, and T. R. Dinger, "The anisotropic nature of the superconducting properties of single crystal $Y_1Ba_2Cu_2O_{7-x}$" Proc. Int. Conf. on High Temperature Superconductors, Interlaken, Switzerland, *Physica C*, Vol. 153–155, Superconductivity, June 1988, pp. 32–37.

Wu, M. K., J. R. Ashburn, C. J. Torng, P. H. Hor, R. L. Meng, L. Gao, Z. J. Huang, Y. Q. Wang, and C. W. Chu, "Superconductivity at 93 K in a New Mixed-Phase Y-Ba-Cu-O Compound System at Ambient Pressure," *Phys. Rev. Lett.*, Vol. 58, No. 9, March 2, 1987, pp. 908–910.

Wu, X. D., D. Dijkkamp, S. B. Ogale, A. Inam, E. W. Chase, P. F. Miceli, C. C. Chang, J. M. Tarascon, and T. Venkatesan, "Epitaxial ordering of oxide superconductor thin films on (100) $SrTiO_3$ prepared by pulsed laser evaporation," *Appl. Phys. Lett.*, Vol. 51, No. 11, September 14, 1987, pp. 861–863.

Wu, P. H., Q. H. Cheng, S. Z. Yang, J. Chen, Y. Li, Z. M. Ji, J. M. Song, H. X. Lu, X. K. Gao, J. Wu, and X. Y. Zhang, "The Josephson Effect in a Ceramic Bridge at Liquid Nitrogen Temperature," *Jap. J. Appl. Phys.*, Vol. 26, No. 10, October 1987, pp. L1579–L1580.

Xiao, G., F. H. Streitz, A. Gavrin, and C. L. Chein, " Magnetic characteristics of superconducting $RBa_2Cu_3O_{6+y}$ (R = Nd, Sm, Eu, Gd, Dy, Ho, Er, Tm and Yb)," *Solid State Commun.*, Vol. 63, No. 9, 1987, pp. 817–820.

Xiao, G., F. H. Streitz, A. Gavrin, Y. W. Du, and C. L. Chein, "Effect of transition-metal elements on the superconductivity of Y-Ba-Cu-O," *Phys. Rev. B*, Vol. 35, No. 16, June 1, 1987, pp. 8782–8784.

Xiao, G., F. H. Streitz, A. Gavrin, M. Z. Cieplak, J. Childress, Ming Lu, A. Zwicker, and C. L. Chein, "Flux pinning and critical current density in $YBa_2Cu_3O_{6+y}$ and $EuBa_2Cu_3O_{6+y}$ superconductors," *Phys. Rev. B*, Vol. 36, No. 4, August 1, 1987, pp. 2382–2385.

Xu, Y., and W. Guan, "Superconductivity and paramagnetism in $ErBa_2Cu_3O_{7-y}$," *Appl. Phys. Lett.*, Vol. 53, No. 4, July 25, 1988, pp. 334–336.

Yamamoto, A., O. Araoka, Y. Doi, K. Hosoyama, T. Inagaki, K. Kawasno, S. Kurokawa, T. Mito, S. Mitsunobu, T. Sato, T. Shintomi, Y. Suzukik, M. Tadano, M. Taino, M. Takasaki, and H. Hirabayashi, "A Superconducting Secondary Beam Line in the 12 GeV Proton Synchrotron at KEK," *Nuc. Instrum. & Methods*, Vol. A257, 1987, pp. 105–113.

Yan, L. G., Y. J. Yu, Z. K. Wang, Z. Y. Kao, Z. X. Ye, C. L. Xue, P. Ye, Y. L. Cheng, X. M. Li, Q. W. Kong, S. S. Song, H. L. Nan, Y. M. Dai, and H. T. Tang, "A laboratory superconducting high gradient magnetic separator," *IEEE Trans. Magn.*, Vol. 25, No. 2, March 1989, pp. 1873–1876.

Yan, M. F., R. L. Barns, H. M. O'Bryan, Jr., P. K. Gallagher, R. C. Sherwood, and S. Jin, "Water interaction with the superconducting $YBa_2Cu_3O_7$ phase," *App. Phys. Lett.*, Vol. 51, No. 7, August 17, 1987, pp. 532–534.

Yang, K. Y., H. Homma, R. Lee, R. Bhadra, M. Grimsditch, S. Bader, "Phase diagram and oxygen stoichiometry of Y-Ba-Cu-O thin films," *Appl. Phys. Lett.*, September 7, 1988.

Ye, Hong-juan, Wei Lu, Zhi-yi Yu, and Hue-chu Shen, "Superconducting energy gap of $YBa_2Cu_3O_{9-\delta}$ measured by far-infrared reflection," *Phys. Rev. B*, Vol. 36, No. 16, December 1, 1987, pp. 8802–8803.

Yntema, G. B., "Niobium Superconducting Magnets," *IEEE Trans. Magn.*, Vol. MAG-23, No. 2, March 1987, pp. 390–395.

Yoshida, K., T. Hashimoto, T. Nagatsuma, and K. Enpuku, "Josephson Analog Amplifier with High Current Gain," *IEEE Trans. Magn.*, Vol. MAG-23, No. 2, March 1987, pp. 723–726.

Yuling, Z., L. Jingkui, X. Sishen, H. Jiuqin, R. Guanghui, C. Xiangrong, L. Hongbin, Z. Dongming, and Q. Shunliang, "The preparation, superconductivity and thermal characteristics of the Tl-Ba-Ca-Cu oxide superconductor," *J. Phys. D*, Vol. 21, 1988, pp. 845–847.

Zappe, H. H., "Memory-Cell Design in Josephson Technology," *IEEE Trans. Electron Devices*, Vol. ED-27, No. 10, October 1980, pp. 1870–1882.

Zappe, H. H., and B. S. Landman, "Experimental investigation of resonances in low-Q Jospehson interferometer devices," *J. Appl. Phys.*, Vol. 49, No. 7, July 1978, pp. 4149–4154.

Zeng, X., X. Jiang, H. Qi, D. Pang, N. Zhu, and Z. Zhang, "Relationships between superconductivity and lattice distortion in Y-Ba-Cu-O compounds," *Appl. Phys. Lett.*, Vol. 51, No. 9, August 31, 1987, pp. 692–693.

Zhou, X. Z., A. H. Morrish, Y. L. Luo, M. Raudsepp, and I. Maartense, "Bulk high-temperature superconductivity in the Tl-Ca-Ba-Cu-O system," *J. Phys. D*, Vol. 21, No. 7, July 1988, pp. 1243–1245.

Zimmerman, J. E., J. A. Beall, M. W. Cromer, and R. H. Ono, "Operation of a Y-Ba-Cu-O rf SQUID at 81 K," *Appl. Phys. Lett.*, Vol. 51, No. 8, August 24, 1987, pp. 617–618.

Zirngiebl, E., J. O. Willis, J. D. Thompson, C. Y. Huang, J. L. Smith, Z. Fisk, P. H. Hor, R. L. Meng, C. W. Chu, and M. K. Wu, "Magnetic and thermal measurements on high-T_c $(La_{0.9}Ba_{0.1})_2CuO_{4-y}$," *Solid State Commun.*, Vol. 63, No. 8, 1987, pp. 721–724.

zur Loye, H. C., K. J. Leary, S. W. Keller, W. K. Ham, T. A. Faltens, J. N. Michaels, A. M. Stacy, "Oxygen Isotope Effect in High-Temperature Oxide Superconductors," *Science*, Vol. 238, December 11, 1987, pp. 1558–1560.

INDEX